GERANIUMS IV

GERANIUMS

IV

The Grower's Manual

Fourth edition, edited by

John W. White

Published by

 Ball Publishing

Batavia, Illinois, USA

Sponsored by

Pennsylvania Flower Growers

Ball Publishing
335 North River Street
Batavia, Illinois 60510 USA

Printed in the United States of America
99 98 97 96   5 4 3 2

Photos: Within the text and color insert, except where noted, the chapter authors provided photos.
Certain other photos appeared in *Geraniums III*. Color insert: Grateful acknowledgment is given to
Professional Plant Growers Association (C1, C11), Richard Craig (C2-4, C6-9, C12) and Ball Seed Co.
(C23-42).

Pennsylvania Flower Growers Inc. hopes that users of this book will find it useful and informative.
While the authors have endeavored to provide accurate information. Pennsylvania Flower Growers
Inc. asks users to call its attention to any errors. The authors have attempted to obtain information
included in this book from reliable sources; however, the accuracy and completeness of this book and
any opinion based thereon are not guaranteed. No endorsement is intended for products mentioned,
nor is criticism meant for products not mentioned.

Library of Congress Cataloging in Publication Data

Geraniums IV : the growers manual / edited by John W. White.—4th
ed.
    p. cm.
  Includes bibliographical references and index.
  ISBN 0-9626796-5-8
  1. Pelargoniums. 2. Geraniums. 3. Floriculture. I. White, John
W. (John William). 1993- . II. Title: Geraniums 4. III. Title:
Geraniums four.
SM413.G35G47 1993   92-21778
635.9'33216—dc20   CIP

Cover photo of Americana Coral by Kurt Reynolds.

DEDICATION

Geraniums IV is dedicated to The Pennsylvania State University for its significant commitment to Pelargonium research and education for almost 35 years. As a commercial crop, geraniums were virtually extinct in the late 1950s; today, they are one of the world's most important floricultural products. This renaissance, initiated at Penn State, is a testimonial to many university and industry researchers' efforts.

Penn State is one of the leading public research universities in the nation. Several departments in the College of Agricultural Sciences, under the auspices of the Pennsylvania Agricultural Experiment Station and Pennsylvania Cooperative Extension, conduct geranium research and technology transfer activities. Penn State's commitment to acquire and disseminate knowledge about geraniums involves faculty and students in the Departments of Horticulture, Plant Pathology, Entomology, Agricultural Engineering, Agricultural Economics and Biology.

Penn State researchers have achieved many notable breakthroughs— the first commercially important seed geranium Nittany Lion, protocols for culture-indexing and virus-indexing, fast crop culture, pioneering growth regulator studies, knowledge of environmental factors' effects on growth and flowering, and improved zonal and Regal cultivars.

Although this dedication is to an institution of higher learning, Penn State research and educational contributions resulted from individual faculty and student commitments. The following faculty members have completed their careers: John W. Mastalerz, Darrell E. Walker, John W. White, James F. Tammen, W. Robert Fortney, Robert A. Aldrich, Peter B. Pfahl, Ernest L. Bergman, Lester P. Nichols, Alvi O. Voigt, Richard F. Stinson and James K. Rathmell Jr.

The current generation of researchers and educators are using the most modern physiology, genetics and molecular biology tools in their quest to develop improved geraniums for the 21st century.

The Pennsylvania Flower Growers' long association with Penn State has provided moral and financial support through the Charles Dillon Research Fund. In addition, PFG has been instrumental in the publication of the 1961, 1971, 1982 and 1993 editions of *Geraniums*.

Richard Craig

CONTENTS

PREFACE

G*eraniums IV* is a step-by-step guide offering specific instructions on how to breed, grow, market and use geraniums, and many chapters are applicable to other greenhouse-grown crops. It's a fascinating story of the past, present and future of geranium culture, breeding, physiology, entomology and pathology.

Geraniums IV has 52 chapters and subchapters, including major revisions and subjects not found in *Geraniums III*. Many new and revised chapters were written by younger authorities who present a fresh approach, challenging and updated information and thought provoking ideas.

You will enjoy the thorough insight in Integrated Management of Insects and Mites and in Whitefly Biology and Management, the uniqueness of Tissue Culture, the nostalgia of tracing the History of Geraniums from the 16th through 20th century, the provocative excursions into the 21st century in Open Markets/Open Minds and in Breeding Geraniums for 2000 and Beyond. Those with environmental concerns will find the intriguing world of new technology in Host Plant Resistance, Beneficial Microbes, Growing Media and Irrigation. The classicists will appreciate Vegetative Propagation and Tree Geraniums.

Traditional zonals are sharing space with seed-grown hybrids, ivies, Regals, and novelty and scented geraniums. Color photos of new cultivars are eye catching and tantalizing. Consumer Preferences may help you decide which types and cultivars to grow. If you discover a new cultivar and want to patent it, Intellectual Property Protection is a must-read. Need to solve a problem? The color photos on nutritional, physiological and pathological disorders may give you a clue. But to pinpoint the problem and its remedy, you will be drawn into the chapters on Diseases, Fertilization, Insects and Other Pests, Physiological and Environmental Disorders, and Clean Stock Production, Growth Regulating Chemicals, and Light and Temperature. Concerned about plant quality and costs of production? See Quality Production and Production Costs for guidelines.

Geraniums IV may not replace Michener for bedtime reading, but it should be on the reference shelf of every grower, marketer, breeder, student and hobbyist of geraniums and other greenhouse-grown crops.

ACKNOWLEDGEMENTS

It should come as no surprise that my long time friend and colleague Dick Craig was the instigator of *Geraniums IV* and the International Geranium Conference in Denmark. We all owe Dick a tremendous thank you for being the spark plug and catalyst in maintaining an active network among growers, breeders, educators and researchers.

Andrea Grazzini did a fantastic job of keeping manuscripts organized, typed and retyped, flowing through the pipeline to Ball Publishing and adding her own editorial changes when I missed something. Rick Grazzini was a big help in pruning down a large manuscript at a time when I was overloaded. Andrew Riseman did a yeoman's job of photography, sorting through *Geraniums III* negatives and numerous errands. Sara Craig and B.J. Myers were a big help in assisting Andrea with typing and printing.

A special thanks to Margery Daughtrey and Richard Lindquist for helping select and encourage authors of the pathology and entomology chapters, respectively, to write their subchapters. And my utmost appreciation to all of the authors for accepting their assignments and doing such excellent writing on such short notice.

Pennsylvania Flower Growers has had a long history of publishing quality crop manuals, including *Geraniums I, II* and *III,* and continue this tradition by sponsoring *Geraniums IV*. They have a significant new partner in Ball Publishing. We owe sincere gratitude to the board of directors, executive secretary and membership of PFG for their foresight and leadership and to John Martens and his staff, especially Peg Biagioni, for bringing crop production manuals to a new level of finished quality.

CONTRIBUTORS

Scott T. Adkins, Ohio State University, Columbus, Ohio

Allan M. Armitage, University of Georgia, Athens, Georgia

Mark E. Ascerno, University of Minnesota, St. Paul, Minnesota

James R. Baker, North Carolina State University, Raleigh, North Carolina

Kenneth F. Baker, retired, USDA-ARS, Albany, Oregon

James E. Barrett, University of Florida, Gainesville, Florida

Bridget K. Behe, Auburn University, Auburn, Alabama

Louis M. Berninger, Speedling Incorporated, Sun City, Florida

Charles L. Bethke, Michigan Peat Company, Williamston, Michigan

Richard L. Biamonte, W.R. Grace and Company, Allentown, Pennsylvania

Robin G. Brumfield, Rutgers University, New Brunswick, New Jersey

David W. Burger, University of California, Davis, California

William H. Carlson, Michigan State University, East Lansing, Michigan

Richard Craig, Penn State University, University Park, Pennsylvania

Margery Daughtrey, Cornell University, Riverhead, New York

Robert S. Dickey

Arthur W. Engelhard, University of Florida, Bradenton, Florida

John E. Erwin, University of Minnesota, St. Paul, Minnesota

Gerard W. Ferrentino, Cornell University, Ithaca, New York

Ralph N. Freeman, Cornell Cooperative Extension, Riverhead, New York

Richard A. Grazzini, Penn State University, University Park, Pennsylvania

Glenn G. Hanniford, Ohio State University, Columbus, Ohio

Mary K. Hausbeck, Michigan State University, East Lansing, Michigan

Charles F. Heidgen, Shady Hill Gardens, Batavia, Illinois

Royal D. Heins, Michigan State University, East Lansing, Michigan

E. Jay Holcomb, Penn State University, University Park, Pennsylvania

R. Kenneth Horst, Cornell University, Ithaca, New York

Ronald K. Jones, North Carolina State University, Raleigh, North Carolina

Mehrassa Khademi, Iowa State University, Ames, Iowa

Michael J. Klopmeyer, Ball FloraPlant, West Chicago, Illinois

David S. Koranski, Iowa State University, Ames, Iowa

Roy A. Larson, North Carolina State University, Raleigh, North Carolina

Linda J. Laughner, PanAmerican Seed Co., West Chicago, Illinois

J. Heinrich Lieth, University of California, Davis, California

Robert G. Linderman, USDA-ARS, Corvallis, Oregon

Richard K. Lindquist, Ohio State University, Wooster, Ohio

Murdick McLeod, South Dakota State University, Brookings, South Dakota

Gary W. Moorman, Penn State University, University Park, Pennsylvania

Ralph O. Mumma, Penn State University, University Park, Pennsylvania

Stephen T. Nameth, Ohio State University, Columbus, Ohio

Terril A. Nell, University of Florida, Gainesville, Florida

Paul E. Nelson, Penn State University, University Park, Pennsylvania

Eugene J. O'Donovan, Goldsmith Seeds, Gilroy, California

Wendy A. Oglevee-O'Donovan, Oglevee Ltd., Connellsville, Pennsylvania

Charles C. Powell, Ohio State University, Columbus, Ohio

Andrew L. Riseman, Penn State University, University Park, Pennsylvania

Marlin N. Rogers, retired, University of Missouri, Columbia, Missouri

Joann L. Rytter, Penn State University, University Park, Pennsylvania

John P. Sanderson, Cornell University, Ithaca, New York

Jay Sheely, Oglevee Associates, Connellsville, Pennsylvania

David R. Smitley, Michigan State University, East Lansing, Michigan

Catherine Anne Whealy, Ball Seed Co., West Chicago, Illinois

John W. White, Penn State University, University Park, Pennsylvania

Robert L. Wick, University of Massachusetts, Amherst, Massachusetts

Dennis J. Wolnick, Penn State University, University Park, Pennsylvania

CHAPTER 1

Status of the Industry

Louis M. Berninger

Geranium sales have proven to be the backbone of the bedding plant industry for many decades. While geraniums often have been taken for granted in some respects with the spotlight focused on new vinca and impatiens introductions, geraniums still command major attention from consumers in virtually all sections of the country.

The wholesale value of vegetatively and seed propagated geraniums increased almost 60% from 1979 to 1990. Wholesale value of geraniums exceeded $160 million at the start of this decade, representing 17.3% of all U.S. wholesale bedding plant sales.

Available statistics come from the annual U.S. Department of Agriculture survey of growers in 28 states, which constitute the major producing areas for floricultural crops. Report accuracy cannot be verified given the voluntary aspect of information contributed by producers. One can, however, identify leading states and gauge trends relative to geranium sales growth.

TABLE 1-1

Leading geranium-producing states in 1990

States	Pots (cuttings)	Pots (seed)	Flats
California	$ 3,949	—	$2,417
Illinois	4,084	$2,094	356
Michigan	6,052	8,572	1,763
North Carolina	2,551	3,161	1,321
New York	12,418	5,506	2,947
Ohio	6,387	3,284	3,022
Pennsylvania	6,520	2,626	471
Texas	3,604	1,874	7,309

Source: 1990 Floriculture Crops Summary, Agricultural Statistics Board, USDA, April 1991.

Competition has played a major role in bringing new enthusiasm and vigor to what had been viewed as a mature industry. The first half of the 1980s featured numerous hybrid seed geranium introductions. At least four commercial breeders aggressively competed for a share of the rapidly growing market for this product. Hybrid seed geraniums have been produced in both pots and flats, and over 90% of potted material has been produced in pots under 5 inches (13 cm) in diameter. Potted material was valued at $45.5 million and flats at almost $30 million.

Reacting to this strong competition, cutting geranium breeders initiated programs that had positive results in the second half of the decade. One firm dominated the cutting industry in the early 1980s; currently, two firms aggressively compete for market share and two more are ready to seek substantial market share in the 1990s. Cutting geraniums, at wholesale, were valued at almost $86.5 million in 1990. We can assume that virtually all plants were sold in pots with more than 80% under 5 inches in diameter.

Producers are reacting to changing consumer purchase patterns by offering a wide selection of geranium flower and foliage colors. Red, long the dominant flower color, now shares the market with pastel colors. New players and new varieties indicate that geranium sales will grow dramatically in the 1990s.

TABLE 1-2

1990 geranium statistics

Category	Potted cuttings	Potted seed	Flats
Producers	4,148	2,040	1,554
Area square footage	26,735,000	14,428,000	6,219,000
Quantity sold:			
Flats			3,545,000
Less than 5-inch	48,899,000	48,943,000	
More than 5-inch	11,150,000	2,719,000	
Total	**60,049,000**	**51,662,000**	**3,545,000**
% of wholesale sales	72%	84%	79%
Wholesale price			
Flats			$8.46
Less than 5-inch	$1.23	$0.77	
More than 5-inch	$2.34	$2.21	
Value of crop			
at wholesale	$86,455,000	$45,534,000	$29,995,000

Note: It is presumed that virtually the entire quantity consists of seed geraniums.
Source: 1990 Floriculture Crops Summary, Agricultural Statistics Board, USDA, April 1991.

CHAPTER 2

Growing Media

Charles L. Bethke

A geranium can be grown in nearly any growing media if the plant gets the proper environment, water, attention and the proper balance of essential elements containing nothing toxic or harmful [24]. A growing medium must supply:

(1) plant support;
(2) water;
(3) nutrients in available forms;
(4) gas exchange to the roots [17]; and
(5) an environment to maintain a biological balance [11].

With careful engineering of the growing media, minimal management can produce optimum performance. Growers often work too hard to maintain a balance or find the best options (a "happy medium"), using a growing mix that has several less-than-desirable properties. It's much easier to grow geraniums in a medium with the proper physical properties, the best chemical characteristics and an optimal environment to promote a biological balance. In engineering a growing medium, first build in the proper physical properties. Next, adjust the chemistry to approach the proper nutritional balance. Then, try to maintain a desirable biological environment. Finally, make sure there is no mishandling or damage to the media in the process.

Physical properties

In engineering media for optimum physical properties, consider the plant's particular requirements and crop handling and management. The first thing to engineer is the physical environment. No single medium has the best properties for all applications. Geraniums require a balance of *bulk density, air porosity, water holding ability and resilience*. The blend must be easy to plant into. It must take up water readily and adapt to either surface watering or subsurface irrigation. It must be uniform throughout all containers and be easily reproducible from batch to batch in any season.

Bulk density

Bulk density provides physical support for the plant and gives the containers stability against wind, water, shipping and handling. Under current management practices the preference is to have a media that is neither too heavy (because of handling and transport), nor too light. Bulk density is defined as the weight of the media, per unit volume, at a certain moisture content. You can compare the bulk densities or components of blends on a dry weight basis. But comparing those values in moist media is of little use unless you know the moisture content. Commonly, saturated media's bulk density can range from 15 to 40 pounds per cubic foot (239 to 639 kilograms per cubic meter).

Table 2-1 shows typical bulk densities of growing media components when dry. Dry bulk densities from 10 to 18 pounds per cubic foot (160 to 288 kilograms per cubic meter) support a large geranium plant in a 4-inch (10-cm) pot. At the same time, the plants are neither too heavy to handle nor too costly to ship.

TABLE 2-1

Bulk density of media components

	Dry weight			
Component	Pounds per cu. ft.[1]	Ounces per 4" pot[2]	Ounces per 6" pot[2]	Ounces per 11½ by 23" flat[2]
Bark, pine	15–28	4.8–9.0	14–26	48–90
Bark, fir	14	4.5	13	45
Bark, redwood	8	2.6	7.5	26
Rockwool	4	1.3	3.8	13
Sphagnum peat	6.5	2.1	6.1	21
Sedge peat	14	4.5	13	45
Perlite	6–8	2.0–2.6	5.6–7.5	20–26
Polystyrene	0.5–2	0.2–0.6	0.5–1.8	2–6
Pumice	30	9.6	28	96
Rice hulls	6.5	2.1	6.1	21
Sand	100	32	30	32
Saw dust	12	3.8	11	38
Vermiculite	5–7	1.6–2.2	4.7–6.6	16–22
Water	62	19.8	58	198

Source: Bunt [5].
1. Multiply pounds by 453.59 to get grams. Multiply cubic feet by 28.32 to get liters.
2. Multiply ounces by 28.35 to get grams. Multiply inches by 2.54 to get centimeters.

The blend's total bulk density is influenced by its components' bulk density. Adding heavy materials like sand, gravel or calcined clay, increases bulk density substantially. Lighter weight materials like rockwool, perlite or vermiculite help reduce bulk density. Large percentages of fine textured components in the blend tend to increase bulk density and decrease porosity. Excessively blending media, especially when moist, can cause the components to break down and the fine particle percentage to increase. This increases bulk density and decreases porosity.

Growers influence media density by the way they fill containers. Heavy packing, excessive mixing time or shaking, vibration, stacking containers in a nested fashion and watering with a heavy pounding stream greatly increases compaction and bulk density and decreases media porosity.

Porosity

A very important physical property of the growing media, porosity influences gas exchange and the water availability. Sufficient porosity eases water management and increases the flexibility needed to control moisture levels while scheduling applications of fertilizers, fungicides and other drenches. Geraniums require air porosities between 5% and 15% by volume for gas exchange to occur [4].

Total porosity

Total media porosity is split into two groupings: The *macropores*, which provide the air-filled pore space or air porosity, and the *micropores*, the water-filled pore spaces at saturation or at container capacity.

Macropore space (air porosity) volume can be measured at saturation in a container the same size as that intended for growing the crop. Water that drains by gravity from fully saturated media represents the macropores at saturation in that size container.

If the roots are to grow and do the work of taking up water and nutrients, they must take in oxygen while giving off carbon dioxide and other respiration products. These gases must diffuse in and out of the medium. Gas diffusion is much faster through air than through water. To increase the air-filled pore space, add coarse particles to the media.

Porosity is influenced by the media's particle size distribution and components. A high percentage of large particle sizes increases air porosity. An increase in the percentage of smaller size particles increases the relative amount of water-filled pore space. The available water is held in these micropores [12]. A wide distribution of all particle sizes decreases total porosity, which is what occurs in the manufacture of cement. A wide range of particle sizes, sand, fine gravel, coarse gravel and stones are blended with mortar to produce cement. In a growing media, with an insufficient percentage of large particles (or a threshold percentage of a given particle diameter), the aggregates "float" in the blend as they do in cement. With similar diameter particles touching one another, however, a bridging effect occurs and pores develop between larger particles.

Threshold percentages of particles with similar diameters usually occur at about 30% to 40%. Therefore, if perlite or vermiculite are blended into a very fine-textured media, at least 30% of similar diameter particles is required to increase air porosity and drainage. By adding up all similar diameter particles from the blended components, you can estimate whether a given additive will increase or decrease porosity. For example, adding sand to a peat-perlite blend doesn't increase porosity, but it does increase bulk density. Sand fills the macropores, decreases the air porosity and gas exchange and decreases drainage. Adding uniformly sized coarse sand to field soil can, however, increase porosity and drainage.

Therefore, particle size distribution is most influential in determining the proportion of macro- and micropores, which influences air porosity and drainage. Additions to or alterations of media components influence their relationships and the way air and water move in the media [9].

In very young seed-propagated geraniums, a lack of gas exchange to the roots can cause a condition called blindness (Color Fig. C69; also see Chapter 12, Growing From Seedlings). This condition can result from saturating or flooding the roots and lack of gas exchange. The symptoms begin in the apical meristem, where the shoot tip loses orientation and apical dominance. The plant then produces many deformed, club-shaped leaves with severely multibranched growth. This delays seedling development. Usually, this problem occurs during a very early development stage. First, the seed leaves, or cotyledons, are deformed, and the first or second true leaves are misshaped or butterflylike in shape. The apex or center shoot is stunted and many axillary shoots begin to develop. The phenomenon is a flood response and is remedied by having sufficient air porosity and avoiding excess watering at young stages.

Blindness is seldom seen in geraniums propagated from cuttings. Occasionally, blindness occurs in poorly rooted cuttings immediately after transplanting, before roots develop to the container edges. The problem is usually recognized several days or weeks after flooding when the physiological damage has occurred. Having sufficient air-filled pores and/or avoiding long periods of excess saturation can remedy the problem. When roots are exposed to the air in the media on the container edge or bottom, gas exchange takes place and prevents blindness.

Water holding ability

Particle size distribution and absorption of the components that make up the media influence its water holding ability. The water available to the plant is the portion of total container capacity that can be used by the plant as it dries from container saturation, after drainage to wilt point. Nonavailable water is that water held in the media so tightly that it's not available to the plant. The total available and nonavailable water make up the container capacity, which is the water held in the media after gravitational water has drained.

Geraniums have the ability to pull water from media under fairly high moisture tensions, up to about 15 bars (15 atmospheres or a 4.2 pF), the permanent wilt point. Therefore, media that have a larger water reserve held at higher moisture tension yield crops with a longer shelf life and better storage ability in containers. This is why field soils and finer textured media show longer shelf life in keeping quality studies. Components like vermiculite, moderately decayed peat, Bacctite (a reed-sedge peat granule) or field soil tend to increase the water quantity held at higher moisture tensions and extend shelf life. Geranium plants will adapt or acclimate to their soil environment based on the moisture stresses they have endured. Wilting a geranium (Color Fig. C64) or exposing it to nearwilt conditions increases its tolerance to drying. This is why geraniums make such good outdoor plants and have been used extensively in cemeteries and places where water is limited.

Many growing media shrink substantially when dried. This can present a significant problem in crop management. If the medium shrinks and pulls away from the container excessively, then rewetting is very difficult. Water runs down the container edge and the medium may become hydrophobic (water "hating" or water repellent).

One way to avoid excessive shrinkage is to avoid media components in the mixture that shrink substantially. Fresh sphagnum that has never dried, super absorbent polymers and some clay materials can shrink significantly. If sphagnum peat has been previously dried, it won't shrink as much as when it was freshly harvested and held in a moist state. If sphagnum peat has never been dried, it will shrink as much as 20%. Once it has been dried down and remoistened, it will shrink only about 2% to 6%.

Another way to avoid excessive mixture shrinkage is to add coarse aggregates. They provide bridging effects that prevent organic fibers from collapsing on themselves, thus reducing shrinkage. The aggregates also create pores or cavities in the media, increasing porosity and decreasing shrinkage.

Wetting agents

Mix wettability is influenced by the media's characteristics and the water chemistry. Uniform distribution of large pores (micropores) helps optimize water absorption throughout the mix. Excessive drying causes organic substances to become hydrophobic. Most organic matter becomes hydrophobic when moisture levels are less than about 30%. Field soil, vermiculite, perlite and sand seldom become hydrophobic. Water has the property of hydrogen bonding, which gives it familiar characteristics. Among these properties is surface tension, the formation of a droplet instead of a thin film on smooth, flat surfaces. Adding a wetting agent or surfactant helps reduce surface tension and, therefore, increases water dispersion throughout a blend. Adding wetting agents or surfactants to media encourages optimum wetting. Contrary to what is often thought, wetting agents don't increase mix wetness or saturation. In fact, they help increase drainage and air porosity.

Most surfactants have a limited useful life. They lose effectiveness as a result of media absorption, leaching losses, microbial decay and chemical oxidation or break down with ultra violet lights. A surfactant should break down over time to be environmentally compatible. Many surfactants are toxic at levels exceeding recommended rates. In blending your own mix, do trial blending with the recommended surfactant concentration rate before using it in full-scale production.

Resilience

A high quality media must withstand handling and hold up during cropping over time. Growing media shouldn't deteriorate too quickly and mustn't shrink readily. They must be easy to plant into. Media are resilient against handling if they can withstand some overblending. They are resilient against compaction if they have some "spring" or solid characteristics in their texture. Resilient media are less likely to be affected by excess packing or nested packing.

Chemical properties

In the last decade many changes have occurred in media and their chemical properties. More and rapid changes are anticipated over the next decade. Understanding and adapting to these changes and continuously growing successful crops requires knowing the fundamental chemical characteristics of growing media.

The individual chemical properties that influence nutrition can be divided into four areas: *buffering ability* (the quantity factor); *nutrient availability* (the availability factor); *nutrient balance* (the nutritional factor); *soluble salts levels* (the intensity factor). All are influenced by the blend components, water and fertilizer used and the plant's growth rate.

Buffering ability

The quantity of available nutrients in the media is more important today than in the past. Continuously applying and leaching nutrients is costly and environmentally unsound. The practice of thoroughly watering to a 20% leaching level at each fertilization is rapidly diminishing. Post production performance is becoming more important. Growers are being required to manage more crops; therefore, the media's nutrient reserve or buffering ability plays an important role.

The cation exchange capacity (CEC) is often referred to as the soil's buffering ability. Each media component contributes a different amount of CEC. The CEC is a measure of the ability to hold positively charged ions called cations. Cations are calcium (Ca), magnesium (Mg), potassium (K), ammonium (NH_4), aluminum (Al), sodium (Na) and hydrogen (H). They are listed in the order of retention in the media from greatest to least. Hydrogen and aluminum cations are called the acids or acidic ions. The others are called the bases or basic ions. They all carry a positive charge, are absorbed on various surfaces of media particles and balance the negative charges (the CEC) on those particles. Organic matter has negatively charged exchange sites provided by carboxyl and hydroxyl radicals. The carboxyl group (R-C-O-O-H) provides sites on which the hydrogen ion can be exchanged for the other cations. Phenolic, enolic and cyclic nitrogen structures also provide CEC in organic substrates. The CEC is generally reported in milliequivalents per 100 grams of soil. The pH has a strong influence on the CEC of organic matter. To be useful, CEC must be expressed at a certain pH in organic substrates. For container growing the CEC per unit weight has little value because of the light weight and difference in bulk density of substrates used. But the CEC per unit volume is a useful measure. Therefore, to properly report the CEC of container media, express it as milliequivalents (meq) per 100 cubic cm at a given pH. When testing CEC, the pH level should be between 5.5 and 6.5 to most closely represent cropping conditions. Some typical CEC values are expressed in Table 2-2. Note that highly decayed, organic materials and clay contributed significantly to CEC.

Blends with low CEC have lower nutrient reserves. Many growers have supplemented these mixtures with slow release fertilizers. The slowly soluble materials provide nutrients over a longer time. They work well either in combination with or in addition to the media's buffering properties.

Growing media low in CEC require frequent testing and careful attention to irrigation and fertilization. With lower buffering levels, nutrition levels fluctuate widely and pH is difficult to control.

TABLE 2-2

Cation exchange capacities of common mixes and media components

	Cation exchange capacities	
Components and mixes	Meq* per 100 grams	Meq* per 100 cm
Loam soil	12	14
Sphagnum peat	100–120	8–10
Humus peat	200–300	60–80
Woody peat	70–100	12–16
Perlite	1–2	0
Vermiculite	150	16
Pine bark	40–60	10–15
Sand	2–5	4–10
Clay (montmorillonite)	100	150
50% sphagnum, 50% vermiculite	140	32
50% sphagnum, 50% sand	8	5
50% sedge peat, 50% sand	21	17
66% pine bark, 33% perlite	24	5

*Milliequivalents
Sources: Bunt [5], Lucas [15], Mastalerz [17].

Nutrient availability

Growing media pH is the primary influence on nutrient availability. The media pH is defined as the negative logarithm of the hydrogen ion concentration. It represents the amount of hydrogen ions (acidic ions) in relation to hydroxyl ions (basic ions) present in the growing media.

In many ways a growing media's nutrient balance resembles soup. With high concentrations of hydrogen ions or acid ion availability the soup becomes more acidic and has a lower pH. That acidity influences ion availability. With low pH comes increased iron and manganese solubility and decreased phosphorous, potassium, calcium, magnesium, boron, copper, zinc and molybdenum availability. High pH levels decrease nitrogen availability; slightly limit phosphorus, calcium and magnesium; and severely limit iron, manganese, boron, zinc and copper. Carefully balancing the blend's pH is the most important part of engineering a medium (Fig. 2-1 and 2-2) [14, 19].

The proper pH level is influenced by the blend components. Each media component has a percentage base saturation that is the percentage of basic cations compared to the total CEC. The degree of base saturation influences the media's pH stability. For balanced nutrition

TABLE 2-3

Iron and manganese symptoms in geraniums (Pinto Red)

Media* starting pH	Symptoms observed under various watering practices in 4" (10 cm) pots		
	Allowed to dry severely and then thoroughly watered to leaching	Watered as needed, so plants never dried significantly	Kept on wet side by watering daily with some leaching
4.5	Smallest plants, slow growth, no toxicity symptoms	Smaller sized plants; severe toxicity symptoms on most leaves	Smaller sized plants, no toxicity symptoms
5.0	Small plants with slow growth, no toxicity symptoms	Small plants, toxicity symptoms on most leaves	Medium plants, no toxicity symptoms
5.5	Small plants, no unusual symptoms	Normal plants, toxicity symptoms on some older leaves	Normal plants, no toxicity symptoms
6.0	Small plants, no unusual symptoms	Normal plants, few older leaves showing toxicity	Large plants, fully normal
6.5	Smaller plants, no foliar symptoms	Normal plants, few showing scattered toxicity symptoms	Large plants, fully normal
7.0	Smaller plants, many deficiency symptoms	Smaller plants, no toxicity symptoms	Large plants, some chlorotic leaves at shoot tip

*Blend of 65% sphagnum (vonPost 3), 20% perlite, 15% vermiculite.

optimum base saturations are 50% to 85% calcium, 6% to 12% magnesium, 2% to 3% potassium, with as little sodium and ammonium as possible [22]. Optimum pH requires a balance of acidic and basic ions. Usually, if 80% to 95% of the CEC is saturated with bases, the pH will be in the 5.5 to 6.5 range. As a result each substrate has either a liming requirement or a liming ability. This liming property depends on its buffering ability and the amount of other free acidic or basic cations.

The influence of two liming materials on two peat types is illustrated in Fig. 2-3. The pH of a sphagnum peat and a reed-sedge peat is influenced differently by each liming material. The reed-sedge peat has a higher buffering and requires more lime to reach an optimum pH

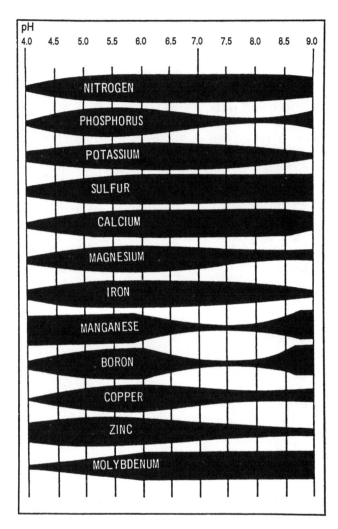

Fig. 2-1. Plant nu-
trient availability
at different pH lev-
els for an organic
field soil [14].
(Source: [19])

(between 5.8 and 6.2) than the sphagnum peat. Liming material A has a greater effect on raising pH than material B. The influence of one component, rockwool, on pH is illustrated in Fig. 2-4. Five rockwool brands were added to sphagnum peat at different rates. Their effects on pH are quite different. Note that rockwool A has a liming effect that raises the pH above optimum with only about 20% in the blend. This limits its use in blending. You can use as high as 60% rockwool B, C and E without exceeding the optimum range.

For sphagnum peat alone, common lime requirements are 2 to 10 pounds of finely ground lime per cubic yard (1.2 to 6 grams per liter) of peat, but this amount varies depending on the buffering ability [5, 6, 21,

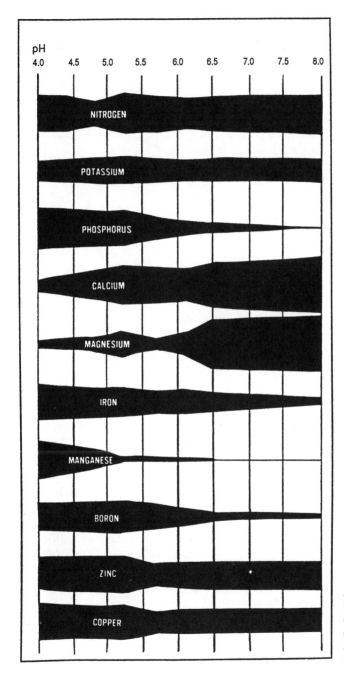

Fig. 2-2. Plant nutrient availability at different pH levels for W.R. Grace Metro Mix 300 [19]. (Source: [19])

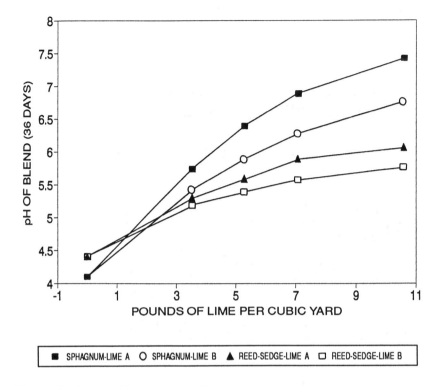

| ■ SPHAGNUM-LIME A | O SPHAGNUM-LIME B | ▲ REED-SEDGE-LIME A | □ REED-SEDGE-LIME B |

Fig. 2-3. Influence of lime type on pH in two types of peat.

23]. Do trial blending and incubate the mix at greenhouse temperatures for at least two weeks to determine the equilibrium pH. Do this well in advance of making up your final blends. If the pH runs high, reduce the lime. If it runs high because of components in the blend, add iron sulfate at a rate of 1 to 3 pounds per cubic yard (0.6 to 1.8 grams per liter) to reduce the pH.

Media pH is extremely important in geraniums. Recently it has been demonstrated [3] that low pH increases the incidence of iron (Color Fig. C58) and manganese toxicity. The symptoms are speckling or stippling of older, mature leaves. The symptoms usually appear shortly after a period of high light or stress. Iron and manganese excesses result from low pH levels and certain management practices. We know that a one unit pH drop increases iron solubility by as much as 1,000 times. It increases manganese solubility about 100 times [13]. To further complicate the problem, growing crops—especially geraniums—excrete hydrogen ions, carboxyl groups and acidic organic substances from the roots [18]. A drop of two pH units can occur in the root zone [16]. Therefore, growing media in direct contact with the root can have a lower pH than other media in the container.

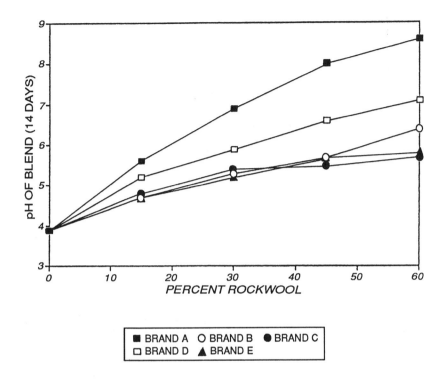

Fig. 2-4. The pH of rockwool and sphagnum at different blend ratios.

In Chapter 5, Fertilization, Biernbaum recommends pH ranges of 6.0 to 6.3. Some growers have observed iron and manganese toxicities in growing media with pH levels in the range of 5.8 to 6.2. Those growers tended to be very careful in their watering practices. Studies were conducted on growing media with a wide range of pH levels under three watering regimes, as described in Table 2-3. In the growing media that was watered as needed, iron and manganese toxicities were the most extensive. But in those treatments watered thoroughly and frequently, or watered only when the plants needed water severely, there was no iron or manganese toxicity. It's likely that ion exchange or ion flushing immediately around the roots prevented the problem. A word of caution: Subirrigation and more attentive watering enhance the problem.

If a growing media's pH is low, leach with neutral or alkaline tap water, or leach with a calcium nitrate solution at about 100 parts per million nitrogen concentration. A thorough leaching usually requires three thorough waterings with an hour wait between each watering.

Proper liming is very important for geraniums. Adjust pH according to blend components, water quality and fertilizer. For geraniums, make every effort to achieve a pH of 6.0 to 6.5.

Nutrient balance

Optimum nutrient levels are relatively easy to maintain. An experienced grower once said, "Geraniums are magnesium and phosphorous pigs." Geraniums consume magnesium and phosphorous at high levels while taking up other nutrients at normal rates. By maintaining a nutrient balance the plants will grow appropriately. Table 2-4 shows the optimum ion balance achieved with proper base saturation. Levels are presented as the percent of soluble salts from a saturated paste water extract. It's best to keep ammonium levels low and sodium and chloride levels extremely low. The exact levels depend on the lab you use and the extraction procedures they apply. Guidelines in Table 2-4 can serve as a general reference. Maintaining these balanced nutrient levels allows geraniums to approach optimum growth.

To reach optimum nutrient levels or to make necessary adjustments, refer to Table 2-5, which lists the different fertilizers and rates needed to alter a blend's test levels.

TABLE 2-4

Nutrient balance desired (from saturated media extract)

Nutrient	Total soluble salts %
Nitrate - Nitrogen	8–10
Ammonium - Nitrogen	<3
Potassium	11–13
Calcium	14–16
Magnesium	5–8
Sodium	<10
Chloride	<10
Phosphorus	1–4

Source: Warncke and Krauskopf [23].

Geraniums particularly are sensitive to magnesium and phosphorous deficiencies. Phosphorous deficiency yields slow growth (Color Fig. C45 and C46). Magnesium deficiency is revealed by interveinal chlorosis on the newly mature and older leaves (Color Fig. C48, C49 and C50).

Nitrogen is best supplied in the nitrate form for geraniums: Geraniums respond best to at least 75% of the nitrogen as nitrate. Calcium nitrate and potassium nitrate are preferred fertilizers.

TABLE 2-5

**Fertilizer needed to increase saturated
paste analysis test levels**

Source of nutrients		Ounces to add per cubic yard to increase nutrient				
Fertilizer	Analysis	N 10 ppm	P 2 ppm	K 25 ppm	Ca 25 ppm	Mg 50 ppm
Potassium nitrate	13-0-44	2.3	-	3.75	-	-
Calcium nitrate	15-0-0	2.0	-	-	7.0	-
Ammonium nitrate	33-0-0	0.9	-	-	-	-
Urea	46-0-0	0.7	-	-	-	-
Potassium sulfate	0-0-50,17%S	-	-	3.25	-	-
Superphosphate	0-20-0, 19%Ca,14%S	-	14.4	-	6.7	-
Triple superphosphate	0-46-0, 14%Ca,4%S	-	6.4	-	10.2	-
Bone meal	2-20-1	>10	12.0	-	-	-
Gypsum	22%Ca,18%S	-	-	-	5.8	-
Magnesium sulfate	9%Mg,13%S	-	-	-	-	4-8
Calcitic lime	40%Ca	-	-	-	>4.0	-
Dolomitic lime	26%Ca,12%Mg	-	-	-	>6.0	>40
20-20-20	20-20-20	1.5	14.4	8.0	-	-

Source: Warncke and Krauskopf, [23].

Soluble salts levels

Soluble salts levels reflect the cumulative level of nutrients and other salts. Total soluble salts are measured as conductivity in a soil-water solution and are commonly referred to as millimho (mMho) or millisiemen (mS). All soluble ions contribute to a growing media's conductivity (soluble salts level). Commonly accepted levels of total soluble salts readings are listed in Table 2-6. Typical media-to-water ratios are also listed in the table. The saturated paste extract is commonly used in laboratory analysis. One part media to two parts distilled water is commonly used in greenhouse monitoring. Dilutions of one part media to five parts distilled water are commonly used when total salts are thought to be high.

When the tap water is high in nonnutrient soluble salts, it's essential that the blend has a well-balanced nutrient charge. If soluble salts continue to climb, leaching is necessary. Again, good porosity is required for efficient leaching. Sodium saturation of the media can occur when tap water contains high sodium levels. When this happens, the growing media won't perform and reacts poorly to additional fertilizer.

TABLE 2-6

Guidelines for conductivity interpretations (for different media:water ratios)

Total soluble salts	Conductivity reading			Nutritional status
	Saturated paste	1 media: 2 water	1 media: 5 water	
Very low	<0.7	<0.3	<0.1	Lack of sufficient nutrients; crops need fertilizer.
Low	0.7–1.5	0.3–0.6	0.1–0.3	Acceptable for growing with steady feeding or subirrigation.
Desirable	1.5–3.0	0.6–1.0	0.3–0.5	Good nutrition level except where tap water contains salts.
Moderately high	3.0–4.5	1.0–1.5	0.5–0.7	Higher than desirable, but acceptable for rapid growth.
High	4.5–5.5	1.5–2.0	0.7–1.0	Expect slower growth and reduced productivity; do not dry out.
Very high	>5.5	>2.0	>1.0	Severe injury from salts is likely: stunting and marginal leaf burn.

Sources: Bunt [5]; Warncke and Krauskopf [23].

It's best to begin a geranium crop in a blend that contains moderate to low levels of balanced nutrients. Monitor nutrient levels on a regular basis. Fertilize at a lower concentration where subirrigation is used or if growing media contains a high level of fine texture components. The current tendency is to maintain balanced nutrition at lower concentrations rather than to approach luxury mineral nutrition at higher concentrations. This decreases fertilizer expense and reduces environmental runoff concerns.

Competitive inhibition often occurs in greenhouse crops where one nutrient inhibits another's uptake or availability. Excessive potassium can inhibit or prevent absorption of other positive ions like calcium or magnesium. A common problem occurs when excess sodium exists in the water or in poor quality fertilizers. The sodium is supplied in excessive amounts in relationship to potassium, calcium or magnesium. Sodium then blocks or inhibits the calcium, magnesium or potassium uptake. Make trial blends and/or get an analysis before growing a crop to prevent this problem.

Other chemical properties

Take care not to introduce excessive toxic substances into the media. In today's environmentally aware community we are exposed to new and alternate growing media components and additives. Often these amendments are a byproduct or waste product of another process. In these products it's common to find excess substances like cadmium, lead, zinc, iron, manganese, copper, sodium, boron or chloride. All can act as toxic materials to geraniums. Therefore, when a new, "creative" additive comes along, conduct a thorough analysis and trial blending.

Watch out for other substance contaminations. Pesticides, herbicides or other potentially damaging chemicals may be present. Investigate the safety and purity of a new additive with chemical analysis and bioassay.

Soils from fields or composting processes are often contaminated with some herbicide or other organic chemical. Assume that they are contaminated until proven otherwise by bioassay. Always use reliable products from a reliable manufacturer. It's best to use products you are familiar with. Try new things on a small scale first.

Biological properties

Growing media's biological properties often have been overlooked, but today's environmental concerns are making them more important. Media must have the proper biological balance (see Chapter 3, Beneficial Microbes) to maintain an optimum environment for the plant and plant protection.

Clean products

Begin with clean products and keep them clean and uncontaminated throughout the growing process. For geraniums the growing media must be free of any undesirable insects, nematodes, diseases, weeds, toxins and pesticides.

pH level

The pH level influences microbial activity [1], because many soil microorganisms are very sensitive to pH. In addition to keeping proper pH for mineral nutrition, keep the pH in the proper range for biological reasons. Extremely high or low pH inhibits microbial activities. Nitrification, essential for converting urea and ammonia to the usable nitrate ions, occurs best at pH levels below 7. Proper pH also enhances many beneficial fungi's development and growth. Most beneficial fungi thrive at pH levels near neutrality (pH 7.0). Lower pH levels tend to inhibit fungal growth and encourage pathogenic behavior. A normal pH range (6.0 to 6.5) provides a balance of beneficial and inhibiting microbes [18].

Excessive moisture

Poor aeration and excessive moisture encourage water molds and other disease-causing microorganisms. A continually wet environment encourages surface algae growth. Keep the media at a moderate to dry moisture level to reduce algae growth.

Geraniums are quite tolerant of dry conditions. When excessively dried, however, stress or damage to root or shoot tissues encourages disease development.

Biological suppression

Biological suppression has recently received considerable attention [11]. Composted bark can significantly suppress microorganisms and diseases in containerized media. Diseases like Pythium, Phytophthora, Rhizoctonia, Thielaviopsis root rot, Fusarium wilt and many nematode diseases can be suppressed when using at least 50% properly composted material in the growing media. Suppressive action works for the first few weeks in the greenhouse. Changes from compost pile to greenhouse containers can influence the biological populations that provide the inhibition, substantially reducing suppressive action over time.

Heating up or composting

Composting bulk blended media piles occurs when moisture levels are between 40% and 60%, oxygen is present and the pile is large enough to trap the heat. Thermophilic composting can occur when the pile is left undisturbed for periods longer than four weeks. Inside these piles temperatures can reach as high as 160 F (71 C). Organic matter begins to break down, soluble salts levels increase, pH changes and the nutrient balance is altered. Using mix rapidly, filling containers or frequently turning the bulk pile reduces or prevents this composting activity. Avoid any composting after blending, as this type of composting doesn't benefit prepared mixes.

Contamination

Contaminated growing media primarily results from improper storage and handling. Often, reusing contaminated soil, dirty equipment or containers is the disease source. Contaminated plant material and soil on the hands or clothing are major transmission factors of soil-borne diseases. Dust and dirt in the soil mixing and potting area also harbor diseases that can later damage the crop.

The major transmitters of biological contaminants into the growing media are insects, animals, wind, water, weeds, dust and people [20]. Therefore, store media in a well-protected area under a roof, away from traffic and equipment and away from pets and people. Observe sanitation in the storage area at all times.

Management and handling

Properly constructed growing media with optimal physical, chemical and biological properties may be destroyed by improper handling. Excessive blending damages the media's physical structure. In most blenders uniform distribution occurs in only one to three minutes of mixing. Overmixing rounds off corners of angular aggregates and breaks down softer particles. It shortens fibers, often destroying air porosity and drainage. Different sized particles can migrate to different sections of the blend after a longer mixing period. Coarse particles tend to come to the top of a blend and come out of a blender first, followed by the finer particles. In filling containers, the coarse particles fill one container while mostly fine particles fill another container. When the bulk densities of the components are substantially different, segregation is more extreme. For example polystyrene beads readily separate from sandy soil, filling entire pots with one component. This makes crop management very difficult.

Excessive blending

Because it separates media from additives like fertilizers and lime, excessive blending alters the chemical makeup from pot to pot. Uniform media among pots and within pots is extremely important for high quality. Subirrigation systems won't work well with media variations.

Overcompaction

Guard against overcompaction when filling or handling containers. Be certain containers are uniformly full and remain full while you handle and grow the crop. In plug production excessive compaction and empty cells are two main problems. In seed-grown geraniums excessive compaction tends to remove air, stop drainage and lead to a condition called blindness. To produce quality plugs, the media must be permeable to oxygen. Both overcompaction and underfilling make water and nutrient management very difficult [8].

Heavy watering

Strong, beating water flooding down onto a dry blend causes finer particles to migrate and coarse particles to float, altering the dry blend's physical structure. It changes porosity and drainage. Use low pressure and fine-textured misting, frequently and lightly, on dry blends. Moisten the blend before filling containers to "knit" the fine components to the coarse components and preserve air porosity and drainage. Containers filled with dry media settle and components separate. In some cases dry media sifts out the holes in the container bottom. Stacking containers into one another or overfilling containers and stacking on top of one another compacts blends excessively.

Media storage

Store media in a clean, dry place as if it were food. Unlike food, it doesn't need refrigeration, but like food it's easily contaminated. The storage area should be well away from trash or chemical supplies. Protect media from excess water, blowing dust, weeds and weed seeds. Prevent contamination by wildlife, livestock or pets. Cleanliness in the immediate storage area is the best defense against diseases and an economical way to prevent losses.

Mushrooms

Usually mushrooms aren't harmful, but they're a nuisance. Mushroom development suggests that the growing medium has been maintained in a moist condition and hasn't dried sufficiently for several days. To avoid mushroom development, use fungicidal drenches or keep the media dry.

In summary, excessive blending, overcompaction, heavy watering and poor storage are the primary sources of media damage. Be sensitive to the qualities and characteristics that exist in a growing mix. Too often, a good quality media purchased at a good price is damaged as a result of poor management and handling.

References

[1] Agrios, G.N. 1978. *Plant Pathology*. Academic Press.
[2] Baker, R.F., ed. 1957. The U.C. system for producing healthy container grown plants. *California Agr. Experiment Sta. Manual* 23,332.
[3] Biernbaum, J., W. Carlson and R. Heins 1988. Low pH causes iron and manganese toxicity. *Greenhouse Grower* 6:3 (March).
[4] Boodley, J.W. and R.S. Sheldrake 1972. Cornell peat-lite mixes for commercial plant growing. *Cornell University Plant Science Information Bulletin* 43.

[5] Bunt, A.C. 1988. *Media and Mixes for Container-Grown Plants.* Unwin Hyman Ltd.

[6] Conover, C.A. and R.T. Poole. 1977. Characteristics of selected peats. Reprinted from *Florida Foliage Grower.* Vol. 14, No. 7.

[7] Elliott, G.C. 1987. Avoid toxicity problems: Be aware of medium-fertilizer interaction. *Greenhouse Grower* 5:11 (October).

[8] Fonteno, W.C. 1988a. Know your media, the air water and container connection. *GrowerTalks* 51(11):110-111.

[9] ——— 1988b. How to get 273 out of a 273 tray. *GrowerTalks* 52(8):68-76.

[10] Foth, H.D. 1984. *Fundamentals of Soil Science.* John Wiley & Sons.

[11] Hoitink, H.A.J. 1980. Composted bark, a lightweight growth medium with fungicidal properties. *Plant Disease* 64:142-7.

[12] Kramer, P.J. 1983. *Water Relations of Plants.* Academic Press Inc.

[13] Lindsay, W. 1979. *Chemical Equilibria In Soils.* John Wiley & Sons.

[14] Lucas, R.E. and J.F. Davis. 1961. Relationships between pH values of organic soils and availability of 12 plant nutrients. *Soil Science* 92:172-182.

[15] Lucas, R.E. 1982. *Organic Soils (Histosols).* Cooperative Extension Service, Michigan State University.

[16] Marschner, H. 1986. *Mineral Nutrition of Higher Plants.* Academic Press Inc.

[17] Mastalerz, J.W. 1977. *The Greenhouse Environment.* John Wiley & Sons Inc.

[18] Paul, E.A. and F.E. Clark. 1989. *Soil Microbiology and Biochemistry.* Academic Press Inc.

[19] Peterson, J.C. 1981. Modify your pH perspective. *Florists' Review* 169:34,35,92,94.

[20] Pirone, P.P. 1978. *Diseases and Pests of Ornamental Plants.* John Wiley & Sons Inc.

[21] Puustjarvi, V. 1966. On the standards of garden peat. *Acta Horticulturae* 4:153-155.

[22] Tisdale, S.L. and W.L. Nelson. 1967. *Soil Fertility and Fertilizers.* The Macmillan Company.

[23] Warncke, D.D. and D.M. Krauskopf. 1983. *Greenhouse Growth Media: Testing and Nutrition Guidelines.* Cooperative Extension Service, Michigan State University.

[24] White, J.W. 1970. Growing media. *Geraniums* 2d ed. Ed. J.W. Mastalerz. University Park, Penn.: Pennsylvania Flower Growers.

CHAPTER 3

Beneficial Microbes

Robert G. Linderman

Commercial geraniums are grown in soilless media, as are many other greenhouse-grown crops, primarily because such media are lightweight and generally have good physical and chemical properties. Growers assume they need control only irrigation and fertilization in soilless media to produce good, healthy plants, although most recognize the need for mixes to have good aeration, drainage and capacity to hold nutrients and be free from insects, pathogens and weeds as well as harmful chemicals. Generally overlooked, however, is the role played by nonpathogenic microorganisms in plant growth and development.

Managing microorganisms in the soil or planting medium and on plant and root surfaces involves eliminating many microorganisms from the growth medium, especially unwanted pathogens, insects and weed seeds. The medium or plant surface microflora is then reconstituted with selected microorganisms known to benefit plant growth and health. Critical to this plan's success, of course, is planting pathogen-free plant material, whether from seed, cuttings or tissue-cultured plantlets.

The rhizosphere phenomenon

Plants grow when roots and shoots interact with soil and air, respectively. The rhizosphere is the soil surrounding roots that is influenced by materials that exude from the roots. Microbial activity is enhanced in that region because of the higher nutrient level from root exudates.

The most numerous and probably most important microorganisms in the rhizosphere are either neutral, beneficial or deleterious bacteria or fungi. Deleterious microbes (pathogens) retard plant growth or kill plants by destroying roots; beneficial microbes impact plant physiology by their metabolic activities, either directly by affecting plant tissue, or indirectly by antagonizing or competing with deleterious microbes. Pathogens' effects on plants are usually obvious, but beneficial microbes' effects are not. Plants grown in soilless media may be devoid of beneficial microbes, and thus their inoculation into the medium or onto plant surfaces is desirable. Inoculating geraniums with three types of microbes—mycorrhizal fungi, biocontrol agents and plant growth-promoting bacteria—could improve growth and health.

Mycorrhizal fungi

Roots of most plants grown in soil form a symbiotic association with specific fungi called mycorrhizae [6]. Geraniums form so-called vesicular-arbuscular (VA) mycorrhizae, as do most agricultural crops. VA mycorrhizal (VAM) fungi colonize the root cortex and establish a direct contact between the host plant cell cytoplasmic membrane and the fungal cell wall. This interface is the exchange site between plant and fungus: Carbon nutrients are transferred from plant to fungus, and mineral nutrients, absorbed by fungal hyphae that grow into the soil, are transferred from fungus to plant. The root colonization by mycorrhizal fungi and the plant tissues' altered nutritional status changes a plant's physiology in ways that can greatly affect its growth and health. For example, under nutrient-deficient conditions, especially with immobile elements like phosphorus, copper and zinc, mycorrhizal fungi can enhance nutrient uptake and overcome the deficiency, thereby aiding plant growth (Fig. 3-1). Other stress situations alleviated by mycorrhizae are drought, manganese toxicity, salt stress, transplant shock and root rot diseases [6].

Fig. 3-1. Geranium seedling growth enhancement under phosphorus-limiting conditions by inoculation with vesicular-arbuscular mycorrhizal fungi (right) compared to nonmycorrhizal control plants (left).

Studies by Biermann and Linderman [1] with geranium seedlings indicated that plant growth enhancement was strong if VAM-inoculated plants were grown under low phosphorus conditions, but the effect disappeared when adequate phosphorus was provided. Even though the plants with or without mycorrhizae were generally the same size, mycorrhizal plants were more uniform than nonmycorrhizal plants. Mycorrhizae formation in soilless medium was generally reduced because of the high fertility level needed to grow the plants resulting from the medium's low nutrient binding properties. Adding some soil or other nutrient-binding ingredient to the medium generally corrected that inhibition.

Biermann and Linderman [1] also demonstrated that mycorrhizae formed on seedlings in seed pots persisted after transplant. This would presumably be true also in the case of plug transplants. Most striking, however, was that pretransplant inoculation increased subsequent geranium growth over that of nonmycorrhizal control plants when both were transplanted into soil, even though the plants were not visibly different at transplant time. The mechanism of improved transplantability isn't known.

Beneficial rhizobacteria

Root exudates stimulate soil microbes, especially bacteria, to multiply in the rhizosphere. Most of those bacteria are neutral saprophytes, but others are deleterious, and a few are potentially beneficial. The proportion of neutral, deleterious or beneficial bacteria depends primarily on the proportions in the soil to start with and each one's competitive capacity. Beneficial microbes usually are a relatively small proportion, often only 1% to 2%. Their low numbers preclude their having a significant impact on plant growth due to microbial competition. Inoculation to increase their numbers and proportion of the total bacterial population could, however, result in enhanced growth or health. This has been demonstrated with several greenhouse crops, but not geranium [3]. A high population of a single bacterium has been shown to increase the plant vigor and growth for reasons not yet determined conclusively. Three hypotheses to explain growth enhancement are: (1) production of growth-regulating phytohormones, (2) production of antibiotics or other substances that could inhibit the activity of deleterious microbes, or (3) alteration of plant nutritional status by some means. Plant-growth-promoting rhizobacteria (PGPR) frequently enhance plant growth in the absence of apparent or known pathogens [4]. Those bacteria that inhibit known fungal pathogens' growth and development are considered potential biocontrol agents (BCAs). Many bacteria can inhibit fungal pathogens under controlled cultural conditions, but their numbers in soil and activity against those pathogens again may be precluded by their relatively low numbers. Root disease control or suppression could be expected only if their numbers could be increased at the appropriate time and site of pathogen ingress into host tissue.

Selecting beneficial microbes

The beneficial microbes discussed above—mycorrhizal fungi, PGPR and BCAs—are indigenous to most soils. Mycorrhizal fungi associate with plant roots and produce large, resistant spores that reside in soil until they germinate and colonize a new host plant's roots. They are so large they can be sieved from soil and used to inoculate plants. Their efficacy on geranium, or any other plant for that matter, is a matter of testing by inoculation in the soil or medium in which the plant will be cultured. Mycorrhizal fungi may be inhibited in soilless media [1], but some have been shown to function well in that environment.

Selecting beneficial rhizobacteria, PGPR and BCAs may be a matter of screening many organisms. The selection process can be streamlined, however, by isolating from soils or media where diseases are thought to be naturally suppressed [7]. The first selection stage is to isolate bacteria, either by soil dilution plating or by direct isolation from roots. The first effectiveness test is to show pathogen inhibition on artificial media, but then, more importantly, to show disease reduction under greenhouse conditions. For geranium, these tests can be by inoculating medium, cutting or seed with the candidate bacterial BCAs and comparing growth and disease incidence of inoculated and noninoculated plants in pathogen-infested soil. Those bacteria that perform best in the first plant screen are ready for larger scale testing. Tests can involve inoculation with a single bacterium or bacteria combinations to demonstrate their compatibility with mycorrhizal fungi [5].

Managing rhizosphere microbes

The formula for managing rhizosphere microbes for greenhouse-grown crops like geranium involves two major steps: (1) treating growth medium with heat or chemicals to eliminate pathogens, insects and weeds, and (2) planting pathogen-free plants into soil or growth media that have been reconstituted with beneficial microflora. Sufficiently high populations of desired microbes, namely mycorrhizal fungi and bacterial biocontrol agents or other PGPR, should be used to ensure their dominance. In some cases, if not most, BCAs are more efficiently introduced directly onto plant tissues—cuttings, tissue-cultured transplants, or seeds—which puts them where they need to be to block a pathogen's entry. Inoculation timing is critical, and the earlier the better. Establishing mycorrhizae when seedlings are first germinating is ideal because the inoculum amount is minimal, and mycorrhizae are established by the time the seedling is transplanted. Managing rhizosphere microbes must also take into account pesticides' sometimes deleterious effects, high fertilizer levels and media components themselves. If the introduced microbes function as planned, then lower fertilizer and pesticide levels will be required to grow a healthy crop.

References

[1]   Biermann, B. and R.G. Linderman. 1983a. Effect of container plant growth medium and fertilizer phosphorus on establishment and host growth response to vesicular-arbuscular mycorrhizae. *J. Amer. Soc. Hort. Sci.* 108:962-971.

[2]   ———— 1983b. Increased geranium growth using pretransplant inoculation with a mycorrhizal fungus. *J. Amer. Soc. Hort. Sci.* 108:972-976.

[3]   Broadbent, P., K.F. Baker, N. Franks and J. Halland. 1977. Effect of *Bacillus* spp. on increased growth of seedlings in steamed and in nonsteamed soil. *Phytopathology* 67:1027-1034.

[4]   Burr, T.J. and A. Caesar. 1983. Beneficial plant bacteria. *Critical Rev. Plant Sci.* 2:1-20.

[5]   Linderman, R.G. 1988a. Mycorrhizal interactions with the rhizosphere microflora: The mycorrhizosphere effect. *Phytopathology* 78:366-371.

[6]   ———— 1988b. VA (vesicular-arbuscular) mycorrhizal symbiosis. *ISI Atlas of Science: Plants and Animals* 1:183-188.

[7]   ————, L.W. Moore, K.F. Baker and D.A. Cooksey. 1983. Strategies for detecting and characterizing systems for biological control of soilborne plant pathogens. *Plant Dis.* 67:1058-1064.

CHAPTER 4

Irrigation

J. Heinrich Lieth and David W. Burger

Water is an essential resource for greenhouse-grown plants' growth and development. It functions as a solvent and is necessary for all plant metabolic processes. Water is taken up through the roots, moves through the plant and evaporates from the leaves in a process called evapotranspiration. This process is vital since it is the driving force responsible for water movement throughout the plant, and as water evaporates from the leaves, it also removes heat. In the absence of evaporation, heat would build up in the plant whenever the sun shone on it, resulting in temperatures high enough to kill plant tissues. The rate at which water leaves the plant is related to leaf surface area, wind speed and temperature. Light is also a factor in that it generally affects leaf temperature and stomatal opening. Water application should parallel water loss through transpiration to avoid water stress.

Water stress

There is no clear point where plants are said to be under water stress. This term is generally used to describe the state where plants experience decreased growth rates or reduced cell expansion rates due to reduced water availability. Under light and moderate water stress, there are no visible symptoms that water stress is occurring; severe water stress involves wilting. Geranium water stress symptoms range from curling leaves and changes in leaf color and sheen to lower leaf yellowing. (See Color Fig. C64.)

Water quality

Water quality can be a determining factor in geranium production success or failure. Several characteristics describe water quality. These consist of pH (a measure of acidity), alkalinity (hardness) and electrical conductivity (EC). Irrigation water pH can, over a long time, affect the container medium pH. A basic pH (>7 on a 0 to 14 scale) is more often the problem than an acidic pH (<5). Water with a high pH may indicate the presence of ions possibly detrimental to plant growth [Na+ (sodium), HCO_3- (bicarbonate)]. The optimum pH for water and soil is between 5 and 7. High quantities of calcium and/or magnesium carbonates (hard water) in the water can increase pH. Hard water can be improved with ion exchange resins (deionizing systems) or reverse osmosis, but these processes are expensive. Using acidifying fertilizers (NH_4SO_4) and/or incorporating iron sulfate or elemental sulfur into growing media can reduce high pH problems. The EC of water and soil, measured with a solubridge, can be a useful measure indicating the total soluble salts in the solution with an acceptable value between 0 and 2 mmhos per square centimeter.

Additives to municipal water supplies such as chlorine or fluoride are generally at concentrations with no known harmful effect on geranium production. It has been shown, however, that geranium growth is diminished by chlorine levels of 2 ppm and above [3].

Leaching

Irrigation can remove unwanted substances from the root zone. Most irrigation solutions aren't perfectly matched to the plant's needs so that as the plant removes water and nutrients and water evaporates from the soil surface, salts and ions dissolved in the water are left behind. These may build up unless the container is flushed (leached) at regular intervals. The elapsed time between flushings depends on water quality, the degree to which the irrigation solution is matched to the plant's needs, and the plant's response to the EC of the irrigation and soil solution. This can be monitored by measuring the salt concentration of the water in the root zone (by measuring the leachate or a saturated paste of a soil sample). If the EC is above 1.5 mmhos, then flushing with clear water is in order. ECs above 2 mmhos per square cm are excessive.

Soil-water-air relationships

Irrigation's primary objective is to supply water to the medium in which the plant is growing so it's available when needed. Water is distributed throughout the root zone by various processes including mass flow (downward movement due to gravity) and capillary action (upward and lateral movement). The soil medium's physical characteristics determine how much water remains in the container after irrigation (container capacity), how much water is freely available and how the water is distributed. All container media consist of a solid component (peat moss, sand, bark, soil) and a nonsolid (void) component (water and air). As the quantity of one void component increases, the other decreases. A moisture-release curve relating moisture content (percent by volume) to soil moisture tension (measured in kilopascals, kPa, or millibars, mbar) shows the water quantity in a given volume of medium at container capacity and the available water. A typical moisture release curve for UC Mix (1:1:1 by volume of sand, redwood sawdust and peat moss) is shown in Fig. 4-1. At container capacity (nearly 0 kPa tension), approximately 75% of the void volume is water. Moisture content decreases (air content increases) with tension (as the water is removed by the plant or by evaporation) so that at 5 to 7 kPa, most of the readily available water has been removed. Above 10 kPa this medium (as well as most container media) has very little available water remaining (despite the fact that water can still be sensed by touch).

Most container media for greenhouse crops are designed to simultaneously have a high water-holding capacity and high porosity. Oxygen is necessary for water and nutrient uptake from the soil solution. Porous media (large particle sizes) holds less water and requires more frequent

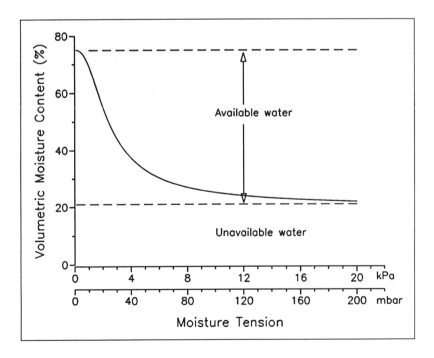

Fig. 4-1. Typical moisture release curve for UC Mix.

irrigations than media having smaller particle sizes. The container medium's ability to hold water and provide air affects how the plants are irrigated. With geraniums, container media should provide both water and air. Waterlogging conditions (soil medium holding too much water or too much water added too frequently) should be avoided. Geranium wilting can occur in response to overwatering as well as underwatering. If adding water to wilting plants growing in a moist medium doesn't reduce wilt, the problem is most likely overwatering. If the geranium's internal water content gets too high, small, pimplelike blisters appear on the underside leaf surface. This physiological disorder, called edema, is due to water uptake exceeding water loss through transpiration [1]. In commercial production close spacing, low light intensities, poor ventilation and high relative humidity are conducive to edema onset. With geranium stock plants used for cutting production, overwatering results in longer internodes and softer cuttings. Moderate watering results in higher quality cuttings that root readily. If watering methods keep the medium too dry, however, the cuttings may be too hard and be more difficult to root. Excessive water in the root zone may also result in root rot problems if pathogens are present.

Irrigation methods

As a general rule, geraniums are more susceptible to overwatering than underwatering. Therefore, regardless of how water is applied, it's a good idea to allow the root zone to dry down before rewetting.

Uniformity in water application is of utmost importance in irrigating any container-grown crop. The goal is to apply the same quantity of water to each container. This requires careful attention during hand watering and careful design and construction when using automated irrigation systems.

Hand watering

Growers use manual irrigation if labor rates are low or if they have found it difficult to make an automated system uniform. In such cases it can be used for touchup watering. Pot plants to be watered manually so that adequate space remains below the pot lip so that all needed water can be applied without having to wait for water to soak into the soil mix. Apply enough water so some water drains from the container bottom.

Overhead

Geraniums grown in plug and liner production, as well as those spaced pot-tight in small containers (≤4-inch[10 cm]) are generally watered with overhead irrigation. This is a convenient irrigation method; however, it can be wasteful and conducive to disease spread. Water is wasted if the overhead irrigation system is poorly designed or constructed, leading to nonuniform water application. Allowing water to stand on the foliage for long, cool periods, may cause disease.

Drip

Drip irrigation systems are well suited for potted geraniums that are not spaced pot tight. These systems involve a supply line that routes irrigation solution to the proximity of each pot. Weighted drip lines then send a slow water flow to each pot. For larger containers (>2 gallon; 8 l), spray emitters may be better since they distribute water more evenly across the top of the soil surface. Drip systems are very efficient water delivery systems since they place water only in the container. Water waste can still occur if the water is applied for too long or if the individual drip lines don't emit similar water quantities.

Some labor is required when setting up a drip irrigation system and when spacing a crop; little labor is needed during crop production.

Subirrigation

Subirrigation provides another way of irrigating container-grown plants. Subirrigation can be accomplished with capillary mats or with watertight benches, trays or floors. With capillary mats, apply water directly to a mat on a solid, sloping bench. Place pots directly on the mat; irrigation solution flowing through the mat is absorbed through holes in the pot bottom. This system has the drawback that since water is typically entering the pot only from below, salts will accumulate at the soil surface as water evaporates and leaves materials dissolved in the water behind. An additional irrigation system may be needed to flush out these salts. Watertight benches or floors tend to be more expensive than the other methods, particularly when coupled with recycling systems.

Irrigation system optimization

Optimizing any irrigation system to deliver water to container-grown plants depends on two major characteristics: (1) the system's capacity to deliver adequate water volume and (2) the delivery of nearly equal water volumes to every container. Capacity is the capability to deliver the needed water amount to all plants on an ongoing basis. Water application should be uniform so that a uniform crop can be produced.

An irrigation system's capacity involves the water delivery rate as well as the water pressure in the irrigation system. If the pressure is too low, the system may be unable to deliver water to all plants, resulting in some plants receiving no water. Excessive pressure can damage the irrigation system by breaking emitters, forcing drip lines out of supply lines and possibly even breaking supply lines.

Pressure variations can affect system uniformity. Therefore, pressure regulators should be designed into any irrigation system so the system pressure will always be the same, assuming the supply pressure is above some minimum level.

To assure that each pot receives approximately the same amount of water, test drip and overhead systems regularly (at least annually) to determine the Coefficient of Uniformity (CU) [2]. To do this, use the following procedure.

(1) Arrange empty, water-tight cans (all should be the same size) in a grid pattern over the irrigated area to catch applied water from overhead sprinklers or place drip emitters into individual cans.
(2) Operate the irrigation system for a known period of time under normal conditions.
(3) Measure the volume of water in each can (x_i).
(4) Determine the mean of all these measurements ($\bar{x}$).
(5) Sum the absolute values of the deviation of each measurement from the mean ($\sum |\bar{x}-x_i|$).
(6) Then calculate the CU for the irrigation system by:

$$CU = 100\left(1.0 - \frac{\sum |\bar{x}-x_i|}{\bar{x}\,n}\right)$$

where n is the number of observations (cans).

CU values below 80 indicate poor system performance that will lead to the nonuniform and wasteful water application. System redesign or maintenance may be required to improve a CU below 80. Nonuniform irrigation systems are likely to result in nonuniform crops where a decision has to be made between losing a percentage of the crop (pots where delivery rates are low) or generating excessive runoff (from pots with the highest delivery rates). Clearly, the more uniform the delivery system, the more economical and precise it is.

Irrigation technology

While all delivery systems can be controlled manually, all except manual irrigation can be automated easily. This involves using solenoid valves to turn water on and off and a controller to schedule the irrigation. Controllers range from simple timers to full-scale greenhouse automation systems using computers. There are several important considerations when programming the controller. Drip and overhead systems require some minimum pressure to operate uniformly and efficiently. With each additional irrigation circuit turned on, the pressure decreases. Thus, care must be taken so the total system pressure is known as well as the pressure-drop occurring with each circuit. It's important that the supply system (pumps or municipal water supply) be rated at a delivery rate adequate to maintain the nursery needs on a hot summer day when water-use rates are highest.

In the past, irrigations have generally been scheduled based on time and fixed durations. The times and durations are generally set to match the maximum use possible to ensure against underwatering. While this methodology ensures against underwatering, it can lead to significant water and fertilizer waste. Computerized, environmental-control technology makes it possible to use sensed data to decide when to irrigate and for how long. Moisture-sensing devices, such as tensiometers, can be used in the root zone of potted plants and benches of stock plants to automate irrigation. Such systems have been under investigation for container-grown crops and are now becoming commercially available [4]. Also, measured light and temperature data can help predict evapotranspiration so the summation of these estimates approximates the total water amount used by the plant. Irrigations are then done when most available water in the root zone is predicted to have been removed.

While all these automated control methods are environmentally sensitive and work well most of the time, grower input is still needed to make sure the system performs as expected. For example, sensors must be serviced and locations evaluated so that the sensors represent what the whole crop is experiencing.

Future prospects

Automation is currently changing the way container-grown plants are being irrigated. The trend is for this technology to become more refined, more powerful and less expensive. As this happens more growers will find that irrigation costs can be reduced by robotics and computer-controlled booms. Computer vision, mathematical models and artificial intelligence are areas where irrigation could be tailored to individual plant needs.

The economics of any automation equipment will depend largely on water and labor costs. Also, in the future, environmental concerns may result in financial pressure on growers through fines and discharge permits. The payback period for equipment that minimizes runoff while optimizing plant water delivery is shortest where these costs represent a significant part of the production cost.

References

[1]   Balge, R.J., B.E. Struckmeyer and G.E. Beck. 1969. Occurrence, severity and nature of Oedema in *Pelargonium hortorum* Ait. J. Amer. Soc. Hort. Sci. 94:181-183.

[2]   Christiansen, J.E. 1942. Irrigation by sprinkling. *University of California Bulletin* No.670.

[3]   Frink, C.R. and G.J. Bugbee. 1987. Response of potted plants and vegetable seedlings to chlorinated water. *HortScience* 22(4):581-583.

[4]   Lieth, J.H. and D.W. Burger. 1989. Growth of chrysanthemum using an irrigation system controlled by soil moisture tension. *J. Amer. Soc. Hort. Sci.* 114(3):387-392.

CHAPTER 5

Fertilization

Richard L. Biamonte, E. Jay Holcomb and John W. White

The essential elements for plant growth are carbon (C), hydrogen (H), oxygen (O), nitrogen (N), phosphorus (P), potassium (K), calcium (Ca), magnesium (Mg), sulfur (S), iron (Fe), molybdenum (Mo), boron (B), copper (Cu), manganese (Mn), zinc (Zn) and chlorine (Cl).

Carbon and oxygen are provided in the atmospheric gas carbon dioxide that enters the plants through small openings in the leaves called stomata. Enriching the atmosphere with carbon dioxide is similar to fertilizing the soil by adding mineral elements. A low carbon dioxide level can reduce geranium growth rate, especially in very tight greenhouses. Hydrogen enters the plant primarily through the roots in the form of water. Water also carries all other essential elements that enter the plant primarily in root absorption.

The essential elements, in conjunction with chlorophyll and sunlight, are raw materials required for plant growth and development. The plant is analogous to a factory whose output of goods depends on each worker, in this case the essential elements. Insufficient levels of one or more essential elements may limit plant performance or potential.

Major elements

Nitrogen. Nitrogen compounds comprise from 40% to 50% of plant cells' living substance. It's required in relatively large quantities in connection with all growth processes in plants. Nitrogen is absorbed by plant roots as ions in the nitrate (NO_{3-}), ammonium (NH_{4-}) or as an organic urea form in amide (NH_{2+}).

Most floriculture plants absorb the largest nitrogen quantities as nitrate. Nitrate is easily transported throughout the plant and is primarily stored in root and shoot cell vacuoles. The plant converts nitrate to ammonia by an enzymatic process called nitrate reduction,

39

which requires molybdenum as a catalyst. This leads to the formation of amino acids and proteins of which nitrogen is a major constituent.

Ammonium nitrogen may be absorbed by roots where it's transformed into organic compounds, a process that requires carbohydrates. Using high ammonium fertilizers may result in lower quality plants, especially during the darker and cooler growing seasons. Cool and overly saturated/low oxygen growing media result in higher ammonium levels and reduce conversion to nitrate.

Certain nitrogen compounds are very mobile within the plant. When a nitrogen deficiency occurs, symptoms appear in oldest leaves first.

Phosphorus. This element is closely related to the plant's vital growth processes. Like nitrogen, it's part of the nucleic acids and proteins that form the protoplast's structural framework. As a catalyst in energy transfer, it may be involved in converting starch to sugar. Phosphates act as buffers to maintain satisfactory acidity and alkalinity conditions in plant cells. Phosphorus, absorbed by plant roots as phosphate ($HPO_{2=}$ or H_2PO_{4-}), is especially important in seed germination, seedling metabolism and root development. In fertilizer, phosphorus combines with many cations to form soluble, slightly soluble and insoluble compounds. By law, the analysis on the fertilizer bag is usually expressed in equivalent phosphorus pentoxide (P_2O_5) units. Compared to nitrogen, phosphorus isn't used by the plant in large quantities, but a constant supply of this nutrient is essential.

Phosphorus and nitrogen functions in the plant are closely related. Available soil nitrogen depresses phosphate uptake and when phosphates are readily available, nitrogen uptake is depressed. It also has been reported that phosphates are more rapidly absorbed by plants when nitrogen is added in organic forms such as urea than when nitrogen is available in other forms.

Phosphorus may be blended in the growing medium prior to planting, utilizing nonsoluble sources, such as single or concentrated superphosphate. An advantage of single (20%) superphosphate over concentrated superphosphate (40%) is that it contains gypsum. The gypsum in single superphosphate provides calcium (Ca^{++}) and sulfates ($SO_{4=}$). A possible attribute of gypsum is the capability to tie up ammonium ions (NH_{4+}) and reduce ammonium toxicity potential.

Phosphorus is also applied with N-P-K water soluble fertilizers. The most common phosphorus source is ammonium phosphate. Potassium phosphate may also be used, but its high cost makes it less common. Another phosphorus source is phosphoric acid, which is used primarily in water treatment to lower alkalinity.

The growing medium pH can influence phosphorus availability, especially as a dry additive such as superphosphate. In general, phosphorus combines with calcium at high pH levels and with aluminum and iron when the pH is low. These situations result in the formation of insoluble compounds. Thus, it's important to regularly apply phosphorus in a fertilizer program to ensure its availability for plant use.

Potassium. Potassium doesn't chemically combine with the plant's building materials. It's probably involved with carbohydrate and protein synthesis, in regulating water conditions within the plant cell and water loss by transpiration, as a catalyst and condensing agent of complex substances, as an accelerator of enzyme action and a contributor to photosynthesis, especially under low light intensity.

Potassium is absorbed and translocated as an ion (K^+). Potassium nitrate is the primary source of potassium used in fertilizers. Potassium salts of chloride and sulfate are used but to a much lesser extent since they contribute to relatively high soluble salt levels in the growing medium. Potassium analysis in fertilizers is usually expressed as percent soluble potassium oxide (K_2O). In some fertilizers (depending on the state) and in most growing medium and leaf analyses, potassium is reported in the elemental form (K). Potassium translocates readily from older to younger tissues; therefore, deficiency symptoms appear on oldest leaves first.

Calcium. One of calcium's main functions is as a cell wall constituent. The middle lamella that glues cells together consists largely of calcium pectate. Calcium has a vital role in the growing points (apical meristems) and is especially important in root development. Calcium may also be important in regulating potassium activity and in nitrogen absorption. Calcium isn't very mobile in plant tissue and, therefore, a deficiency often occurs in shoot and root tips. Calcium also neutralizes acid soils or growing media and helps to improve soil structure by increasing colloid formation.

Magnesium. Magnesium is important for protein synthesis and is the main chlorophyll molecule constituent. When magnesium is deficient, reduced chlorophyll in chloroplasts results and the common symptom is chlorosis. A low magnesium level may also reduce root growth. Magnesium is very mobile within the plant and a deficiency appears first in older leaves.

Calcium and magnesium are usually present in the soil as ions (Ca^{++}) (Mg^{++}) and in fertilizers as salts of sulfates and chlorides and sometimes as salts of carbonates, bicarbonates and nitrates. Calcium is often applied to the soil in form of calcitic or dolomitic limestone, or as gypsum. Magnesium sources include Epsom salt ($MgSO_4 \cdot 7H_2O$), Mag-Ox, emjeo, Kieserite, K-Mag, magnesian limestone, PRO/MESIUM, Sul-Po-Mag and Mag-Amp.

Sulfur. Sulfur is usually absorbed by roots as the sulfate ($SO_{4=}$) ion. Sulfate in fertilizer is combined with many cations and is the primary carrier for many elements such as iron, manganese, copper and zinc. Sulfur, like nitrogen and phosphorus, is a constituent of the amino acids, cystine, cysteine and methionine, three of the compounds from which proteins are made, and is also a constituent of thiamine and biotin, important plant hormones. Sulfur usually is applied to the soil as flowers of sulfur, single superphosphate, gypsum or sulfur.

pH. The pH scale indicates acid and basic values in the growing medium. The pH level of 7, midpoint in the scale from 0 to 14, is neutral and the values below this point are acidic and those above are basic. Most crop plants don't grow when the pH level is below 4 or above 8.

Geraniums are influenced by the growing medium's pH (see Chapter 2, Growing Media). Koranski [7] suggested that hybrid geraniums would not flower at a pH below 5.5 and that the leaves would develop brown spots. Biernbaum et al [2] reported that at a pH of less than 5.8, geraniums were susceptible to iron toxicity and possibly manganese toxicity. Biamonte [1], working with hybrid Red Elite geraniums, observed iron toxicity symptoms of speckling in the interveinal areas in older leaves at a pH of 5.45 or less and none at a pH level of 5.84. R.P. Vetanovetz and J.F. Knauss [13] indicated that saturated media extract (SME) pH levels less than 6.0, iron of 1.0, and iron-to-manganese ratio greater than 3.1 may lead to iron toxicity. Biernbaum et al [2] recommended the pH range of 6.0 to 6.3 for geraniums.

Trace elements

Plants use trace elements in very small amounts. Many function as constituents of specific enzymes working in a catalyst role. Although required in very small amounts, a specific trace element deficiency or excess can damage plant growth. Iron, manganese, boron, copper, zinc and molybdenum are trace elements that most commonly cause problems in plants.

Iron. The role of iron appears to be that of a catalyst in chlorophyll formation but not as a constituent of it. As a result, chlorosis is a common symptom of iron deficiency. Since iron is relatively immobile in plant tissue, deficiency symptoms are first noted in the younger tissue. Iron is usually absorbed as physiologically available ferrous (Fe^{++}) or ferric (Fe^{+++}) ion. The ferrous (Fe^{++}) can oxidize to the less available ferric (Fe^{+++}) form in leaves. Foliar analysis doesn't necessarily distinguish between these iron forms. Iron chlorotic plants may have a sufficient but physiologically unavailable form of iron when compared with normal plants. Iron is applied as a salt ($FeSO_4 7H2O$) or as a chelate (disodium ferrous versenate, iron EDTA) or a frit (ground glass) but may be present in the soil as ferric oxide (Fe_2O_3). Iron is seldom lacking in soil but is often present in an unavailable form. Iron may become deficient as a result of being tied up with phosphates, carbonates or hydroxides. The latter two anions are associated with overliming or high pH conditions in the soil. Water saturated/low oxygen soil conditions may also inhibit iron uptake.

Iron toxicity. Geraniums appear to be efficient in their iron uptake and accumulation. Iron toxicity symptoms in geraniums include chlorotic to necrotic speckling in interveinal areas of mature leaves. R.L. Biamonte [1] grew Red Elite seedling geraniums and determined that iron toxicity symptoms weren't present when the iron-to-manganese ratio was 1:1 and no greater than 2:1 in the growing medium and the pH level was 5.8.

Manganese. Manganese is involved in chlorophyll formation and acts as a catalyst in oxidation reduction reactions and an enzyme activator. Manganese deficiency results in lessened photosynthetic activity. It's also closely associated with iron in the plant and the two elements may show antagonistic effects. If manganese is present in excessive quantities, iron solubility may be decreased to such an extent to cause a deficiency resulting in chlorosis. Conversely, an excessively high iron level may cause a manganese deficiency to develop. Manganese is absorbed in the form of a manganous (Mn^{++}) ion and applied as manganese sulfate ($MnSO_4 H_2O$) or as chelated manganese EDTA.

Manganese toxicity. Using soilless growing media in place of soil has eliminated the need, in most cases, for steam pasteurization. Therefore, manganese toxicity, as a result of steaming, is less likely to occur in soilless growing media. Manganese, however, may be found in pine bark in a soluble form. Where growers are still using soil, especially in stock plant production, steam treating soil at 212 F (100 C) for several hours resulted in the conversion of manganese into potentially toxic forms [14]. Other methods that could produce high levels of available (water soluble and exchangeable) manganese are reducing pH, creating areas of high moisture and low oxygen (poor drainage) and using large quantities of ammoniacal nitrogen fertilizer. Limestone soils may contain large quantities of naturally occurring manganese in unavailable forms. Continued use of complete fertilizers, particularly the more acidifying formulae, which contain appreciable manganese quantities, greatly contribute to the total supply. Thus, in time, soil manganese can reach levels that will be toxic given the above conditions. Lower steaming temperatures (140 F to 160 F; 60 C to 70 C), using steam-air mixtures, routine leaching, maintaining near neutral pH (7.0), using only nitrate forms of nitrogen and using adequate calcium sulfate (6 ounces per 1 bushel) (70 g per 35 l) helps prevent this problem. The symptoms are chlorosis of the terminal and young leaves and root loss, which eventually lead to either wilting or severe growth stunting.

Holcomb and Fortney [4] grew geranium stock plants in perlite fertilized with nutrient solutions containing varying manganese concentrations. With a manganese level in solution of 50 ppm (100 times recommended manganese solution level), growth and cutting production were similar to the recommended level. Cuttings taken from high manganese fertilized plants, which had foliar manganese levels as high as 2,000 to 3,000 ppm, rooted as rapidly as cuttings from control plants.

Boron. Boron has been ascribed many roles. It's necessary for sugar translocation and may enhance respiration; it's involved in nucleic acid synthesis and pollen germination. It also tends to keep calcium in a soluble form within the plant and may act as a regulator of potassium/calcium ratios. Furthermore, boron may be involved with nitrogen metabolism and oxidation equilibria in cells. Boron, applied as sodium borate ($Na_2B_4O_7$ $10H_2O$), is absorbed as an ion ($BO_3^=$). Two major boron sources are borax and boric acid. Some irrigation waters contain toxic quantities of boron. Using fertilizers with low boron levels would be advisable under such circumstances.

Copper-zinc-molybdenum. These elements are required in such small quantities that a deficiency seldom occurs in greenhouse-grown geraniums. They are absorbed by plants as ions: (Zn^{++}), (Cu^{++}) and ($MoO_4^=$).

Copper and zinc are provided in water soluble fertilizers in chelated form as copper EDTA and zinc EDTA. These nutrients are also available as copper sulfate, $CuSO_4$ $5H_2O$ and zinc sulfate, $ZnSO_4$ $7H_2O$. Fungicides may contain copper and/or zinc. Molybdenum sources are sodium molybdate, Na_2MoO_4 $2H_2O$ and ammonium molybdate [$(NH_4)_2MoO_4)$]. Most water soluble fertilizers use sodium molybdate because of lower cost.

Fertility management

Determining fertility requirements

A geranium crop's nutritional status can be determined by using information obtained from the analyses of growing media, plant tissue and irrigation water and by visual plant observation. These tools are available to growers who wish to properly manage fertility programs.

Growing media testing. Growing medium analysis, conducted on a regular schedule, is an excellent way for the grower to maintain the proper nutrient level and balance plant uptake. Although the extractable nutrients indicate their presence in the growing medium, that presence doesn't guarantee that root absorption will occur. Factors such as high or low pH and high soluble salts may be involved and inhibit nutrient uptake. Tissue analysis, however, reveals the nutrients and the levels taken up. The combination of growing medium and tissue analysis aids in interpreting and diagnosing plant nutritional disorders.

The procedure utilized in extracting nutrients from the growing medium can differ among the various university and commercial laboratories. It's a good idea for growers to become familiar with the procedure utilized by their service laboratory. This is necessary for the proper data interpretation. For example, the laboratories that utilize a weak acid compared with the laboratories that use distilled water in extracting nutrients from the growing medium show results with different extractable nutrients levels.

One of the most popular procedures is the saturated medium extract procedure (SME). Distilled water is the extractant, and it removes the readily available nutrients from the growing medium. The saturated medium extract procedure is used primarily with soilless growing media. An attribute of SME is that it compensates for water holding capacity differences among growing media. The normal range for extractables using the SME is found in Table 5-1.

TABLE 5-1

The extractables using the saturated Medium Extract Procedure and their normal range

Test	Normal range*
pH	5.2–6.3
Soluble salts (mmhos/cm)	0.75–3.5
Nitrate ppm NO_3	35–180
Ammonium ppm NH_4 - N	0–20
Phosphorous ppm P	5–50
Potassium ppm K	35–300
Calcium ppm Ca	40–200
Magnesium ppm Mg	20–100
Boron ppm B	0.05–3.0
Iron ppm Fe	0.30–3.0
Manganese ppm Mn	0.02–3.0
Copper ppm Cu	0.001–0.5
Zinc ppm Zn	0.30–3.0
Molybdenum ppm Mo	0.01–1.0

*The normal range is general for most crops. The preferred pH range for geraniums is 5.8 to 6.5.
Source: Grace-Sierra Testing Laboratory.

Water analysis. Water quality may differ from one region to the next. The grower whose range depends on more than one water source may find qualitative and quantitative differences among them. Although water quality is not expected to change drastically over time, growers should have their irrigation water analyzed at least once each year. Knowing the water quality is useful information for designing the fertilizer program. Aspects of water quality are found in Table 5-3 for alkalinity and Table 5-4 for elemental content. Growers should assess water quality before changing fertilizer formulations.

Tissue analysis

Leaf analysis or foliar analysis gives the grower an additional tool to assess a crop's nutrient status. It's better to *prevent* than to *correct* a nutritional problem. By the time a nutrient disorder has resulted in a deficiency or toxicity, it has also affected plant growth.

Tissue analysis standards for zonal, seed, ivy and Regal geranium leaves are listed in Table 5-2. The most recently mature leaves were sampled for developing these standards. Factors affecting plant tissue nutrient content are plant age, leaf position on plant, cultivar, time of day of leaf harvest, season, fertilization methods, number of days after

TABLE 5-2

Tissue analysis standards

Nutrient	Zonal geranium	Seed geranium	Ivy geranium	Regal geranium	Normal range
Nitrogen (%)	3.8–4.4	3.7–4.8	3.4–4.4	3.0–3.2	3.3–4.8
Phosphorus (%)	0.3–0.5	0.3–0.6	0.4–0.7	0.3–0.6	0.4–0.67
Potassium (%)	2.6–3.5	3.3–3.9	2.8–4.7	1.1–3.1	2.5–4.5
Calcium (%)	1.4–2.0	1.2–2.1	0.9–1.4	1.2–2.6	0.81–1.2
Magnesium (%)	0.2–0.4	0.2–0.4	0.2–0.6	0.3–0.9	0.2–0.52
Iron (ppm)	110–580	120–340	115–270	120–225	70–268
Manganese (ppm)	270–325	110–285	40–175	115–475	42–174
Zinc (ppm)	50–55	35–60	10–45	35–50	8–40
Copper (ppm)	5–15	5–15	5–15	5–10	7–16
Boron (ppm)	40–50	35–60	30–280	15–45	30–280
Molybdenum (ppm)	–	–	–	–	0.2–5.0

Sources: GrowerTalks August 1989. Assembled and updated by R.E. Widmer (1985) and updated by Dole and Wilkins (1988), University of Minnesota. Geranium normal range information provided by the Grace Sierra Testing Laboratory.

fertilization, media pH, growth regulators, physiological condition of tissue (healthy or aberrant and presence of diseases and insects) and temperature.

Tissue standards are general guidelines and not absolute values. They are best used in combination with media analysis values, visual observations and records of fertilization procedures and rates. Sampling at regular intervals, especially for stock plants used for propagation, helps establish a track record so you can compare variations over time.

Major elements—visual symptoms

Nitrogen deficiency. The most obvious symptoms are reduced growth plus a slightly lighter green color, particularly on the youngest leaves. With moderately severe nitrogen deficiency the venation area becomes markedly redder and red pigments show clearly on major veins. (See Color Fig. C43 and C57.)

Phosphorus deficiency. The first symptoms are reduced leaf size. Necrosis appears in large blotches at older leaf margins. Margins tend to curl. As the deficiency progresses, older leaves die, although younger leaves remain green but small. A pronounced, purplish red pigment is usually visible in the severely affected leaves on the lower plant portion. (See Color Fig. C45, C46 and C57.)

Potassium deficiency. First symptoms are a slight, interveinal chlorosis and dull, rather than glossy leaves followed by a general, moderate chlorosis somewhat like nitrogen deficiency. The leaf veins remain slightly greener than the rest of the leaf. Reddish blotches develop on leaves. Young leaves lose their luster early and never attain the size of leaves on healthy plants. When the deficiency is severe, cells elongate but don't divide. (See Color Fig. C44, zonals and C59, Regals.)

Calcium deficiency. At low calcium concentrations, roots become flaccid and, in the extreme case, turn black within a few days, or become stubby and thick with the meristem apparently dying. With slight deficiency, top growth is reduced but general plant color lacks gloss, there is a slight, interveinal chlorosis and some leaves are smaller than usual and have chlorotic margins. With severe deficiency, top growth is greatly stunted and dieback occurs, while new leaves stay small. There may be some chlorosis, but major veins remain green. Zonation is normal. (See Color Fig. C47 and C51.)

Magnesium deficiency. The first symptoms are reduced growth and smaller leaf size. Leaf margins tend to curl down. A pronounced chlorosis develops, which may have reddish tones. The pattern takes on several forms, depending on leaf age. The margins sometimes become chlorotic with a tendency for veins to remain green. Where the deficiency develops gradually, chlorosis first appears at the margins of older leaves spreading inward. These leaves sometime become completely necrotic, while the upper leaves appear normal in color but smaller. Various degrees of chlorosis and necrosis may be evident between these extremes. With severe deficiency, roots are stubby with much branching, somewhat dark and slightly flaccid. In general, cell elongation isn't reduced under magnesium deficiency. (See Color Fig. C48, C49 and C50.)

Trace elements—visual symptoms

Iron deficiency. Younger leaves first turn chlorotic with veins remaining green, then younger leaves turn almost completely white. Zonation is at first prominent, then absent. Leaf size is reduced. Chlorosis begins at younger leaves' outer margin. The apical meristem may die. Under severe conditions younger leaves become necrotic. (See Color Fig. C55, zonals and C61, Regals.)

Iron toxicity. Mature leaves develop chlorotic to necrotic speckling in their interveinal areas. (See Color Fig. C58.)

Manganese deficiency. A light, mottled chlorosis of recently matured and younger leaves beginning first at margins and then spreading over the entire leaf becomes evident before growth is significantly reduced. Younger leaves' zonation is markedly reduced in early symptoms. As the deficiency becomes more pronounced, a very distinctive chlorotic pattern develops where the veins remain green, but the interveinal region has a spotted appearance. A "leopardlike" pattern eventually develops on all leaves. An advanced symptom is chlorosis in the veins near the margins. Numerous, small chlorotic spots about 0.3 inch (2 or 3 mm) in diameter, more or less uniformly distributed over the leaves, begin developing when symptoms are moderately pronounced. The spots remain about the same size but become increasingly chlorotic. Manganese deficiency is usually associated with high pH or excess liming. (See Color Fig. C56.)

Boron deficiency. Boron is relatively immobile within the plant. Symptoms first become evident on the youngest plant parts. Kofranek and Lunt [6] observed first symptoms on recently matured leaves, which were lighter green. These leaves displayed a general light chlorosis. Being lighter, venation in the color zone stands out clearly. On the underside of nearly mature leaves are elongated translucent specks that appear to be somewhat raised and develop first on top of secondary and tertiary veins. As symptoms become more pronounced on these leaves, just the major veins become chlorotic with blotchy, interveinal chlorosis developing, often most pronounced near the petiole. In the early stages the growing tip seems normal, but the youngest leaves show mild, mottled chlorosis. Subsequently, the growing tip dies, flower buds abort and the youngest leaves are deformed and become puckered. Leaf petioles are loosely held to stem and leaves may fall off with slight downward pressure. (See Color Fig. C53 and C54, zonals and Fig. C60, Regals.)

McFadden [11] reported boron deficiency symptoms when young leaves had accumulated less than 27 ppm boron and toxicity when above 325 ppm boron. He suggested that boron use should be restricted to instances where boron deficiency symptoms have actually been diagnosed. He recommended as general guides for application of boron either Solubor at ½ to 1 ounce (15 to 30 milliliters) or household borax at 1 to 2 ounces (30 to 60 milliliters) per 100 gallons (379 l) irrigation water.

Boron toxicity. On older leaves a thin (about 1 mm) black, necrotic zone develops at leaf margins. The condition is barely noticeable at first and advances quite slowly. (See Color Fig. C52.)

Copper deficiency. The first symptoms are a mild chlorosis with reddish tones near middle and younger leaf edges. The zonation becomes more pronounced as green intensity is reduced and a slight, copper-colored tint develops in the zonation. As the symptoms advance, older leaf margins become marked with pinkish red pigmentation. Some necrosis starts at leaf margins; most of the leaf is pinkish red except a small green zone near the petiole, which is marked by green veins extending into the pink zone.

Zinc deficiency. Deficiency symptoms show first on middle and subsequently on upper leaves. The symptoms are quite similar to copper deficiency except that the chlorosis and bronzing of the leaves is more general. Mild chlorosis tends to develop near the leaf edge. Later, the leaf edge sometimes becomes chlorotic with a reddish pigmentation developing near the margin. Upper leaves become small, and terminal growth develops slowly.

Sulfur deficiency. Deficiency symptoms are uncommon, but resemble nitrogen deficiency. Shoot growth is restricted, though less severely than with nitrogen deficiency, and a general yellowing or paleness ("tea yellows") appears on all leaves. Sulfur in soil comes from atmospheric SO_2, especially in industrial centers, and from sulfate fertilizers such as gypsum, Epsom salts, iron sulfate and single (20%) superphosphate. Sulfur is largely available due to bacterial activity in the media, and therefore could be slow in conversion in soilless media when the media is cool. Some plants obtain atmospheric SO_2 through the leaves, but this phenomenon hasn't been reported in geraniums. (See Color Fig. C62, Regals.)

Color Fig. C43 through C62 aid the diagnostician in identifying some of these nutritional deficiencies and toxicities, but you shouldn't rely completely on visual diagnosis in determining corrective treatments. Use growing media and tissue test results as supportive diagnostic tools. Even when using all three tools, a good measure of common sense is necessary. If in doubt, try corrective treatments on a small plant group before treating an entire crop. The best control method for nutritional problems is to never allow them to develop. Routine growing media testing combined with a regular, planned fertilization program is the best approach to fertility management.

Maintaining plant nutrition

When following a regularly scheduled fertilizer program, it's possible to maintain the proper nutrient level and balance in the growing medium necessary to produce high quality geraniums. Apply water soluble fertilizers with each application of water to provide a relatively low nutrient dosage per application.

In keeping with constant feeding is controlled release fertilizer, which may be topdressed at planting time or preincorporated in the growing medium. This fertilizer releases nutrients over time through openings in the surface of resin-coated prills. One benefit of controlled release fertilizers is nutrient runoff reduction and elimination.

Nutritional program

A most effective nutritional program is combining water soluble fertilizer and controlled release fertilizer. The combination creates better plant performance than when either fertilizer type is applied solely. The combined fertilizer program also offers the advantage of using a lower dosage of each fertilizer.

Parts per million. The concentration of nutrients in a solution of water soluble fertilizer is expressed as parts per million (ppm). Parts per million is essentially a ratio of the nutrient weight added in water and the weight of water. When 1 ounce (30 milliliters) of fertilizer is added in 100 gallons (378 l) of water, this is equivalent to 75 ppm. In using the number 75, it's possible to calculate ppm for any nutrient in the fertilizer. Although several nutrients are available in a fertilizer, it's customary to use ppm of nitrogen as the reference for the dosage concentration when fertilizing plants, including geraniums.

In the example that follows, 200 ppm nitrogen will be calculated.
Step 1. Multiply the percentage of nitrogen in the fertilizer by 75.
Step 2. Divide the result from Step 1 into the desired ppm of nitrogen. A 15% nitrogen fertilizer will be used.
Example: Step 1. 75 (.15) = 11.25 ppm N;
Step 2. 200 ppm N ÷ 11.25 ppm N = 17.8 ounces (.5 l). The 17.8 ounces added in 100 gallons (378 l) of water is equal to 200 ppm N. That number of ounces of fertilizer would also be added per gallon of concentrate with a 1:100 or 1% dilution ratio injector.

Injector dilution ratio. Fertilizer injectors provide the advantage of preparing small concentrate quantities and diluting them to a required level of fertilizer nitrogen. A 1:100 dilution ratio was used in calculating ppm N. This refers to the capability of making 100 gallons of dilute fertilizer from one gallon of concentrate. The ounces of fertilizer added per gallon of concentrate is influenced by the dilution ratio. Changing the dilution ratio in the previous example from 1:100 to 1:200 would require doubling the amount of fertilizer per gallon of concentrate in order to maintain the 200 ppm N. Water soluble fertilizers have limits of solubility that may be reached or exceeded with high dilution ratios and result in a precipitate in the stock concentrate.

Monitoring the fertilizer solution. Water soluble fertilizers are composed of nutrient salts that separate in solution into the component ions and contribute to electrical conductivity. Increasing the amount of fertilizer in solution should increase the conductivity. Therefore, a most effective way to monitor the fertilizer solution is to measure its electrical conductivity. Portable units are available for this purpose, as are electrical conductivity meters that can monitor the continuous flow of dilute solution from the injector. Fertilizer electrical conductivity tables are available. It's important to subtract the electrical conductivity of the irrigation water to obtain the net fertilizer's conductivity. Note: Urea is chemically organic and does not contribute to conductivity.

Water quality. Plant nutrients and other elements in the irrigation water may influence plant growth. Growers are encouraged to have their water sources tested since fertilizer programs should be based on water quality. High alkalinity or bicarbonates can tie up trace elements and also increase the growing medium pH. It is necessary to employ mineral acid, such as phosphoric acid, at alkalinity levels above 180 milligrams calcium carbonate per liter (180 ppm). There may also be a need to supplement the fertilizer program with calcium and/or magnesium when these nutrients are at low levels in the irrigation water.

Selecting the water soluble fertilizer. Fertilizers most often used by geranium growers are 15-15-15, 15-16-17 and 20-10-20, since they provide from 48% to 60% nitrate nitrogen. These higher nitrate-containing fertilizers are preferable in producing crops in soilless growing media to prevent an ammonium nitrogen buildup. When ammonium is converted to nitrate in the growing media, however, it may eventually lower the pH below the desired range of 5.8 to 6.5. Plant roots, when absorbing ammonium, release H^+ ions in the growing medium, thereby contributing to a lower pH. In taking up nitrates, plant roots release bicarbonates (HCO_{3-}), which tend to keep the pH level higher.

Nitrogen dosages. The irrigation method used in producing crops can influence the nitrogen dosage to be applied. Geraniums are fertilized at

200 to 250 ppm nitrogen during constant feeding when a top surface application is made, such as in drip irrigation. When applying fertilizer by subirrigation, such as ebb and flood, 150 to 200 ppm nitrogen is sufficient. There is less leaching with subirrigation and, therefore, a higher nutrient level in the growing medium.

In combinations of water soluble fertilizer and controlled release fertilizer, the suggested nitrogen level for geraniums is 100 to 150 ppm with top surface applications of fertilizer solution. With ebb and flood, the nitrogen dosage is 100 ppm. Osmocote-type (resin coated) fertilizers for geraniums, such as 14-14-14, 13-12-11 and 15-10-10, are incorporated at 6,7 and 8 pounds per cubic yard (2.5 to 3.5 kilograms per cubic m) respectively when used in combination with water soluble fertilizers.

Calcium and magnesium application. Levels of 60 ppm calcium and 30 ppm magnesium allow constant feeding. Some irrigation waters may provide these nutrient levels. If so, there's no need to supplement the fertilizer program. In addition, limestone incorporated in the growing medium should provide both nutrients, especially dolomitic limestone, although extractable calcium and magnesium is often insufficient.

Water soluble sources of calcium and magnesium should supplement the fertilizer program whenever levels are low. Calcium sources such as calcium nitrate and fertilizers that include calcium nitrate are available. In order to provide 60 ppm calcium, add 4.25 ounces per 100 gallons (120 g per 378 l) of 15.5-0-0 (Calcium nitrate), 7.25 ounces per 100 gallons (185 g per 378 l) of 15-0-15, or 13.3 ounces per 100 gallons of (377 g per 378 l) 20-0-20. Adding 15-5-15 at 18 ounces per 100 gallons (510 g per 378 l) provides 66.7 ppm calcium and 26.7 ppm magnesium.

TABLE 5-3

Water alkalinity guidelines

(Pot diameter size impacts alkalinity effects)

Intended use	Normal range (Mg Ca CO$_3$/L)	Level of concern	
Plugs	60 to 100	< 40	> 120
Small pots/ shallow flats	80 to 120	< 40	> 140
4" to 5" pots/ deep flats	100 to 140	< 40	> 160
Pots: 6" or more/ long term crops	120 to 180	< 60	> 200

Source: Grace-Sierra Testing Laboratory.

Although water soluble, these calcium sources shouldn't be added with most N-P-K fertilizers in the same stock concentrate. The phosphorus supplied as ammonium phosphate forms insoluble calcium phosphate. Therefore, separate fertilizer concentrates are required to prevent this undesirable formation. Use a twin headed injector or alternate feeding with calcium.

The most common magnesium source is magnesium sulfate or Epsom salts. In order to provide 30 ppm of magnesium, add 4 ounces (113 g) of Epsom salts per 100 gallons (378 l) of water. Avoid adding this highly soluble material in the same concentrate with calcium, since insoluble calcium sulfate, or gypsum, will occur. Epsom salts may be added in the

TABLE 5-4

General water quality guidelines for growing in soilless media

| Parameter | (parts per million except where noted) | | |
	Normal range	Low	High
Soluble salts (mmhos/cm)	0.3–1.0	<0.2	>1.3
Major nutrients*			
Nitrate nitrogen (NO_3-N)			>10
Ammonium nitrogen (NH_4-N)			>10
Phosphorus (P)			>10
Potassium (K)			>10
Calcium (Ca)	40–75	<25	>100
Magnesium (Mg)	30–50	<15	>50
Trace nutrients			
Manganese (Mn)			>1.5
Iron (Fe)			>2.00
Copper (Cu)			>0.20
Boron (B)			>0.50
Zinc (Zn)			>0.40
Molybdenum (Mo)			>0.20
Other elements			
Sodium (Na)			>50
Chloride (Cl)			>70
Fluoride (F)			>1.0
Aluminum (Al)			>1.0

*N, P and K levels are usually low. Levels greater than 10 to 20 ppm may indicate nutrient runoff into water source.

Source: Grace-Sierra Testing Laboratory.

N-P-K stock concentrate. Magnesium nitrate at 4 ounces (113 g) per 100 gallons (378 l) provides 27 ppm magnesium. This nutrient source is compatible with calcium nitrate. A commercially available product consists of 10-0-0 plus 9% magnesium.

References

[1]   Biamonte, R.L. 1990. Grace-Sierra Hort. Products Co. Unpublished photo-copy.
[2]   Biernbaum, J.A., W.H. Carlson, C. Shoemaker and R.D. Heins. 1988. Low pH causes iron and manganese toxicity. *Greenhouse Grower* 6 (3):92–97.
[3]   Boodley, J.W., C.F. Gortzig, R.W. Langhans and J.W. Layer. 1977. Fertil-izer proportioners for floriculture and nursery crop production manage-ment. *Cornell Univ. Info. Bul.* 129.
[4]   Holcomb, E.J. and R. Fortney. 1981. Effect of manganese on geranium cutting production and rooting of cuttings. *HortScience* 16(3):415-416.
[5]   Holcomb, E.J. and J.W. White. 1982. Fertilization. *Geraniums, 3d ed.* Ed. J.W. Mastalerz and E.J. Holcomb. University Park, Penn.: Pennsylvania Flower Growers.
[6]   Kofranek, A.M. and O.R. Lunt. 1969. A study of critical nutrient levels in *Pelargonium hortorum* cultivar Irene. *Jour. Amer. Soc. Hort. Sci.* 94:204-207.
[7]   Koranski, D.S. 1979. Cultural techniques for growing hybrid geraniums. *Minn. State Florists Bulletin* 28(1):11-13.
[8]   Laurie, A. and D.C. Kiplinger. 1980. *Commercial Flower Forcing.* 5th ed. New York: McGraw-Hill Book Company.
[9]   Link, C.B. and J.G. Seeley. 1954. House plants bring nature's beauty indoors. *The Care and Feeding of Garden Plants.* Washington, D.C.: Amer. Soc. Hort. Sci. and Natl. Fert. Asso.
[10] Marschner, Horts. 1986. *Mineral Nutrition in Higher Plants.* Academic Press.
[11] McFadden, L.A. 1970. Geranium problem solved by New Hampshire plant scientists. *Flor. Rev.* 145(3768):31, 63-64.
[12] Post, K. 1950. *Florist Crop Production and Marketing.* Orange Judd Publishing Co. Inc.
[13] Vetanovetz, R.P., J.F. Knauss. 1989. Iron toxicity: what you should know. *Benchmarks* Vol. 4 No. S. Grace-Sierra Horticultural Products Company.
[14] White, J.W. 1971. Interaction of nitrogenous fertilizers and steam on soil chemicals and carnation growth. *Jour. Amer. Soc. Hort. Sci.* 96:134-137.
[15] White, J.W. and A. Wick. 1968. Fertilizing seedling geraniums with slow release fertilizers. *Penn. Flower Growers Bul.* 204:1-5.

CHAPTER 6

Light and Temperature

John E. Erwin and Royal D. Heins

Although geranium is a major floriculture crop, relatively little is known quantitatively about how light and temperature affect its growth. Most geranium growers rely on personal experience and limited research for information.

Both seed and zonal geraniums are popular. Although production methods differ, they are both classified taxonomically as *Pelargonium* x *hortorum* Bailey. Zonal geraniums' relative popularity decreased somewhat during the late 1970s and early 1980s when seed geraniums' popularity increased substantially. The increase in seed geranium popularity was due to several factors including low production cost, ability to uniformly flower large plant quantities and their tolerance for adverse environmental conditions. In particular, seed geraniums were more tolerant of high temperatures and drought stress. New zonal geraniums, introduced during the late 1980s, were more tolerant of adverse environmental conditions.

This chapter concentrates on how light and temperature affect geranium growth, presenting environmental research conducted on both seed and zonal geraniums. Some research on seed geraniums was conducted by Allan Armitage and Douglas Hopper while at Michigan State University and by John W. White and I. Warrington at the DSIR-PPD Phytotron in New Zealand. Recent research results from the University of Minnesota and the University of Hanover on environmental effects on zonal geranium growth are also presented.

Light

Light quantity. Light is the energy that drives plant growth through photosynthesis. Insufficient light or excessive light and heat can limit plant growth.

Light can be limited through low light intensity and/or a short photoperiod. In contrast, supraoptimal light levels (>600 umol s-1 m-2)

TABLE 6-1

A comparison of photometric and radiant energy measurements*

Photometric		Radiant energy (400 to 700 nanometers)			
Footcandles	Lux	umol s-1 m-2		Wm-2	
		Daylight	HPS	Daylight	HPS
300	3,240	60	39.5	13.1	7.9
400	4,320	80	52.7	17.5	10.6
500	5,400	100	65.9	21.9	13.2
600	6,480	120	79.0	26.3	15.8
700	7,560	140	92.2	30.6	18.5
800	8,640	160	105.4	35.0	21.2
900	9,720	180	118.5	39.4	23.8
1,000	10,800	200	132.7	43.8	26.6

*Conversion factors:
1. Footcandle to lux: Multiply footcandle reading by 10.8.
Lux to footcandle: Multiply lux reading by 0.926.
2. umol s-1 m-2 (uE s-1 m-2) to Wm-2 PAR (400 to 700 nanometers): Divide umol s-1 m-2 reading by the following factor for each lamp or light source: (umol s-1 m-2 is micromols per second per square meter; uE s–1 m–2 is microEinsteins per second per square meter; Wm-2 is watts per square meter).

sun and sky, daylight	4.57
high pressure sodium	4.98
metal halide	4.59
cool and white fluorescent	4.59

3. Lux to umol s-1 m-2 PAR (400 to 700 nm): Divide lux reading by the following factor for each lamp or light source:

sun and sky, daylight	54
high pressure sodium	82
metal halide	71
cool and white fluorescent	74

Source: R.W. Thimijan and R.O. Heins [14].

for 16 hours can be detrimental to geranium growth because geranium leaves' high heat-absorbence capacity raises leaf temperatures, especially on terminal leaves, which then turn white. According to John White, if the leaves could be kept cool, we don't know what the upper limits of light might be. The importance of light quantity to geranium growth depends on the plant's development stage.

Geraniums are day-neutral. In other words, photoperiod length has no effect on flower initiation. Although photoperiod length doesn't affect geranium flower initiation, light quantity does [1, 2, 17]. In general, initial flower initiation occurs earlier as the amount of light that seedlings receive daily increases. For instance, flower initiation was hastened in Red Elite hybrid seed geraniums as the daily light integral

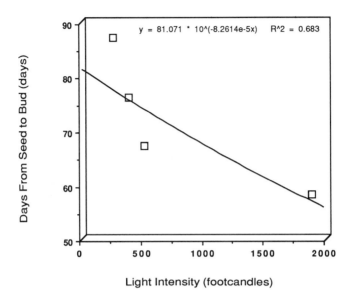

Fig. 6-1. The effect of daily light integral on the time from germination until anthesis in hybrid seed geranium Red Elite.

that plants were grown under increased from 3.3 to 13 moles per day per m-2 (Fig. 6-1). Flower initiation didn't occur at daily light integrals below 3.3 moles per day per m-2. Increasing the daily light integral above 17 moles per day per m-2 didn't significantly hasten flower initiation [17].

After initial flower initiation, light quantity has little effect on the flower development rate [1, 15, 16]. In contrast to light effects on flower initiation, light quantity has little effect on the geranium leaf unfolding rate [17].

Light intensity. Irradiance affects geranium morphological development. For example, Red Elite plant height and stem diameter decreased as irradiance increased up to 600 umol s-1 m-2 (3,000 footcandles or 32.4 klux) for 12 hours [9]. Further increasing light intensity above 600 umol s-1 m-2 generally doesn't result in shorter plants (Fig. 6-2). Conversely, low irradiance [<100 umol s-1 m-2 (<500 footcandles)] can reduce branching. It is, therefore, critical to maintain stock plants and cuttings under light intensities sufficient to ensure compact, stocky growth (> 2,000 footcandles or 21.6 klux).

Other morphological characteristics affected by irradiance are petiole length, pedicel length, leaf size, leaf thickness, flower number and flower size [16, 9]. The petiole is the stem section connecting the leaf blade to

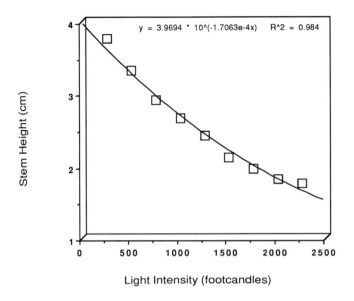

$$y = 3.9694 * 10^{(-1.7063e-4x)} \quad R^2 = 0.984$$

Fig. 6-2. The effect of irradiance on plant height at flower of hybrid seed geranium Red Elite.

the stem. The peduncle is the stem section connecting the inflorescence to the stem. As light intensity increases, petiole and peduncle length decrease. Similarly, leaf expansion and flower number and size decrease as light intensity increases (Fig. 6-3).

Light intensity also affects plant color. Leaf zonation, leaf color and flower color are more intense or darker at higher light intensities.

Zonal geranium fresh weight can determine both cutting and plant quality. In general, cutting and plant dry weight increase as daily light integrals increase from 4 to 26 moles per day per m-2.

Light quality. Light quality or color greatly affects plant appearance [12]. Red and far red light greatly affect the "state" of phytochrome, a bluish pigment in plant leaves that reacts to photoperiod and light quality [12]. In general, an increased proportion of far red to red light decreases plant quality. Far red light increases stem elongation, increases leaf size, decreases leaf thickness, increases pedicel, petiole and peduncle length, decreases branching and reduces leaf and flower color intensity [12].

A plant's phytochrome state is artificially altered by supplemental lighting. Incandescent light is high in far red light. Fluorescent light is high in red light. This is the basis for the shorter, darker green, well-branched growth habit of plants grown under fluorescent lights and the taller, lighter green, sparsely branched growth habit of plants grown under incandescent lamps.

A plant leaf acts as a light filter—the leaf absorbs red light but lets far red light pass through [12]. Therefore, leaves below a canopy are exposed to more *far* red light than red light. This shading results in stretching and poor shoot growth in the lower internodes and leaf axils. Space stock plants to avoid shading.

Temperature

Temperature affects plant development rate, appearance and quality. The average daily temperature and the way temperature is delivered during a 24-hour period affects plant development.

Plant development rate. Plants' developmental responses to temperature follow a similar pattern. Below a base temperature no growth occurs. As temperatures increase above the base temperature, plant development rate increases almost linearly to some maximum rate. As temperatures increase above this point, the plant development rate decreases.

The plant development rate depends on the temperature to which it is exposed at any given moment. In general, geraniums are maintained at temperatures in the "linear range" of the growth response curve to temperature. In this range, the plant development rate is strongly correlated with the average daily temperature in the growing area during a 24-hour period.

Leaf unfolding rate. For instance, hybrid seed geranium Red Elite leaf unfolding rates increased almost linearly as average daily temperature increased from 48 F to 81 F (9 C to 27 C) [17]. Similarly, Veronica zonal geranium's leaf unfolding rate increased as average daily temperature increased from 54 F to 76 F (12 C to 24 C) [6].

Leaf unfolding rate decreased as temperatures increased above 76 F to 81 F (24 C to 27 C) on hybrid seed and zonal geraniums. Therefore, increasing temperature above 76 F (24 C) doesn't hasten geranium flowering but probably delays flowering and/or leaf unfolding rate [1 and Erwin, unpublished data).

Understanding how average daily temperature affects geranium leaf unfolding is an important tool in producing geraniums of a given size; every leaf that unfolds represents a potential branch or inflorescence (Fig. 6-3).

Flower development rate. Geranium flower bud development appears to have a lower maximum temperature than leaf unfolding. Flower

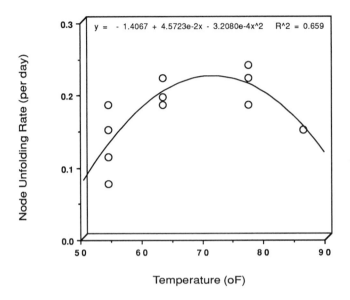

$$y = -1.4067 + 4.5723\text{e-}2x - 3.2080\text{e-}4x^2 \qquad R^2 = 0.659$$

Fig. 6-3. The effect of average temperature on leaf unfolding rate per day of seed or zonal geraniums.

bud development rates increase as temperatures increase from 50 F to 72 F (10 C to 22 C) (Fig. 6-4). Increasing temperature above 72 F doesn't increase the flower bud development rate [17, 11]. In fact, increasing temperature much over 72 F probably slows flower bud development.

Plant appearance. Stem elongation is affected by *how* temperatures are segmented during a 24-hour period, while flower number per inflorescence, flower and inflorescence size and flower dry weight are influenced more by average daily temperature.

Plant height. Geranium stem elongation increases as the day temperature under which plants are grown increases [6]. Conversely, geranium stem elongation decreases as night temperature increases [6]. Recent research on several plant species shows that the effects of day and night temperature on stem elongation can be best described by the difference (DIF) between day and night temperature rather than absolute day and/or night temperature within the 50 F to 80 F (10 C to 27 C) temperature range [4, 5, 7, 10]. Geranium plant height and internode length increased as the DIFference (day temperature minus night temperature) increased [8, 6]. As long as the difference between the day and night temperature is the same between two plants, the internode lengths are the same, regardless of the absolute temperature. (See Color Fig. C66).

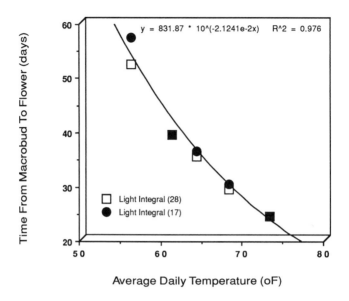

Fig. 6-4. The effect of average daily temperature on the time from macrobud to flower on hybrid seed geranium Red Elite.

Geranium stem elongation is most sensitive to cool temperatures during the first three morning hours [5]. Dropping plant temperature during the first three morning hours to below the night temperature reduces stem elongation substantially. Using this morning drop in temperature technique decreases stem elongation almost as much as growing plants with a cooler day than night all day. This technique is most useful in greenhouses with poor temperature control or for southern growers who can't grow plants with an equal or cooler day temperature than night temperature.

Pedicel, petiole and peduncle elongation are also strongly influenced by DIF [6, 7]. As with stem elongation, petiole, peduncle and pedicel elongation increase as DIF increases.

In contrast to stem elongation, flower number and size are affected by the average daily temperature under which zonal geraniums are grown [6]. For instance, Veronica zonal geranium's flower numbers per inflorescence decreased from 52 to 15 flowers as the average daily temperature increased from 54 F to 86 F (12 C to 30 C). Although the relationship isn't strong, flower diameter decreased slightly as the average daily temperature increased from 54 F to 86 F [6]. Neither average daily temperature or DIF affected individual flower dry weight with temperatures between 54 F and 86 F [6]. Actual temperatures also affect leaf and flower color. In general, as the day and night temperatures decrease, leaf and flower intensities increase.

Plant quality. Plant quality is judged in many ways. A common belief is that plant quality increases as plant fresh or dry weight increases. Zonal geranium fresh weight at flowering decreases as temperature increases. Fresh weight is greatest when plants have been grown at 54 F (12 C). Day temperatures over 82 F (28 C) reduce zonal geranium quality. Similar day and night temperature, or a slightly cooler day than night, tends to produce the highest quality plants [11].

Inflorescence dry weight decreases as the average daily temperature under which plants are grown increases [7]. For instance, Veronica zonal geranium's inflorescence dry weight decreased exponentially from approximately 1.1 grams to .2 grams per inflorescence as the average daily temperature increased from 54 F to 86 F (12 C to 30 C).

References

[1]   Armitage, A.M., W.H. Carlson and J.A. Flore. 1981. The effect of temperature and quantum flux density on the morphology, physiology and flowering of hybrid geraniums. *J. Amer. Soc. Hort. Sci.* 106:643-647.

[2]   Craig, R. and D.E. Walker. 1963. The flowering of *Pelargonium* x *hortorum* Bailey seedlings as affected by cumulative solar energy. *Proc. Amer. Soc. Hort. Sci.* 83:772-776.

[3]   Erickson, V.L., A. Armitage, W.H. Carlson and R.M. Miranda. 1980. The effect of cumulative photosynthetically active radiation on the growth and flowering of the seedling geranium, *Pelargonium* x *hortorum* Bailey. *HortScience* 15:815-817.

[4]   Erwin, J.E., R.D. Heins and M.G. Karlsson. 1989a. Thermomorphogenesis in *Lilium longiflorum*. *Amer. J. Bot.* 76(1):47-52.

[5]   Erwin, J.E., R.D. Heins, B.J. Kovanda, R.D. Berghage, W.H. Carlson and J.A. Biernbaum. 1989b. Cool mornings can control plant height. *GrowerTalks* 52(9):75.

[6]   Erwin, J.E. 1991. The effect of day and night temperature on zonal geranium flower development. *Minn. Flower Growers Bul.* 40(2):16-19.

[7]   Erwin, J.E. and D. Schwarze. 1991. The effects of day and night temperature on zonal geranium flower and peduncle dry weight. *Minn. Flower Growers Bul.* 40(3):1-3.

[8]   Heins, R.D. , J.E. Erwin and M.G. Karlsson. 1989. Use temperature to control plant height. *Greenhouse Grower* 6(9):32-37.

[9]   Hopper, D.A. 1986. "Light and temperature effects on hybrid seed geranium development." Master's thesis. Michigan State University.

[10]   Moe, R., R.D. Heins and J.E. Erwin. 1991. Stem elongation and flowering of the long day plant *Campanula isophylla* Moretti in response to day and night temperature alterations and light quality. *Scientia Hort.* (in press).

[11]   Pytlinski, J. and H. Krug. 1988. Modeling *Pelargonium zonale* response to various day and night temperatures. Proc. International Symposium on

models for plant growth, environmental control and farm management in protected cultivation. *Acta. Hort.* 248:75-84.

[12] Smith, H. 1986. The perception of the light environment. *Photomorphogenesis In Plants.* Eds. R.E. Kendrick and G.H.M. Kronenberg. Boston: Martinus Nijoff Pub.

[13] Smith, H. and D.C. Morgan. 1983. The function of phytochrome in nature. *Encyclopedia of Plant Physiology,* New Series 16B, Photomorphogenesis, Eds. W. Shropshire Jr. and H. Mohr. Berlin: Springer-Verlag.

[14] Thimijan, R.W. and R.O. Heins. 1983. Photometric, radiometric and quantum light units of measure: a review of procedures for interconversion. *HortScience* 18:818-824.

[15] White, J.W. and I.J. Warrington. 1984a. Effects of split-night temperatures, light and chlormequat on growth and carbohydrate status of *Pelargonium* x *hortorum. J. Amer. Soc. Hort. Sci.* 109:458-463.

[16] ——— 1984b. Growth and development responses of geranium to temperature, light integral, CO_2 and chlormequat. *J. Amer. Soc. Hort. Sci.* 109:728-735.

[17] ——— 1988. Temperature and light integral effects on growth and flowering of hybrid geraniums. *J. Amer. Soc. Hort. Sci.* 113:354-359.

CHAPTER 7

Growth Regulating Chemicals

James E. Barrett and E. Jay Holcomb

Chemical growth regulators are frequently used in the greenhouse industry. These natural or synthetic organic compounds promote, inhibit or otherwise modify a plant's physiological processes. These chemicals make geranium culture easier, increase life and make geraniums aesthetically more pleasing for consumers.

There are five generally accepted growth regulator classes: (1) Auxin or synthetic auxin is used primarily on vegetatively propagated geraniums to induce root formation on cuttings. Auxin is usually available as indolebutyric acid (IBA), indoleacetic acid (IAA), or naphthaleneacetic acid (NAA). (2) Cytokinins increase plant cell division, but there's been little interest in using cytokinins on geraniums since geraniums seem to branch well naturally. (3) Ethylene affects rooting and causes branching, flower bud abortion and flower shattering. Ethylene is a gas in nature but is also available as the commercial product Florel. (4) Gibberellins, either natural or synthetic, increase stem and petiole elongation so they can be used for creating special plant forms like tree geraniums and enlarging inflorescences. Generally GA_3 or GA_{4+7} are the commercial gibberellins formulations used on geraniums. (5) Inhibitors such as antigibberellins and antiethylene compounds also have been used on geraniums.

To make a geranium aesthetically pleasing, generally the height must be controlled. Since both seed geraniums and cutting geraniums can grow too tall, antigibberellin growth regulators, called growth retardants, are applied to restrict stem elongation and keep plant height in proportion to the pot. Several compounds, including chlormequat (Cycocel), ancymidol (A-Rest), paclobutrazol (Bonzi) and uniconazole (Sumagic), restrict geranium height.

The antiethylene compound silver thiosulfate (STS) inhibits ethylene activity, thus reducing single geranium flower shattering.

Auxins

Auxins, naturally occurring plant hormones, are the most important root-promoting substances used in plant propagation. IBA is most frequently available as either a powder or liquid. A 0.01% powder induces root formation on geranium cuttings. Many growers routinely use a root-promoting substance. Other growers see no need to apply a root-promoting substance on geranium cuttings since they root quite readily without one. The decision of whether to use a root-promoting substance is up to you. Additional information on root-promoting substances appears in Chapter 9, Vegetative Propagation.

Auxin also has an effect on branching through apical dominance. The growing tip produces auxin, which inhibits auxiliary shoot development, thus permitting a single terminal shoot to continue growing. Carpenter and Carlson [8] found that IAA didn't improve branching compared to plants that were pinched. Currently, there doesn't seem to be any benefit from applying auxin to geranium foliage.

Gibberellic acid

This hormone affects cell elongation and interacts with other plant hormones to control growth and flowering. When GA was applied alone, treated plants were taller and often had slightly larger flowers.

GA was first used as a spray to increase stem elongation and thus produce tree geraniums more rapidly. Information on the techniques involved appears in Chapter 13, Tree Geraniums.

Because GA interacts with other plant hormones, GA affects flowering. Armitage [1] demonstrated that GA applied as a foliar spray delays geranium flowering, in part, because elevated GA levels retard flowering initiation. GA may also delay development of geranium flower development slightly. Armitage [1] reported that Cycocel counteracts GA's effect to some extent, suggesting that GA application could reverse excessive Cycocel effects.

This reverse effect has been demonstrated for some crops. Cox [9] showed it's possible to use very high GA concentrations (100 ppm) to reverse effects of very high paclobutrazol spray and drench treatments. He noted that the best GA treatments created abnormally large leaves as a result of GA sprays. Fig. 7-1 shows results from a study using uniconazole at rates slightly above the optimum, followed by GA at 5 to 15 ppm. The 5-ppm GA spray reversed the uniconizole effects, but the GA-treated plants were growing faster than the plants not receiving uniconazole or GA. In most commercial situations GA rates of 5 to 10 ppm have proven more effective for reversing growth retardant effects; however, the response can be unpredictable. Although it's possible to reverse a growth retardant's effect with GA, growers shouldn't do this because if too much GA were applied, the effect would be so completely reversed that the plants might become too tall. Since appropriate rates of growth retardants and GA haven't been established, growers shouldn't try to reverse growth retardant effects with GA.

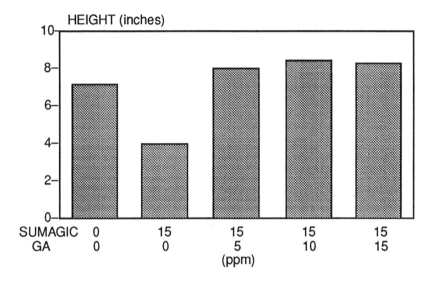

Fig. 7-1. Using GA to reverse effects of Sumagic (uniconazole). Red Elite seed geranium plugs were planted on March 16, 1990, into 4-inch (10 cm) pots. Sumagic was sprayed on April 6, GA was sprayed on April 17 and final plant height was measured on May 1. (Courtesy J.E. Barrett and T.A. Nell.)

Some geranium cultivars normally have short internodes, limiting the number of cuttings of a given length that can be taken from a stock plant. Adding GA to the growing media increases stem length, thus increasing the number of cuttings that can be harvested from a stock plant [2]. If the rate of GA is too high, cuttings would have unsuitably long internodes with very thin stem diameter. The correct GA concentration for slow growing cultivars should increase stem length without adversely reducing stem diameters. The correct concentration hasn't been determined for most cultivars. Although findings reported by Arteca et al. [2] resulted from applying GA to the growing media, the same responses should be obtainable from applying GA as a spray.

Ethephon (Florel)

Ethephon is a commercially available ethylene formulation. Ethylene can cause shorter internodes, suppressed flower bud growth and early maturation and flowers and fruit abscission.

Carpenter and Carlson [8] demonstrated that ethephon causes more branching on geraniums if applied two weeks before pinching. Because greater branching results, many growers sprayed Florel on geranium stock plants. Florel caused shorter stems, smaller leaves, flower bud abortion and better branching. Tsujita and Harney [15] reported that Florel increased cutting yield on geranium stock plants by 12% over non-Florel controls. Moorman and Campbell [13] determined that Florel concentration was cultivar dependent within a range of 500 to 1,000 ppm for the cultivars they treated. Because Florel restricts plant height, it could be used as a growth retardant. It hasn't often been used that way commercially, probably because it also causes flower bud abortion, which increases time to flower.

In general, for geranium stock plants, apply Florel as a spray. If the spray tank solution pH is too high and the time between mixing the spray application is too long, Florel's effectiveness may be reduced. A pH that's too high causes ethylene to be released to the air. To combat this problem use tap water with a neutral pH and apply the spray as soon as possible after mixing.

One concern is what effect Florel would have on rooting cuttings from treated stock. Tsujita and Harney [15] reported that cuttings from Florel-treated stock plants had more roots than cuttings from untreated stock plants. Maleike [11] reported that spraying ivy geranium cuttings with Florel increased rooting. It's unlikely that growers are using Florel to induce rooting on geranium cuttings, but limited research results indicate that Florel might be usable for that purpose.

Growth retardants

Using chemical growth retardant effectively controls plant size and allows faster production of both seed and cutting geraniums at closer spacing. Most published research studies of growth retardants use on geraniums have dealt with seed geraniums, probably due to the generally greater need for chemicals to control size in seed varieties and growth retardant effects on flower initiation in seed varieties.

The most commonly used chemical is Cycocel (chlormequat) at a rate of 1,500 ppm. Cycocel is often applied about one week after transplanting and again as needed one to two weeks later. Actual application rates and timing can vary widely with individual production situations and grower preferences. Plugs are generally more sensitive to Cycocel and other growth retardants and require lower rates. Cycocel is known for promoting early flower initiation when applied two weeks after seeding [12], but application timing is important. White and Warrington [16]) observed that Cycocel treatments 40 days after seeding didn't hasten flowering. Cycocel's effect on flower initiation isn't unique: Earlier flower initiation was also reported for A-Rest-treated geraniums by Miranda and Carlson [12]. Schekel and Blau [14] found that Bonzi reduced flowering time in Sprinter Scarlet by 20 days but treatments at or after the sixth leaf stage had no effect on flowering time. Growth retardants' effects on earlier flowering aren't as important in varieties with short crop times.

A major problem with Cycocel use is the phytotoxicity that often results. Symptoms are light green to yellow spots that appear on newly expanded leaves within three to seven days after treatment. (See Color Fig. C63.) In worst-case situations Cycocel can cause tissue death and necrotic areas. The chlorotic spots usually regain green color over time and are only slightly noticeable after three to four weeks. This phytotoxicity reduces the ability to use Cycocel near the end of a crop and prevents using higher rates that would be more active. Growers commonly opt for using lower Cycocel rates and more applications to avoid or reduce leaf spotting. Rates of 800 ppm or less are commonly used. Many theories give steps to reduce the spotting problem; they include spraying when it's cool and cloudy, spraying when humidities are low and the plants dry faster, spraying in the evening, using an additional wetting agent, using only light spray volumes, mixing B-Nine with the Cycocel and not spraying plants under drought stress. Little is really known, however, about how these and other factors influence the spotting that results from Cycocel sprays. Researchers at the University of Florida found that Cycocel causes a chloroplast breakdown in expanding leaves and that the more height control activity obtained from the Cycocel treatment, the greater the injury. Therefore, using lower rates reduces injury level, as does using lighter spray volumes to reduce the tendency for drops to run together and collect at leaf margins and in the leaf center. The longer the leaf stays wet after a spray, the more Cycocel it takes up and the greater the injury. So, less injury usually results from morning spray when the sun dries the leaves. Drench applications don't cause this problem unless a very high rate is used.

A-Rest is also used on geraniums. It's effective as either a drench or spray and isn't phytotoxic at recommended rates of 100 to 200 ppm. Its relatively high cost per plant prevents more frequent use.

B-Nine (daminozide) is commonly used on other bedding plants, but isn't often used on geraniums. It has little activity except in cool climates on seedlings and is sometimes used in combination with Cycocel as a tank mix similar to the way it's used on poinsettias.

Bonzi is a triazole growth retardant with very high activity on geraniums as shown in Fig. 7-2. This activity is its major advantage and disadvantage. Some growers have been successful in using Bonzi on geraniums, but there have also been problems. Sumagic is a new triazole chemical similar to Bonzi, and the results in Fig. 7-3 illustrate that single Sumagic applications can provide considerable height control.

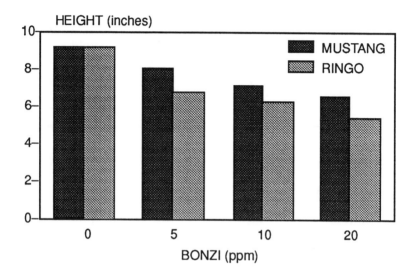

Fig. 7-2. Mustang and Ringo Deep Scarlet seed geranium cultivars' response to Bonzi (paclobutrazol). Plugs were transplanted into 4-inch pots (10 cm) on March 17. Bonzi was sprayed on April 14 and final height was determined on May 24. (Courtesy J.E. Barrett and T.A. Nell.)

Bonzi and Sumagic don't cause leaf spotting, delay flowering or reduce flower size. Geraniums are very sensitive to Bonzi and Sumagic, however, and too high concentrations can cause extreme stunting from which the plants recover slowly. Slightly high concentrations can cause an undesirable reduction in leaf size. Also, excessive Bonzi or Sumagic can greatly affect garden performance. It's important to ensure proper rates to avoid these problems. Barrett and Nell [6] found, in direct comparisons, Sumagic was more active than Bonzi, which is reflected in the optimum spray rates on geraniums: 5 to 15 ppm for Bonzi and 1 to 5 ppm for Sumagic. Optimum rates depend on spray techniques and often vary for different growers.

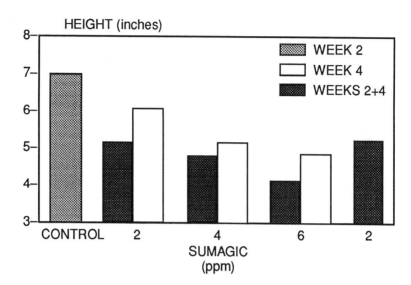

Fig. 7-3. Time of application of Sumagic (uniconazole) on Red Elite seed geranium in 4-inch (10 cm) pots. Plugs were planted on January 19 and final height was measured on March 16. Plants were sprayed once with Sumagic rates of 2, 4 or 6 ppm at either two weeks or four weeks after planting. One group of plants received Sumagic at 2 ppm at both two and four weeks. (Courtesy J.E. Barrett and T.A. Nell.)

Bonzi and Sumagic are different from Cycocel. Their activity is from the chemical taken up through the stems; the chemical that runs off the plant into the media is taken up by the roots [3,4,5]. The spray volume applied is important because higher volumes result in more chemical entering the media and being taken up through the roots. A Bonzi or Sumagic optimum rate at one volume would prove to be too high if applied at a higher volume. The generally recommended volume is 2 quarts per 100 square foot (0.2 liters per square meter), but most growers are using spray volumes between 1 and 3 quarts per 100 square foot (0.1 and 0.3 liters per square meter). Light Cycocel sprays are about 1 to 1.5 quarts (liters). At volumes below 2 quarts (liters), too little Bonzi or Sumagic reaches plant stems and the results usually vary. University of Florida trials have shown that Bonzi and Sumagic usually provide more uniform, predictable responses when applied at about 3 quarts per 100 square feet (0.3 liter per square meter), and many growers successfully using Bonzi on geraniums apply low rates in large volumes to achieve a heavy drench effect.

Each growth retardant, except B-Nine, is active in either drench or spray applications, but growers usually prefer sprays because of their easy application. Drenches provide longer, more uniform control and cause fewer problems with phytotoxicity than spray applications but require more labor. Drenches may be more convenient on larger container sizes and hanging baskets. Cycocel drenches, used by some growers for years, aren't as active as others and are considerably more costly. Bonzi and Sumagic are the least expensive drench options. The amount of Bonzi or Sumagic needed per gallon for drenches is very small and may be difficult to measure in fluid ounces. They are much easier to measure in milliliters using a graduated cylinder.

Apply drench applications to moist media to ensure even distribution; a good procedure is to water the plants one day and apply the drench the next day. The amount of solution applied should also allow for good distribution throughout the media. Four fluid ounces (118 milliliters) is good for 6-inch (15 cm) pots and is recommended on the A-Rest, Bonzi and Sumagic labels. The Cycocel label recommends 6 fluid ounces (178 ml). For larger size containers, increase the amount proportionally to container volume.

Many cultural and environmental factors influence geranium plant growth and affect growth retardant use. When geraniums are watered frequently, receive higher fertilizer and have higher temperatures, the growth rate increases and more chemical is necessary to produce a desired-size plant. Growers try to maximize space efficiency and many want to grow geraniums at very close spacing, which requires greater dependency on growth retardants to prevent overcrowding. Plant age is another important factor. Fig. 7-3 illustrates that Red Elite geraniums treated with different rates of Sumagic were more responsive two weeks after transplant than four weeks. One of the most important factors affecting chemical requirements is the growth characteristics of different cultivars. Many of the newly released cultivars are much less vigorous and less chemical is needed in production. There can be small differences in cultivars as shown by the fact that Bonzi at 8 to 20 ppm had greater effect on Ringo than Mustang (Fig. 7-2).

Silver thiosulfate

Shattering or rapid and severe petal abscission occurs on geraniums, with single-flower types being particularly sensitive to shattering. One reason is that pollination can cause shattering, and the single types are readily pollinated by insects. The cutting types, generally semidoubles, seem less susceptible to shattering than seed-propagated singles. Another factor associated with shattering is ethylene. Elevated ethylene levels can cause rapid geranium flower shattering. Cameron and Reid [7] demonstrated that STS applied as a spray to seedling geraniums before the first floret opened prevents flower shattering. STS even prevented shattering when the plants were exposed to low ethylene levels. Cameron and Reid [7] reported that STS applied two to three weeks before sale prevented petal shattering during transport and sale. (See Chapter 18, Use and Care Advice and Chapter 32, Physiological and Environmental Disorders.)

Healthy appearing seedling geraniums treated with STS soon showed signs of having Pythium root rot. Hausbeck et al. [10] reported that STS predisposes geraniums to Pythium root rot. (See Chapter 29, Vascular Wilt Diseases.) If STS is going to be part of your geranium production program, you need excellent Pythium control.

References

[1]   Armitage, A.M. 1986. Chlormequat-induced early flowering of hybrid geranium: The influence of gibberellic acid. *HortScience* 21(1):116-118.

[2]   Arteca, R.N., E.J. Holcomb, C. Schlagnhaufer and D. Tsai. 1985. Effects of root applications of gibberellic acid on photosynthesis, transpiration and growth of geranium plants. *HortScience* 20(5):925-927.

[3]   Barrett, J.E. and C.A. Bartuska. 1982. PP333 effects on stem elongation dependent on site of application. *HortScience* 17:737-738.

[4]   Barrett, J.E. and T.A. Nell. 1990. Factors affecting efficacy of paclobutrazol and uniconazole on petunia and chrysanthemum. *Acta Horticulturae* 272: 229-234.

[5]   ——— 1991. Efficacy of uniconazole altered by application methods. *HortScience* 26(6):770 (abstr.).

[6]   ——— 1992. Efficacy of paclobutrazol and uniconazole on four bedding plant species. *HortScience* 27:(in press).

[7]   Cameron, A.C. and M.S. Reid. 1983. Use of silver thiosulfate to prevent flower abscission from potted plants. *Scientia Hortic.* 19:373-378.

[8]   Carpenter, W.J. and W.H. Carlson. 1972. Improving geranium branching with growth regulator sprays. *HortScience* 7(3):291-292.

[9]   Cox, D.A. 1991. Gibberellic acid reverses effects of excess paclobutrazol on geranium. *HortScience* 26(1):39-40.

[10] Hausbeck, M.K., C.T. Stephens and R.D. Heins. 1987. Variation in resistance of geranium to *Pythium ultimum* in presence or absence of silver thiosulfate. *HortScience* 22(5):940-943.

[11] Maleike, R. 1978. Ethylene and adventitious root formation. *International Plant Propagators Society Combined Proceedings* 28:519-525.

[12] Miranda, R.M. and W.H. Carlson. 1980. Effect of timing and number of applications of chlormequat and ancymidol on the growth and flowering of seed geraniums. *J. Amer. Soc. Hort. Sci.* 105:273-277.

[13] Moorman, F. and F.J. Campbell. 1980. Florel's effects on three geranium cultivars. *Florists' Review* 166(4297):16, 58-60.

[14] Schekel, K.A. and S. Blau. 1987. Influence of paclobutrazol and gibberellic acid on flowering of geranium. *HortScience* 22:1155 (Abstr.)

[15] Tsujita, M.J. and P.M. Harney. 1978. The effects of Florel and supplemental lighting on the production and rooting of geranium cuttings. *Journal of Hort. Sci.* 53(4):349-350.

[16] White, J.W. and I.J. Warrington. 1984. Effects of split night temperatures, light and chlormequat on growth and carbohydrate status of *Pelargonium* x *hortorum*. *J. Amer. Soc. Hort. Sci.* 109:458-463.

CHAPTER 8

Stock Plants

Eugene J. O'Donovan

Despite substantial changes in the business climate and technology, there still exists a need for growers to produce cuttings from their own stock plants. This task must be performed "conscientiously and well." Proficient and profitable cutting production from stock plants results from a series of deliberate decisions and timely actions. It doesn't just happen but is the result of good management.

Before scheduling any bench area, ordering product or planting cuttings, you must make a series of initial decisions. These decisions are the first steps in a good management system that delivers consistent quality production.

Certified stock

The single factor that most limits stock production success is disease. Combined with the relatively long-term crop cycle, the exposure to risk associated with disease is high. For that reason, the program must start clean and be kept clean.

Using certified stock, cuttings that are directly generated from plants that have been culture-virus indexed, is imperative. This reduces potential exposure to fatal geranium disease that results in crop failure. To further reduce exposure to other fatal geranium diseases, isolate stock plants from other crops in the production program. Additionally, if you use cuttings generated from multiple sources, maintain and manage these products as separate crops. Renew this stock on an annual basis. Further, don't allow crop overlap from one season to another. This overlap poses a very high risk of exposure to potential problems. Plant the new stock only after all geraniums from the prior year have been removed from the premises. This includes the prior year's cuttings, finished plants and stock plants.

Using stock that has been certified free from systemic bacterial, fungal and viral diseases is the start. But it's only as good as the downstream systems that are put in place to keep this stock free of these problems. Maintaining the stock plants in a "disease-free" state is the first rule of successful stock production.

Planting dates

The key to profitable stock production is achieving increased yields per unit area under cover. Examine all components in this equation before each season. In addition to strategies to increase the number of stock plants under cover, the stock planting date is important in determining final productivity. Stock plant age and cutting productivity are directly related. Simply, older stock plants are able to produce more cuttings per unit of time than younger plants.

Table 8-1 shows the relative effect of various planting dates on productivity. In general, for every month's delay in planting, expect approximately 10% reduction in cutting production.

TABLE 8-1

Effect of planting date on estimated yield

Month	Yield index
June	100
July	90
August	80
September	65
October	50
January	20

Not only will younger stock plants yield fewer cuttings, but they will also be smaller. Because of their reduced size, more stock plants fit on a unit area of bench space. This increased plant density per unit area will help offset reduced productivity. Simply divide the index into 100, subtract 1 from the result and multiply that result by 100, to determine additional plants required. For example, stock plants established in August produce approximately 80% of those planted in June. $[(100 \div 80)-1]$ x 100 = $[(1.25)-1]$ x 100 = 25%. An August planting will require an additional 25% more stock to approximate older stock plant production.

This planting density is critical for improved productivity. The number of stock plants grown per unit area depends on the variety and planting date. Manage density to provide maximum yield and quality plus easy pest and disease management. Often the rule of one plant per square foot is acceptable, but with the advent of newer, more compact varieties and later planting schedules, this rule may need modification.

Variety selection is also important. Significant productivity differences do exist. Options within a given color class and plant habit are available. The stock plant size doesn't dictate final productivity. Rather, the manner in which the framework (scaffold) of the stock plant was developed affects actual productivity. The concept of a framework for a stock plant is critical and often overlooked. Framework could be defined as the combination of primary, secondary and tertiary branches, with various stages of cutting development in progress. It's important that at all times the main framework, the primary, secondary and tertiary branches, be filled with maximum dormant and started eyes and cuttings. The quantity of these eyes (dormant and started) and cuttings drive the eventual production. If one were to visualize the space above the rootball as a cube, the framework would fill this cube. The various stages of cutting development ensure continuous cutting production during the required time frame.

The planting date decision drives many other considerations, including planting density, container and media. The earlier the stock plant is established, the greater the productivity. This increased productivity requires additional labor and resources.

Containers

The key to good top growth (productivity) is good root growth. Geranium root systems perform best in situations where the water table can be managed to provide an optimum amount of substrate air space. Container height and adequate drainage characteristics are very important. Containers provide additional height and both side and bottom drainage holes are preferable. Containers that have at least a three-fourths profile (azalea) are preferred to one-half profile (pan) types.

The growing container size is linked to the planting date. Stock plants grown on a long-term crop cycle (more than nine months) produce very large plants and require additional substrate for easy management and productivity. Stock plants grown on a short-term crop cycle (less than five months) produce a much smaller plant and require less substrate for continued growth and productivity.

For ease of management it's important that the container be uniform in characteristics and size.

Often, the additional cost of a larger container and additional substrate can be offset by the increased production or the ease of managing nutritional factors.

The style of the production container, florist pot, nursery container, plastic or clay is a choice that should be decided by available market opportunities. In recent years demand for instant color items have

TABLE 8-2

Container size related to crop cycle

Crop cycle	Container (diameter)
9 to 11 months	7" (18 cm) or greater
4 to 8 months	6" to 8" (15 to 20 cm)
less than 4 months	4" to 6" (10 to 15 cm)

created a market for large geraniums. Many growers have successfully forced their stock plants into flower to fill that need. The proper container can improve the finished product's appearance and result in increased value and revenues that often offset additional cost incurred.

Substrate considerations

Substrate selection is fundamental to the program success. Select the substrate based on crop cycle length, container size, existing water/nutritional conditions on site and grower preference as the primary criteria. Substrate cost should be the last, not the first factor in the decision process.

Because most stock programs are grown for most of a year and maximum productivity is necessary at the cycle's end, the substrate should maintain its structure and physical properties without substantial degradation throughout the entire crop cycle.

Prior to planting, steam pasteurize the substrate, and analyze it for nutritional status to ensure trouble-free performance.

System considerations

Insect and disease control. *The ability to keep the stock plants disease free is directly linked to the program's ultimate success.* Properly planning and executing these systems can result in continued trouble-free production. Isolating stock plants from other crops that represent a disease risk is vital. Where possible, grow geranium stock in separate facilities with limited access. Limiting access controls disease transmission risk. Entry into these facilities should require all to pass through a wash station, where body parts that should come in contact with plant material can be disinfected. At this place, any clothing that has come in contact with other plant material should be covered with a clean, protective garment to avoid potential disease and insect transfer.

Grow stock plants on raised benches. The height of the raised benches should place the crop above splashing water. Give attention to harvesting ease. These benches should permit the leachate to freely drain away from the container and not accumulate around the plant base. Where possible the bench construction shouldn't permit the leachate from plants to touch another plant. For this reason, open bottom benches, those constructed with wire mesh, expanded metal or similar products, are preferable. These open bottom benches also aid air movement through the canopy, helping to reduce the threat of Botrytis and other airborne fungal diseases. The bench width, although not a consideration for disease control, should be limited to 6 feet wide. This will aid in harvesting.

Use trickle irrigation in all geranium stock production. These systems provide water and liquid fertilizer directly to the substrate and avoid splashing water and/or substrate particles from one container to the next. In addition, they keep the irrigation solution from the foliage, helping to reduce the risk of Botrytis and other foliar fungi. These systems are also very useful in providing timely irrigations, reducing moisture and soluble salt stress.

Screen greenhouse openings. This includes vents and doors used for ventilation. In combination with screened openings, implement an ongoing insect monitoring program. Place sampling cards at plant height randomly throughout the facility. Read these cards daily. Post the results and observe trends. Change the cards frequently, at least once a week.

Provide separate utensils and protective clothing for exclusive use on this crop and facility. Sterilize the utensils prior to work each day and as frequently as possible during the day. Practical solutions to utensil sterilization include resterilizing them between cultivars and/or changing benches. These practices help to reduce disease spread throughout the crop.

Light. Grow geranium stock in facilities that provide the greatest light transmission during the winter months. Grow no crops over head in the stock range. Overhead crops reduce the light that reaches the stock plants and increase the disease threat. Cutting productivity and quality are linked to the quantity and quality of light the stock plant receives.

For stock plants established in the summer and early fall, limit light intensities to the range of 5,000 to 6,000 footcandles (54 to 65 klux). Do this by shading the greenhouse covering. Be sure to remove this shade when the ambient light, temperature and daylength start to lessen substantially. For many growers this occurs early to mid September. From this point until late spring, prior to the selling season, grow geranium stock in full ambient light conditions.

Supplemental assimilation lighting programs to increase cutting productivity have found limited commercial application. Although increases in productivity can range from 20% to 50% or more, many times these increases aren't sufficient to provide ample return on investment.

Fertility management. Geranium stock production requires high fertility levels. This statement sounds rather simple, but often is one of the most limiting factors in quality production.

For long-term production, constant liquid fertilization programs provide greater flexibility, faster response and easier management than other options. Manage the fertility program to maintain optimum elemental nutrition levels in the stock plant tissues. Monthly substrate and foliage analyses are necessary to determine the proper fertilizer levels to be applied. Post these results to provide a tracking mechanism for future reference.

Pay strict attention to the pH and electrical conductivity (EC) levels (see Chapters 2, Growing Media and 5, Fertilization) in the substrate. With high fertilizer levels required and crop cycle length, these tend to fluctuate to undesirable levels during periods of peak harvest requirements. These fluctuations result in productivity and/or quality loss.

Injecting carbon dioxide during fall/winter and early spring enhances cutting production and quality. Maintaining supplemental levels of this compound at 750 to 1,000 ppm can result in 10% to 20% increased production. This increased production, when compared to installation and operation costs provides a substantial return on investment.

Cultural practices

Prior to planting, clean and sanitize the planned facility. Chemically disinfect the structure; sterilize benches with heat or chemicals. Clear the walkway and areas under the benches of any weed or foreign plant material. Sterilize all utensils that come in contact with the stock plants with heat or chemicals and properly store them to avoid possible reinfection. Clear the area outside the greenhouse of geraniums, weeds and tall grasses that might harbor insects and/or disease.

Getting started

Planting depth is very important. In the case of a geranium stock plant, planting too shallow results in a mature plant that tends to be unstable and subject to damage during harvest. This damage can range from damaged roots to a broken main stem. Transplant the cutting so the lowest breaks are just above the soil line after the substrate is watered in. Cuttings transplanted too deeply will be delayed in developing top growth. Uniform transplant height is very important for uniform canopy development.

The first seven days after transplanting are very important in the plant's long-term development. To achieve the most from the cutting, at the end of this time frame the cutting should have established new, active root growth in the substrate. Active, white roots should be visible on the substrate surface and on the outside diameter of the substrate mass within seven days after transplant for at least 10% of the cuttings. If this development isn't present, conduct chemical analysis of the substrate. Based on these analysis results, take corrective measures.

Should existing conditions require fungistatic substrate applications to control water- or soil-borne diseases, make these applications after the root system is initiated, and then as conditions or symptoms dictate; consult manufacturers' labels prior to use.

During the first few weeks of development, until the cutting has developed an adequate root system, mist foliage during stressful periods when the cutting might wilt. These mistings help the plant recover from extreme water loss. These cycles also encourage the development of started eyes on the lower cutting portions. At all times the cutting should go into night (dark) time with dry foliage.

Initial framework

Timing the first pinch depends on the root system development rate. In most cases the first pinch for framework development can be made from three to five weeks after transplanting. Removing the apical growing point promotes lateral branching. Any difference that would result from a soft pinch or removing a cutting has little impact on the plant's overall timing or productivity. After the tip has been removed, the remaining stem should range in length from 4 to 6 inches (10 to 15 cm) with at least four internodes. Additional internodes can remain on the stem as long as the stem length isn't increased.

These two criteria are very important in developing a compact, productive framework that stands up to the abuses of harvesting during peak demand. Practices that leave a longer stem length behind the pinch result in a taller framework. This framework is more likely to be damaged during harvesting and reduces active, ongoing production from the lower plant. This taller plant doesn't result in additional production. Practices that result in a shorter stem length behind the pinch create stock plants with overall lower productivity.

Intermediate framework

The next framework development phase should be conducted continuously until peak harvest requirements begin. The procedures in this phase continue until the final height is obtained. At that point, modify procedures slightly to maintain maximum height. The stock plant's maximum height should be between 12 and 15 inches (30 to 38 cm) from the substrate surface. The objective in this development phase is to produce as many potential cuttings as possible (started and dormant eyes as well as actual cuttings) in the prescribed space.

To accomplish this task, schedule and perform work on a regular basis. Maintaining optimum fertility programs is vital to this production phase. Low fertility or high electroconductivity levels reduce potential cutting production. Provide adequate light (3,000 to 5,000 foot-candles; 33 to 54 klux) to the entire plant framework to increase potential cuttings.

Each main stem in the framework must receive adequate light. The amount of light the plant receives determines eventual productivity. Higher irradiance helps keep the internode distance as compact as possible. This is very important and shouldn't be overlooked. The more compact the internodes, the more potential cuttings (dormant eyes, started eyes and apical tips) will be housed on the framework. Grow this crop in the best available light in the greenhouse. *Once every month, remove all fully expanded and mature leaves that aren't part of a usable cutting.* Removing these leaves does many things. First, it opens the framework (canopy), permitting light to reach all levels of the plant. This helps promote started eyes in the lower framework. These started eyes provide continued production later in the season. This open canopy permits air currents to move through the crop, increasing the CO_2 concentration in the canopy and reducing potential problems with Botrytis and other airborne fungi.

At regular intervals, possibly as often as every two weeks, remove the largest cuttings from the plant. Removing these cuttings helps keep the canopy open and enhances the production of started eyes behind the cutting on that main stem. Leave two or three nodes on the plant. Simple arithmetic shows that for every cutting removed from the stock plant, at least one additional cutting is being developed on the framework. That is, two breaks behind the cutting, subtract one cutting removed, resulting in a net gain of one cutting. The number of cuttings removed at a given time depends on the relative framework size and the plant's growth rate. Remove cuttings to shape the plant for maximum future production, creating a well-shaped plant that fills the space allotted.

Should the plant reach maximum height prior to the start of peak harvest requirements, continue the above practices with an emphasis on limiting the framework height to the prescribed limits.

Peak season harvest management

At the start of peak season, the basic rules for harvesting change to those that produce the most cuttings. But it's important that this harvesting practice not reduce future production. During peak season the first cuttings are normally removed from the outer framework. It's important that those potential cuttings (dormant and started eyes) located on the inner framework remain on the plant. For each cutting removed prior to the last month of peak harvest requirement, one node should remain. This practice helps ensure late season production.

Although removing fully expanded, mature leaves that aren't part of a cutting would benefit production, in most cases there isn't ample labor to carry out this task. Hence, remove just the fully expanded, mature leaves that threaten the crop's overall health and continuity.

Other production alternatives

For some, a traditional stock plant may not service the production plan needs. Literature has cited the use of tree geraniums for stock plants (see *Geraniums III*, Chapter 9, for details). Using tree geraniums as stock plants offers several advantages. First, this program permits large batches (flushes) of cuttings to be harvested in a narrow time frame. Next, it provides unique postproduction market opportunities.

This method calls for establishing the stock plant early in the production cycle, June or July, for peak harvest in January, February and March. The mature tree geranium requires a large growing container to physically support the plant and prevent the plant from falling over. A minimum container size of at least 10 inches (25 cm) in diameter provides ample moisture and fertility reserves. At the same time it should provide sufficient mass to maintain the plant vertically. Stock plants need substantial space, often 3 to 5 square feet (0.3 to 0.5 square meters) per stock plant.

To achieve maximum productivity, train the plant on a trellis or stake it to help support the main stem and avoid breakage. Applications of gibberellic acid in late summer and early fall help the plant obtain the required height. Once the majority of height is obtained, discontinue gibberellic acid applications and remove the apical growing point. From this point to one month prior to the first scheduled peak harvest, prune and shape these plants frequently and remove large leaves that aren't part of cuttings.

Tree geraniums produce more cuttings per plant, but due to their relatively large space requirements, the yield per square foot is essentially the same as for more traditional methods, but concentrated in a much narrower time frame.

Using growth regulators for stock production can enhance productivity. One growth regulator labeled for use on geraniums is Florel (active ingredient Ethephon). Properly utilized, this chemical can result in increased cutting production of at least 15% to 20%. Stock plants treated with Ethephon are more compact and have closer internodes. Ethephon also eliminates or reduces flowering for several weeks after application. Cuttings from stock plants treated with Ethephon have a shorter stem length and a reduced caliber with smaller leaves and elongated petioles.

Application rates vary, dependent on variety. Range of rates commonly used in production are 300 to 500 ppm active ingredient. Stock plants treated with one application of Ethephon produce more cuttings than those not treated. Stock plants treated with two applications produce more cuttings than those that have received one. There doesn't appear to be a maximum number of applications. Discontinue chemical applications one month before the first scheduled peak harvest. This practice permits the plant to grow out of the growth regulator effect and provide a flush of cuttings when desired. Supplemental injections of CO_2 further enhances the quality of cuttings from treated plants and increases productivity.

Using horizontal air flow (HAF) has proven very beneficial in stock production. This constant air movement through the canopy reduces Botrytis in the canopy, helping to increase yield.

Cutting quality

Over recent years the industry has seen the size of geranium cuttings offered for sale decrease. This has led to much speculation about proper cutting size. This reduced size is the result of new cultivars, changing markets and increased ease in shipping and propagation.

The harvesting procedure shapes the framework for maximum productivity. Remove cuttings from the stock plant without regard for size. A large cutting can be tailored to the proper requirements at a later time. Instead, focus on shaping the plant during harvest. Where possible, *snap or break cuttings from the stock plant*. A well-grown cutting snaps cleanly from the main stem. Cuttings that don't snap freely are the first indicators of either nutritional or environmental disorders. In some situations, cuttings must be removed from the stock plant with knives. Should this be the practice employed, implement measures to ensure that the utensils used are sterile and haven't been used for other purposes in the greenhouse. Sterilize utensils frequently to avoid disease spread. Never harvest cuttings with scissors or shears. These tend to crush the cutting stem and the stock plant, increasing losses.

For ease in propagation, all cuttings must have approximately the same surface area (leaves). Leaf area dictates the transpiration rate and strongly influences the amount of mist required. The less mist required in propagation, the fewer problems encountered. Therefore, a cutting with a uniform and minimal surface area will require less mist and have fewer problems. It's less expensive to grow a small cutting into a larger unit in the propagation area than to perform this task on the stock plant.

An ideal cutting should have an active apical growing point, and at least two expanded leaves below the growing point. It's not necessary to have more leaves. Additional leaves require additional mist. Stem length can vary, depending on the stock plant's internode spacing. In most cases a stem length of 1½ to 2½ inches (3.8 to 6.3 cm) is sufficient. Prior to sticking, square the cutting's basal end. Enter cuttings into the propagation sequence as soon as possible (see Chapter 9, Vegetative Production and Chapter 10, Growing from Vegetative Propagules).

Vegetative Propagation

Tissue Culture

Roy A. Larson

Thirty years ago bacteria, fungi and viruses made profitable vegetative geranium production a venturous enterprise. Some specialist propagators were so aware of the diseases their material probably possessed that they routinely purchased an extra 10% cuttings. Then plant pathologists such as Dimock at Cornell University and Tammen and Nelson at The Pennsylvania State University developed techniques to detect pathogens and used culture indexing to produce disease-free cuttings. Some growers believed that indexed geraniums were disease resistant and continued to obtain diseased cuttings to supply some needs. That practice resulted in disease problems among growers who were convinced that their plants would remain invincible to pathogens. The errors of such thinking soon became apparent, and healthy, vigorous, disease-free cuttings almost eliminated the poorer cuttings from the market. Cultured cutting costs were substantially higher than their predecessors' costs because of the labor and facilities required, but they produced superior plants when pathogens were absent.

Seedling geraniums became available commercially in the mid-1960s, and these seedlings were relatively disease free in the early production stages. Again, they weren't immune to diseases.

This brief history, discussed in greater detail in other sections of this book, illustrates the concern geranium propagators have had about the disease organism transmission from their stock plants to cuttings to finished products. They have been willing to invest in innovations that should reduce disease. The brief history also reveals that the geranium would be an obvious choice for tissue culture micropropagation. Much information already was available about how the tissue would respond under aseptic conditions, in agar, and how much labor would be required to accomplish various laboratory and greenhouse procedures.

Plant material propagation by tissue culture isn't a recent idea. Haberlandt, a German plant physiologist, postulated in 1902 that one should be able to extract a cell from a living plant and end up with a complete plant [10, 25, 26]. The process required the optimum medium, one that contained a source of carbohydrates (sucrose), a proper balance of auxins and cytokinins and balanced nutrient elements.

Originally one medium was used to obtain callus formation when the tissue was first removed from the original plant, another affected shoot growth and another promoted root development. The ideal situation was to have one medium that could be used throughout the micropropagation phase, and such a medium has been developed for some species.

Tissue culture as a means to propagate plants has both benefits and limitations. In most instances the benefits outweigh the disadvantages, but some plant species don't exhibit enough benefits to make tissue culture profitable.

Benefits

A primary benefit of tissue culture is clean plant material production [3, 20, 21], assuming that the parent stock indexes disease-free. That advantage appears if the correct tissue amount is removed from the parent plant. Orchids were among the first floricultural crops to be commercially reproduced by tissue culture, but virus problems became worse, rather than better, with tissue culture. The plants supposedly were micropropagated by meristem culture or mericloning, but much more than the meristem was removed. The extreme apical cells of the shoot tip most likely would be virus free, but shoot tips at least 10 mm long often were used and viruses were present. Tissue culture, combined with heat therapy, has resulted in essentially disease-free plant material. Reuther [21] found that shoot tips removed from geranium plants were more likely to survive if they were 0.5 to 1 mm in length. Smaller shoot tips were less likely to be involved in the transmission of disease organisms, but the chances for successful propagation were much reduced. Tissue culture can result in plant material that indexes disease free, but the lesson should have been learned from culture-indexed geraniums that such plant material isn't disease resistant.

Another major micropropagation benefit is that thousands of plantlets can be reproduced from a small amount of material. Callus in one flask could eventually produce as many plants as would result from many large stock plants. These larger stock plants would have consumed time, space and labor and been subjected to many pests. The potential for a vast increase in plant population is amplified by the rapid rate at which this increase can occur. A new plant could be quickly replaced if acute shortages suddenly occurred. *Plantlets would be identical to the original material, unless a mutation occurred.*

In many regions stock plant production is restricted by climate, but micropropagation enables a propagator to produce propagules indoors throughout the year. The ideal environmental conditions, optimum nutrient medium and disease-free tissue culture production also are satisfactorily controlled. Desirable stock material can be maintained for a long time. Geranium stock plants from which terminal cuttings are removed for spring and summer sales usually are discarded at the conclusion of the propagation season, because such stock plants require large amounts of water, fertilizer, space and pest control.

Some plant species aren't easily propagated by more conventional ways, while tissue culture is successful. However, most *Pelargonium* species aren't difficult to propagate from vegetative cuttings.

Limitations

Disadvantages of in vitro or micropropagation do exist. A primary handicap is cost. Building and equipping modern laboratory facilities are expensive, and procedures used in the tissue culture laboratory are labor intensive. Skilled employees are required [4]. According to Donnan [6], the labor cost can range from 60% to 85% of the total cost. Some laboratories have been relocated to countries where labor costs are lower, but most tissue culture laboratories still are located in countries with high standards of living and high wages. The handicap of costly geraniums from tissue culture will be intensified as more cuttings are imported into the United States from Mexico.

Chu [3] believed that the inability to make technology developed by university researchers into a profitable venture was an unexpected limitation. Crop scheduling has often been difficult, and customers haven't always been able to obtain plantlets they ordered when they wanted them. Many growers don't know how to handle the plantlets when they arrive, so losses often are greater than they would be with conventional cuttings. The need for special care instructions was emphasized by Klein-Cox [13]. Janick [12] has emphasized the need for propagule acclimatization.

Though micropropagation has been credited with the rapid multiplication of many identical clones, it also has been criticized because genetic instabilities occasionally occur [3]. Detecting mutants often requires considerable time, effort and expense.

Geranium micropropagation has had the advantages and disadvantages listed above, but this also has been true for many other plant species. Considerable research specifically restricted to geraniums has provided additional technology.

Micropropagation technology

The first use of micropropagation on geraniums was to eliminate disease. Paludan [17] noted that cuttings that were free of tomato ringspot virus were more vigorous and had larger, glossier leaves than infected cuttings. Horst et al. [11] reported that cuttings that were indexed to be virus-free and had been derived from meristem tip culture rooted 10% faster than cuttings that had not been cultured. Forty *Pelargonium* x *hortorum* cultivars, 11 *P. peltatum* and five seedling lines were included in the study. Cultivars showed pronounced differences in shoot and root regeneration success. Oglevee-O'Donovan [14, 15] has been instrumental in developing virus-free, cultured cuttings. Reuther [21] used tissue culture methods to eliminate the bacterium *Xanthomonas pelargonii* from geranium plantlets. Shoot tips 0.5 to 1 mm long were used by Reuther as tips smaller than that were less likely to survive. More bacteria could be found at the base than at the shoot tip. The ability of tissue to regenerate declined if Pelargonium callus was subjected to long-term tissue culture, a barrier not encountered with carnations and chrysanthemums.

Geraniums' economic value and popularity have prompted many tissue culture experiments on the genus but only a few examples will be cited here. Different tissues have been tried, under various environmental conditions and with different nutrient media. Ward and Vance [23], working with *P.* x *hortorum* and *P. zonale*, found that pith callus could give rise to shoots and roots, while stem tips produced shoots. Callus from the stem tips were virus-free. They found that small explants were less likely to survive than large explants, and more shoots were produced on the larger explants. Blue, green, red and white (full spectrum) light and darkness were evaluated by Ward and Vance. The greatest amount of callus tissue was produced under white or blue lights. Pillai and Hildebrandt [18] failed to obtain growth from pith callus unless vascular tissue was present. They later reported [19] that shoots were induced in six to eight weeks from undifferentiated callus tissue if a 16-hour light, 8-hour dark cycle was used. Roots required another eight to 10 weeks. They found that callus that had been subcultured more than three times often didn't produce the desired shoots and roots, a fact confirmed by Reuther [21] almost 20 years later.

Light intensity and date of explantation were studied by Hammerschlag [8]. She used the geranium cultivars Sprinter (diploid), Sincerity (tetraploid) and Kleine Liebling (haploid) in her experiments. More callus formation was obtained from Sprinter and Sincerity at a light intensity of 926 footcandles (1 klx or approximately 13.5 micromols per second per square meter of PAR [400 to 700 nanometers] from cool, white, fluorescent lamps), while Kleine Liebling was most prolific at 5,000 footcandles

(5 klx). All three cultivars responded poorly at 10,000 footcandles (10 klx). In a later study Hammerschlag and Bottino [9] showed that successful micropropagation of the seedling cultivar Sprinter depended on light quality and intensity, season and plant material age. More callus production and more roots were obtained when tissue was derived from young seedlings. Dunbar and Stephens [7], trying to develop seedling cultivars resistant to bacterial blight, worked with three cultivars in the Orbit series. They found that callus was initiated when tissue was taken from seedlings 10 to 14 days old, and an appropriate medium was used. Five to 50-shoot primordia were produced per explant when the Murashige-Skoog medium was used, supplemented with 2 mg per liter zeatin and 1.9 mg per liter of indoleacetic acid. Callus was initiated more readily for Regal Pelargoniums when the medium was supplemented with 2 mg per liter 6-Benzylamino purine and 2 mg per liter naphthalene acetic acid.

Cassels and Minas [1] studied the influence of season, light intensity and duration, temperature, relative humidity and water on the shoot tip culture. They also tried to improve the efficiency of tissue culture handling systems and to obtain greater success at less cost. They found they could store isolates at low temperatures, increasing the flexibility of micropropagation.

Plant production from protoplasts has been considered. Yarrow, Cocking and Power [24] predicted that new genotypes for geranium breeders would be possible with protoplasts obtained from cultured callus. The chances for regeneration of P. x hortorum protoplast were improved only 10% on most media to 60% when a liquid shake culture regimen was used.

Chen and Galston [2], working with Pelargonium pith cells, found they could produce a whole plant if they used White's medium, with auxin and cytokinin added, to get cell division. Cell division was best maintained on the Murashige-Skoog medium, while cell differentiation was best on a semisolid medium. Further differentiation reportedly proceeded most rapidly on liquid culture, while growth and development were enhanced on a semisolid medium.

Growth medium

The necessity of using the proper medium is emphasized by almost everyone working with tissue culture. Some basic components are always used, and then slight modifications are made. A prominent medium used for almost 30 years was developed by Murashige and Skoog [14] (Table 9-1). Other media are listed by Dodds and Roberts [5].

The optimum medium and laboratory environment have received most attention of researchers studying micropropagation. The post-laboratory phase seems to have been somewhat neglected, yet improper

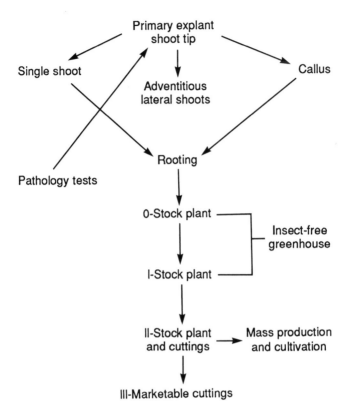

Fig. 9-1. Production schematic. (Source: Reuther [20].)

conditions in the greenhouse or nursery can nullify all the knowledge available on laboratory technology. Plantlets, though perhaps disease-free and vigorous, often have not been properly acclimated to conditions outside the laboratory. The mortality rate of the plantlets can be unnecessarily high. Zumwalt [26] advocated a 2- to 4-week pretransplant stage during which the plantlets would be acclimated, primarily by increasing light intensity. Yarrow, Cocking and Power [24] have studied all phases of Pelargonium tissue culture, and they suggest a 16-hour daylength, light intensities ranging from 10,000 to 20,000 footcandles (10 to 20 klx), and temperatures of approximately 72 F to 80 F (20 C to 27 C). Adequate water and sanitary greenhouse or nursery conditions are necessary. The disease-free advantage can be quickly lost if sanitation and pest control are neglected.

Though scientific research has enabled tissue culture propagation to reach its present level, Yeoman [25] stated that experience, rather than theory, has been largely responsible for the tissue culture's commercial

TABLE 9-1

Prominent tissue culture medium

Ingredients	Amount (mg per l)
$(NH_4) NO_3$ (Ammonium nitrate)	1,650
KNO_3 (Potassium nitrate)	1,900
$CaCl_2 2\text{-}2H_2$ (Calcium chloride)	440
$MgSO_4\text{-}7H_2O$ (Magnesium sulfate)	370
$FeSO_4\text{-}7H_4O$ (Iron sulfate)	27.8
Na_2-EDTA (Sodium-EDTA)	37.3
$MnSO_4\text{-}4H_2O$ (Manganese sulfate)	22.3
$ZnSO_4\text{-}7H_2O$ (Zinc sulfate)	8.6
H_3BO_3 (Boric acid)	6.2
KI (Potassium iodide)	0.83
$Na_2Mo_4\text{-}2H_2O$ (Sodium molybdate)	0.25
$CuSO_2\text{-}5H_2O$ (Copper sulfate)	0.025
$CoCl_2\text{-}6H_2O$ (Cobalt chloride)	0.025
Myo-inositol	100
Nicotinic acid	0.5
Pyridoxine-HCl	0.5
Thiamine-HCl	0.1
Glycine	2.0
Sucrose	30,000
Kinetin	0.04 to 10
IAA (Indolacetic acid)	1.0 to 30
pH	5.7 to 5.8

Source: Murashige and Skoog [14]

use. Thorpe [22] estimated that at least 300 commercial firms are involved in producing plantlets in 1990, and many firms have "in house" tissue culture laboratories that might not have been included in Thorpe's original estimate.

The success of tissue culture in geranium propagation has been proven, particularly in reduced severe losses caused by diseases. The procedure's economics will be challenged in the future with new cuttings sources, but tissue culture benefits should exceed limitations.

References

[1] Cassels, A.C. and G. Minas. 1983. Plant and in vitro factors influencing the micropropagation of Pelargonium cultures by bud tip culture. *Sci. Hortic.* 21:53-65.

[2]   Chen, H.R. and A.W. Galston. 1967. Growth and development of Pelargonium pith cells in vitro. II. Initiation of organized development. *Physiol. Plant.* 20:533-539.

[3]   Chu, I.Y.E. 1986. The application of tissue culture to plant improvement and propagation in the ornamental horticulture industry. *Tissue Culture as a Plant Production System for Horticultural Crops.* Ed. R.H. Zimmerman, R.J. Griesbach, F.A. Hammerschlag and R.H. Lawson. Martinus Nijhoff Publishers, Dordrecht, The Netherlands.

[4]   Dixon, G. 1986. Pricing micropropagation to keep ahead of the competition. *Grower* 105(24):22-24.

[5]   Dodds, J.H. and L.W. Roberts. 1982. *Experiments in Plant Tissue Culture.* Cambridge Univ. Press, Cambridge.

[6]   Donnan Jr., A. 1986. Determining and minimizing production costs. *Tissue Culture as a Plant Production System for Horticultural Crops.* Ed. R.H. Zimmerman, R.J. Griesbach, F.A. Hammerschlag and R.H. Lawson. Martinus Nijhoff Publishers, Dordrecht, The Netherlands.

[7]   Dunbar, K.B. and C.T. Stephens. 1989. Shoot regeneration of hybrid seed geraniums (*Pelargonium* x *hortorum*) and Regal geraniums (*Pelargonium* x domesticum) from primary callus cultures. *Plant Cell, Tissue and Organ Culture* 19:13-21.

[8]   Hammerschlag, F.A. 1978. Influence of light intensity and date of explantation on growth of geranium callus. *HortScience* 132:153-154.

[9]   Hammerschlag, F.A. and P. Bottino. 1981. Effect of plant age on callus growth, plant regeneration, and anther culture of geraniums. *J. Amer. Soc. Hort. Sci.* 106(1):114-116.

[10]   Hartmann, H.T., D.E. Kester and F.T. Davies Jr. 1990. *Plant Propagation. Principles and Practices.* Prentice Hall, Englewood Cliffs, NJ.

[11]   Horst, R.K., H.T. Horst, S.H. Smith and W.A. Oglevee. 1976. In vitro regeneration of shoot and root growth from meristematic tips of *Pelargonium* x *hortorum* Bailey. *Acta Hortic.* 59:131-141.

[12]   Janick, J. 1986. *Horticultural Science.* New York: W.H. Freeman and Co.

[13]   Klein-Cox, C. 1985. Tissue culture marketing requires care. *Amer. Nurserymen* 162(6):59-61.

[14]   Murashige, T. and F. Skoog. 1962. A revised medium for rapid growth and bioassays with tobacco tissue culture. *Physiol. Plant.* 18:100-127.

[15]   Oglevee-O'Donovan, W. 1982. Culture-indexing for vascular wilting. *Geraniums*, 3d ed. Ed. John W. Mastalerz and E. Jay Holcomb. University Park, Penn.: Pennsylvania Flower Growers.

[16]   ——— 1986. Production of culture virus-indexed geraniums. *Tissue Culture as a Plant Production System for Horticultural Crops.* Ed. R.H. Zimmerman, R.J. Griesbach, F.A. Hammerschlag and R.H. Lawson. Dordrecht, The Netherlands: Martinus Nijhoff Publishers.

[17]   Paludan, N. 1976. Virus diseases in *Pelargonium hortorum* specially concerning tomato ringspot virus. *Acta Hortic.* 59:119-130.

[18]   Pillai, S.K. and A.C. Hildebrandt. 1968. Origin and development of geranium callus from stem tips, internode pith and petiole in vitro. *Phyton* 25:89-95.

[19]   ——— 1969. Induced differentiation of geranium plants from undifferentiated callus in vitro. *Amer. J. Bot.* 56(1):52-58.

[20]   Reuther, G. 1985. Principles and application of the micropropagation of ornamental plants. *In Vitro Techniques, Propagation and Long Term Storage.* Ed. A. Schafer-Menuhr. Dordrecht, The Netherlands. Martinus Nijhoff/Dr. W. Junk Publishers.

[21] ——— 1988. Problems of transmission and identification of bacteria in tissue culture propagated geraniums. *Acta Hortic.* 225:139-152.

[22] Thorpe, T.A. 1990. The current status of plant tissue culture. *Plant Tissue Culture: Applications and Limitations. Developments in Crop Science 19.* Ed. S.S. Bhojwani. Elsevier, Amsterdam.

[23] Ward, H.B. and B.D. Vance. 1968. Effects of monochromatic radiation on growth of Pelargonium callus tissue. *J. Expt. Bot.* 19:161-174.

[24] Yarrow, S.A., E.C. Cocking and J.B. Power. 1982. Plant regeneration from cultured cell-derived protoplasts of *Pelargonium aridum, P.* x *hortorum* and *P. peltatum.* Plant Cell Rep. 6:102-104.

[25] Yeoman, M.M. 1986. The present development and future of plant cell and tissue culture in agriculture, forestry and horticulture. *Plant Tissue Culture and its Agricultural Applications.* Ed. L.A. Withers and P.G. Alderson. Butterworth, London.

[26] Zumwalt, G. 1976. Plant propagation by tissue culture moves from academic theory into commercial production. *Grower's Rept. Florist* 10(1):72-81.

Cuttings

Marlin N. Rogers

All commercially grown geranium cultivars were vegetatively propagated by cuttings before true geranium breeding lines were developed. While seed propagated plants may gradually replace vegetatively propagated geranium clones, correct vegetative propagation practices still are of prime importance to ensuring success with clonal cultivars.

Disease control

Pathogen-free stock. Stock plants must be free of all major bacterial, fungal and viral pathogens to ensure a healthy, salable crop. Consequently, cuttings must come from stock plants that have originated as recently as possible from culture- and virus-indexed nucleus blocks.

Pasteurized media. Since geraniums are susceptible to several pathogens that are spread through infested rooting media, it is also important to eliminate this possible infection source. Several diseases cause losses during propagation.

Pythium (blackleg) and Rhizoctonia (damping-off) usually occur because nonpasteurized propagating media are used. The easiest way to prevent such problems is by heat treatment of propagating benches containing rooting media at 140 F to 160 F (60 C to 71 C) for 30 minutes the day before inserting cuttings.

Air-moisture balance. Proper air- and moisture-holding capacities of propagating media result in rapid rooting and aid in controlling pathogens such as Pythium. This fungus thrives best in media having high moisture content and restricted aeration and can cause major losses. Disease incidence is particularly high in media with restricted air-holding capacities when mist systems apply more water than needed.

Sanitation. Meticulous sanitation in all geranium propagation operations is crucial. Cuttings harvested from stock plants in one part of a greenhouse range and carried to the propagation area should be transferred in clean, pathogen-free containers. Many growers use clean polyethylene bags or new flats for this operation. Carrying containers permitted to sit on the ground while cuttings are being harvested shouldn't be placed on the freshly pasteurized propagating bench, or recontamination is almost certain to occur.

Breaking and snapping cuttings off stock plants rather than cutting them off was recommended when stock plants were often contaminated with the bacterial pathogen Xanthomonas. (See Chapter 26, Vascular Wilt Diseases). This practice helped to prevent accidental inoculation of cuttings by bacteria that might be carried on the knife used to remove cuttings, first from a diseased and then from a healthy plant. Using a freshly sterilized cutting knife for each stock plant was also suggested [12].

You can snap cuttings off stock plants if stems have the proper degree of succulence, but it's difficult to remove cuttings easily from tougher, more fibrous stems. With today's availability of Xanthomonas-free stock plants, the danger of spreading bacteria from plant to plant or plant to cutting is very low. As a result, most growers now remove cuttings from stock plants with a sharp knife.

Other important sanitation practices that help prevent recontamination of clean propagating areas include keeping hose ends off the ground and keeping cats and other greenhouse pets off propagating benches [6]. Use disinfectant solutions frequently to wash hands and knives.

Proper bench spacing. The gray mold fungus Botrytis (see Chapter 25, Foliar Diseases), propagated by airborne spores, can cause major losses during cutting propagation. This fungus must have high moisture levels to thrive, making proper spacing of cuttings in the bench for good air movement and ventilation important for preventing losses from this pathogen. Usually, only the lower leaf on stem cuttings must be removed for easy sticking. Keep remaining leaves on the cutting for fastest rooting. Space cuttings adequately to minimize foliage overlap and reduce Botrytis losses.

Geranium cutting types. Many commercial growers think only geranium terminal or tip cuttings are valuable for propagation. These may range from 1 to 2 inches (2.5 to 5 cm) up to 4 to 5 inches long (10 to 12.5 cm), depending on the grower's personal preferences.

Stock plant productivity can be doubled or tripled (see Chapter 8, Stock Plants) by dividing stem portions below the terminal cutting into leaf bud or single-eye cuttings (sometimes also referred to as dormant-eye cuttings). More and more geranium growers use this method to produce additional cuttings from their plants. Such cuttings consist of a stem section long enough to anchor the cutting in the propagating medium (usually about 1 inch or 2.5 cm), an attached leaf and its axillary bud (Fig. 9-2). These cuttings form adventitious roots on the lower main stem segment, while the axillary bud breaks into growth and forms the shoot.

Fig. 9-2. Terminal and leaf bud cuttings of zonal, ivy and Regal geraniums. (Photo courtesy Andy Riseman.)

Axillary buds often begin growing while still on stock plants, particularly after removing stem terminal cuttings. These stems also can be divided into short sections called started-eye cuttings, which have essentially identical performance and reliability as short-terminal cuttings.

A fourth kind of geranium cutting sometimes used is the heel cutting, which consists of a short axillary shoot with a bit of the original geranium stem attached [13]. These also are equivalent to short-terminal cuttings in their potential productivity.

Rapid rooting

Keep each production step to a minimum number of days to reduce total production costs. Ideal propagation conditions save time and money.

Stock plant factors. The environmental and nutritional conditions under which stock plants are grown influence rooting of cuttings taken from them. Haun and Cornell [4] reported that the highest rooting percentage occurred in cuttings from stock plants that had received medium nitrogen levels and high phosphorus and potash levels. Oversucculent stock plant growth reduced rooting percentages and delayed rooting [1, 2]. Moderate moisture levels and temperatures with high light levels are optimum for best stock plant growth and propagation results.

Direct rooting techniques. Many growers prefer rooting cuttings directly in containers to rooting cuttings in propagating benches and then potting. Some growers use cell-paks, Jiffy strips or 2¼-inch pots filled with light soil mixes or inert growing media or preformed propagating units such as rock wool blocks, Oasis Wedge, Rubber Dirt or Jiffy-7 peat pellets (Fig. 9-3). Each container has advantages and disadvantages for particular situations. All types provide excellent results when properly used. Direct sticking eliminates one step—the potting of rooted cuttings into small pots—which is a much slower operation than sticking unrooted cuttings in the same container. Furthermore, direct sticking results in superior growth in most cases.

Some geranium growers also eliminate moving the plant from the small container to the 4-inch (10-cm) pot in which it will be sold by rooting cuttings directly in the salable container. Initially this requires more bench space than the previous method but reduces labor costs, which are often higher than overhead costs of additional bench space. Direct rooting in the final pot has earned increased acceptance during recent years with light weight, inert growing media having a higher aeration level than traditional, soil-containing media once used for geranium culture.

Rooting hormones. Some authorities recommend hormones to hasten geranium cutting rooting [1, 2], while others [15] do not. My experience indicates that geranium cuttings usually root very readily without hormones if they are at the proper degree of maturity and succulence. If using hormones, apply them with a puff duster, rather than dipping the cuttings directly into the powder. This reduces the danger of spreading bacteria from diseased to healthy cuttings through the hormone powder container and uses less material.

Some leaf bud cuttings are slower and more difficult to root than others. Most require longer periods to initiate roots than terminal cuttings. Rooting rates are closely correlated with cutting age and hardness; woody cuttings take longer to root than softer tissue cuttings.

Rooting responses of leaf bud cuttings were evaluated following applications of Hormodin #1, #2 and #3. The only difference between the

Fig. 9-3. Molded growing containers.

formulations was increasing active ingredient concentration as the numerical designation became larger. Leaf bud cuttings of three geranium cultivars, which had been arbitrarily classified as being "soft" wood or "hard" wood, were treated with the three hormone powders, inserted in a sand propagating bench in a 70 F (20 C) greenhouse and removed five weeks later.

Increasing the hormone concentration generally resulted in progressively poorer rooting. Cuttings from softer wood rooted significantly better than those from harder wood; thus, the stem tissue should be as soft as possible when leaf bud cuttings are taken. Don't use hormones stronger than Hormodin #1.

Treating geranium stock plants with foliar sprays of ethephon (another growth regulator) at concentrations of 500 to 1,000 ppm improved the rooting percentage and increased root number on cuttings [8]. Dipping cutting ends into 2,500 or 5,000 ppm solutions of the growth retardant Alar (SADH) also enhanced geranium cutting rooting as much as stem dips of 500 ppm indolebutryic acid [9].

Temperature. Rapid rooting occurred when the cuttings were placed in a 60 F (15 C) night temperature greenhouse and bottom heat temperatures of 70 F to 75 F (but not over 80 F [26 C]) were applied [1, 2, 7, 14, 3]. Bottom heat is especially important if cuttings are rooted under mist, since evaporative cooling reduces the rooting medium temperature.

Moisture and aeration. Twenty years ago using mist propagation for geraniums was a sure route to disaster. If cuttings were infected with bacterial or fungal pathogens, the moist environment provided by mist was ideal for rapid and devastating disease development. Under these conditions, mist propagation wasn't recommended.

Although some research horticulturists and plant pathologists still don't recommend mist propagation for geraniums, many growers are now using mist propagation successfully. The key to successful mist use lies in nearly pathogen-free cuttings. Cuttings come from cultured and indexed stock free from internal pathogens, and growers use hospital-like sanitation practices and regular and frequent applications of Botrytis-inhibiting fungicides to reduce or eliminate external spore loads on their cuttings. When using mist, don't use more water than absolutely necessary to prevent foliage wilt, since geranium foliage appears to be more susceptible to metabolite leaching and senescence than other plant foliage, such as chrysanthemum and poinsettia.

Light. Geranium cuttings need the maximum light they can tolerate without excessive wilting for fastest rooting. After the first few days, cuttings propagated during the late winter won't be damaged by full sunlight in most northern greenhouses. Shade, where needed, should be applied as cheesecloth or saran over the propagating bench rather than as newspaper or other opaque materials laid directly over the cuttings. Such materials confine moisture and humidity around the cutting leaves and promote disease development.

Carbon dioxide. Elevated day temperatures (75 F to 80 F; 24 C to 26 C) and enriched carbon dioxide levels (800 to 1,500 ppm) in greenhouse atmospheres increase rooting speed and early root system growth on geraniums [11] and are recommended for both stock production and rooting areas by Yoder Brothers [13].

Leaf bud vs. terminal cuttings

Detailed observations and measurements of how terminal vs. leaf bud cutting affected the final product's size and timing showed that at four or five months after propagation, plants of both kinds were comparable in size and development [10]. Full flowering, however, was delayed about two weeks on plants propagated from leaf bud cuttings compared to those coming from terminal cuttings.

Special needs of leaf bud cuttings. Small leaf bud cuttings respond more favorably to warmer environments (70 F [21 C] night temperatures) in early growth stages than terminal cuttings and grow very little at air temperatures less than 60 F (15 C). Leaf bud cuttings grown at 60 F (15 C) temperatures during the first six weeks after removal from the stock plants caught up with terminal cuttings grown during the same period at 52 F to 55 F (11 C to 13 C) temperatures, and from that time forward both groups responded almost identically [10].

Growers and researchers have found that leaf bud cuttings are more susceptible than terminal cuttings to infection and loss from Botrytis during propagation. Initial infections usually occur through the wounded tissue at the cutting's top and progress downward (see Figures in Chapter 25, Foliar Diseases). These problems occurred when the best anti-Botrytis fungicides available were the dithiocarbamates and captan—neither very effective against Botrytis. Growers today should be able to achieve much higher rooting percentages with leaf bud cuttings, since more effective fungicides are available for Botrytis control.

Don't insert leaf bud cuttings too deeply into the propagation medium. Optimum rooting occurs when the axillary bud is about level with the medium surface. Deeper insertion gives poorer results.

References

[1]   Boodley, J.W. 1961. Geranium propagation. *Flor. Rev.* 128(3312):37-39, 88-95.

[2]   ——— 1961. Production of geraniums. *N.Y. State Flower Growers Bul.* 181: 1-3.

[3]   Carpenter, W.J., E.N. Hanse and W.H. Carlson. 1973. Medium temperatures effect on geranium and poinsettia root initiation and elongation. *J. Amer. Soc. Hort. Sci.* 98:64-66.

[4]   Haun, J.R. and R.W. Cornell. 1951. Rooting response of geranium cuttings as influenced by N, P and K nutrition of the stock plant. *Proc. Amer. Soc. Hort. Sci.* 58:317-323.

[5]   Heng, D.A. and P.E. Read. 1976. Influence of daminozide on the levels of root-promoting substances in *Pelargonium hortorum* Bailey. *J. Amer. Soc. Hort. Sci.* 101:311-314.

[6]   Juchartz, D.D. 1961. Geraniums. *Mich. Florist* 361:24. *Dis. Reptr.* 53: 412-414.

[7]   Mastalerz, J.W. 1965. Geranium stock plants. *Penna. Flower Growers Bul.* 171:3-6.

[8]   Moorman, F. and F.J. Campbell. 1980. Florel's effects on three geranium cultivars. *Flor. Rev.* 166(3297):1, 6, 58-60.

[9]   Read, P.E. and V.C. Hoysler. 1969. Stimulation and retardation of adventitious root formation by application of B-Nine Cycocel. *J. Amer. Soc. Hort. Sci.* 94:314-316.

[10] Rogers, M.N. 1961. Increasing geranium yield. *Flor. Rev.* 128(3325):21-22, 79-82.

[11] Shaw, R.J. and M.N. Rogers. 1964. Interactions between elevated carbon dioxide levels and greenhouse temperatures on the growth of roses, chrysanthemums, carnations, geraniums, snapdragons and African violets. Part 5. Various flowers. *Flor. Rev.* 135(3491):19, 37-39.

[12] Tammen, J. 1960. Disease-free geraniums from cultured cuttings. *Penna. Flower Growers Bul.* 117:1, 6-9.

[13] Yoder Brothers Inc. 1968. Yoder 40/one geraniums. *Grower Circle News* 64:1-2.

[14] ———— 1969. Propagation. *Grower Circle News* 77:8.

[15] Zoeb, R. 1965. Geraniums—western (Wisconsin, that is) style. *Ohio Florists Assn. Bul.* 424:3-6.

CHAPTER 10

Growing from Vegetative Propagules

Jay Sheely and Richard Craig

The zonal geranium market has expanded rapidly over the past five years due to the introduction of many new cultivars into the marketplace (see Chapter 22, Novelty Geraniums and Chapter 23, New Cultivars). These cultivars offer new flower colors to the consumer and also represent improved genetics in greenhouse and garden performance.

This chapter is devoted to finishing zonal geraniums and is divided into three segments. Phase I examines the relative advantages of various cuttings and preparing the greenhouse for the cuttings' arrival. Phase II reviews all production aspects—media, irrigation, fertility and growth regulator treatments. Phase III discusses preparing plants for sale and some container size and scheduling alternatives.

Phase I—preparation

Clean stock implies that mother plants are indexed free of Xanthomonas and certain viruses. Modern geranium production is possible only if you use clean stock. Assuming that you use clean stock, modern geranium production and profitability result from *superior cultivars* and *proper cultural systems*. If you don't use clean stock, you can expect that your costs will escalate and your profits will disappear.

Thus, even with superior cultivars, superior greenhouses, superior production techniques, systems, superior growers and superior marketing, Xanthomonas can cause a 0% performance. Xanthomonas is a highly contagious disease and you *must* adopt a zero tolerance attitude. Two percent infection, 1% infection, or 0.5% infection is not acceptable. No other geranium disease or virus is as devastating as bacterial blight. Don't let anyone sell you a bill of goods that says that you will always have a little Xanthomonas. The only acceptable definition of clean stock is zero Xanthomonas.

Clean stock maintenance techniques

- Purchase clean-stock cuttings from a reputable propagator.
- Do not carry over geraniums from year to year—zonals, ivies or Regals. Regals (*Pelargonium* x *domesticum*) can be especially troublesome. Xanthomonas-infected plants may not show symptoms, but they can contaminate zonals or ivies.
- Do not mix geraniums from various supply sources. If you buy from several propagators, separate each group and keep the cuttings and pots isolated throughout the production cycle. You can invite serious problems by mixing cuttings from clean stock with those harvested from noncultured stock plants.
- If you suspect a problem, immediately isolate the suspect plant and have the plant material tested by your state agriculture department or a commercial laboratory. If the plant is infected with Xanthomonas, there is no cure.
- Thoroughly clean the greenhouse area prior to cuttings' arrival, being sure to remove debris from previous geranium crops. Disinfect all tools, carts and flats. Spray all surfaces (bench tops, side walls) with a surface sterilant solution (10%).
- Limit entry to geranium production areas to those workers who really need to be there. Train your employees to be aware that a clean stock system is in place and to know and follow sanitation rules.
- Schedule crop rotation so that you are out of geraniums for at least 30 days. This is cheap insurance to avoid carry-over problems from one production year to the next.

Vegetative cuttings

Unrooted cuttings. The demand for unrooted cuttings is increasing as the cutting supply from indexed stock programs continues to develop. Because growers have installed capital improvements to meet the requirements of plug programs and other propagation systems, more growers can now successfully root geraniums. A good bottom heat system to sustain a media temperature of 70 F (21 C) degrees and a mist system to alleviate stress during the first five to seven days encourages quick rooting, rapid initial growth and minimal loss.

Calloused cuttings. Although a smaller market segment, calloused cuttings are still in demand by growers who stick the cuttings directly into the finished container. Using calloused instead of unrooted cuttings decreases the production cycle by seven to 10 days. Relatively low product and freight costs and planting ease drive this buying decision.

Rooted cuttings—plugs. One of the fastest growing market segments over the past five years has been using geranium cuttings rooted in synthetic media. The Oasis Wedge and Oglevee Rubber Dirt are two examples. Used primarily by large growers, the plugs have lower product and freight costs than the traditional 2¼-inch liner. While their handling requirements are more stringent, they offer significant cost savings to the grower.

When handling a plug product, the most essential step in assuring rapid initial growth is irrigating the pots two or three times the first day and daily for several days thereafter. This practice leaches any excess salts and initiates root development in the soil. Roots should be visible at the edge of the pot after one week. If not, leach the soil mix.

Rooted cuttings—traditional 2¼-inch liners. Overall, the traditional 2¼-inch rooted product remains an attractive alternative for many growers. This product commands the highest price, but offers three advantages: least crop time (as little as four weeks—see fast cropping), ability to offer retail customers greater cultivar selection due to lower cultivar multiples and relative growing ease compared to other products.

Receiving cuttings or young plants

Zonal geraniums are among the most difficult of all horticultural products to ship successfully. Any transit delay or high temperature extremes during transit will result in lower leaf yellowing and abscission. While plants outgrow this condition, a longer crop time results. If

Fig. 10-1. Terminal cuttings are on the left and stem or leaf bud cuttings are on the right. A dormant eye cutting is shown upper right. Started leaf bud cuttings (lower right) are equivalent to terminal cuttings in handling and performance.

the plants or pots are shipped in a sealed box, it is essential to remove the field heat from the plants prior to sealing the box.

Remove cuttings from the shipping container immediately upon receipt, space them on a raised bench and irrigate. If this isn't possible, open the boxes and store the cuttings in a cooler at 40 F (4 C). Always attempt to pot cuttings on arrival or at least within 24 hours.

Phase II—growing the crop

Spacing

If you wish to produce high quality geraniums, adequate spacing is essential. While many growers choose to pot the cuttings and hold them pot-to-pot for several weeks, the pots should receive their final spacing during the third or fourth week after potting. This ensures increased branching and compact growth. Crowded plants become leggy, lose bottom leaves and are more susceptible to disease. In general, space the plants so that the lower leaves are barely touching (see Chapter 7, Growth Regulating Chemicals).

Media, irrigation and fertilization

Media. Zonal geraniums may be grown successfully in almost any media as long as it meets certain cultural requirements.
- The medium must be free of soil-borne pathogens.
- The optimum pH for growing zonal geraniums is 5.8 to 6.5 in a soilless medium. If the pH drops below these levels, the plants simply will not grow to optimum size. (*Editor's note:* In Chapter 5, Biernbaum recommends a pH range of 6.0 to 6.3 for geraniums.) Low medium pH (lower than 5.5) combined with high levels of micronutrients (derived from constant feed programs with peat-lite fertilizers and/or incorporated micronutrients) can result in iron/manganese toxicities (see Chapter 5, Fertilization).
- The medium must be well-drained to sufficiently aerate the root system.

Irrigation and fertilization. Quality geraniums grown as a fast crop require excellent fertilization/irrigation regimens, termed fertigation by some experts. Fertigate as soon as possible after potting.

Use a fertilizer solution to irrigate the plants and fertigate at least three times during the first day, if possible. Remember, you want to attain root growth quickly; all growth requires proper nutrition. Subsequent fertigations are less frequent, but always heavy enough to leach at least 10% to 15% of the applied solution from the container; this practice removes undesirable salts.

Many growers use or are planning to use ebb-flow or flood-drain systems. In theory, these systems permit greater labor efficiency, more controlled fertilization and irrigation schedules and generally, if used properly, maximum plant quality. Potentially a problem exists with using these systems with geraniums: disease spread. Two of the most devastating geranium diseases are Xanthomonas and Pythium. Research being conducted with geraniums and other crops in ebb-flow systems is currently focusing on three viewpoints:

- Reported disease spread in these systems in Europe, where they have been used regularly for several years, haven't been serious.
- These systems are recommended only if recirculating solution is sterilized.
- Not enough experimental evidence has been gathered to prove or disprove these systems' effect on disease spread.

Xanthomonas and Pythium can be readily disseminated by water or nutrient solutions. Use extreme caution in producing geraniums with these systems. By all means, use clean stock!

Many fertility programs can succeed equally if you follow several rules. Geraniums require frequent fertilization. Most geranium fertilizers have an equal or nearly equal ratio of major elements such as soluble 15-15-15, peat-lite 15-16-17 or Sierra Geranium Mix 13-11-12. Fertilizers with a high proportion of nitrogen in nitrate form are best for geranium production.

We don't know whether high ammoniacal nitrogen levels are deleterious due to ammonia per se, or due to the lower pH that results when higher ammoniacal ratio fertilizers are used. The bottom line is that most geranium fertilization is accomplished with a high nitrate fertilizer. Use at least 200 to 300 parts per million of 15-15-15 and trace elements or a peat-lite formulation such as 15-16-17 with trace elements for best results.

Slow-release fertilizers can be used completely or in concert with soluble fertilizer. You may incorporate slow-release fertilizer into the media or broadcast it on top of the media; in the latter case, keep the fertilizer away from the plant stem.

Two nutritional problems occur commonly with geraniums. The first is calcium deficiency (see Color Fig. C47); calcium deficient plants exhibit two visible symptoms: abnormal growth of the apical meristem and inflorescence or floret bud abortion. If you observe aborted inflorescences or floret buds, you can be almost certain that the plants are calcium deficient. To avoid the problem, alternate regular fertilizer with one of three high-calcium fertilizers: 15-0-15, 20-0-20 peat-lite, or a mixture of calcium nitrate-potassium nitrate plus magnesium sulfate

(Epsom salts). Apply these alternate fertilizers every third fertigation or as needed, based on regular foliar analysis.

The second major nutritional problem is metal toxicity—usually iron (see Color Fig. C58), manganese or zinc. Our experience indicates that the real problem is due to low pH coupled with generous trace elements. Most metals are much more available to plant roots at low pH. Early symptoms resemble pinpricks in the mature leaves. Symptoms can become more severe and greater foliar damage becomes evident. Ultimately, foliage can be destroyed (see Chapter 5, Fertilization).

Of course, this problem won't occur if you *monitor media pH* and use metals sparingly in the media. By properly amending the media initially so that the pH is 5.8 to 6.2 and by using high nitrate fertilizers, you can control this problem. When it does occur, the only solution is to increase the pH by applying finely ground limestone as a slurry or top dressing. If you diagnose the problem early and raise the pH, you can still produce high quality, salable plants.

Helpful hints. If your zonal geraniums exhibit long internodes, large leaves and are taller than optimum:
- Fertilizer levels in the medium may be too low. If the injection system is working properly, review the application method.
- The cultivar may just be a vigorous grower, and you may need to apply a growth regulator or select a less vigorous cultivar from the same color class.
- The plants may not have adequate spacing, causing them to stretch.

If your zonal geraniums are compact and are smaller than optimum:
- Check the root system. Compact growth may indicate a root system problem caused by excessive substrate salt levels, a soilborne pathogen or insufficient aeration.
- If the plants were treated with a growth regulator, either the rate or the volume applied may have been excessive.
- You may have selected a compact cultivar that you could replace with another cultivar better suited to your management program.

Environmental factors

Temperature. Geraniums can be grown between 55 F and 85 F (13 C and 29 C); however, optimum growth occurs when night temperatures are 60 F to 65 F (15 C to 18 C) and day temperatures are maintained at a minimum 70 F (21 C). Vent at 80 F (27 C). During most growth phases, geraniums can tolerate and grow at much higher temperatures, especially if light and CO_2 aren't limiting.

Light. With respect to flowering, geraniums are day-neutral and don't respond to short photoperiods or to long photoperiods of low irradiance.

Conversely, high irradiance improves geranium growth and flowering. Geraniums are full-irradiance plants; grow them in your brightest, cleanest greenhouses and without shade except late in the spring to protect flowers from burning. Geraniums respond positively to supplemental irradiance from fluorescent or HID lamps. Thus, under low natural irradiance, supplemental lighting is beneficial. It may or may not be cost-effective; you will have to make this determination.

Relative humidity. The relative humidity of most greenhouses is adequate for growing geraniums. Humidity control, however, is critical to prevent many foliar diseases, including Botrytis and rust. Under highly humid conditions, Botrytis can destroy a complete crop. Control humidity by proper plant spacing and adequate ventilation.

Carbon dioxide. Geraniums respond positively to supplemental carbon dioxide. CO_2 deficiencies can occur in greenhouses, especially when ventilation is limited. This often occurs on cold, bright mornings when vents are closed. Deficient CO_2 reduces growth. Exceptional growth can be observed when supplemental CO_2 is used. Normal CO_2 concentration is 350 ppm; supplementing to 800 to 1,000 ppm for at least eight hours per day or until 80 F (27 C) is reached (usually venting temperature) is a cost-effective method of optimizing growth when irradiance levels are high or when using supplemental irradiance.

Pinching and growth regulators

While some growers still apply a soft pinch in 4½- or 6-inch geranium programs, many have abandoned the practice, as it is no longer essential. Pinching adds two weeks crop time, and new cultivars have basal branching capabilities. Many growers apply a hard pinch only to remove a tip cutting from an early planting.

The need for a growth regulator is based on cultivar growth habit, forcing temperature, media salinity levels, and irrigation frequency. As a general rule, compact cultivars require no growth regulators; vigorous cultivars require at least one application; and, anything labeled "moderate or medium" may or may not require growth regulators, depending on the above factors. Cultivars such as Kim, Veronica or Pink Expectations exhibit growth characteristics that are often labeled as medium.

Growth regulators reduce leaf size, increase lateral branching and shorten internode length. Cycocel, Bonzi and Florel all have this effect. Florel, instead of hastening flowering like Cycocel, often aborts flower buds at higher application rates and is useful for programs where flower removal is necessary to enhance vegetative growth.

To produce the best looking plants, apply growth regulators frequently at reduced rates, rather than infrequently at high rates. For example, a geranium sprayed twice with Cycocel at 1,500 ppm (at 10 to 14 days after potting and again 10 days later) will have more uniform leaf size and habit than a crop sprayed once with 3,000 ppm. With some of the newer growth regulators (Bonzi, for example), lower rates also reduce the risk of applying excessive amounts. Phytotoxicity can be a problem if Cycocel is applied at high rates during periods of stress (high temperature or dry media).

The following rules serve as general guidelines when applying growth regulators:

- Not all growth regulators are approved for geraniums, or approved in all states. Refer to your local cooperative extension office for specific information. Those growth regulators mentioned above have been used in research situations; however, their mention does not imply endorsement.
- Apply when the plants are not under stress from high temperatures; early in the day or on a cloudy day is best.
- Don't apply when plants are under moisture stress. Irrigate plants the day before the application is planned.
- Not all cultivars require treatment. Isolate compact cultivars on separate benches or in other greenhouse areas.
- Consider both application *rate* AND *volume*. Keep accurate records of application frequency and rates and carefully note any cultivar sensitivity.

Phase III—realizing the profit potential

Preparation for sale

Following several procedures just prior to sale can improve the product appearance and enhance its shelf life.

- Water geraniums thoroughly to leach any excess salts prior to sale.
- Remove spent blooms. Very often the plants are irrigated overhead in the sales area, and petals on the foliage increase Botrytis risk.
- Remove yellow leaves. If yellowing is severe, check the root system.
- Apply gibberellic acid at 1 to 2 ppm when the first two or three florets open to make the inflorescence larger and increase shelf life in the retail area.
- Apply a light shade during late spring to reduce light levels and prevent flower petal burning (especially for red flowered cultivars) prior to sale.

If the geraniums are delivered by truck to the retail outlet, be sure everyone is aware that the plants should be unpacked and spaced as soon as they arrive. Geraniums packed pot tight in a box for long periods can produce ethylene in damaging quantities and quickly reduce a high quality crop to unsalable junk. Also, consider the temperatures the geraniums will encounter in transit. Is there any air moving through the truck to prevent temperature extremes?

One of the best ways to benefit your customers is to properly label your geranium cultivars. Not only is it a contractual requirement in many cases to properly identify the plant as a patented or proprietary product, but it also aids your customers in defining exactly which cultivar they are buying. With the large number of new cultivars now available, how can the customer select the cultivar that performed so well in their garden last season when all the red flowered geraniums are labeled "red?"

Timing and scheduling

Fast crop techniques in 4-inch programs. To finish a geranium cutting in a 4- or 4½-inch pot in six to eight weeks, use fast cropping (unchecked growth) techniques: First remove all stress conditions and limiting factors and then use higher temperature, nutrient and moisture levels than normal. The fast crop requirements are:
- Well-rooted cuttings from clean stock
- Well-drained medium or peat-lite mix
- Warm temperatures—minimum 62 F to 64 F (17 C to 18 C) night and 70 F to 80 F (21 C to 27 C) day
- Continuous fertilizer and soil moisture supply
- Full light intensity
- CO_2 if available at 1,000 ppm for eight hours daily
- No pinching

Opportunities in larger containers. Growers have been finishing their zonal geranium stock plants for sale as large pots or tubs for years. The final product is determined to a large degree by the starter plant's size and shape. A few suggestions:
- You cannot chop back a tall, leggy stock plant to a reasonable size and end up with a well-developed plant with multiple blooms in a reasonably short crop time of four to six weeks. Ideally, the stock plant should be no taller than six to eight inches above the pot after the last cuttings are harvested.
- Keep any pruning to a minimum. Remove all old, large leaves to allow basal breaks to develop.
- Allow about 10 to 14 days for the plant to recover from any pruning/cleaning and apply Cycocel at 1,500 ppm on vigorous cultivars. If the sales date is at least eight weeks away, consider a Florel application of 500 ppm. This chemical reduces internode length and leaf size, and also stimulates breaks to develop from the plant bottom. For an early

sales date, using DIF temperature control (see Chapter 6, Light and Temperature) may be an alternative to growth regulator applications to control height.
• Choose a cultivar with a medium to vigorous growth habit. A compact habit isn't useful in large containers, since the product's overall height and size often directly relates to selling price.

To finish stock plants, allow six weeks at 64 F (18 C) nights, 8 weeks at 58 F to 60 F (14 C to 15 C) nights.

You may use the following table as a guide:

TABLE 10-1

Container size (in inches)	Number of cuttings/pot	Crop time in weeks	Pinch
6	1	8	yes
6	3	4–5	no
8–10	4–5	10–12	yes

Hanging zonal baskets. Ivy production dominates the geranium hanging basket market (see Chapter 21, Ivy Geraniums), but certainly there's a place for some newer zonal cultivars as well.
• Select compact, early flowering cultivars, such as Victoria, Tango or the dark leaf Eclipse Series. Pot three cuttings in an 8-inch (20 cm) container or four in a 10-inch (25 cm) container. Allow eight to 10 weeks depending on night temperature. Pinching adds two weeks to crop time.
• Watch the pH. Zonals like a pH around 6.0, while ivies can be grown in the 5.0 to 5.5 range. If you're supplementing your feed with iron for the ivies, you need to run a higher pH for the zonals, or you may experience problems with heavy metal toxicity.
• Consider producing a portion of your stock plants in hanging baskets. The baskets finish in five to six weeks from the last harvest and you gain some extra space.

Geraniums in the landscape

Geraniums in the landscape

C13. Appleblossom Rosebud geranium
C14. Scarlet Rosebud geranium
C15. Patricia Andrea tulip-flowered geranium

C16. Poinsettia cactus-flowered geranium
C17. Silver Ruby fancy-leaved geranium
C18. Petals fancy-leaved geranium

*N*ovelty geraniums

C19. Mrs. Cox fancy-leaved geranium
C20. White Mesh ivy-leaved geranium

C21. Chinese Cactus stellar geranium
C22. Pixie Rose stellar geranium

*N*ew cultivars

C23. Ritz

C24. Sassy Dark Red

*N*ew cultivars

C25

C28

C26

C29

C27

C30

C25. Eclipse Red
C26. Elizabeth
C27. Tango

C28. Americana Cherry Rose
C29. Charleston
C30. Satisfaction Salmon

C31. Showcase Dark Salmon
C32. From left to right, Judy, Grace and Marilyn
C33. Pink Expectations

C34. Marilyn
C35. Rio
C36. Risque

New cultivars

C37. Laura
C38. Fox
C39. Americana White

C40. Showcase White
C41. Ben Franklin
C42. Princess Balcon

$p$hysiological disorders

C43. Complete nutrient solution (left) and nitrogen deficiency (right).

C44. Potassium deficiency on young and older leaves. Compare to C59.

C45 and C46. Phosphorous deficiency. Early stage on older leaf (left) and advanced stage on older leaf.

C47. Calcium deficiency, four gradations. *Editor's note:* Plant roots with boron deficiency and low pH look similar, stubby with brown or black tips.

C48, C49 and C50. Magnesium deficiency, several symptom expression forms.

C51. Low pH (4.5). Some calcium deficiency and chloride toxicity phases also look similar.

C52. Boron toxicity. Plant water stress may look similar, especially on older leaves.

_P_hysiological disorders

C53 and C54. Boron deficiency. The photo on the bottom shows internal breakdown of pith—a good diagnosis method. Symptoms could be confused with crinkle virus. Compare to C60.

C55. Iron deficiency, advanced stage. High temperature may cause a similar appearance. Compare to C61.

C56. Manganese deficiency, advanced stage. Could be confused with yellow net vein virus.

C57. Could be phosphorous and nitrogen deficiency or older leaf of a plant grown too cool and too dry.

C58. Iron toxicity. Mature leaves develop speckling with chlorotic and necrotic interveinal symptomology.

physiological disorders

C59. Potassium deficiency on Regals. Compare to C44.

C60. Boron deficiency on Regals. Compare to C54.

C61. Iron deficiency on Regals. Compare to C55.

C62. Sulfur deficiency on Regals. Symptoms uncommon but similar to nitrogen deficiency or high temperature stress.

C63. Cycocel phytotoxicity on newly expanded leaves, three to seven days after treatment. Leaves often regain green coloration.

C64. Moisture stress from low soil moisture. Note dull plant leaf on left under low moisture conditions vs. sheen of turgid leaf on the right.

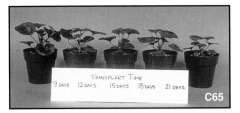

C65. Late transplant delay on seedling geraniums.

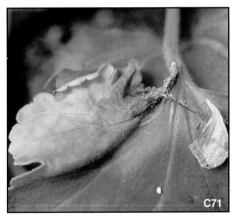

C66. Dramatic DIF results. Geranium on left grown at 25 C day/15 C night temperatures. Plant on right at 15 C day/25 C night.

C67. Tissue culture reproduction.

C68. Mite impeded on trichome.

C69. Seedling blindness. Growing point death and flower bud abortion.

C70. Botrytis flower blight. Brown, soft (water-soaked) areas on petals covered with grayish brown spore masses. Florets matted together.

C71. Botrytis leaf blight from petals. Black, soft lesions on leaves assuming shape of fallen petal.

Geranium diseases

C72. On this stock plant Botrytis stem blight began in the wounded stem and is progressing downward.

C73. Dense canopies limit light and promote lower leaf senescence.

C74. *Botrytis cinerea* readily infects senescent leaves and sporulates.

C75. Botrytis stem blight started from injury infection on this cutting.

C76. Botrytis crown rot. Black, soft stems with white fuzz (spore masses).

Geranium diseases

C77. Pythium black leg. Brown to black water-soaked rot.

C78. Bacterial leaf spot: (left) *Pseudomonas syringae*; (right) *P. cichorii*. Species not distinguishable by symptoms.

C79. Leaf spot caused by *Cercospora brunkii* on Ringo. Includes small brown spots, larger brown spots and large necrotic areas. Chlorosis is usually associated with diseased tissue. The pathogen sporulates on both upper and lower leaf surfaces.

C80. Alternaria leaf spot. Raised, blisterlike, wet-appearing spots, some with white centers on lower leaf surfaces.

C81. Bacterial blight *(Xanthomonas pelargonii)* spread by overhead irrigation.

C82. Wilted, yellow leaf showing systemic infection by *X. pelargonii.*

Geranium diseases

C83. Yellowing and collapsed zonal geranium infected with bacterial blight.

C84. Leaf spots on ivy geranium from *X. pelargonii* infection.

C85. Ivy geranium leaf showing yellowing and brown patches caused by bacterial blight.

C86. Tobacco ring spot virus. Yellow to necrotic spots, rings and line patterns on leaves.

C87. Pelargonium ring spot virus. Symptoms are yellow spot, partial rings and complete ring spots.

C88. Yellow net vein virus. Symptoms are most intense in winter.

Geranium diseases

C89. Cucumber mosaic virus six weeks after inoculation.

C90. Pelargonium flower break combined with cucumber mosaic.

C91 and C92. Rust. Bottom left (C91): Brown spore pustules 13 days after infection. Top right (C92): Concentric rust uredia rings surround a single rust lesion.

C93. Inflorescence and petiole infected with cottony stem blight. White tufts of cottony fungus growth appear on petiole and inflorescence.

C94. Sclerotinia crown rot. Stem infection close-up with white cottony fungus growth and black sclerotium.

CHAPTER 11

Plug Culture

David S. Koranski and Mehrassa Khademi

Single-cell production or plug production has mechanized the floriculture industry. Plug production provides several advantages not found in traditional bedding plant methodology. Mechanization has made seeding, growing and transplanting more efficient: Time, space and labor are optimized. Single-cell plants are more vigorous, resulting in reduced crop production time and allowing two or more crops in one season.

Over the past 10 years plug production has become a major factor in greenhouse growing. The last five years have seen significant refinement of growing techniques and equipment needed to produce healthy, quality plugs, while ongoing research enhances our knowledge of various crops' plug production requirements.

Grow your own plugs or buy them?

A commercial grower has several production choices: producing geraniums in the time-honored method of sowing seed flats and then transplanting the seedlings; producing plugs; purchasing plugs from an outside specialist-producer for finishing; or using a combination of growing and buying.

For many growers producing or purchasing geranium plugs is becoming the favored production method. Producing geranium plugs is a rapid production method because of fast plug germinating and finishing, but this method requires a sizable investment in specialized equipment, a large space to germinate seed (approximately four to five times more space then conventional seed flats) and specialized growing methods. For growers with adequate space, capital to invest and a willingness to learn special plug growing considerations, producing plugs enables them to grow more bedding plants faster and with less labor. Plug production allows more crop rotations in the same space compared to conventional sowing, and crop times are highly predictable. Producing your own plugs also allows you to control your crop.

Producing plugs can be advantageous if you have optimum conditions for the crops you grow. It may be difficult, however, to produce high quality plugs under one environment. In making the decision to purchase or produce geranium plugs, consider availability of a controlled environment. Research work at Iowa State indicates that the optimum geranium plug is produced under high light intensity and/or supplemental lighting. Therefore, if you are going to grow your own geraniums without supplemental lighting, you may produce an inferior product with longer time-to-flowering.

Producing plugs

Plug production is an entire growing system that affects every aspect from sowing to transplanting. A substantial initial investment is necessary for germination facilities, mechanical seeders, trays and other equipment. Germination and growing-on require accurate environmental control devices to provide proper light levels, temperature, moisture and nutrient levels.

Equipment

Automatic seeders. One key piece of specialized equipment necessary to plug production is an automatic seeder. Several different seeders are currently available, and the dollar investment can vary from under $1,000 to over $30,000.

Before purchasing a seeder, evaluate your volume, production methods, financial investment ability and your operation; then, gather as much information as you can about the various seeders available. Write to manufacturers for literature and ask for demonstrations at trade shows or a supplier's business place. Growers who use automatic seeders are also a good information source.

Some features you will want to consider include: Is sowing speed variable or fixed? What types of trays will the seeder sow? How much seed will the seeder hold or require to operate?

Plug trays. Growers can choose plug trays in different sizes and shapes. The most common range from 128 to 800 cells per tray. The plug tray choice depends on plug use. Smaller plugs finish well in flats, while larger plugs produce attractive pots, baskets and planters. Larger plugs also produce a higher quality finished plant in a shorter time period than smaller plugs. For geraniums we suggest the 288 or 128 plug tray. The 288, for some growers, is then transplanted into the 50 plug, which is later transplanted into the 4- or 5-inch (10- or 12-cm) pot. The 128 cell is preferred by many growers since it allows more air movement, usually resulting in less Botrytis, and these plants can be directly transplanted into finishing containers.

Research at North Carolina State University has determined that the deeper the cell, the more oxygen available in the medium. A deeper plug cell, however, might not lead to better seedling production if the proper moisture isn't provided. Providing optimum temperature and moisture is more critical than plug tray depth. Never allow plugs to dry completely, although some drying must occur to allow increased oxygen for optimum germination and growth.

Seed

Beyond different seed classes and varieties, you have a choice among refined, primed and standard seed.

Primed seed. The priming process, used for many years with vegetables, starts some of the germination processes and then stops them before root emergence occurs. The seed is then dried back to its original moisture content. This process results in greater and more uniform germination than even refined seed.

Refined seed. This seed has been mechanically separated by density, size, gravity, weight and/or shape.

Standard seed. Of all the different seed types tested, the standard seed usually provides excellent results. Few difficulties have been encountered with geranium seed. At times, however, the seed has been improperly scarified, a necessary process since it has a very hard seed coat. If the scarification process isn't conducted properly, the radicle (root) may

Fig. 11-1. Plug germination on benches.

be malformed, twisted or gnarled and have a tendency to grow upward instead of penetrating the growing media. The cotyledon may tend to grow downwards and actually penetrate the medium. If this occurs, improper scarification process usually is the cause, and you should notify the seed supplier.

Culture for seed geranium germination

Successful seed germination depends on light, medium, temperature and moisture. Light isn't necessary but may enhance germination. Stages 1 and 2 will be referred to during the discussion of annuals germination culture. Stage 1 is the physiological stage after the root has emerged. Stage 2 is the growth of the root system and cotyledons (the first true leaves).

Media temperature is very important for germination. Maintain uniform temperature throughout germination facilities. Geraniums need a temperature between 75 F and 78 F (24 C and 25 C) for optimum germination (Table 11-1). Temperatures above 78 F may cause seed dormancy, resulting in poor germination. This crop is extremely sensitive to substrate temperatures above 78 F and therefore requires careful monitoring.

TABLE 11-1

Germination counts (No. per 100 seeds) at 10 days after sowing

Cultivar	Temperature (degrees)	
	72 F (22 C)	90 F (32 C)
Smash Hit	48.1	23.3
Hollywood Red	52.9	37.9
Ivy	59.3	35.9
Ringo Scarlet	85.9	56.6

Air circulation is essential for uniform temperature in a facility where lamps supply heat. A recording thermometer or thermograph helps monitor air temperature; medium temperature is approximately two degrees F (one degree C) lower than the air temperature. Bottom heating can provide proper medium temperatures for bench germination.

Plug germination success or failure is directly related to moisture applied to the seed. Too much moisture doesn't allow sufficient oxygen to reach the seed, which can result in seed death. Fine mist with a droplet size of 15 to 30 microns provides ample moisture and oxygen for optimum seed germination. Propagation mist usually used for geraniums and poinsettias has a droplet size of 300 to 500 microns, which is too heavy to allow sufficient oxygen for germination in plug culture.

Temperature and moisture requirements are determined by the crop; refer to Table 11-2 for specific details. Once the seeds have germinated, allow the growing medium to partly dry between irrigations.

A grower installing a new germination facility should consider "zoning" the facility to provide different temperatures and humidities. Zoning should provide the grower with better germination percentages than germinating in only one environment. Monitor humidity with a dew point hygrometer or an aspirated psychrometer.

Iowa State University experiment results indicate that a temperature in the 75 F to 78 F (24 C to 26 C) range during Stage 1 and 70 F to 72 F (21 C to 22 C) and reduced moisture during Stage 2 allow optimum geranium germination.

Germination facilities

TABLE 11-2

Geranium germination requirements

	Stage 1	Stage 2
Temperature	75 F–78 F (24 C–25 C)	70 F–72 F (21 C–22 C)
Moisture	100%, 0.1 vapor pressure deficit	95%, 0.5 vapor pressure deficit
Light	50 fc	350 to 400 fc
Fertilizer	25 ppm $CaNO_3$ and KNO_3, one application (within 1 to 3 days)	

Germination room. Germination rooms are environmentally controlled rooms or chambers used for seeded plug tray germination. They are designed so trays can be stacked vertically on movable carts and rolled out of the room for observation or transported to a growing-on area. Most growth room heat is supplied by cool, white fluorescent lamps. Air conditioning is usually available to cool the chambers.

A mist or fog system in a germination room provides the best germination results. These systems increase the humidity at seed level, which is important for optimum germination. The moisture droplet size should be 15 to 80 microns to allow adequate oxygen to reach the seed. The most uniform moisture distribution is attained with very little air movement; too much air movement can cause uneven distribution and drying in some areas. Don't hand irrigate germination trays because irrigating can wash seed out of the cells or bury the seed in the media.

Germination room walls can be constructed of exterior plywood and insulation. The inside is compartmentalized so carts containing plug trays can be wheeled into each section. Cool, white fluorescent tubes are installed 8 to 10 inches (20 to 25 cm) horizontally or vertically from the trays to provide 200 to 400 footcandles (0.43 klux) when measured at tray level.

Create a uniform temperature by forcing conditioned air through a perforated false ceiling into the chamber. Another method of maintaining the temperature is by installing hot water pipes around the chamber's inside walls. In germination chamber construction, several factors are critical, including minimal air movement for uniform moisture distribution and waterproof equipment.

Sweat chamber. Very similar to germination rooms, sweat chambers are designed for germination only and contain no lights. Seed trays remain in the sweat chamber for only 2 to 4 days or until the radicle emerges. Generally, a 90% relative humidity is recommended with an air temperature of 80 F (26 C), but this should be adjustable for different temperature/relative humidity requirements for different crops. Refer to Table 11-2 for specific recommendations.

Problems can arise, however, when the young seedlings are removed from the dark into a dry, high light intensity greenhouse. To ease the seedlings' adjustment to the greenhouse environment, provide a low light intensity area with controlled moisture and temperature regime before moving the seedlings into the growing-on area.

Germination on benches. A portion of a greenhouse can be devoted to plug germination if proper conditions can be maintained. Germination on benches allows easy access to and close inspection of plug trays. Root zone, perimeter or overhead heating systems should emit approximately 70 BTU per square foot (approximately 700 BTU per square meter) to provide optimum temperatures. Overhead fog or mist systems can supply necessary moisture. It's difficult to use irrigation booms for small seed because larger water droplets inhibit oxygen penetration to the seed. They can be used, however, with a very fine nozzle (50 to 80 microns).

Some growers use capillary mats when germinating plugs on benches, especially if using root-zone heating without fog. The mats act to supply moisture and humidity to the plug trays and reduce the drying effects of root-zone heating. Capillary mats also help "pull" moisture through the plugs, spreading it more evenly. You must watch for roots growing into the mats, however.

One method for germinating on greenhouse benches involves covering the trays with a porous, nonoil-based material such as Agricloth. This material provides an ideal microclimate for seed germination by not allowing moisture droplets to "drown" the seed. Remove this covering after germination, when the cotyledons unfold.

High intensity discharge (high pressure sodium) lamps provide supplemental light energy (at least 400 footcandles [0.43 klux]) for seedling development after germination. Maintaining uniform conditions at plant level is necessary for optimum germination and subsequent seedling growth and development.

Germination medium. One problem frequently encountered in growing plugs is providing a suitable germination medium. Because of the small container size, the medium may fluctuate in moisture content, aeration, pH, soluble salts and nutrient levels. A desirable medium should have a high buffering capacity (resistance to pH change), high water-holding capacity and a broad range of particle size to ensure proper drainage.

If the medium particles are similar in size, they compact when irrigated, preventing sufficient drainage, which can kill seed. Carefully prepare medium and test it for drainage in different sized germination containers to prevent serious problems.

The medium pH is extremely critical to geranium plug seedlings' optimum germination, growth and development. The initial pH should be 6.2 (not above 6.6) and at transplant. Conduct an analysis to determine the medium's soluble salts concentration. The soluble salts concentration for plug grown media should be approximately one-half that of finishing substrate. The soluble salts should be less than 1.0 (saturated paste extract). Commercial mixes contain a variable nutrient charge. Test the medium to determine nutrient content; test the pH weekly, by adding distilled water to the media. A nutrient charge above 1.0 soluble salts might burn seedlings if it's not properly irrigated. Phytotoxicity on lower leaves of geranium seedlings grown below 6.0 pH may result from high concentrations of iron and manganese (see Color Fig. C51 and C58).

Water quality

A top quality germination medium can be of no value if poor quality water is applied. Have your water tested by an independent laboratory. Bicarbonate, fluoride, chloride, sodium and boron levels cause major difficulties if they are excessive. Neutralize bicarbonate levels with sulfuric, nitric or phosphoric acid. Review test results with a water quality expert. High bicarbonate and sodium absorption ratio levels may require reverse osmosis (R/O) or deionized water systems (see Chapter 2, Growing Media).

TABLE 11-3

Optimum water quality for seed geranium germination and growth

pH	5.0–6.5
Soluble salts	0.5–0.75
Calcium	60 ppm
Magnesium	35 ppm
Sodium	Less than 30 ppm
Boron	Less than .5 ppm

Lighting and fertilization

During Stage 2, seedling growth improves with supplemental lighting at approximately 450 footcandles (0.49 klux). Geraniums are most responsive to supplemental lighting from 16 to 18 hours per day. Leaf temperature shouldn't go above 85 F (29 C).

Recent nutrition experiments show that optimum growth and development occur with calcium- and potassium-based fertilizers. Maintain the electroconductivity level at approximately 0.75 to 1.0.

Temperature, moisture, lighting and fertilization for Stage 3

In Stage 3 the first true leaves develop. Medium temperature should be approximately 68 F to 72 F (20 C to 22 C). This will vary by cultivar. Air temperature shouldn't be five degrees F less than the medium temperature. Irrigate seedlings daily, preferably with overhead mist and water heated to 70 F to 80 F (21 C to 26 C). Uniform moisture application determines uniformity of seedling growth and development. Supplemental lighting benefits certain crops in Stage 3. Chlorosis or bleaching indicates too much light or the need for additional fertilization.

Nutrition is especially important in Stage 3. Fertilizer may be applied at every other irrigation for faster and more uniform seedling growth. To avoid Botrytis, keep the plant on the hard or toned side. Take substrate samples at least every two weeks for analysis. The most common problems are excess soluble salts, nitrates and phosphorus. Several thorough irrigations with clear water leach the salts and nitrates. Excessive phosphorus can tie up iron and cause iron deficiency (see Color Fig. C55). Leach excessive phosphorus out of soilless media with several thorough irrigations.

Low pH can tie up magnesium and calcium, causing a deficiency of these elements and releasing more iron and manganese, causing phytotoxicity (see Color Fig. C51). This may vary among cultivars. All seed geraniums seem to be sensitive, although the sensitivity also varies among cultivars. Raise pH above 6.2 to avoid phytotoxicity for Stage 4.

Temperature and fertilization for Stage 4

A Stage 4 seedling is almost ready to transplant. The temperature at Stage 4 is extremely critical. Don't expose geranium seedlings to temperatures below 59 F (15 C) for more than one to one-and-a-half weeks or the flowering process may be delayed. Optimum medium temperature for holding plants is 62 F (17 C) for two weeks.

The fertilization regimen at this stage is mainly using a fertilizer with no ammonium. Use a fertilizer that contains mostly calcium nitrate and potassium nitrate to promote root growth.

Growth regulators

Seedlings growing under low light intensity at temperatures above 75 F (24 C) and high moisture are likely to elongate. Applying chemical growth regulators is necessary when you can't control the environment, such as in late spring.

Timing growth regulator application is a factor in successfully growing quality plug seedlings. For seed geraniums, Iowa State University research indicates that 450 ppm Cycocel gives optimum results and should be applied when approximately two to four leaves are present (see Chapter 7, Growth Regulating Chemicals).

Be certain to read and follow label directions for growth regulators. The most suitable time to apply chemicals is approximately two hours after sunrise on a cloudy day because leaf stomates are open to their maximum and allow greatest chemical penetration. Don't spray to runoff. More frequent applications at a lower concentration may produce more manageable control than one or two applications at a higher concentration. High concentrations may delay flowering or produce smaller flower size, especially when applied late in seedling development. Not all chemicals are cleared for bedding plant usage. Ask your extension specialist (university consultant) or other experts for advice. Because some compounds may inhibit root development, examine the root system to ensure active growth before applying chemical growth regulators. Seedlings also should be uniform size at treatment time to obtain consistent results. Conduct small scale sample tests to determine the response under your particular greenhouse conditions. Follow manufacturer's instructions to apply the correct amount of chemical per given area at a rate that delivers the correct concentration of active ingredient per container.

Timing and scheduling

Flexible timing and scheduling is one of the many advantages of growing from plugs. Variety in plug size helps you develop different time schedules based on different development stages and expands production period. Under continuous light and 70 F (21 C) temperature, geraniums stay two days in Stage 1, seven to 10 days in Stage 2, and 23 to 30 days in Stages 3 and 4. That means it takes five weeks from 400-plug size to 18 or 32 packs. Under the same conditions, 128-plug size would take six weeks to grow into 4-inch (10-cm) pots. Time from transplanting to flower also depends on plug size, differing from seven to eight weeks for 128-plug up to eight to nine weeks for 400-plug. Most geranium cultivars respond the same way, except highly vigorous geraniums like the multibloom series, which are very early flowering. Multiblooms may even go to flower while in the plug tray, around week 5. With 400-plug, two applications of Cycocel (750 ppm), timing the first one at the first true leaf and the second application one week later, keep the plants on time. Growing 128-plug, you will need three Cycocel applications (one week apart) to stay on the schedule. Manipulate timing and scheduling with environmental factors such as temperature, light, nutrition, watering and chemicals. You can do this by different means at different plug development stages. Usually growth should be slowed at Stages 3 and 4. Temperature and light are the two primary factors controlling plug growth. Lowering the temperature lessens the need for feeding and watering, resulting in a tuned plug and longer holding time.

Transplanting

When seedlings have completed their growing-on in Stage 4, they are ready for transplant. Two or three hours before transplant, irrigate the trays thoroughly; this helps in removing the plugs from the flats. Prepare your packs and pots by filling them with moistened substrate and predibbling holes for the plugs. Dislodge the plugs and transplant them into prepared packs and pots. Don't pull the plugs out; they may break. A plug dislodger helps release plugs from the trays. Don't attempt to separate multiple seedlings or all of them may be damaged. Closely space transplanted flats and pots on benches in the greenhouse. Space pots according to normal growing recommendations.

Place healthy, actively growing plugs in a medium that provides 15% to 20% aeration with soluble salts under 1.0 to encourage fast rooting. After transplant, irrigate seedlings thoroughly.

Major problems

Media pH. Levels less than 6.0 can cause excessive levels of some micronutrients, such as manganese, iron, sodium and zinc. Toxic levels of these nutrients are expressed as speckling on the foliage, necrosis and destruction of the growing point (see Color Fig. C58).

Nutrition problems. Poor root development, which makes removing the plugs from the tray difficult, can occur when less than optimum calcium and magnesium levels are in the soil. When a fertilizer such as 20-10-20 is used on a constant basis, large leaves may form causing insufficient air movement around the plant, allowing infection and development of diseases such as Botrytis.

CHAPTER 12

Growing From Seedlings

Allan M. Armitage*

Seed-propagated geraniums have been with us for over 20 years and they certainly aren't the babies they once were. Although the flood of new cultivars has slowed to a manageable stream, the advent of multiflora forms, faster flowering cultivars, new flower colors and the potential for bronze leaf forms demonstrate that geranium breeders are alive and well. New dwarf cultivars and refined growing techniques have also allowed growers to move further away from chemical height control. A few geraniums are still produced "from seed to flower" by individual growers, but most plants are finished from plugs. Plug growers have assumed an even greater importance in the geranium trade than before and, the plug quality dictates the finished product quality. Recognizing the importance of greenhouse treatments near the crop's end to enhance postproduction life, good growers now incorporate "integrated postproduction management" as a cultural and marketing tool.

Germinating plugs

It's estimated that approximately 88% of total bedding plant production is from plugs and seed geranium production is certainly close to that percentage. Growing geranium plugs isn't a mystery; it's no more complicated than growing seeds in an open seed flat. However, you must *pay much closer attention to detail*. Using a germination room allows growers to fine tune the emerging seedling's germination needs. Researchers have coined the term "germination stages" for various germination phases (Table 12-1).

The germination stages concept has been around for many years, but only recently has it been recognized that different stages benefit from different environments, regardless if germinated in a plug or in an open seed tray. Stages 1 and 2 are most critical to success. Improper heat, aeration and moisture can be disastrous during the early germination stages. Stages 2 and 3 also greatly benefit from proper amounts of light, water and carbon dioxide.

---

*This chapter is a highly edited version of the book *Seed-Propagated Geraniums and Regal Geraniums* by A.M. Armitage and M. Kaczperski, Timber Press, Portland, Oregon.

TABLE 12-1

Germination Stages 1 to 4

Stage	Description
1	Root emerges from seed coat.
2	Stem and seed leaves emerge.
3	True leaves grow and develop.
4	Seedlings ready for shipping, transplanting or holding.

Source: Koranski and Karlovich, *Grower Talks* August, 1989.

Selecting the plug flat density is a most important decision. Growing high density plug trays with little soil volume (for example, 648 or 800 plugs cells per flat) should be the sole domain of the wholesale plug producer. High density plugs are less forgiving of errors, dry out more rapidly and require constant attention to detail; however, they turn over more rapidly and are a good size for the finisher who wishes to finish in high density flats (72 flat vs. 48 flat). Problems with high density plugs include low soil volume, more finishing time in the final container and storage intolerance. The 406 and 288 plug sizes are most popular for geranium culture. Few geraniums are produced in 800s.

The subject of germination stages sometimes intimidates growers and is more complicated than necessary. With some bedding species, such as pansies, vinca and begonias, paying attention to stages has proven extremely beneficial. For geraniums, if the seed is fresh and properly treated, and if moisture and temperature of 72 F to 77 F (22 C to 25 C) can be maintained, germination should be well above 90% even if you don't have a clue which stage is which. Germination stage awareness is most useful if additional CO_2 or supplemental light is to be applied. Knowing when to apply such inputs results in a faster, more vigorous seedling at the most economical cost.

Many smaller growers still sow geraniums and other bedding plants in open seed trays. This "traditional method" can result in excellent quality finished material, and if you find it more economical than purchasing plugs, you should continue. The traditional method requires a similar germination environment and often results in fewer losses and more uniform stands. Producing uniform, full plug trays requires paying more attention to detail than germinating in open trays, but you can't be sloppy or lackadaisical with tray culture. Due to greater soil volumes, however, tray culture is a little more forgiving than plug culture.

Growing on plugs

When bringing seedlings out of germination chambers, place them out of full sun and mist them occasionally the first day to acclimate them to their environment. Seedlings require a temperature of 68 F to 72 F (20 C to 22 C) after coming out of the germination chamber. The major concern in plug production is proper moisture control, particularly in early stages. For production Stages 2, 3 and 4, continue to supply light, temperature and CO_2.

Open seed tray transplanting

Transplanting from open trays occurs much earlier than from plugs. Seed trays are generally shallow and contain only a small nutrient reservoir, if any; therefore, prompt transplanting is necessary. Geraniums may be transplanted as early as 10 days from sowing. At this stage only the cotyledon leaves have emerged and the root system is small. Obviously significant damage can result if the seedlings are handled roughly. Transplanting can be deferred until one to two true leaves develop without any significant delay in flowering time, but no longer. Work by Jim Tsujita at the University of Guelph showed significant stunting and flower delay if transplanting from open trays occurs too late. Experience with other annual plants indicates that a two- to three-week transplanting delay significantly delays flowering time and geraniums respond similarly.

Factors affecting flowering

Light

Natural light. The main factor affecting time to flower is light intensity. Geraniums are day-neutral plants, so day length is important only because plants are receiving light for a longer or shorter time. Geraniums are most responsive to light intensity changes when young; therefore, keep greenhouse covers clean and free of shading compounds throughout the crop cycle.

Flower initiation is directly influenced by the cumulative light the plant receives. Flowering is most directly influenced in the first six- to eight-leaf stage of geranium growth. These leaves must be healthy and exposed to as much light as possible because these leaves' photosynthetic activity provide carbon-containing substances (sugars and starches) to the developing meristem [6]. The initial six to eight leaves are produced sooner and are larger when grown in high light compared with shaded conditions; they require more time for the meristem to obtain substances necessary to initiate reproductive growth when light is low.

High light intensity accelerates plant development from a leaf producer (vegetative state) to a flower producer (reproductive state). Light intensity greatly affects the plants' vegetative growth phase. Less time is required in the vegetative stage when light levels are high [7].

Flower parts are made during flower development. These parts' formation also accelerates when plants receive high light levels. Once the bud is visible, however, light is necessary to maintain bud quality (bud number and size), but has little effect on time required for further development [5]. Once the plant has reached visible bud stage, regulating light intensity isn't an effective way to accelerate or slow crop. At this stage, temperature becomes the major controlling factor (see Chapter 6, Light and Temperature).

Other obvious light effects also occur on plant growth. Low light reduces dry weight and flower number, causes thin leaves and results in reduced sugar and starch content in leaves, thus reducing shelf life, regardless of the temperature at which the plants are grown [16].

Supplemental light

- Germination. Light isn't necessary for germination if germinated in heated, moist chambers. Seedlings stretch rapidly and irreversibly under dark conditions, but installing lights in germination chambers limits stretching problems. Light intensity can be as low as 300 footcandles (0.32 klux) at tray level to as high as 1,500 footcandles (1.62 klux). A low light system is adequate if seedlings are removed to the greenhouse at Stage 2. Higher intensities are useful for growth rooms where plugs continue to be grown to the beginning of Stage 4.

- Stage 3 to transplant. Natural light is significantly reduced in winter months. In northern states such as Wisconsin, December light may be as low as 21% of July light and if glass or poly is dirty, winter light may be as low as 10% of summer light intensity [5]. Supplemental lighting of hybrid geraniums significantly reduces time to flower and ensures sturdy, compact seedling growth, early leaf expansion and increased branching. Not only does it enhance plant growth, but the light also provides significant heat, reducing heating bills. Although many light units exist, main supplemental light sources are cool-white fluorescent, high-pressure sodium lamps (HPS) or metal halide lamps (MH). The HPS and MH lamps are better supplemental sources than fluorescent tubes because they block less natural light and are more efficient in utilizing electrical power. Don't use low pressure sodium (HPS) lamps as the *sole* light source because the light is monochromatic.

Stage of lighting. Light all young plants, whether produced in a seed or plug flat. Although some growers light for 24 hours a day, 18 to 20 hours per day for six weeks appears satisfactory. Lighting for longer than 20 hours or more than six weeks doesn't provide enough benefit to warrant the additional expense [3]. Commercial light intensities are between 250 to 750 footcandles (0.27 to 0.81 klux), but higher light levels induce faster flowering. Obviously, the more daylight, the less supplemental light is required. With older, slower cultivars, high light levels reduced time to flower even more than with faster cultivars. Although Northern U.S. growers may derive more benefit from supplemental lighting than Southern U.S. growers (due to less natural light), additional lighting in winter months enhances geranium quality throughout the United States.

HID lamps are available with 400 and 1,000 watt bulbs. The larger wattage lamps are more expensive but fewer are required; therefore, they create less shade. The greenhouse roof must be sufficiently high to suspend them well above crop height.

Supplemental light plus Cycocel. Supplemental light combined with Cycocel accelerates flowering in an additive fashion; that is, the plants flower faster with both supplemental light and Cycocel than with supplemental light or Cycocel alone [2]. Apply light and Cycocel while geraniums are in the plug stage. Lighting after plants are more than six weeks old or after the visible bud stage does little except provide extra heat and extra expense.

Lighting geraniums also allows the plant to better use extra carbon dioxide; supplemental winter lighting and supplemental carbon dioxide are natural partners for more rapid geranium production.

Carbon dioxide

Carbon dioxide (CO_2), present in the atmosphere at a concentration of approximately 330 ppm, is an essential raw material for photosynthesis. On days when ventilation isn't required because of low outside temperatures, CO_2 concentration can fall below 330 ppm. This often occurs during the winter, resulting in reduced photosynthesis and plant growth. Even with open ventilators, CO_2 concentrations in the greenhouse often stay below that of outside air.

Like other crops, geraniums grow more rapidly and are better quality when treated with supplemental CO_2. The most efficient CO_2 supplement level is usually around 800 to 1,000 ppm. Plants are most sensitive to CO_2 enrichment when they are young, similar to the plant response to supplemental light. Fertilizing with CO_2 during the entire plant life gives the greatest growth increase; however, the most economically important time for growth increase is the first four to six weeks of development. Supplemental CO_2 also results in earlier geranium maturation and earlier flowering.

Recent plug research has demonstrated that even at the beginning of Stage 2, CO_2 can be effective. Kessler [12] significantly accelerated begonia plug growth with 800 ppm and work at Cornell [9] used 1,100 ppm CO_2 to reduce impatiens plug time. Both studies showed that light intensities and temperatures must be raised for efficient supplemental CO_2 use. Supplemental CO_2 is best utilized by plants grown at higher temperatures and high light (either naturally high or supplemental light) because as light and temperatures increase, the plant's ability to use the additional CO_2 also increase. If CO_2 is applied under low light levels (winter light), plants will likely stretch. Similarly, using ammoniacal nitrogen isn't recommended when carbon dioxide is used.

CO_2 in the South. Southern growers hesitate to use CO_2 in the greenhouse because they feel vents are opened too early to benefit the crop. Applying CO_2 at 800 to 1,000 ppm from sunrise to open vents is effective and reduces crop time, even in the southern United States. Payback for equipment and gas is slower than in the northern United States, but unless you're venting from 10:00 a.m. until sundown in the winter, CO_2 fertilization can be beneficial.

As growing methods become more intensive, combining supplemental light, supplemental CO_2, higher temperatures and greater fertility will become commonplace in growing geraniums. For the best review of sources, application methods and measuring techniques for CO_2 in the greenhouse, see the excellent book, *CO_2 Enrichment in the Greenhouse* by Peter Hicklenton (Timber Press).

Temperature

Temperature significantly influences growth and flowering time of seed-propagated geraniums. Temperatures below 50 F (10 C) and above 90 F (32 C) retard growth and decrease flowering time due to reduced photosynthetic activity [1, 4] and metabolic processes.

More recent research confirmed the role of night temperature in growth and flowering. Merritt and Kohl [14] grew Red Elite and Cardinal Orbit at 80 F (26 C) day temperatures and either 44 F or 64 F (6 C or 18 C) night temperatures. As expected, flowering time was inhibited by three weeks at cooler temperatures and cool-grown plants were half the height of warmer plants when the latter flowered. Although slower, the cool-grown plants were compact, attractive and excellent horticultural quality. Although a positive DIF occurred in this work, the difference between day and night temperature was so great that it canceled any effect on height. If time isn't a factor in finishing geraniums, cool night temperatures can be used. Seeds may be sown about three weeks early and, more importantly, plants can and should be spaced closer together. Plants grown under cool temperatures produce fewer, smaller leaves and may be spaced closely together for a longer time [14].

For most growers, however, production time is a major concern. Night temperatures of 60 F to 62 F (15 F to 16 C) and day temperatures of approximately 70 F (21 C) have proved successful in producing high quality hybrid geraniums. If the crop is on schedule, reduce night temperatures to 50 F to 55 F (10 C to 13 C) when the first flowers open to increase shelf life. Split night temperatures between 50 F and 60 F (10 C and 16 C) didn't delay flowering [16].

Effect on scheduling. Temperature plays a key role in scheduling hybrid geraniums. The time between the flower buds being just visible (larger than 0.5 cm, the size of a tiny pea) and their opening to full flower is approximately 25 to 30 days for most cultivars under normal greenhouse conditions. This stage is directly influenced by temperature and can be accelerated, slowed down or brought to a virtual standstill by judiciously using heat [5]. You can raise or lower night temperatures to help you meet sales dates. During the spring every one degree F (0.5 degree C) decrease in night temperature results in approximately one day delay between visible bud and flower [10]. As temperatures warm up during late spring, night temperature changes have less effect due to warmer day temperatures and increased light.

Timing and scheduling

Bedding plant growers are often surveyed about the problems experienced in producing crops during the previous year. The leading internal problem is usually crop timing and scheduling. Timing seed geranium flowering for a designated week is demanding at best and virtually impossible without advance planning. (See Color Fig. C65.)

Unfortunately, weather can't be controlled and if the fall and winter are particularly dark and cold, then plants will be late no matter how much planning has been done. Similarly, if warm, bright weather persists longer than usual, crops will be early. Nor can anything be done about the weather during the weekends when plants are to be shipped. If rain decides to hover about each weekend causing retailers to wait, plants must still be in top condition (see holding section).

Flowering times have been presented for many geranium cultivars in trade journals and university and seed supplier publications. Within limits, they're excellent reference sources. The key to scheduling, however, is keeping detailed records so local weather effects can be minimized. Records kept over several years provide written experience to anticipate plant response and manage the crop in your own environment. Simple records, shown in Table 12-2, provide information needed to monitor scheduling from year to year. Notice that the visible bud date (bud approximately the size of a small pea) is included. This is one of the most important pieces of information. Once the flower bud is visible, flowering acceleration or retardation can be directly controlled by regulating temperature. Since it takes approximately 25 to 30 days from visible flower bud to open flower (62 F night, 70 F days; 16 C day, 21 C nights) for most cultivars, the crop may still be fine tuned.

Scheduling flowering

A market date of May 1 (Week 18) may be assumed for a typical schedule for Cardinal Orbit geranium. Based on published data [8], this cultivar flowers in approximately 93 days in East Lansing, Michigan and 85 days in Athens, Georgia. The milder climate, fewer clouds and longer days during the winter in Georgia result in reduced time to flower compared to the cooler climate, cloudy winters and shorter days in Michigan. If you use the Michigan data and work backwards 93 days from May 1, sowing should occur on Week 5. But if an extra seven days cushion is required, sow during Week 4.

Schedule transplanting 10 to 14 days later and apply Cycocel (if neccesary) during Week 9 and Week 10. Flower buds should be visible approximately 3 to 4 weeks prior to May 1 (Week 14 to 15). At that time you may alter temperature to bring the crop in on time (Table 12-2).

This schedule assumes good, natural daylight, a constant liquid feeding program and no insect or disease problems. Most growers have staggered sales dates so planned production becomes more important due to seasonal changes in light intensity, photoperiod and temperature.

As a general rule, the later the sowing date, the faster the crop time because more growing occurs during late spring when days are longer, light is stronger and temperatures are warmer.

TABLE 12-2

Traditional scheduling of Cardinal Orbit[1]

	Sow	Transplant	Growth regulator	Growth regulator	Visible bud	Color in bud	Market
Calendar week	4–5	7	9	10	14–15	17	18
Temperature (F)	75	70 day	alter as necessary				55
Light	low	medium	high[2]		ambient ———→		
Nitrogen (ppm)	0	100–150		250-300 ———→		150	50

1. Based on environment in East Lansing, Michigan.
2. Supplemental light is useful.

Scheduling flowering with plugs

Plugs have already been grown by the specialist propagator for a certain number of weeks prior to your receiving them. It's essential to know how old the plugs are when they arrive. Some plug producers provide prebudded material that may require only two to four weeks to flower while 40 day old plugs may need six to nine weeks on the bench. Most plugs are grown for six to seven weeks prior to shipping; however, if grown in dense plugs (600, 800), they will be only approximately four weeks old and will need additional bench time. Plugs will have received at least one application of growth regulator, fungicide and fertilizer. Be sure to ask the plug supplier or salesperson about the plugs' cultural history. For example, some plug growers may not apply growth regulator, relying instead on reduced water, starvation or DIF to control height.

Holding plugs. You should transplant plants to their final container immediately upon arrival; however, this isn't always possible. Although geraniums aren't as susceptible to plant stretch and overgrowing as impatiens or petunia, don't leave mature plugs on the greenhouse bench longer than a few days. Ignoring plants in the plug tray too long results in flowering plants that are poorly branched and stunted.

Research has demonstrated that plugs may be placed in 40 F (5 C), preferably, but not necessarily, with light [11, 13]. If light can be provided in the cooler, use low light levels from incandescent or special cold fluorescent lamps. Apply a fungicide prior to moving plugs into the cooler, water sparingly and don't fertilize. Plugs may be kept up to three weeks under these conditions. Remove from the cooler during the evening or very early morning and place under shade.

The above scenario falls to pieces when dense plug trays are stored. The smaller the plug, (512, 800), the fewer the days it can remain useful if not transplanted. If geraniums were ordered in dense plugs, be prepared to transplant immediately.

References

[1] Agnew, N.H. and D.S. Koranski. 1987. High temperatures influence seed germination in plug trays. *BPI News* 18(8):10.
[2] Armitage, A.M., M.J. Tsujita and P.M. Harney. 1978. Effects of Cycocel and high intensity lighting on flowering of seed propagated geraniums. *J. Hort. Sci.* 53:147-149.
[3] Armitage, A.M. and M.J. Tsujita. 1979. The effect of supplemental light source, illumination and quantum flux density on the flowering of seed-propagated geraniums. *J. Hort. Sci.* 54:195-198.
[4] Armitage, A.M. and W.H. Carlson. 1981. Hybrid geranium shatter. *BPI News* (April) 5.

[5]   Armitage, A.M., W.H. Carlson and J.A. Flore. 1981b. The effect of tempera-
      ture and quantum flux density on the morphology, physiology and flower-
      ing of hybrid geraniums. *J. Amer. Soc. Hort. Sci.* 106:643-647.

[6]   Armitage, A.M. 1984. Effect of leaf number, leaf position and node number
      on flowering time in hybrid geranium. *J. Amer. Soc. Hort Sci.* 109:233-236.

[7]   ――― and H.Y. Wetzstein. 1984. Influence of light intensity on flower
      initiation and differentiation in hybrid geranium. *J. Amer. Soc. Hort. Sci.*
      19:114-116.

[8]   Derthich, S. and W.H. Carlson. 1988. Geraniums evaluated at pack trials.
      BPI News 18(8):6-7.

[9]   Dreesen, D. and R.W. Langhans. 1989. Research update. *Grower Talks*
      53(8):82-83.

[10] Heins, R.D. 1979. Influence of temperature or flower development of
      geranium Sprinter Scarlet from visible bud to flower. *BPI News* (Dec) 5.

[11] Kaczperski, M. and A.M. Armitage. 1990. Short term storage of plug-grown
      bedding plant seedlings. *HortScience* 25:1094 (Abstr.).

[12] Kessler, R. 1990. "Influence of carbon dioxide on the growth and flowering
      of begonia." Ph.D. thesis, University of Georgia.

[13] Koranski, D., P. Karlovich and A. Al-Hemaid. 1989. The latest research on
      holding and shipping plugs.*GrowerTalks* 53(8):72–79.

[14] Merritt, R.H. and H.C. Kohl Jr. 1989. Crop productivity and morphology of
      petunia and geranium in response to low night temperature. *J. Amer. Soc.
      Hort. Sci.* 114:44-48.

[15] Voight, A. 1991. Sales were strong for the 1990 bedding plant season,
      despite adverse weather conditions. *PPGA News* 22(1), Spec. Suppl.:19p.

[16] White, J.W. and I.J. Warrington. 1984. Effects of split night temperatures,
      light and chlormequat on growth and carbohydrate status of *Pelargonium*
      x *hortorum. J. Amer. Soc. Hort. Sci.* 109:458-463.

CHAPTER 13

Tree Geraniums

William H. Carlson

Geraniums' increased popularity has created demand for novel plant sizes to meet special consumer needs. "Tree-Type (Standard)" geranium production has increased because of its attractive size and distinctive appearance but most growers haven't attempted to produce this geranium because of the long cultural period required for salable plants.

Previously, standard geranium growers required 12 to 15 months to produce marketable tree geranium plants three feet or taller. But when gibberellic acid (GA_3) applications on geranium foliage were reported to promote greater geranium stem and leaf growth, larger inflorescence and longer lasting flower life, the potential existed to greatly reduce the time required [2, 3, 1]. The procedure described in this chapter to shorten tree geranium production time and improve quality was first reported by Pudlo, Carlson and Aung in 1967.

Gibberellic acid application

Plant rooted cuttings or 2¼-inch (6.4 cm) potted geraniums in 4-inch (10 cm) pots containing a standard, well-drained media. After two weeks make the first GA_3 application. Make five consecutive weekly spray applications of 250 ppm GA_3 solutions to the entire plants to runoff. GA_3 concentrations above 25 ppm cause stem elongation. The higher the GA_3 concentration, the longer the internodes, up to 1,000 ppm. Concentrations above 1,000 ppm cause geraniums to become distorted and commercially unacceptable.

You can purchase GA_3 as Pro-Gib, containing 20,000 ppm concentration. To obtain the proper concentration, mix 1¼ parts Pro-Gib to 98¾ parts water to equal 250 ppm.

Cultural practices

Remove lateral branches from GA₃-treated geraniums to develop one, strong terminal stem. Staking is necessary since the stem isn't strong enough to support the plant crown. Plants reach desired height about a month after the final GA_3 treatment. When this height is achieved, soft-pinch the terminal so the plant crown starts to develop. Pinch lateral branches several times to produce the desired crown. After crown development begins, strip all lower leaves from the plant. Also, remove all flower buds until the crown reaches desired size, then allow flowering. GA_3 delays flowering slightly, but this isn't a serious problem since it takes a month to develop the crown and flowering occurs by then. GA_3-treated plants have larger flowers, sometimes twice normal flower size.

Fertilize plants well. In most areas of the country, 720 ppm nitrogen, phosphorus and potassium (3 pounds of 20-20-20 per 100 gallons; 1.35 kilograms per 380 liters) used weekly should be adequate. If your soil is low in trace elements, add these also.

Grow the plant at a minimum 62 F (17 C) night and 70 F (21 C) day. Lower temperatures slow growth and result in a longer growing period.

Transplant tree geraniums from the 4-inch (10-cm) pots to 8-inch (20-cm) pots after the fourth GA_3 application. They will grow to marketable size in this 8-inch (20-cm) pot.

Remember, gibberellic acid is not a panacea, but it elongates stem growth and increases flower size. If used properly it can help produce profitable tree-type geraniums; *however, good growing and greenhouse practices are also necessary to produce a marketable plant.*

Steps for tree-type geranium production

(1) Spray regular 4-inch (10-cm) geraniums for five consecutive weeks with 250 ppm GA_3.

(2) Remove all lateral growth until desired height is obtained.

(3) Stake the weak stem.

(4) Shift from 4-inch to 8-inch (10-cm to 20-cm) pots to finish.

(5) To produce plant crown, pinch terminals at desired height; pinch resulting breaks several times as needed for desired size.

(6) Remove lower leaves when crown starts to develop.

(7) Fertilize well—3 pounds of 20-20-20 per 100 gallons weekly or equivalent (1.35 kilograms per 385 liters).

(8) Grow warm—62 F night, 70 F day (17 C night, 21 C day).

References

[1] Biswas, P.K. and M.N. Rogers. 1963. Effects of gibberellic acid on size and quality of inflorescence in geranium, *Pelargonium hortorum. Proc. Amer. Soc. Hort. Sci.* 82:490-493.

[2] Davis, D., J. Deak and J.W. Rothrock. 1957. The effect of gibberellins on vegetative growth and flowering habit of geranium. *N.Y. State Flower Growers Bul.* 142:3.

[3] Lindstrom, R.S. and S.H. Wittwer. 1957. Gibberellin and higher plants. IX. Flowering in geranium (*Pelargonium hortorum*). *Mich. Agr. Exp. Sta. Quat. Bul.* 40:225-231.

[4] Pudlo, M., W.H. Carlson and L.H. Aung. 1967. A rapid method for standard geranium production with gibberellic acid. *Flor. Review* Vol. 141, No. 3647 10/1:32-33,54-56.

Satisfaction Geranium Trees

Catherine Anne Whealy

Although tree geraniums are considered a specialty item, they continue to increase in popularity and command top prices. Because of their strong vigor and profuse flowering, the varieties in the Satisfaction vegetative geranium series are excellent choices for tree geraniums.

Product specifications

Large trees: Plant three 4-inch (10-cm) plants in the center of a 16- to 20-inch (40- to 50-cm) container with three to five 5-inch (13-cm) geraniums planted at the tree base. At sale, trees should be 3 to 4 feet tall (90 to 120 cm) (including container) depending on container size and in full flower.

Small trees. Produce smaller trees in 10- to 13-inch (25- to 33-cm) containers by planting one 4-inch (10-cm) geranium plant in the center and two to three smaller 4-inch (10-cm) plants around it. Finished height should be proportionally shorter for this product, 2 to 3 feet (60 to 90 cm) tall (including container) depending on container size, and tree should be in full flower at sale.

Production recommendations

Plant rooted Satisfaction geraniums cuttings, one per 4-inch (10-cm) pot in a well-drained, soilless medium.

Cuttings should be established within two weeks of planting; that is, roots should reach the pot bottom.

Fig. 13-1. Satisfaction geranium tree. (Photo courtesy Ball Seed Co.)

Begin gibberellic acid (GA_3) applications at this time. Gibberellic acid decreases production time by increasing the main tree stem and increasing flower head size.

Apply GA_3 as a foliar spray at a rate of 250 ppm to plants that will become the "tree" geraniums. Generally, space applications one week apart; however, if stems appear too thin, allow 10 to 14 days between applications to allow stems to thicken. Depending on the desired final height and environmental conditions, apply three to five times. GA_3 applications should cover the entire plant and be sprayed to "run-off."

(This is a nonregistered use for GA_3, but this use was first reported by Pudlo, Carlson and Aung in *Florists' Review* in 1967, further described by William Carlson in *Geraniums*, 3d ed. (Pennsylvania Flower Growers) and used in practice, therefore establishing it as an appropriate use.)

As GA_3-treated geraniums grow, remove lateral stems to maintain one central terminal stem. Stake after the second or third application if stems become weak as they elongate.

After all GA_3 applications, remove lower leaves. Grade treated plants and plant three plants of equal size and proportion per container in the center of the large container. Provide support to the tree's stem with a wooden stake in the center.

Once planted, soft-pinch all three tree geraniums to increase the tree's canopy. A subsequent pinch two to three weeks later increases canopy size and improves canopy shape.

Remove all flower buds until desired shape and size is attained. When canopy is at desired size, permit flowering.

Plant flowering 4-inch (10-cm) pinched plants at the tree base to provide a base of color.

Throughout the production phase, fertilize plants and water well. Fertilizer should come from a balanced fertilizer source such as 20-10-20 in combination with calcium nitrate.

You have two fertilizer options. Apply soluble fertilizer continuously at a rate of 240 ppm nitrogen from a 20-10-20 source with 60 ppm nitrogen from calcium nitrate with a clear water leaching after every third fertigation.

Or you can fertilize once a week at a rate of 576 ppm from a 20-10-20 source with 144 ppm nitrogen from calcium nitrate. The majority of the nitrogen, 60% to 70%, should be in the nitrate form with 30% to 40% in the ammoniacal form. Increasing the nitrate-to-ammoniacal ratio enhances overall plant quality and increases the tree's postproduction longevity. Adding calcium nitrate to the fertilizer program increases leaf size, stem strength and root development.

Maintain night temperatures at 62 F to 65 F (17 C to 19 C); day temperatures should range from 75 F to 90 F (24 C to 32 C) depending on light conditions. Lower night temperatures result in increased production time and reduced growth.

Geraniums require high light, so 5,000 to 6,000 footcandles (53.8 to 64.5 klux) is best for optimal growth.

Depending on desired final shape and size, environmental conditions, cultural practices and number of GA_3 applications, Satisfaction tree geraniums can be produced from rooted cuttings in 18 to 24 weeks.

CHAPTER 14

Quality Production

Roy A. Larson

Assessing geranium quality is difficult because there are no grades and standards to use as guidelines. Desirable traits or merits listed below appeared in a flower judging manual [1]:

- Flowers and flower bud potential provide a symmetrical display.
- Individual flowers are uniform in size.
- Leaves are a proper size.
- Plant and pot sizes are in proper proportion and balance.
- Foliage is dark green.
- Flowers are conspicuous above foliage.
- Plant form is symmetrical.
- Plants are free of insect and disease damages.
- Flower color is clear and intense.

Major faults would be faded, concealed flowers and asymmetrical plants that are either too tall or too short.

Characteristics mentioned above don't result in specific grades or standards. This lacking of standards is a common feature among flowering potted plants in the United States and is in contrast to most horticultural products. Pot size is more often used to assess geranium plant value than is plant quality.

Environmental effects

Some described merits can be attributed to a cultivar's inherited characteristics, but several are consequences of environment or cultural practices. Cultivar characteristics are discussed in another chapter, so the major emphasis here is on environmental factors and cultural practices that affect geranium quality.

Light intensity

A key factor in producing high quality geranium plants is light intensity. It affects both vegetative growth and flowering. Geographic location and climate largely determine natural light intensity. In some areas, such as in South Florida, light intensity can be too high and young foliage can be light yellow to almost white, as chlorophyll synthesis is impaired. Inadequate light intensity can cause undesirable shoot elongation, yellowish leaves and diminished plant quality. Delayed flower initiation, early development and fewer flowers may also occur. If natural light is inadequate, the grower can install artificial lighting systems.

Sometimes low light conditions and poor quality result when growers space plants too closely. Excessive spacing is uneconomical because of decreased greenhouse population, but excessive crowding could be even more costly as many plants are rendered unsalable or acceptable only at reduced prices. Disease problems such as Botrytis are more severe when plants are crowded because poor air circulation and relative humidity can promote disease development.

Temperature

Geranium quality is affected by temperature. The interaction of high light intensity and high leaf temperatures has already been mentioned, and it's difficult to separate the effects of temperature and irradiation. High quality geraniums can be produced at night temperatures ranging from 60 F to 70 F (16 C to 21 C). Day temperatures are often coupled with light intensity. Lower temperatures if the light intensity is low to avoid the consequences of high respiration rates.

Carbon dioxide

Injecting carbon dioxide into the greenhouse atmosphere can be effective in regions where ventilators and fans can remain inoperative for an adequate duration each day. Rates of 1,000 to 1,500 ppm can improve plant quality [2]. The largest plants are produced when CO_2 is injected throughout crop production, but the first four to six weeks are most important.

Irrigation

Plant and leaf size are proportional to water supply. Geraniums can withstand dry conditions, but excessive stress adversely affects photosynthesis and it never regains normal levels [2]. Gear watering practices to improve plant acclimation to outdoor conditions.

Nutrition

Both inadequate or excessive nutrients lessen plant quality. The nitrogen source can also affect plant quality, since some leaf burn can occur if ammoniacal nitrogen is too high and peat-lite mixes are used. Stop fertilization before placing plants in the marketing channels. Although continued fertilization is often recommended, terminate fertilization prior to sale to increase post-greenhouse longevity. High soluble salts accompanied by dry conditions can damage plants placed outdoors.

Media

The potting medium used in geranium culture often is more suitable for greenhouse production than for retaining high plant quality when geraniums are planted in the landscape or placed in urns or patio containers. The consequences are particularly pronounced when watering is inadequate.

Essential information has long been available for producing high quality geranium plants with proper greenhouse practices. Information has been lacking, however, on those practices effects on plant performance after the greenhouse phase is over.

References

[1]   Anonymous. 1987. *A Manual for Flower Judging,* 7th ed. Phi Alpha Xi National.
[2]   Fonteno, W.C. 1992. Geraniums. Ed. R.A. Larson. *Introduction to Floriculture,* 2d ed. Orlando, Florida: Academic Press.

CHAPTER 15

Production Costs

Robin G. Brumfield

How much does it cost to produce a flat of seedling geraniums or a cutting geranium in a 4-inch pot? Is your selling price for geraniums high enough to cover all your costs plus a reasonable profit? Should you produce your own cuttings from stock plants or buy cuttings? Should you produce geraniums from cuttings or seeds? If you produce seed geraniums, should you use barerooted seedlings or plugs? Are geraniums unprofitable and due for a price increase? If you don't think your customers will accept a price increase, should you discontinue production and buy prefinished or finished plants from another producer? Are geraniums more profitable than other crops, and can you increase geranium sales? If, like most growers, you produce geranium plants in several sizes and a myriad of other crops, it's difficult to keep track of each crop's production costs. Tax records give you a good picture of your business's overall profitability from a tax accounting viewpoint, but tax records don't provide information on the profitability of geraniums or any other crops you produce.

One useful and relatively simple technique that provides answers to these questions and guides business growth is cost accounting. Cost accounting allows you to allocate business costs to specific crops and provides an information base for making decisions.

You can determine each crop's production costs with simple cost accounting using existing records usually kept for tax purposes and simple cost allocation techniques. A good starting point for gathering necessary information is the income statement and balance sheet. The first step is to allocate as many of the costs listed on these two statements as possible. Then, add the remaining costs and allocate then to specific crops on a per-square-foot-week basis.

Information from previous studies is used here as an example of cost accounting. These studies reflect industry averages in Northeastern United States. Except for heating costs, these data can serve as a baseline for other parts of the country as well.

Remember that *everyone's costs are different! Don't assume that your costs are the same as those given in the examples!* Every greenhouse firm faces unique circumstances and unique costs. Costs vary from one greenhouse to another because of such factors as location, size, managerial skill and style, market channel, time of year, space utilization, price and availability of permanent and part-time labor, greenhouse construction, heating system and age and condition of facilities and equipment. Comparing your costs to the industry average will point out areas where your costs are too high or other areas where your costs are low and you have a market advantage. *But looking at industry averages is no substitute for doing your own cost accounting.*

In the following tables, each cost is categorized into overhead and variable types because these need to be treated differently. Variable costs vary as the number of units produced changes. Overhead costs don't vary directly with the number of units produced but are incurred regardless of output.

Overhead costs

Total overhead costs, or fixed costs, remain constant regardless of what crop or how many units are produced. On the other hand, overhead costs per unit decrease as more units are produced. Thus, overhead costs are difficult to allocate to a particular crop. Overhead costs include managerial salaries, depreciation, interest, insurance, repairs and other items that can't be allocated to a specific crop and must be allocated on some other basis such as cost-per-square-foot-week. Even though heating fuel costs vary depending on the greenhouse temperature, it's difficult for most managers to calculate the heating fuel consumed to produce a geranium plant, so it's left as an overhead cost in these examples.

The annual income statement includes all overhead costs as well as variable costs. To calculate overhead costs, first allocate as many costs as possible to geraniums and to other crops. After you have subtracted all variable costs that can be allocated to a specific crop, treat the remaining costs as overhead costs. Divide that number by the number of weeks you use the greenhouse for production to give an annual overhead cost per week. Next, divide that number by the square footage of greenhouse space actually utilized to determine the cost per square foot-week. This has been done for a typical, 20,000 square foot (1,858 square m), double-layer, polyethylene greenhouse producing in the mass-marketing channel (Table 15-1). Overhead costs presented here were derived from previous studies [3, 2]. In these studies, the average greenhouse used 77% total floor area as production area. Overhead costs were updated using Producer Price Indices and Employment Cost Indices.

These overhead costs include salaries for managers, salespeople and office staff as well as unemployment insurance, workmen's compensation and social security insurance for these positions. Hourly wages will be added later as a variable cost. If you don't know how much time is spent on each production task, leave those wages as an overhead cost and allocate them on a per-square-foot-week basis. Overhead costs also include depreciation, interest, insurance, repairs and taxes on greenhouses, buildings and equipment. Other overhead expenses include utilities, advertising, dues and subscriptions, travel and entertainment, office expenses, professional fees, truck expenses, land use cost, contributions and bad debts. In Table 15-1, overhead costs per square foot-week of bench space are calculated by dividing the total overhead costs by 52 weeks and multiplying by 77% of the floor space to determine a cost of $0.194 per square foot-week of bench space.

TABLE 15-1

Overhead costs*

	Annual total	Cost per sq. ft.-week of bench
Overhead labor & benefits	$60,000	$0.075
Utilities		
Heating fuel (13,420 gal.)	13,420	0.017
Electricity	1,843	0.002
Telephone	1,212	0.002
Depreciation	23,887	0.030
Interest	14,848	0.019
Insurance	4,621	0.006
Repairs	11,818	0.015
Property taxes	732	0.001
Advertising	682	0.001
Dues & subscriptions	328	0.000
Travel & entertainment	1,616	0.002
Office expenses	758	0.001
Professional fees	960	0.001
Truck expenses & equipment rental	13,333	0.017
Land rental	1,490	0.002
Contributions	379	0.000
Miscellaneous	2,651	0.003
Bad debts	631	0.001
Total	**$155,208**	**$0.194**

*20,000 sq. ft., double layer, polyethylene greenhouse.

Variable costs

Variable costs, or direct costs, can be allocated to each unit produced based on information from invoices. Total variable costs are directly related to the number of units produced and increase as the number of units produced increases. But the cost per unit stays the same except for quantity discounts. Material costs, which include the costs of flats, inserts, seedlings, media, labels and chemicals are direct costs and are among the easiest variable costs to allocate to each unit produced. For example, the total amount spent for flats divided by the number of flats purchased yields the cost of each flat.

Production labor is more difficult to allocate to each unit, but you can do it with some simple record keeping. Many jobs are done for one size container over a period of several hours. Simply note the number of people performing the operation and when they start and finish. Then count the number of units they finish in that time period. From this you can calculate the time per unit and multiply it by the wage rate, including benefits, to obtain the cost of that labor task per unit. Even though this method is not exactly precise, you will be amazed at how consistent the time per flat will be if you time the same operation on several different days. We are all creatures of habit and tend to do things basically the same way at the same speed. So once you time an operation, you can use that time figure in your cost accounting with confidence. This method works with an experienced crew. A new crew will, of course, become more efficient with experience and you must take this into account with a relatively new crew. Labor that is difficult to allocate to a specific crop can be included in overhead costs and allocated on a per-square-foot-week basis like all other nonallocated costs.

Variable costs were derived from previous studies [1, 2, 5, 6] and were updated using Producer Price Indices and Employment Cost Indices and by contacting supply companies to obtain 1991 costs (Table 15-1).

Production labor costs for this example were estimated using a wage rate of $8.00 per hour. This includes a base wage of $6.50 per hour and $1.50 per hour for benefits including social security, unemployment insurance, worker's compensation and paid holidays and sick days. This labor cost doesn't include overhead salaries paid to managers, growers, sales personnel and office staff, which are included in overhead costs.

Interest on fixed assets, such as greenhouses, are included in overhead costs, but you also have interest on production expenses. You must purchase supplies and pay for labor before you sell plants and collect money, so you must calculate an interest cost for materials and labor. First, determine the annual interest rate and divide it by 52 to obtain the weekly interest rate. Then, multiply that by the number of weeks that your money is tied up for geraniums. The interest rate is assumed to be 9% for the example here.

Loss

Not all plants are salable. Some plants are lost due to diseases, insects or other cultural problems, some may be poor quality, and others just can't be sold. To calculate the cost of losses, add all other costs per unit and multiply by the number of unsold units. Then divide by the number of plants sold to assign the cost of losses to each unit sold. The losses in these examples were assumed to be 5% of plants produced. For example in Table 15-3 the sum of variable costs ($1.214) and overhead costs ($0.824) is $2.038. If you start with 100 plants and do not sell five of them, multiply $2.038 cost per plant by 5 unsold plants and divide by 95 sold plants to yield a loss of $0.107 to be assigned to each finished plant.

Four-inch pots from cuttings

Geraniums may be produced from cuttings or seedlings. While seed geraniums are growing in popularity, the demand for geraniums from cuttings is still strong, so most growers produce both types. In this example, 12 cuttings were produced from each stock plant. The stock plants were produced from purchased rooted cuttings. The labor inputs and other costs of producing the stock plants and finished geraniums

TABLE 15-2

Labor inputs*

Task	Seconds per pot	Dollars per pot
Planting	20.7	$0.045
Irrigate and fertilize	19.0	0.042
Spray Florel	0.8	0.001
Clean plants and apply insecticide	24.0	0.052
Harvest and treat for Botrytis	70.9	0.154
Total	**135.4**	**$0.292**

*Time: 16 weeks; spacing: 6" by 6"

from cuttings taken from the stock plants were obtained from studies by Brumfield et al. [2] and Jenkins [5] and Brumfield [1] and were updated for this book. These studies synthesized budgets for a 5-acre (2-hectare), double-layer polyethylene-covered greenhouse selling flowering geraniums in 4-inch pots to the mass market (Table 15-2).

About half the labor involved in producing cuttings from stock plants was devoted to harvesting cuttings, and the other half was devoted to producing stock plants. Stock plants were produced from purchased rooted cuttings that were then potted directly into 6-inch (15-cm) pots. Cuttings were the largest variable cost, followed by labor (Table 15-3). At a cost of $0.194 per bench square foot-week, for the 16 weeks of production plus one week of harvest, the overhead costs of producing one stock plant are $0.824. On average, 12 cuttings were produced per stock plant. So the cost per cutting is $0.179 under the assumptions made for this example. The unrooted cuttings then become an input in the production of finished geraniums from cuttings (Tables 15-4 and 15-5).

The unrooted cuttings were stuck into Oasis rootcubes and kept on a propagation bench for four weeks. The rooted cuttings were then potted in 6-inch (15 cm) pots and finished at 6-inch by 6-inch spacing.

TABLE 15-3

Geranium cutting production costs[1]

	Dollars per pot
Variable costs	
Rooted cutting	$0.559
6" plastic pot	0.160
Fertilizer	0.037
Root medium	0.109
Insecticide	0.011
Fungicide	0.003
Labor	0.300
Interest on variable costs	0.035
Total variable costs	**$1.214**
Overhead costs	0.824
Loss allocation[2]	0.107
Total cost per plant	**$2.145**
Total cost per cutting	**$0.179**
(12 cuttings per plant)	

1. Time: 16 weeks; spacing: 6" x 6"
2. Based on 5% of total production.

Approximately 20% production labor was devoted to each of the following operations: sticking cuttings, potting and watering using constant
liquid feed (Table 15-4). About one-third of the labor was allocated to
harvesting. The largest materials cost was the unrooted cutting (Table
15-5). Variable costs accounted for about two-thirds of the total costs.

TABLE 15-4

Labor inputs*

Task	Seconds per pot	Dollars per pot
Stick cuttings	15.0	$0.033
Pot	5.5	0.034
Irrigate and fertilize	14.6	0.032
Apply pesticide	3.3	0.008
Harvest	23.7	0.052
Total	**62.1**	**$0.159**

*Per 4" pot from unrooted cuttings.

TABLE 15-5

Production costs[1]

	Dollars per pot
Variable costs	
Unrooted cutting	$0.240
Oasis Rootcube	0.023
4-inch plastic pot	0.074
Fertilizer	0.013
Rooting medium	0.037
Insecticide	0.001
Fungicide	0.003
Labor	0.159
Interest on materials and labor	0.025
Total variable costs	**$0.575**
Overhead costs	0.257
Loss allocation[2]	0.044
Total per plant	**$0.876**

1. Per 4" pot from unrooted cuttings.
2. Based on 5% of total production.

Producing seedling geraniums

Plug production has been adopted rapidly by bedding plant producers in the past few years, probably because it promises several economic benefits over conventional, barerooted seedling production. Plug production allows the seeding operation to be mechanized. Transplanting is faster than conventional production due to ease of handling plugs as opposed to barerooted seedlings. Plug production offers managerial flexibility because plugs can be held longer than barerooted seedlings in the seedling flats. Cropping time is reportedly less than in conventional production. The cost estimates of producing seedling geraniums were derived from studies by Jenkins [5], Nelson and Brumfield [6] and Brumfield [1].

Many systems exist for producing geranium seedlings, and costs vary depending on the system, size of firm and technology level. In the example shown in Tables 15-6, 15-7 and 15-8, small greenhouses are 20,000 square feet (1,858 square m), medium are 100,000 square feet (9,290 square m) and large greenhouses are 400,000 square feet (37,160 square m).

Production labor inputs for the seedling and finished flat stages were collected for conventional greenhouses producing barerooted seedlings and plugs. In all greenhouses, the flats were filled and seeds sown in a central work area or headhouse and then moved into the greenhouse until they were ready for transplant. The seedling flats were moved back to the central work area where they were transplanted into finishing flats. Some large greenhouses used systems similar to greenhouses in the other size categories for transplanting. The operations were staged in different locations within the headhouse but weren't automated. In large automated greenhouses, transplanting took place on a conveyor belt so that all operations took place smoothly in an assembly line with considerably reduced labor.

Plants were hand watered an average of four times per week. Growth regulator was applied once and pesticides, twice.

In general, the labor inputs per flat were reduced as the greenhouse size increased (Table 15-6). Moving required more time in the medium-size plug producing greenhouses than in small or large greenhouses. Many medium-size greenhouses began smaller and didn't plan to grow. Since they didn't plan for expansion, their expansion was often in a manner that didn't result in large greenhouse blocks conducive to labor efficiency. Labor inputs were less in greenhouses that used plug technology than in greenhouses of the same size that used barerooted seedlings. Plug technology's labor saving advantage is reduced transplant time.

TABLE 15-6

Labor inputs for seedling geranium production

	Time (seconds per flat)						
	Conventional			Plug			
	Small	Med.	Large	Small	Med.	Large	
						Nonauto	Auto
Seedling stage							
Fill flat	35	37	22	24	49	47	47
Seed and move to							
germination area	144	162	85	103	140	81	81
Move to greenhouse	16	40	33	36	29	14	14
Irrigate	14	14	6	14	7	5	5
Move to work area	40	30	67	46	51	25	25
Total	**249**	**283**	**213**	**223**	**276**	**172**	**172**
Finished flat stage							
Fill flat	45	120	35	36	44	66	4
Transplant and move							
to greenhouse	280	240	256	100	252	183	144
Irrigate	99	78	62	84	66	53	53
Spray growth regulator	14	10	2	14	10	2	2
Spray pesticide	14	2	2	14	2	2	2
Total	**452**	**450**	**357**	**248**	**374**	**306**	**205**

The number of seeds per bareroot seedling flat varies but is assumed to be 1,000 in the example. Plugs can be produced in trays containing 200 to 648 plugs per tray, but this example assumes 273. A germination rate of 80% is assumed for all greenhouse sizes and technologies. Since seeds are the largest cost involved in seedling production (Table 15-7), an increase in germination rate would impact greatly on the cost per seedling. Seedling production costs decreased with increased greenhouse size. This resulted from quantity discounts for variable costs and economies of size for overhead costs. The cost to produce a plug seedling was greater than to produce a barerooted seedling because the plug required five weeks and the barerooted required only three weeks.

The seedling cost becomes an input price in the production of finished flats (Table 15-8). One advantage of plug production is reduced production time. Jenkins [5] reported that the time from transplant to finish was shorter for finished flats using plug technology than for conventional production, but that the plugs took longer in the seedling stage. Thus, both technologies resulted in the same total production time.

TABLE 15-7

Seedling production costs

	Conventional			Plug			
	Small	**Med.**	**Large**	**Small**	**Med.**	**Large**	
						Nonauto	**Auto**
Seeds[1]	$85.00	$73.67	$46.55	$23.20	$19.72	$18.10	$18.10
Tray	0.46	0.34	0.25	0.40	0.32	0.31	0.31
Medium	0.58	0.42	0.32	0.13	0.10	0.10	0.10
Fertilizer	0.01	0.01	0.01	0.01	0.01	0.01	0.01
Labor	0.55	0.61	0.46	0.48	0.60	0.37	0.37
Interest on variable costs	0.78	0.68	0.43	0.22	0.19	0.17	0.17
Overhead costs[2]	0.95	0.69	0.68	1.59	1.15	1.13	1.13
Total per flat	**$88.33**	**$76.41**	**$48.70**	**$26.03**	**$22.08**	**$20.19**	**$20.19**
Total per seedling	**$0.11**	**$0.10**	**$0.06**	**$0.12**	**$0.10**	**$0.09**	**$0.09**

1. 1,000 seeds planted per conventional flat and 273 per plug flat. Germination rate is assumed to be 80%.
2. Overhead costs are calculated at $0.194 per square foot-week x 1.64 square feet per flat x 3 weeks for conventional flats and 5 weeks for plug flats.

Geraniums took 13 weeks from transplant to finish using barerooted seedlings and 11 weeks using plugs, so the total time was 16 weeks in both cases.

As in the seedling production, costs declined as greenhouse size increased. Unlike the seedling stage, however, production cost for a finished flat from plugs was lower than for a flat finished using barerooted seedlings.

At current geranium prices it appears that geraniums are profitable using any of the above scenarios except for small growers using barerooted seedlings. Costs vary from one producer to another because of market conditions, labor supply, age and greenhouse condition, managerial skill and many other factors. The above examples can serve as a guide for the industry, but don't reflect the actual costs of any particular greenhouse. Compare your costs to these to determine your strengths and weaknesses, but doing your own cost accounting gives a true picture of your actual costs.

TABLE 15-8

Finished flat production costs

	Conventional			Plug			
	Small	**Med.**	**Large**	**Small**	**Med.**	**Large**	
						Nonauto	**Auto**
Variable costs							
Seedlings[1]	$1.99	$1.72	$1.10	$2.15	$1.82	$1.67	$1.67
Flat	0.46	0.36	0.25	0.46	0.36	0.25	0.25
Insert	0.34	0.20	0.19	0.34	0.20	0.19	0.19
Rooting medium	0.32	0.26	0.18	0.32	0.26	0.18	0.18
Label	0.08	0.06	0.05	0.08	0.06	0.05	0.05
Fertilizer	0.03	0.02	0.02	0.03	0.02	0.02	0.02
Growth regulator	0.03	0.02	0.02	0.03	0.02	0.02	0.02
Pesticide	0.01	0.01	0.01	0.01	0.01	0.01	0.01
Labor	0.98	0.98	0.77	0.53	0.81	0.66	0.45
Interest on variable cost	0.19	0.16	0.12	0.18	0.16	0.14	0.13
Total variable costs	**$4.43**	**$3.79**	**$2.70**	**$4.13**	**$3.72**	**$3.18**	**$2.96**
Overhead costs[2]	4.14	2.98	2.94	3.50	2.52	2.48	2.48
Loss allocation[3]	0.45	0.36	0.30	0.40	0.33	0.30	0.29
Total per flat	**$9.02**	**$7.13**	**$5.94**	**$8.03**	**$6.57**	**$5.97**	**$5.74**

1. 18 seedlings per finished flat.
2. Overhead costs are calculated at $0.194 per square foot-week x 1.64 square feet per flat x 13 weeks for conventional flats and 11 weeks for plug flats.
3. Based on 5% of total production.

References

[1] Brumfield, R.G. 1991. Production costs. *Tips on Growing Zonal Geraniums*, 2d ed., Ohio State Coop. Ext. Ser.
[2] Brumfield, R.G., P.N. Walker, C.R. Jenkins, C.A. Frumento, L.R. Heard and J.M. Carson. 1990. Economic feasibility of conventional and reject water-heated greenhouses. *Rutgers Cooperative Ext. Bul.* E135.
[3] Brumfield, R.G., P.V. Nelson, A.J. Coutu, D.H. Willits and R.S. Sowell. 1980. Overhead costs of greenhouse firms differentiated by size of firm and market channel. *N.C. Agr. Res. Tech. Bul.* 269.
[4] Bureau of Labor Statistics. U.S. Department of Labor. Producer Price Indices, Data for July 1979 to August 1991.

[5]   Jenkins, C.R. 1987. "Economic analysis of Pennsylvania greenhouse pro-
      duction, bedding plant costs by firm size and production method: a linear
      programming model of output and income in a conventionally heated and
      a reject-water-heated greenhouse." M.S. thesis, The Penn. State Univ.
[6]   Nelson, P.V. and R.G. Brumfield. 1982. Production costs. *Geraniums,* 3d
      ed. Ed. J.W. Mastalerz and E.J. Holcomb. University Park, Penn.: Pennsyl-
      vania Flower Growers Association.

CHAPTER 16

Open Markets, Open Minds

Dennis J. Wolnick

Whether you're the owner-manager of an established geranium business or an entrepreneur considering geraniums as a future enterprise, you must continually assess the geranium market in order to plan geranium production and sales strategies.

There's no doubt that the geranium will continue to capture worldwide interest, and that overall sales will continue to climb. At the same time, it's simplistic to assume that anyone can enter the geranium business successfully, and that all geraniums can be grouped together as one big market with a share for anyone. In fact, true success in the geranium business beyond this decade will come only to individuals who can anticipate significant changes likely to occur in the marketplace and in business operations and who can reposition their firms to take best advantage of those changes.

As we now consider how geranium production and marketing will change, keep these three ideas in mind:

- Every product (in this case geraniums) has a lifecycle that extends from the time it begins to its decline and replacement by other products;
- the competitive edge goes to those producers who can maintain the position of least-cost suppliers;
- the seeds for geranium production and marketing changes have already been sown, and many have already begun to germinate.

It's useful for discussion to group factors that will influence change in business activity, but please understand that in reality these factors are mixed together in complex ways, and all of them are changing and evolving concurrently, differing mainly by location and rate of change.

The plants

Nearly 300 naturally occurring species of Pelargonium exist today. In addition, there are hundreds of named cultivars. These species and cultivars are grouped into categories and called by such names as zonal, Regal, ivy and scented-leaf. What exists now strongly flavors our concepts about what will continue. Nevertheless, geranium architecture will change. Plants will be genetically programmed for height, from very small to very tall, and each will have its own particular niche. Plants will be free branching from the plant base, and low-growing types may be used for ground covers as commonly as geraniums are used for specimen plants today.

Flower color

Geraniums will undoubtedly be available in the primary colors of blue and yellow besides the present red, and each primary color will be available in many tints and shades. Today's marketplace shows no strong, consistent demand for plants that don't "look like" geraniums, and aren't colored as geraniums "should be" colored. Tomorrow's open market knows no such boundaries and clever entrepreneurs will seize opportunities to create new markets for new plants. Geranium cultivars will be available in such an array that the geranium products will be sold and used year-round, like today's chrysanthemum.

Genetic resistance

Tomorrow's geranium will be genetically resistant to pests and diseases that plague it in various locations and may even possess a "sphere of influence" that suppress undesirable plant (weed) growth nearby. Future geraniums will have increased drought tolerance and won't require abundant water to maintain a vivid outdoor display.

In tomorrow's marketplace it will be woefully inadequate to label geraniums with the current familiar terms zonal, ivy, cascading, seed-grown, cutting-grown, red, pink and white. Many geranium "products" will be available. Each will have unique product characteristics and marketing requirements. Plant characteristics that provide the geranium with a long, healthy display life will also guarantee the geranium a long product life cycle. At the same time rapidly changing consumer interests and demands will move any new geranium cultivar quickly from the introductory stage to maturity of its product life, fueling a continual demand for improvements and changes in cultivars to maintain profitability and consumer interest.

Brand names

Brand names associated with geranium cultivars will probably become more important in marketing strategies, but brand name recognition is likely to be less long-lived than at present, because less time will be available for the necessary slow, steady cultivation of a widely recognized brand name.

Compressed market maturity

The time from the introduction of any new geranium cultivar into the marketplace to the plant's product maturity will be significantly compressed when compared to that process today. Why? For several reasons:

- The knowledge base necessary to bring about desired genetic changes in plants is expanding exponentially, along with skills needed to achieve the task quickly, so new plants will reach the market with greater speed.

- The development timetable for plant changes will be accurately simulated by computer, then developed through rapid biotechnological processes.

- Information about new cultivars will move in great amounts and with great speed to potential consumers who have been specifically identified as likely users.

- And finally, competition among firms developing improved plants will increase.

All these changes and improvements will cost money. New geranium cultivar developers will expect a return for their investment, and these costs will pass along the distribution channel from producers to consumers. Producers and retailers will need to build the extra costs of new plants into their production and merchandising strategies.

Geranium production

Changes in the geranium plant will bring radical changes in production. Production inputs such as chemical growth regulators, fertilizers, growth media, containers and labor will be optimized or minimized. Some inputs, such as applied chemical growth regulators and pesticides, will be eliminated altogether, replaced with genetic, physical and biological controls. Computers will optimize production, simulating and overseeing the entire production operation. Computers will drive "smart" production systems that electronically sense, diagnose and remedy or prevent plant disorders, automate plant propagation, production, grading and packaging and plan, monitor and adjust product distribution systems. Electronic and automation costs will be evaluated against more traditional labor as a production input to be optimized.

Production facility locations

Future open markets such as the free trade markets envisioned between the United States and Canada, or Mexico, within the European Economic Community or from Asia to Europe and the United States will partially dictate the geranium production facility locations. Computer simulations will include all production inputs and one or more optimized production solutions will determine facility locations, either by competitively optimizing inputs at a firm's present location or by suggesting competitive solutions at other locations to which the firm can move.

In an open market a production firm will also be able to easily import materials needed for plant production from far-flung sources in order to remain competitive without relocating. Such widespread resources and potential competition may dictate that the production firm further specialize, that it import geranium plants at various development stages to further process then pass on, or that it target its production to markets more distant or different from traditional markets.

Societal concerns

Societal interests will also influence geranium production. Geranium plants will be produced in recyclable or compostable containers. As natural materials such as soil and peat moss dwindle or become protected, potting media will be created from a wide variety of recycled and/ or composted materials. In some cases geranium plants will be grown hydroponically (that is, without potting media or container) through most of the production period, potted only as the plants are being readied for market. It's also likely that the term timed-release fertilizer will take on new meaning as new chemical "containers" are developed that meter fertilizers and water properly to plants not only throughout production, but also through the marketing process and well into the display and maintenance period at the consumption point.

These profound changes in the nature and location of production activities will certainly influence supply and technical firms dedicated to developing new products for the grower such as containers, media and chemicals. Future supply and development firms will be located all over the world, probably specializing in what they do best, but forming and reforming into loose confederations with other development, production and marketing firms on a project-by-project basis.

In the future geranium development and production environment, information will be king, perhaps the only king. The entrepreneur who can quickly and creatively process the vast amount of information related to geranium development, to changing and expanding distribution systems and to changing market structures and strategies will be the consistent profit and reputation leader.

Societal and consumer attitudes and behavior

Issues that affect societal behavior in general or plant consumers in particular will continue to exert a major influence over geranium production and marketing. In fact, sensitivity to consumer attitudes and needs will be greater.

Society will be very much concerned about waste control: resource management, including recycling all materials, conserving scarce natural materials, particularly water, and environmental preservation. Geraniums will be valued for their durability and longevity, their minimum consumption of natural resources such as water and nutrients and their contribution to environmental "greening." Plants grown in recycled, composted or "waste" materials will be in demand, as will plants in containers that disappear with no environmental impact.

Geraniums and other bedding plants will become institutionalized. That is, they will be commonly accepted and even expected in public and private settings, thus automatically increasing demand, with a swing in the market favoring larger commercial customers ranging from landscape contractors to municipalities and other government entities. Consumers will maintain a very positive opinion about geranium producers as long as they perceive that the producers are environmental stewards who act responsibly to minimize environmental contamination, provide positive solutions to environmental problems and speak clearly and consistently about production technology's effects.

An example that illustrates the possibly delicate relationship between societal issues and the greenhouse industry may help stimulate future thinking. Currently our society is bombarded in the media with the concept that accumulating greenhouse gases are the major cause of the earth's average atmospheric temperature increase, otherwise known as global warming. Fatalists believe the world will be destroyed by rising ocean waters, exceedingly hot summers and major droughts that will destroy crops. Clearly, global warming causes are complex, and neither the scientific community nor the average consumer may understand this issue clearly. Consider, then, the phrase greenhouse gases. Are greenhouse operators major polluters, or do they nurture and properly utilize natural elements that make our planet inhabitable? Public opinion may swing either way on such issues but is more likely to swing against growers who remain silent and uninvolved. Future geranium growers will need to stay informed about consumers' concerns and regularly present themselves as having special skills that repair and enhance the environment rather than degrade it. Such image-building will be as much a part of future marketing strategies as advertising, packaging, pricing and display activities and will depend on early, accurate awareness of societal concerns followed by positive image-building strategies.

Bigger international consumer segments

Numerous consumer segments will continue to be interested in geraniums. If anything, consumer segments will increase as geranium cultivars are developed for more specialized needs. In a future open market atmosphere, a firm's targeted consumer segment(s) may have members not just in the local market area but throughout the nation and even the world. A geranium marketer focusing on a niche market of specialized customers may be marketing geraniums to a worldwide customer base—still specialized, but on a much larger scale.

Product benefits

A multinational or worldwide customer base concept naturally follows from the idea of open markets and can certainly influence business strategies. Highlighting geranium benefits to consumers through promotion will remain an important marketing strategy, but the specific benefits highlighted may need to be adjusted and perhaps broadened to cross national boundaries. At the same time a worldwide niche market may have such a clear, universal need that benefit promotions may cross national boundaries virtually unchanged, reaching many additional customers.

Future competition

Competition to supply geranium products to the market place will inevitably increase as trade barriers are dropped and aggressive importers hungrily eye the American market. The best approach is to recognize that increased competition will exist and to create an aggressive, creative action plan, rather than resist change and hope the problem evaporates. We tend to overlook opportunities for exporting products and processes to other countries with potentially large future markets.

The worldwide market

Many future plant businesses will buy and sell plant products worldwide. These firms will market products electronically and import supplies, products and information where necessary to enhance and expand their product lines. These firms will also export products and information when opportunities exist. The average firm will be more fluent with international trade terms and impatient with government restrictions on free trade. Business administrators will likely adapt to free trade more quickly than their employees who may see international trade as a threat to job security rather than a benefit.

Transparent distribution channels

With lowering trade barriers and multiple "centers" throughout the world marketing, new and perhaps competitive technologies, and distribution channels for plants and production supplies are likely to become more complex. This increased complexity, however, will be transparent to consumers because products will move through the channel easily with little delay and no value loss.

Open markets, open minds

Many firms strongly oriented toward a local market will continue to develop and operate as we move into the 21st century, but it's likely that even these firms will have a continuing strategy to tap a worldwide information base for developments affecting local consumers. In other words the future's successful geranium entrepreneur will face the new, open markets with an open mind, actively positioning the firm to seize far-reaching marketing opportunities.

CHAPTER 17

Marketing and Merchandising

Bridget K. Behe

Planning geranium marketing and merchandising is as important as planning production. Not having a strategy or ignoring marketing and merchandising makes you noncompetitive with the many retailers who have a geranium sales strategy.

It may be helpful to first understand how geraniums get from producer to consumer. Understanding this distribution channel helps you plan marketing activities. The distribution channel is a map showing various firms that help the product, in this case geraniums, reach the consumers. As Fig. 17-1 shows, product suppliers are at the top of the distribution channel. The seed suppliers, propagation material suppliers, growing medium and related hardgood suppliers are first in the distribution channel. They supply products and services directly to the commercial producer or grower who puts the products together to "make" the geranium plants for sale or indirectly through brokers and distributors. The producer may sell the plants to a broker or wholesaler or retail directly to the consumer. The grower may choose to ship plants to a retailer directly, perhaps a garden center, mass merchandiser or landscape contracting firm. The retailer sells the plants to the consumer or commercial end-user for planting.

One important concept in the distribution channel is the concept of functions. Throughout the channel, each member performs certain functions. These functions include growing plants, packaging them, labeling them, putting assorted varieties together, pricing plants, attaching care and planting instructions, delivery, making credit terms and installation. It's important to understand functions because the functions *must* be performed by someone in the channel. If one member doesn't perform, someone else will have to do that function. Functions may be shifted but they're never taken away.

For example, the delivery function from supplier to consumer is an important one. Someone must assume this function and its associated cost. The landscape contractor may deliver the plants to the commercial site, but often the customer pays for that delivery or the landscape contractor absorbs that cost. Similarly, the retail garden center may

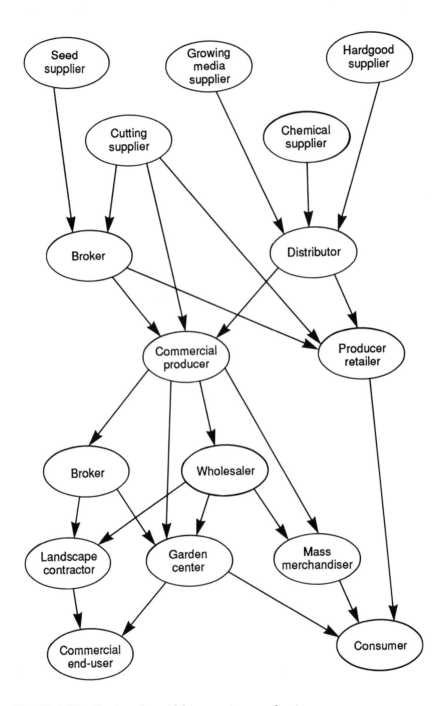

Fig. 17-1. Distribution channel for geranium production.

choose to deliver a plant order, including geraniums, to a consumer and charge that consumer for delivery. The consumer pays for this convenience. Most mass merchandisers would never deliver plant material. The delivery function doesn't go away; rather, the consumer must perform that function for him/herself and indirectly pay for this inconvenience. The mass merchandiser saves the delivery cost, but passes the function onto the consumer. Understanding which functions your business will perform for its customers will help you plan your marketing.

If you are a geranium producer, you may decide to perform the functions of producing, packaging, labeling and delivering. Each function has a cost to your firm, which should be accounted for in pricing the product. It's the inability to account for several of these functions (which many retailers expect you to perform) that can mean the difference between profitable and unprofitable geranium production. Once you understand your role in the distribution channel, you're ready to prepare a market plan.

Market plan

Many business managers plan their vacations or Christmas shopping better than they plan their product marketing. Not having a plan means you're willing to accept things however they happen. Having a plan means that you have established goals and a direction and purpose for geranium sales. Planning can mean the difference between making a profit and taking a loss.

The market plan components can be stated in five questions: Which products? What sizes? How many? What time? What price? Answer these five questions and you have a basic market plan for your geranium crop. Plans don't need to be long or complex; the shorter and simpler, the better. You could fit your market plan on only a few pieces of paper.

Which products? You need to select some geranium varieties, and perhaps even some geranium species you plan to market. Information on consumer preferences may help you determine which varieties you'll produce. Consider past purchases in the market, since they often best indicate future purchases. Plan to add a small percentage of new or unusual varieties as some customers will likely be looking for the new and unusual. Determine which varieties you will market.

What sizes? The product size can vary tremendously from plug trays to planted barrels of mixed plants. Determining this size mix is important. Some plants may be purchased in the plug size and transplanted into cell pack for flat sales. Other geraniums produced from cuttings are often sold in 4- to 8-inch (10- to 20-cm) pots. Consider, too, any mixed planters or window boxes you can market as these increase in popularity.

Customers may demand color throughout the season and want larger plants as the spring progresses.

How many? Most managers use historical data (last year's production and sales records) to plan the number of plants produced or marketed in the coming year. Realizing that the weather and economy may affect sales, plan for the number of plants to market. Presale orders may account for a large percentage of market plan numbers. Some growers produce only a limited percentage on speculation while others grow entirely on speculation. Predetermining the number to market is essential to planning for supplies and space needed.

What time? There are several "windows of opportunity" for geranium marketing, but most producers consider only one: spring. Most geranium sales occur in the spring. Depending on your geographic market area, spring may begin in March or May. The more Southern U.S. markets begin bedding plant sales several weeks before the more Northern U.S. market areas. Sales may continue through the end of May but could continue through the summer by offering alternative geranium products. One large, concentrated market window occurs in Northern markets for Memorial Day or Decoration Day where geranium plants traditionally decorate cemeteries. There may be a great demand for products early in the season if it's a warm spring. Planning for early marketing may bring more profit if the weather is favorable, but profits can be lost by being too early for the market. Timing is critical.

What price? Price is an important component of any market plan. One part of determining price is direct costs associated with geranium production. This should not, however, be the only factor in determining price. Profits must be added to costs, along with indirect or overhead costs. Consider, too, some of your products' perceived value. Some geraniums, either new varieties or species, may be worth more to consumers because they're different. While the costs, both direct and indirect, may add up to a given price, the perceived value may add even more to that price. Competitors must be careful not to consider the competition's lowest prices, as they may not have fully accounted for their costs or may be using the products as "loss leaders" to get customers into the store to buy other merchandise. Consider the cost of marketing your geraniums at a loss (to match the competition) vs. not growing them at all. Is it better for the competition or for you to lose money on sales?

Consider these five factors simultaneously when creating your market plan. The factors depend on each other. Variety may dictate size and availability may dictate price. Create a plan incorporating these five factors, but be prepared to alter it along the way.

Merchandising

Consumers can purchase geraniums and other bedding plants from various retail establishments. As shown in Fig. 17-1, the distribution channel can include several nontraditional or nonhorticultural retail outlets. Consumers *expect* more from horticultural firms than from nonhorticultural firms. Consumers look more often to horticultural firms for new varieties, innovative plant uses, assistance with problem diagnosis and solutions and consistently high plant quality.

Retail geranium merchandising, the in-store geranium display and promotion, can affect plant sales. Properly displayed and promoted products can be profitable while improperly displayed and unpromoted plants can be left in the store to deteriorate and not generate profits. The easier you and your displays make geranium purchasing, the more geraniums consumers will purchase.

Most customers prefer to purchase plants with some flower color showing. Display colorful geranium blooms to help generate sales. Display plants in a neat, orderly manner. Keep shelves fully stocked at all times, usually restocked once during the day and then at day's end for the next morning sales. Full shelves or displays give a positive impression while nearly empty display areas create the impression that the customer is "too late" for the best plants. Integrate plants fully in bloom with plants just showing bud color to help increase sales of plant material not yet in full bloom.

Display height and width is critical. Most customers resent reaching any distance to select plant material. Position the displays at waist height or approximately 3 feet (.9 m) above the ground. Most customers resent having to bend over to select plants. Tiered shelves or displays use more vertical space, but the plants in the middle will be selected most often. Display shelves should be no deeper than 4 feet (1.2 m). Customers can't reach into a shelf more than 1 or 2 feet. A 4-foot width accessible from both sides would accommodate many plants, yet is easily accessible for most customers. Position your display at the proper height, 3 feet, and in the proper width, 2 feet (.6 m) from any side, and the plants will more likely be purchased.

Geraniums are living plants and will hopefully be displayed only a short time before they are sold. To keep them in the best possible condition for customer purchases, remember these horticultural considerations. Water plants once or twice a day. Plan shelves that have adequate drainage so customers aren't walking in puddles, and plants aren't standing in puddles. Avoid stacking shelves vertically: Light can't reach the plants in the middle shelves, customers can't see them to buy them and watering is very difficult. Single layer or tiered shelves make the best display for the customer and the retail employee charged with caring for them.

Locate at least one display near a high traffic area in the retail store to create impulse sales. Geraniums are recognized by many customers, more so than other blooming plants, and high visibility can stimulate impulse purchases. Keep this highly visible display well stocked at all times. You might consider featuring some newer varieties in this display area and have signs directing interested customers to another store area where more geraniums are located.

Include as much information as possible on the plant label. Today's consumers demand more information concerning the products they purchase. Vital information includes variety, description, planting instructions, price and anticipated blooming time. Variety name should be prominent on the label along with a picture if the plant isn't in flower. Some consumers seek specific varieties that were successful for them in previous years. The plant description should include its mature height for incorporation into landscape planning. Novice consumers may prefer to have some planting instructions while more experienced gardeners may already know planting basics. It doesn't hurt to remind customers that geraniums require a sunny location and benefit from a layer of mulch, adequate irrigation and regular fertilization.

More sophisticated merchandisers may offer additional information in displays or demonstrations. Displays promoting new geranium uses, such as creatively planted in a wooden barrel with other uncommon bedding plants, stimulates the gardener's imagination and may stimulate sales. Be sure to have the demonstration model priced for immediate sale so that you don't waste valuable sales time if someone wants to purchase the finished product. Demonstration gardens show the customer which geranium varieties, all clearly labeled, perform best in their locale. The display garden is often viewed throughout the summer months. It's an attractive sales tool that can enliven the store exterior.

Use the add-on sales technique in your geranium-related product displays. Near the geraniums, include gardening gloves, trowels, fertilizers and decorative planters. This display may stimulate the consumer to recall the need for one or more add-on products and can be an effective means of reducing inventory.

Developing and executing a plan for marketing and merchandising geraniums is as important to profitable sales as developing and executing a plan for fertilization and pest management. Lack of planning can lead to lack of profit. Make the plan early in the season. Evaluate last year's plan immediately after the season is over, which is also the best time to begin next year's plan. While spring is a busy time, it's very important to record numbers and sizes sold so that you can incorporate that information into next year's marketing and merchandising plan.

CHAPTER 18

Use and Care Advice

Terril A. Nell

Geraniums are versatile plants used in landscapes and hanging baskets and as potted plants. Zonal geraniums, both seed-propagated hybrids and vegetatively propagated cultivars, dominate landscape use. Some are sold as specialty potted plants, especially for Mother's Day in the United States, but the inflorescence life is quite short and plants are generally transplanted into the landscape. Regal (Martha Washington) geraniums are used primarily as potted plants because of their large, showy flowers and superior interior performance. Regals offer a wide range of flower colors. Ivy leaf geraniums are commonly grown for hanging baskets due to their natural, vinelike growth and weeping appearance.

Seed-propagated geraniums, when originally introduced in the early 1970s, were thought to be more profusely flowering and more heat tolerant than vegetatively propagated cultivars. Today, however, landscape performance appears to depend on cultivar rather than propagation method. Seed geraniums are more susceptible to petal shattering than vegetatively propagated cultivars. All geraniums, however, need proper care and handling.

Avoid storage/shipping stress

Storing and transporting geranium cuttings, seedlings (plugs) or flowering plants often result in leaf yellowing and petal shattering. Several factors may affect the degree of injury that results from storage and shipping, including darkness, transit vibration, exposure to ethylene and temperature. While all these factors may contribute to petal shattering and leaf yellowing, ethylene and warm temperatures are the most serious problems encountered in geranium storage and shipping.

Geraniums are one of the most ethylene-sensitive flowering plants. Plants exposed to ethylene exhibit petal shattering and/or leaf yellowing even without the detrimental effects of darkness and transit vibration. Ethylene injury worsens as storage or shipping temperatures increase. Thus, the most effective way to reduce ethylene injury on geraniums is temperature control. A carnation is 1,000 times more sensitive to ethylene at 65 F (18 C) than at 35 F (2 C) and you can anticipate a similar response with geraniums. Cool cuttings, plugs and flowering plants to 35 F to 40 F (2 C to 4 C) from the time they are removed from the production facility until they are planted at the production facility or delivered to the retail market. Cool temperatures reduce plant respiration, which may lead to leaf yellowing. Temperature control appears to be the only effective method to avoid leaf yellowing in cuttings and plugs, but another alternative is available to minimize petal shattering.

Silver thiosulfate

Marketing plants in the bud stage reduces petal shattering, since open flowers are more susceptible to ethylene. Applying anti-ethylene compounds, such as silver thiosulfate (STS), also reduces petal shattering. Plants are sprayed with silver thiosulfate as the first floret on an inflorescence is beginning to open at a concentration of approximately 1 millimol STS. Applying STS at a later development stage reduces STS effectiveness. Several commercial STS solutions are available and growers generally obtain better results with these prepared compounds than with solutions mixed in-house. Consult appropriate country, state and local ordinances regarding STS use and disposal before use. Most local ordinances restrict disposal of heavy metals such as the silver contained in STS. For the same reason, countries such as the Netherlands don't allow foliar sprays of this compound.

The presence of ethylene in retail display areas has become a major concern in the United States and Europe as more plants, especially Regal geraniums, are sold in supermarkets where ethylene levels are high and temperatures can't be adequately controlled. Applying silver thiosulfate is the primary method available to minimize ethylene injury under warm retail conditions. Unfortunately, STS applications make geranium plants more susceptible to Pythium root rot.

Geraniums do well as landscape plants or interior plants if handled properly throughout shipping and retail display. To summarize, follow these recommendations to maximize geranium performance by reducing petal shattering and leaf yellowing:
- Maintain cuttings, seedlings and flowering plants between 35 F and 40 F (2 C to 4 C).
- Apply STS as the flowers begin to show color.
- Market plants in the bud stage since mature flowers are more sensitive to ethylene.

CHAPTER 19

Consumer Preferences

Bridget K. Behe

Consumer preferences capture the attention of marketers and business managers; consumer preferences for flowers and plants are no exception. Understanding which varieties, colors and plant shapes consumers prefer enables managers to better plan the plants they grow and sell.

Much available information on consumer preferences for geraniums is obtained from historical data or sales figures. For example, a manager may notice that in previous years 75% of the geraniums sold had red flowers. From that historical data, the manager may conclude that most consumers prefer red-flowered geraniums. The problem with this conclusion is that consumers may have been exposed only to red geraniums and were unaware that other geranium colors were available.

The average geranium flower color mix produced by commercial producers or sold in retail establishments is approximately 75% red and 25% other colors. The typical breakdown for the other colors is 15% salmon and pink, 5% white and 5% other novelty colors. Red is a traditional or customary color for geraniums, much like dark blue is a traditional color for a suit. Red may be popular because that's the color people expect geraniums to be. More progressive markets with consumers looking for different geraniums would likely sell more nonred colors.

Consumer preferences for flowers haven't been well-documented; only a few studies have been published. One of the first consumer behavior studies with floral products was conducted in 1956 at Ohio State. Sherman, Kiplinger and Williams [2] determined that consumers preferred red carnations over yellow and white carnations. Hutchison and Robertson [1] determined that men preferred red roses over other colors and that women preferred salmon-colored roses over red roses. The also found that both groups preferred pink roses least of any color.

Only two studies have been published concerning consumer preferences for geraniums. Michigan State researchers determined that there were significant differences in the geranium flower color preferences of consumers and producers [4]. Their results showed that consumers preferred salmon and pink geranium flowers, while growers preferred

red. Using paired comparisons, Wolnick [3] demonstrated that consumers preferred red flower color over pink when only these two color choices were presented. Wolnick also found that people preferred cultivars with highly variegated leaves over nonvariegated leaves.

Red flowers are sold most often in the U.S. market. This may be due to a "push" strategy where growers produce a high percentage of red flower colors, so consumers purchase them. As consumers become more familiar with other bloom and foliage colors, their preferences may shift from red to salmon or pink to other novelty colors. Consumers may be reluctant to try new varieties because they're unfamiliar with them. Education is the key to successfully introducing and marketing nonred geraniums.

Individual horticultural businesses can determine, to some extent, their own customers' preferences by looking at historical data. Gradually, they could increase the percentage of nonred varieties. The annual increase should be very small, only 1% to 2% per year. Teaching customers about the availability, use and care of nonred geraniums may reduce their reluctance to try new kinds. Consumers must be aware of the tremendous geranium varieties to purchase them.

References

[1]   Hutchison, N.R. and J.L. Robertson. 1979. Consumer demand analysis for roses. *J. Amer. Soc. Hort. Sci.* 104:303-308.
[2]   Sherman, R.W., D.C. Kiplinger and H.C. Williams. 1956. Consumer preferences for cut roses, carnations and chrysanthemums. *Ohio Ag. Exp. Sta. Res. Circ. 31.* Ohio Ag. Exp.
[3]   Wolnick, D.J. 1983. Consumer preference studies with zonal geraniums. *Florist Review* 162:31-33.
[4]   Zehner, M.D., L.J. Nelson and W.H. Carlson. 1982. Color preferences for hybrid seed geraniums. *Research Report 441.* Michigan State Univ.

CHAPTER 20

Regal Geraniums

Glenn G. Hanniford and Andrew L. Riseman

Regals (*Pelargonium* x *domesticum*) are the showiest Pelargoniums. Their large, azalea-like flowers with bright colors, intricate petal markings and profuse flowering delight consumers of all ages. Whether given as a housewarming present, get well pick-me-up or holiday gift, a Regal Pelargonium will be enjoyed day after day.

For the grower, the advent of precise scheduling, improved cultivars and a better understanding of the production process have made Regals a dependable and profitable crop. High quality plants can be produced year-round by carefully managing the environment during each production stage and following cultural guidelines presented in this manual.

Ancestry

Pelargonium x *domesticum* is one of three major categories of cultivated Pelargoniums that has resulted from numerous interspecific crosses within the genus. This hybrid species has a complex ancestry and considerable genetic variability, but with recognizable plant and flower characteristics which distinguish these plants as a separate Pelargonium class. The major contributing species include:

P. cucullatum, which was introduced to Europe around 1690, is a naturally large flowering species and is believed to be the parental species exerting the most influence on plant phenotype. It's probably responsible for the flower form, predominating lavender flower color and source of the colored markings on the upper two flower petals.

P. angulosum (new taxonomic designation for *P. cucullatum* subsp. *cucullatum*; Van Der Walt, 1985), which was introduced to Europe in 1724, is believed to contribute the purple flower color and strongly influence plant form.

P. fulgidum, which was introduced to Europe in 1732, is primarily responsible for the red color pigments in the flower.

P. grandiflorum, which was introduced around 1794, may be the source of the white flower color, although *P. cucullatum* has a white form in the wild.

Other species such as *P. betulinum, P. capitatum, P. cordifolium* and *P.* x *ignescens* are believed to have contributed to the domesticum class development, but with minor influence on the plant's characteristics.

Regal development began with the introduction of the species and their five-petaled florets, with the upper two petals larger than the lower three petals. Gardeners and hybridizers, fascinated by the diversity within the genus, began making crosses—some planned, others by chance. Around the 1830s a recognizable plant class with common plant and flower characteristics emerged. These early domesticum plants were called Pelargoniums. Early hybridizers were content to concentrate their efforts on improving flower size and color. The first improvements included enlarging lower petals. As improvements continued, the early domesticums became known as show, or fancy, Pelargoniums.

Around 1850 petal markings appeared on lower petals and hybridizers began to develop these markings and intensify the color. Petal ruffling was developed in the 1870s and the blotchy petal markings gave way to intricate feathering. Also around this time florets with six petals appeared. Coinciding with these improvements, the name royal or Regal Pelargoniums was used and quickly ascribed to this Pelargonium class.

Between 1870 and 1900 Regals were introduced into the United States. They became very popular in the early 1900s and were commonly called Lady or Martha Washington geraniums due to the popularity of a cultivar introduced under that name. In 1930 L.H. Bailey assigned the botanical name *Pelargonium* x *domesticum* to this cultivared Pelargonium class. Regals in this era seemed to grow best in cool climates and cultivation was primarily on the Northeast and Western coasts of the United States.

Early cultivar development in the United States was quite secretive; however, breeding efforts emphasized improved blooming quality, plant compactness, number of petals per floret and petal ruffling. In the 1920s, Carl Faiss of Germany introduced several cultivars to the United States, including Grandma Fischer (Grossmama Fischer) Mackensen (Feldmarschall von Mackensen), Spring Magic (Fruhlingzauber), Pink Gardeners Joy (Marktgaertners Fruede), Easter Greetings (Ostergruss) and Marie Vogel (Frau Marie Vogel). Faiss' cultivars featured greatly improved blooming qualities and are still important in Regal development, not only as popular cultivars but also as part of continued Regal Pelargonium improvement by hybridizers and Pelargonium enthusiasts. Some of today's leading cultivars can be traced to original Carl Faiss introductions.

Environmental flowering control

Understanding environment's impact on growth and flowering control is crucial to making intelligent production decisions. Regal Pelargoniums are much more sensitive to environmental conditions than zonal geraniums. Knowing the effects of temperature, irradiance and photoperiod during growth stages will help you produce a high quality crop.

Temperature

The role of temperature can be divided into two parts. Temperature level is the amount of heat energy to which plants are exposed during their growing period. Temperature duration is the length of time that plants are exposed to a particular level during any particular stage.

The first published information on the effects of temperature level on flowering were made by Post [22]. He observed that plants kept at a minimum 60 F (15 C) night temperature failed to form flower buds during the spring, while those grown at 50 F (10 C) night temperature formed buds and flowered normally.

Research [8, 18, 12] indicates that temperature is the most important environmental factor affecting Regal Pelargonium flowering. A cold temperature induction cycle of constant 45 F (7 C) hastens flowering by an average 37 days for the cultivars Parisienne and Rapture as compared to using a minimum temperature of 60 F (15 C) (Table 20-1). Lavender Grand Slam didn't flower when grown at temperatures above 70 F (21 C), illustrating the vernalization requirement necessary for flower induction [13]. For some cultivars, 65 F (18 C) is low enough to cause initiation; other cultivars require 50 F (10 C), which should be used to ensure bud initiation in all cultivars.

Temperature exposure duration as well as the temperature level affects the Regal Pelargoniums' flowering response. Crossley (Table 20-1) found time to flowering was advanced for each cultivar as the temperature treatment duration was increased from one to two months. Table 20-2 illustrates the effect of temperature level and duration on the flowering of the Regal cultivar Fruhlingzauber (Spring Magic). At 48 F and 53 F (9 C and 12 C), 100% of the plants flowered after six weeks of temperature treatment; however, as the temperature increased to 60 F (15 C), 100% flowering occurred only after nine weeks of treatment. At 65 F (18 C), flowering didn't occur in many plants, even after nine weeks of temperature treatment.

Research [12, 20] substantiated the cold temperature requirement and indicated no different response to flowering using a cooling temperature of 1, 3 and 5 degrees F (2, 7 and 10 C). As little as two weeks of cold

TABLE 20-1

Effect of temperature on Regal flowering

Mean number of days of subsequent treatment[1] to anthesis and percentage of flowering plants

Photoperiod										
Temp. (degrees C)	Treatment (hours)	Time (months)	Aztec days	%	Grand Slam days	%	Parisienne days	%	Rapture days	%
15	8	1	—[2]	0	—	0	115	100	134d	100
15	8	2	—	0	—	0	90	100	106c	100
7	8	1	83b[3]	100	77b	100	80	100	88b	100
7	8	2	63a	100	63a	100	61	100	69a	100

1. Photoperiod 16 hours; temperature 15 C.
2. Anthesis not attained when experiment was terminated 147 days following pretreatment completion date, but 40% of Aztec and 20% of Grand Slam had initiated flower buds.
3. Within columns, means followed by the same letter are not significantly different at the 5% level.
Source: J.H. Crossley [8].

TABLE 20-2

Effect of temperature on Fruhlingzauber flowering

Temperature (degrees C)	Length of treatments (weeks)*		
	3	6	9
9	0%	100%	100%
12	10%	100%	100%
15	0%	30%	100%
18	—	—	0%

Source: J.H. Nilsen [18]

treatment could induce flowering, but the response was greater (increased number of inflorescences and a faster flowering rate) as the temperature duration treatment was increased up to eight weeks. Acceptable flowering was obtained with four weeks of cold treatment.

After floral initiation, temperatures higher than 60 F (15 C) can hasten flower development considerably; however, higher temperature can also cause flower bud abortion. This response varies depending on the degree and duration of the high temperature and the irradiance levels during the high temperature exposure [12, 24]. In a study of the effects of photoperiod and finishing temperature on floral development, Hanniford et al. found a decreased percentage of flowering plants, a reduced number of inflorescences and increased final plant height as the temperature increased from 50 F (10 C) to 77 F (25 C).

Lavender Grand Slam plants finished at 65 F (18 C) produced an acceptable number of florets per inflorescence with high total irradiance; however, with low irradiance the number of florets decreased by half. Plants finished at 55 F (13 C) didn't show a decreased number of florets produced at any irradiance level, indicating an interaction between finishing temperature and total irradiance on floret production [24, 13]. There appears to be considerable cultivar variation with respect to irradiance and finishing temperature interactions.

Total irradiance

Exposure to different irradiance amounts has a significant impact on floral initiation and flower development in Regal Pelargoniums. Investigations [12] suggest that high irradiance may partially substitute for the low temperature requirement necessary for flower initiation. Oglevee and Craig (in a process patent [21]) have proposed using high irradiances and warmer temperatures to induce flowering and produce quality Regals on a year-round basis. Supplemental lighting or high natural irradiance during the floral initiation period provides several alternatives to using temperature alone to trigger flower initiation.

Total irradiance exposure is also very important in controlling flower development rate. Regals respond to high irradiance by flowering faster. Lavender Grand Slam plants finished at 55 F (13 C) and high daily total irradiance flowered 21 days earlier than plants finished at the same temperature but low daily total irradiance [24]. Finishing Regals using high irradiance allows growers to increase finishing temperature to hasten flower development without substantial florets loss.

Photoperiod

Initial observations had indicated that *Pelargonium* x *domesticum* was a short day/long day plant requiring short days for flower initiation and long days for flower development [32]. Subsequent investigations [18, 12, 10] found daylength not to be important in flower initiation, but extended daylength hastened flower development. When cooled, plants forced at 60 F (15 C) in long days (incandescent night lighting of 20 watts per square m for eight hours) flowered two to three weeks earlier [9].

Experiments conducted by Hanniford et al. in 1983 compared the effects of three photoperiod treatments at three forcing temperatures on flower development. By comparing mean number of days to flower after vernalization, we found that long days provided by either night-break or extended daylength considerably shortened the number of days to flowering. Higher forcing temperatures also shortened flowering, but there was a decrease in flowering plant percentage and an increase in the number of aborted inflorescences as temperatures increased from 50 F (10 C) to 77 F (25 C). Flowering response to methods of long day application didn't differ.

Light and temperature interactions

Higher forcing temperatures can hasten flower development; however, the response depends on the level and duration of the forcing temperature and irradiance during high temperature exposure. Under high forcing temperatures and low irradiance, the number of developing flower buds can sharply decrease.

Plant breeders are utilizing genetic variation related to the temperature and light interaction as a criterion for selecting new cultivars. New cultivars selected have a much wider tolerance in their response to warm temperatures and increased light levels, which ameliorates the vernalization process required for floral initiation.

The flowering process in *Pelargonium* x *domesticum* can be separated into two distinct growth stages, each having its own set of optimal environmental conditions. Floral initiation in each cultivar requires a critical minimum temperature, isn't affected by daylength, but is affected by irradiance. High irradiance can at least partially replace vernalization in most cultivars. Too warm a finishing temperature in conjunction with low irradiance conditions can cause bud abortion and stretching during the flower development stage.

Production phases

The production cycle for Regal Pelargoniums can be divided into four phases: propagation, floral initiation, forcing and postproduction. Each phase incorporates its own environmental criteria and management decisions. The major choice is determining where in the developmental cycle you wish to begin. In general, the earlier in the cycle you become involved, the less expensive the plant material, but the more complex the process. The production alternatives fall into two categories—complete production control from start to finish or the purchase of plant material at a specific developmental stage for process completion.

If you grow Regals from propagation to finishing, you'll need to understand stock plant management, rooting, floral initiation and floral development and have the appropriate facilities to conduct each phase. In addition, if you choose to propagate patented or protected varieties, you'll be responsible for the appropriate licensing agreements and royalty payments.

Beginning later in the production cycle generally requires proportionally fewer resources, shorter time to maturity and less knowledge and skill to produce a successful crop, but your initial per-plant costs may be higher. The alternatives include buying unrooted cuttings from a specialist propagator, purchasing rooted single stem or multi-stem (budded and nonbudded) cuttings from a specialist propagator or purchasing prefinished potted plants for the final forcing stages from a specialty grower. In making your decision, consider your knowledge of each production phase, your ability to control the environment and each alternative's costs.

Propagation phase

Regal Pelargonium propagation is strictly by vegetative, herbaceous cuttings. Maintaining pathogen-free stock plants is critical.

Stock plants

The importance of a pathogen-free stock program cannot be overemphasized. *Pelargonium* x *domesticum* is reported to be resistant to *Xanthomonas campestris* var. *Pelargonii* (bacterial blight) in that it doesn't exhibit the typical wilt disease symptoms; however, it can be a disease carrier. This can lead to very serious problems if you also grow zonal geraniums as part of your product mix. Using culture-indexed plants for stock is critical and strongly recommended if you plan to produce you own cuttings (see Chapter 29, Clean Stock Production).

Likewise, virus infections can also be problematic if you produce your own stock. Viruses that infect stock plants are carried through vegetative propagation cycles to infect new propagules. Using virus-indexed stock can be a starting place for minimizing the impact of viruses on Regal production (see Chapter 29, Clean Stock Production). Additionally, strict control of insect pests such as whitefly, aphids and thrips is important in preventing infection and virus spread in Regal stock (see Chapter 31, Insects and Other Pests).

Regals grow best in a well-drained, porous media that doesn't compact over time. Using vermiculite in a stock plant media is discouraged due to its compaction characteristics. The media should have a high cation exchange capacity, hold water well and be pathogen-free. Commercially available peat-based soilless media can provide excellent results as can media containing 10% to 20% pathogen-free soil. Various media can be used successfully in growing Regal stock as long as the fertilization and irrigation practices match the medium. Adjust the medium pH initially to 6.0 and maintain it between 5.8 and 6.2 during the stock plant life cycle. In addition, incorporate phosphorous, magnesium and trace elements prior to planting unless they are adequately supplied by the fertilization protocol. (*Editor's note:* See Color Fig. C58 through C62 for physiological disorders of Regals.)

Fertilization should consist of balanced liquid fertilizers, using 150 to 250 ppm nitrogen. Concentration should vary depending on seasonal environmental conditions. Proportionately less frequent fertilization is appropriate during low light periods. Alternating 15-16-17 and 20-0-20 can work well in a stock plant fertilization program. Both formulations provide micronutrients necessary for healthy growth. In addition, the 20-0-20 formulation supplies calcium.

Stock plants require approximately three to three and one-half months to develop, depending on the time of year. It's important to begin stock production well before cuttings are required. Pinch stock plants several times to develop the scaffold needed for maximum cutting production. Success in Regal rooting and unchecked growth depends on starting with high quality cuttings. Nutritionally deficient cuttings root slower and produce poorer finished plants.

European research [18] has indicated that optimum temperature for stock plant production is around 65 F (18 C). Temperatures below 65 F delayed vegetative growth and decreased cutting production. Temperatures higher than 65 F (18 C) produced sufficient numbers, but longer cuttings and taller finished plants. Slightly higher day temperatures are best. Cutting production can be enhanced during low irradiance seasons by the addition of HPS or HID supplemental lighting at 1,850 to 3,700 footcandles (2 to 4 klux).

A comprehensive pest control program is important in producing Regals and especially critical in the stock plant production phase. The major insect pest which infests Regals is the greenhouse whitefly, although aphids, mealybugs and thrips have been reported as potential problems. These insects also can act as vectors for viruses. Botrytis and Pythium control is a major component of successful stock plant management. Refer to Chapters 25, Foliar Diseases, and 27, Stem and Root Diseases, for identification, prevention and control strategies.

Rooting

The rooting process begins with a high quality cutting, harvested from a well-managed stock plant. Use softwood terminal cuttings, approximately 2 to 3 inches (5 to 8 cm) long, which have been trimmed to two fully expanded leaves and the terminal growing tip. Cuttings can be broken from the plant if the stems are turgid or cut from the stock plant using a knife. Sterilize knives between plants to prevent disease spread. Protect harvested cuttings from desiccation by removing them from the greenhouse quickly and placing them in the propagation area as soon as possible after harvesting. Each cutting's base can be treated with a rooting hormone containing 0.1% IBA to improve uniformity and rooting rate. The medium used for propagation should be well-drained, porous and sterile and have a low soluble salt level.

To reduce the chance of Botrytis infection on stock plants when harvesting cuttings, cleaning the plant or pruning, treat the plants with a fungicide the day prior to and the day after cutting. Fungicide rotation is important since Botrytis can develop resistance to some fungicides.

The rooting environment should consist of high humidity (usually a mist or fog system), media temperature of 70 F to 75 F (21 C to 24 C), moderate light levels of 1,500 to 2,000 footcandles (1.6 to 2.2 klux) and nighttime air temperature around 65 F (18 C). In general, misting will be frequent for the first three days, followed by a gradual decrease for two weeks. Cuttings should be completely rooted in approximately three to four weeks. Rooting may take slightly longer during certain seasons. Discontinue rootzone heat after two weeks to prevent plant stretching. Fertilize with a balanced fertilizer such as 15-16-17 at 100 ppm nitrogen when roots have formed.

Pinching

If a pinched (multi-shoot) plant is desired, pinch after propagation and prior to floral initiation phase. Pinching will vary depending on cultivar, but leave a minimum five nodes above the soil line. Grow plants for two to three weeks following pinching using a minimum 65 F (18 C) day temperature, minimum 60 F (15 C) night temperature and irradiance levels of 3,000 to 4,000 footcandles (3.2 to 4.3 klux). Carbon dioxide enrichment of 1,000 to 1,500 ppm can benefit lateral break formation. Irrigation should be frequent since Regals are sensitive to dry growing conditions. Grow multi-shoot plants under this cultural regimen until the lateral shoots are ¾ to 1 inch (2 to 2.5 cm) long, when they begin the floral initiation phase. Pinching delays the crop two to six weeks, but results in a larger plant. Many newer cultivars have been selected for improved branching and don't require pinching to produce well-branched plants. The choice to pinch or not to pinch will be determined by the cultivar selected, your environmental conditions and market preference.

Floral initiation phase

The second critical phase of Regal production is floral initiation. Regals differ significantly in this characteristic from zonal and ivy geraniums in that a special floral initiation treatment is required. Historically flower bud initiation was induced by exposure to a low temperature treatment (vernalization). Plants were treated with a minimum 35 F (2 C) and maximum 50 F (10 C) with optimum temperature 48 F (9 C).

The recommended low temperature duration was a minimum 10 hours per day, preferably 24 hours per day, for a minimum four-weeks.

Experimental evidence has shown that high irradiance levels during floral initiation can at least partially replace the low temperature vernalization requirement for some cultivars. High irradiance during the floral initiation phase could provide a method for year-round induction of Regals for potted plant production in conjunction with cultivar selection [21]. Various temperature, irradiance and duration combinations might induce floral initiation at different times of the year. The key to success is understanding the threshold requirements for cumulative irradiance and temperature level and duration. Continued cultivar improvement for efficient flower production under higher temperatures would also increase the flexibility of conditioning plants for flowering.

At present there are several alternatives for providing the necessary environment for flower induction. After rooting, plants can be placed in a cooler at 48 F (9 C) using minimum irradiance for a period of five to six weeks. Plants alternatively could be initiated in a cooler at 48 F (9 C) using high irradiance provided by HID lights for four weeks. Initiation might also be accomplished in a greenhouse using natural irradiance and low outside air temperatures for a four to six weeks. (This approach would be possible only in certain geographic locations and at certain times of the year). A fourth alternative would be to initiate flowers in a greenhouse using natural irradiance, plus supplemental HID irradiance and outside air temperatures (potential year-round production in some geographic locations).

Forcing phase

Whether you have propagated and induced your own plants or purchased prebudded plants from a specialist propagator, the forcing phase requires good irrigation, fertilization, temperature, photoperiod and pest management. Transplant plants into 4-inch (10-cm) or 6-inch (16-cm) azalea pots depending on cultivar and whether single- or multi-stem plants are being grown. The media should be well-drained and porous with the pH adjusted to 5.8 to 6.2. Many commercially prepared peat-based media or media high in organic matter (such as 1:1:1 peat:soil:perlite) can be used successfully. Incorporate a source of phosphorous, magnesium and trace elements prior to planting if soil test results and expected fertilization program indicate a need.

Irrigate plants several times following transplanting to ensure thorough wetting of the media and good root/media contact. Fertilize with a constant liquid feed program. Fertilization levels vary from 150 to 350 ppm nitrogen depending on cultivar, environment and geographic location. Regular soil and foliar tests can help you tailor your fertilization program. Using a balanced peat-lite formulation, such as 15-16-17, and alternating fertilizations with a high calcium 20-0-20 formulation helps maintain adequate calcium and micronutrient availability in the medium. Slow release fertilizers can be used successfully if proper amounts of calcium, magnesium and micronutrients are present in the medium. A typical application of a 14-14-14 formulation would be ½ teaspoon per 6-inch (15-cm) pot at potting and again four weeks later.

You may use growth regulators to control Regal height where necessary. Application rate and frequency varies with cultivar, time of year, forcing temperature, desired final plant height and application method. Cycocel applied at active ingredient rates of 500 to 3,000 ppm has successfully controlled Regal height when applied two to five weeks after transplanting. Higher concentrations and/or more frequent applications are generally used during low irradiance months. Under high irradiance conditions, height control isn't usually required. Growth regulator applications are discussed in Chapter 7, Growth Regulating Chemicals.

Photoperiod control is critical for flower development. Most cultivars respond to long days by flowering more quickly and maintaining higher bud counts. Regals respond to extended daylength (18 hours) or night-break lighting using a minimum 5 watts per square foot (54 watts per square m). This irradiance can be obtained by stringing one line of 100 watt incandescent bulbs on 4-foot (1.2 m) centers, 6-feet (1.8 m) above the benches. This provides sufficient irradiance for two 4-foot benches with an 18-inch (.5 cm) aisle between them. Apply photoperiod lighting from September to April, from the time plants are potted until flower buds show color. Natural day irradiance shouldn't exceed 5,000 footcandles (5.4 klux) to obtain high quality plants.

Maintain forcing temperatures at 58 F (14 C) night temperature and 68 F (20 C) day temperature to produce high quality plants. Forcing Regals at higher temperatures can result in excessive stretching, decreased bud count and flower bud abortion. Growing Regals at cooler daytime temperatures won't adversely affect flower quality, but will lengthen the time to flower. Adequate ventilation is critical, particularly near the end of the forcing cycle. Supplemental irradiance during low natural light periods can help maintain plant quality.

Final plant spacing varies with pot size. In general, space 6-inch (15-cm) pots one plant per square foot (929 square m) and 4-inch (10-cm) pots at two and one-half plants per square foot. Spacing may vary with cultivar selection and growing method. The time to flower depends on the cultivar, forcing environment and whether you are growing a pinched plant. Nonpinched plants should flower in eight to 11 weeks from the beginning of forcing. Pinched plants may take from three to six weeks longer to flower.

Major pest problems during the forcing phase include Botrytis and the greenhouse whitefly. Controlling these pests is crucial to maintaining high plant quality (see Chapters 25, Foliar Diseases, and 31, Insects and Other Pests).

Postproduction phase

No published information is available on how management practices during forcing affects Regal Pelargoniums' keeping quality. Common sense would dictate that good management practices with proper environmental conditions and cultural practices would maximize each cultivar's post-harvest potential. It's important to realize that finishing plants under lower light conditions and/or higher temperatures than recommended decreases quality. Plants finished under these conditions are generally taller, have lower bud counts and shorter floret longevity. This decreased quality may not be evident at harvest but results in poor post-harvest performance, which could affect future sales.

Regals are sensitive to ethylene and petal abscission is the typical response to ethylene exposure. Silver thiosulfate (STS) reduces petal abscission caused by ethylene when applied in concentrations of one millimole to foliage when flower buds begin to show color. Spray STS to runoff. Adding a drop of surfactant aids active ingredient contact with the leaf surface (see Chapter 7, Growth Regulating Chemicals).

Cultivars

Only cultivars available as clean stock are listed here. Four distinct Regal groups are available to growers: traditional cultivars, Vavra cultivars, the Elegance series cultivars and the Royalty series cultivars.

Traditional cultivars, which were originally selected by Oglevee Ltd. from their evaluation of over 300 cultivars for greenhouse production, represent the best of the older types (Table 20-3). These cultivars are generally large plants and are more sensitive to warmer temperatures with respect to flowering. They typically need four to six weeks at 48 F (9 C) for floral initiation and 10 to 12 weeks for finishing. They are best suited for spring production in 6-inch (15-cm) pots.

The Vavra cultivars, originally developed by the Austrian breeder Wolfgang Kirmann working for Mirko Vavra, are available through both Oglevee and Fischer Geraniums (Table 20-4). These compact cultivars offer excellent plant habit in a wide range of flower colors; however, they require cool temperatures throughout the production cycle to produce a quality plant. They typically need four weeks at 50 F (10 C) for floral initiation and eight to 10 weeks for finishing. They are best suited for production as nonpinched plants in 5-inch (12 cm) pots. Year-round production is possible, but extreme care is necessary to ensure that floral initiation and development progress normally.

The Elegance series, developed by Richard Craig, Glenn Hanniford and Leon Glicenstein at The Pennsylvania State University, represents the most advanced Regals on the market (Table 20-5). They are currently available exclusively through Oglevee Ltd. These plants offer significant improvements in growth habit, flowering response and production ease. Self-branching and free-flowering qualities are two major improvements over traditional cultivars. They also have been bred to tolerate higher temperatures during both floral initiation and finishing phases. The Elegance series can be initiated in four weeks at 58 F (14 C) if given

TABLE 20-3

Traditional cultivars

Country Girl	Melissa*
Granada	Olga*
Grand Slam	Parisienne*
Lavender Grand Slam	Virginia*
	White Glory*

Note: Cultivars marked with (*) may no longer be available as clean stock.
Available through Oglevee Ltd.

TABLE 20-4

Vavra cultivars

U.S. market name	European market name	U.S. patent number/status
Micky	Micky	6080
Vicky	Gottweig	6029
Josy	Joseph Haydn	6019
Jenny	Belvedere	Patent pending
Mary	Jasmin	6027
Shirley	Silvia	6015
Dolly	Muttertag	6028
Lilly	Valentin	6022
Honey	Mikado	6030
Macy	Rubin	6012
Peggy	Jupiter	6020
Rosemary	Aquarell	Patent pending
Candy	(Frueh. Gruss)	6023
Sally		6021
Lucy		6072

Available in 1991 through Oglevee Ltd. and Fischer USA or Fischer Pelargonien.

supplemental irradiation and can be finished in eight to 10 weeks depending on the season. They can be scheduled year-round as either a 4-inch or 6-inch crop (10-cm or 15-cm).

The Royalty series, developed by Ernest Walters in England, is being marketed by Oglevee Ltd. (Table 20-6). These plants show vigorous, upright growth habit and continuous flowering response. They have been noted for good postproduction quality due to reduced flower shattering. The Royalty cultivars require a slightly longer finishing time than other Regals, approximately 10 to 14 weeks; however, they produce a good display of numerous small flowers. They are suitable for production as either a 4-inch or 6-inch crop.

TABLE 20-5

Elegance series

U.S. market name	U.S. patent number/status
Allure	7467
Crystal	7343
Fantasy	7538
Splendor	7656
Majestic	7387
Flair	7620

Available through Oglevee Ltd.

TABLE 20-6

Royalty series

U.S. market name	U.S. patent number/status
Duke	Patent pending
Duchess	Patent pending
Prince	Patent pending
Princess	Patent pending

Available through Oglevee Ltd.

References

[1]   Ameele, A. 1975. The *Pelargonium domesticum. Geraniums Around the World* 23 (3):3-4.

[2]   Bailey, L.H. 1950. *The Standard Cyclopedia of Horticulture.* New York: The Macmillan Co.

[3]   Bode, F.A. 1966a. Character study of a queen. *Geraniums Around the World* 13(4):76-81,95.

[4]   ——— 1966b. Character study of a queen. Part II. *Geraniums Around the World.* 14(2):28-35.

[5]   Chritensson, H. 1991. *Engelsk pelargon—odling av belysta kulturer.* Swedish University of Agricultural Sciences Research Information Centre ISSN 1101-3753.

[6]   Clifford, D. 1956. The parents of the geranium. *Royal Hort. Soc. J.* 81:20-22.

[7]   ——— 1958. *Pelargoniums, Including the Popular Geranium.* London: The Blanchford Press.

[8]   Crossly, J.H. 1968. Warm vs. cool short days as preconditions for flowering of *Pelargonium domesticum* cultivars. *Can. J. Plant Sci.* 48:211-212.

[9]   Duffet, W.E. 1965. Personal communication. Long day treatment for earlier flower formation in *Pelargonium* x *domesticum.* Translated from *Viola* 71:3(65):3.

[10] Falasca, R. 1966. Unpublished experiments. The Pennsylvania State Univ.

[11] Geissinger, D. 1966. Fred Bode on geraniums. *Geraniums Around the World* 14(1):4-5, 17.

[12] Hackett, W.P. and J. Kister. 1974a. Environmental factors affecting flowering in *Pelargonium domesticum* cultivars. *J. Amer. Hort. Sci.* 9(1):15-17.

[13] Hackett, W.P., J. Kister and A.T.Y. Tse. 1974b. Flower induction of *Pelargonium domesticum* Bailey cv. Lavender Grand Slam with exposure low temperature and low light intensity. *HortScience* 9(1):63-65.

[14] Hall, O.G. 1968. The response of potplants to nightbreak light: Calceolarias, Regal Pelargoniums and cinerarias. *Shinfield Progr.* No. 13:26-29.

[15] ——— 1969. The response to night-break light: Regal Pelargoniums and calceolarias. *Shinfield Progr.* No. 14:27-32.

[16] Holcomb, E.J. 1979. Effect of growth regulators on non-cooled and cooled *Pelargonium* x *domesticum. HortScience* 14(3):280-281.

[17] Hosmer, T.L. 1957. Lady Washington geraniums capture the spotlight. *Horticulture* 35:318.

[18] Nilsen, J.H. 1975. Factors affecting flowering in Regal Pelargoniums (*Pelargonium* x *domesticum* Bailey). *Acta Horticulture* 51:299-309.

[19] O'Donovan, J. 1981. Personal communication. Oglevee Associates Inc., Connellsville, Penn.

[20] Oglevee, E. 1977. Personal communication. Oglevee Floral Co., Connellsville, Penn.

[21] Oglevee, J.R. and R. Craig. 1990. Precision flowering of Regal Pelargoniums (*Pelargonium* x *domesticum*). Process Patent Number 4,897,957 issued February 6, 1990.

[22] Post, K. 1942. Effects of daylength and temperature on growth and flowering of some florist crops. *Cornell Univ. Agr. Expt. Sta. Bul.* 787:1-70.

[23] Post, K. 1949. *Florist Crop Production and Marketing*. New York: Orange Judd Publishing Co., Inc.

[24] Powell, M.C. and A.C. Bunt. 1978. The effect of temperature and light on flower development in *Pelargonium* x *domesticum*. *Scientia Horticultura* 8:75-79.

[25] Ross, M.E. 1961. The royal geranium. *Geraniums Around The World* 9(2): 39-41.

[26] Runger, W. 1981. Untersuchungen uber bluhreaktion and streckungswachstum einiger sorten pelargonium grandiflorum hort. Fachbereich Landschaftsentwicklung der Technischen Universitat Berlin zur Verleihung des akademischen Grades Doktor der Agrarwissenschaften genehmigte Dissertation.

[27] Schmidt, W.E. 1955. Pelargonium breeders I have known. *Geraniums Around the World* 3(3):5-6, 15.

[28] —— 1956a. *Pelargonium domesticum*—Originators, varieties and dates of introduction. *Geraniums Around the World* 4(1):8-10.

[29] —— 1956b. On hybridizing. *Geraniums Around the World* 4(3):6-8.

[30] —— 1974. Regals in California prior to World War II. *Geraniums Around the World* 22(3)3-7.

[31] Tappeiner, J. 1959. Lady Washingtons in pots. *Geraniums Around the World* 12(2):18-19.

[32] Van De Veen, R. and G. Meijer. 1959. *Light and Plant Growth*.Eindhoven, Netherlands: N. V. Philips Gloeilampenfabrieken.

[33] Wood, J. 1975. The Regal Pelargonium. *Geraniums Around the World* 22(4):11.

CHAPTER 21

Ivy Geraniums

E. Jay Holcomb and Eugene J. O'Donovan

Ivy geraniums, once thought a novelty, today are a mature component in the market, satisfying the need for increased diversity. This product provides continual performance in sheltered garden locations. Predominantly used in hanging baskets, today's cultivars flower profusely with a trailing habit.

Pelargonium peltatum, or the ivy geranium, was first discovered in Southern Africa in the Cape Province growing in sheltered locations. Apparently *P. peltatum* in nature is quite variable in plant habit, forcing Clifford (1970) to conclude that *P. peltatum* is a hybrid with several species contributing characters. *P. peltatum* has been cultivated since 1701. One very early cultivar that we may see today is Galilee, which dates back to 1882.

Today's ivy geranium is a very broad product category. Variety exists in foliage color, size and variegation, bloom color and types and growth habit; today's consumer can enjoy more than 75 different commercial cultivars (see Chapter 23, New Cultivars, and Color Fig. C42). These cultivars provide both grower and consumer the change necessary to meet today's gardening demands.

Most of today's ivy geraniums are asexually propagated by specialist propagators who provide rooted plants for most growers. Unrooted cuttings are also available and tend to root with ease in porous, well-drained substrate maintained at 65 F to 70 F (18 C to 21 C) with intermittent mist. Seed-propagated ivy geraniums are also available and germinate under similar conditions as *P. x hortorum* hybrids.

Cultural information

To obtain optimum crop performance, the successful grower must recognize ivy geranium's unique cultural demands. These unique demands differentiate this crop from garden or zonal geraniums and are the keys to eliminating many maladies that affect this crop's timely performance. Interacting environmental factors (light, temperature and relative humidity), cultural practices (irrigation, fertilization and pest management) and cultivar differences directly affect the crop's overall quality and profitability.

Light, temperature, humidity

Ivy geranium's environmental requirements need special attention when you plan this crop. Light, temperature, relative humidity and the ability to control these factors in the greenhouse deserve top priority. Light is one of the most critical factors in commercial ivy geranium production. Often ivy geranium is incorrectly compared to *P. x hortorum* (zonal or garden geranium), for which higher light equates to increased performance. Instead of light-temperature relationships, leaf temperature and light energy and their effect on plant physiology are more precise. This discussion utilizes less precise terms to provide more acceptable commercial application.

Generally, the ivy geranium requires lower, more moderate light intensities for optimum performance. Although light isn't an independent variable in plant growth, and in this case is closely related to other factors such as temperature and relative humidity, normal light intensity levels for this crop should range between 2,500 and 4,000 footcandles (2.7 and 4.3 klux). The interaction of temperature and relative humidity have a direct relationship to the light intensity at which the crop should be grown. Under higher temperature (greater than 85 F; 29 C) and relative humidity regimes, this crop's maximum light intensities should approach the lower limits of this range. Under regimes of moderate (65 F to 80 F; 18 to 26 C) temperature and relative humidity, this crop's maximum light intensities can approach the upper limit.

If leaf temperatures can be maintained at lower, more optimum temperatures, then light intensity can be increased. Individual cultivar requirements also impact the amount of light that can be used. Use Table 21-1 as a guideline:

Growing substrate

The substrate should provide adequate moisture retention, soil air space at full saturation, good drainage and buffering capacity. With this in mind, many suitable alternatives are available. Carlson [1] recommended a 1:1:1 or a 2:1:1 (soil:peat:perlite) substrate. A substrate that contains 75% long-fibered sphagnum peat moss, 15% mineral soil, 10% coarse aggregate and trace elements also provides acceptable performance and increased drainage. Many commercial peat based substrates are also acceptable. Each component in the substrate has a function. The mineral soil provides two functions: weight and a micronutrient source.

TABLE 21-1

Light tolerance

Highest light cultivars (range: 3,000 to 3,500 footcandles; 3.2 to 3.8 klux)

Balcon Princess	Minicascade Lilac
Bright Cascade	Minicascade Pink
King of Balcon	Minicascade Red
Lilac Cascade	Sophie Cascade

Medium light cultivars (range: 2,500 to 3,000 footcandles; 2.7 to 3.2 klux)

Beauty of Eastbourne	Nanette
Cornell	Nicole
Double Lilac White	Salmon Queen
Galilee	Solidor
Mexicana	Snow Queen

Lowest light cultivars (range: 2,000 to 2,500 footcandles; 2.2 to 2.7 klux)

Amethyst	Sugar Baby
Barock	Sybil Holmes
Butterfly	Yale

The mineral soil amount can vary, but reducing mineral soil in the substrate generally improves drainage and increases dependency for micronutrient nutrition in the liquid fertilizer. This improvement in substrate drainage is very important in controlling edema.

The growing medium's pH is very important and should be maintained within close tolerances to avoid nutritional disorders. The pH depends on the ambient level of nutrients provided by the existing water source and growing substrate. Generally ivies require a substrate pH approximately 0.5 pH units less than utilized on existing zonal programs (see Chapter 5, Fertilization). (*Editor's note: In Chapter 5, Biernbaum recommends a pH range of 6.0 to 6.3 for zonal geraniums; in Chapter 10, Sheely and Craig recommend a pH range of 5.0 to 5.5 for ivies.*) Maintaining this pH is very important for the entire crop cycle. Fluctuations cause decreased quality, nutritional disorders and/or increased edema.

Fertilization

Balance is the operative word in a functional fertilizer program for ivy geraniums. The liquid fertilizer program should complement the existing water quality conditions and growing substrate. A constant liquid fertilizer program with a complete fertilizer at a concentration of 250 to

350 ppm nitrogen at each irrigation provides a good start. Ivy geraniums grown on long-term crop schedules tend to deplete normal magnesium and iron levels near the end of the program. Elevated levels of phosphorus, magnesium, iron and nitrogen prove beneficial in the development of tissue more tolerant to edema. One common foliar problem associated with this crop is interveinal chlorosis, which is often associated with lower levels of iron and/or magnesium. In both cases these elements must be maintained at higher minimum levels in ivy tissue than in their zonal counterparts (see tissue analysis standards in Chapter 5). Cultivars more sensitive to this disorder include Amethyst and Sybil Holmes.

Management of electroconductivity (EC) levels is very critical; ivy geraniums are more sensitive to elevated EC levels than are zonal geraniums. Proper irrigation practices that provide adequate leachate and uniform moisture levels are essential. Excessive drought, moisture or wide fluctuations between drought and high moisture often hasten root loss. Irrigate this crop only when required. Leachate should be managed to 10% to 15% of the applied amount to ensure a proper EC level. Irrigations with less leachate will result in a rising EC; irrigations with more leachate will result in lower EC levels. Often, removing saucers from hanging baskets enables more consistent EC levels management and provides unrestricted percolation.

Edema

One major problem with ivy geraniums is edema, a physiological disorder characterized by many corky spots, varying in size predominantly on older leaves' undersides. This disorder results from a series of environmental and cultural misjudgements that often create severe plant stress. The stress leads to ruptured plant cells and the formation of scar tissue (corky spots) on leaves' undersides (see Chapters 25, Edema, and 32, Physiological and Environmental Disorders).

Light, temperature and relative humidity interaction are critical in controlling edema, with moderation the key. Keep air temperatures to a maximum 75 F to 85 F (24 C to 30 C) and relative humidity between 60% and 70%. As light intensity increases towards the maximum, maintaining temperature and relative humidity within the prescribed limits becomes critical. On the other hand as temperature approaches the maximum, maintaining moderate light intensities and relative humidity regimes is of foremost importance. Controlling these three variables as an integrated system is necessary to optimize ivy performance.

Selecting suitable growing substrate and associated fertility program is very important to crop success. A light, well-drained mix that has a high cation exchange capacity is essential. A well-drained substrate matched to the container enables easier management of the water table, promotes stronger roots and helps reduce the relative humidity in a dense foliar canopy. The fertility program should provide balanced nutrition, stable pH and low EC levels.

TABLE 21-2

Six keys to reduced edema

1. Lower substrate pH (0.5 pH lower than used for zonal geraniums).

2. Adequately drained substrate.

3. Proper nutritional balance and levels.
   Increased nitrogen and iron aid in reducing incidence.

4. Moderate light intensities, 2,500 to 4,000 footcandles (2.7 to 4.3 klux); level depends on variety and temperature.

5. Cooler day temperatures.

6. Irrigation to maintain moderate moisture and EC levels and avoid artificially high water tables.

Craig[3] tested 18 ivy geranium cultivars and grouped them by edema susceptibility. If a grower has traditionally had problems with edema and the cultural practices haven't eliminated or reduced these problems, it may be necessary to select cultivars more edema resistant.

Growth regulators

Sometimes ivy geranium's natural growth habit isn't suitable to produce a top quality marketable plant. There's a greater interest in producing ivy geraniums in 4-inch (10-cm) pots than in the past. Since the plants

TABLE 21-3

Susceptibility to edema

Most susceptible	Intermediate	Most resistant
Amethyst	Madeline Crozy	Sugar Baby
Yale	Cornell	Double Lilac White
Balcon Princess	Spain	Salmon Queen
King of Balcon	Pascal	Sybil Holmes
Balcon Imperial	Rigi	Galilee
Balcon Royale	Rouletta	
Beauty of Eastbourne		

naturally have a trailing habit, they don't look good in 4-inch pots without a growth retardant. Even hanging baskets have a more desirable form when treated with a growth retardant. Holcomb [4] determined that the response of ivy geranium cultivars to growth retardants was cultivar dependent. Plants in the Balcon series didn't seem to respond to growth retardants, while most other cultivars did. In general, increasing chlormequat (Cycocel) spray applications from one to three resulted in shorter plants. Ancymidol as a spray generally produced shorter plants than those untreated.

Pest problems

Ivy geraniums aren't bothered by many pests. The ivies are not affected by most foliar diseases but may be affected by root rots. Use a pasteurized substrate, reasonable care to prevent infection and preventative drenches to limit root rot problems.

Mealybugs can be very troublesome but are controllable. Red spider mites can cause brown spots on leaf undersides, which look very much like edema. Cyclamen mites can cause severe apical tip distortion and stunting. Unless you are sure you *don't* have two-spotted spider mites or cyclamen mites, a miticide spray would be beneficial (see Chapter 31, Insects and Other Pests).

Scheduling

To flower a 10-inch (25-cm) hanging basket for Mother's Day, Carlson [1] recommends the following schedule for a 60 F (15 C) forcing greenhouse.

If sales are planned for late May or early June, you can delay planting two weeks. The cultivar will influence the scheduling. For 4-inch (10-cm) pots, add two weeks to the schedules used for zonal geraniums.

Cultivars

With more than 75 different cultivars available in today's market, choices are available in color, form and application. Today's cultivars can

TABLE 21-4

Flowering schedule

Time planted	No. of cuttings	Pinching
December 1	1	When ready until mid-March
February 1	3	Two to three times
March 1	5	No pinching

service a wide range of applications from accent plants in sheltered locations to full sun accent or bedding plants.

The more popular cultivars are best suited for production and application in a sheltered garden location. These cultivars provide the grower with a well-balanced specimen hanging basket with many flowers. With improved genetics, specimen baskets production is becoming easier. These new cultivars also provide increased performance with lower maintenance for the consumer.

Often overlooked but increasing in popularity are those cultivars that not only fill expectation in traditional applications, but provide high performance and low maintenance in high light settings. Developed in Europe and introduced in North America, these cultivars are characterized by easy production, low maintenance and abundant single flowers. Known in the trade as Balcon or Balcon-types, this cultivar family is available in three distinct habits:

Vigorous. Balcons, available in several colors—red, light pink, rose, lavender—are best suited for long, trailing applications and in large areas of ground cover or bedding.

Moderate. Cascades or Decoras feature variegated stems and foliage appropriate in baskets and as a groundcover or bedding plant. They require moderate to high light to obtain maximum flower production.

Compact. Minicascades, a series containing three colors—red, pink and lavender—make outstanding hanging baskets and excellent controlled ground covers.

Craig [3] developed the following classification system:

Traditional ivies. Have large, thick leaves, large semidouble or double flowers and usually produce fewer but more showy flowers. Examples are Sybil Holmes, Yale and Tavira.

Balcon types. Of European origin, the leaves are smaller and thinner than traditional cultivars, flowers are usually single but are produced in profusion; plants produce many branches. They are available as green-leaved or variegated forms. Examples are the Balcons, Cascades and Decoras.

Dwarf forms. Similar to balcony types but with dwarf, compact plant habit and many smaller flowers. Examples include Sugar Baby and Mini Cascades.

Ivy-zonal hybrids. Leaves and flowers are closer to zonals but they exhibit a modified, vining habit. Flowers are semidouble and flower number is similar to traditional ivies or zonals. Examples are Pascal, Madeline Crozy and Schone von Grenchen.

Craig [3] reported the number of flowers produced by each cultivar per 10-inch (25-cm) basket when four cuttings were planted into each basket. In the following table cultivars are ranked from lowest to highest based on number of flowers.

These cultivars, although nontraditional in appearance, provide unexcelled, perpetual high performance with relative low maintenance in high light conditions. In short, an ivy geranium can be placed next to zonals in the garden and thrive!

TABLE 21-5

Flower production

Cultivar	No. of flowers	Cultivar	No. of flowers
Pascal	18.5	Spain	31.2
Madeline Crozy	20.0	Cornell	36.2
Yale	23.0	Beauty of Eastbourne	36.5
Rigi	25.0	Rouletta	44.0
Amethyst	28.0	Sugar Baby	60.2
Galilee	28.0	Balcony Imperial	79.2
Sybil Holmes	30.3	Balcon Royale	83.2
Salmon Queen	30.5	Balcon Princess	85.6
Double Lilac White	30.6	King of Balcon	95.3

Marketing

Ivy geraniums are traditionally sold in hanging baskets but they needn't be limited to this application. Diverse commercial cultivar choices can fill many markets and niches. New genetics combined with increasing production knowledge result in a product with added perceived quality.

An emerging growth segment of this market is ivy geraniums produced in "small" containers. This product offers opportunity in the production cycle and a new customer base in the marketplace. In the production cycle, this product can be produced in a shorter time frame than a hanging basket, thus utilizing a later planting date when cuttings are readily available. Compact cultivars can be grown at very population densities, which results in improved revenue potential. Small container ivies permit the consumer to enjoy this plant at a more competitive price and provide garden applications other than hanging baskets.

Other opportunities exist for color combinations in the same container. Many ivy geraniums are sold as a single color in a container.

Mixing colors with compatible shades, tones and growth habits expands possibilities. The many new cultivars in the market make the variation innumerable and increasingly easy to produce.

Education will continue to play an increasing role in successfully marketing this crop. Consumers need more information about caring for this crop than they need for other crops so the plants can achieve their full potential. Whenever possible, label products with accurate care information and suggestions for garden applications.

Regardless of how ivies find their way to the market, a well-produced geranium can fill many needs in today's garden, providing perpetual high performance with relatively low maintenance.

References

[1]   Carlson, W.H. 1973. Ivy geraniums: a growing demand. *Grow Magazine* 1:6-7.
[2]   Clifford, D. 1970. *Pelargoniums, Including the Popular "Geranium."* Great Britain: Blandford Press.
[3]   Craig, R. 1991. Personal communication.
[4]   Holcomb, E.J. 1985. Effect of growth retardants on ivy geranium. *HortScience* 20(4):771-772.

CHAPTER 22

Novelty Geraniums

Charles F. Heidgen

Geraniums have evolved over a period of 350 years from approximately 200 different Pelargonium species.

In this evolutionary process many species, hybrids and cultivars have been discovered and developed that were novel, but not quite up to large scale commercial production standards. For lack of a better name we call them "novelty" geraniums. Novelty geraniums refer to unusual flowers, foliage and growth habit.

Rosebud geranium

Rosebud-flowered geraniums are mentioned in English horticultural literature from as far back as the 1870s. The rosebuds' flowers have so many petals that they resemble tiny rambler rosebuds. (See Color Figs. C13 and C14.) A single floret often has 30 to 40 petals. Although rosebud geraniums are very spectacular in full bloom, they frequently don't bloom well, especially in rather hot weather. Appleblossom Rosebud is one of the most photogenic geraniums in existence. Sybil Holmes is an excellent example of a rosebud ivy geranium.

Tulip-flowered geranium

The tulip-flowered geranium is frequently referred to as a rosebud geranium. Actually, the tulip geranium has only six to nine petals, but because of their incurved form they look like tulips. (See Color Fig. C15.) The incurved petals rarely open completely, and the large, full heads are thought by some to resemble popcorn balls. The first tulip geranium was discovered in the mid- to late 1960s by Frank Andrea in his greenhouses in a Boston, Massachusetts suburb, and was named after his wife, Patricia Andrea.

He marketed this new geranium at his greenhouses and at flower shows all over the United States. A remarkable quality of the tulip geranium is its ability to bloom all winter long in the greenhouse, even during a dark, cloudy winter.

In addition to the unique flower form, the tulip geranium also has unique foliage that resembles very crinkly lettuce. Occasionally, a "sport" branch will come off the main plant, usually right at the soil line; the "sport" will have usual zonal geranium foliage and salmon flowers much like Salmon Fiat.

Cactus-flowered geranium

The cactus-flowered geraniums are so named because they have long, narrow, pointed petals, resembling cactus spines. (See Color Fig. C16.) The single forms display the petals better, but usually the double forms are more attractive. The cactus geraniums occur in all colors from white to pink to salmon to purple.

Carnation-flowered geranium

The carnation-flowered geraniums date back to the 1880s. The usually single flowers have delicately serrated petal edges. Some say the serrated edges are caused by a virus, while others say that it's the result of a chimera. It's quite common for a carnation-flowered geranium to have a "sport" or "reversion" to a different color and standard flower.

Bird's egg-flowered geranium

The bird's egg-flowered geraniums have flower petals speckled with a contrasting color, definitely resembling a bird's egg. Most bird's egg geraniums have single flowers, though some are double. The speckled flowers are far more noticeable on the single flowers than on the doubles.

Phlox-flowered geranium

The phlox-flowered geraniums are all single flowered, having broad, flat petals, giving the flower a very round appearance. A darker color band runs across each petal about midway and looks like a halo encircling the flower's center, much like garden phlox.

Fancy-leaved geranium

Fancy-leaved geraniums are zonal geraniums with extraordinary markings and unusual coloring and zonation. (See Color Figs. C17, C18 and C19.) Foliage can be various shades of green, yellow or gold; some with a simple white border, others combining cream, crimson and brown. When a dark zone crosses a variegated leaf portion, an exotic, tricolor effect results.

These fancy-leaved geraniums have been mentioned in geranium journals since the mid-1700s, and many of the varieties propagated today date back to the 1850s to 1870s. Some varieties seem to be flower color sports of the same variegated leaved plant, such as Petals (Color Fig. C18), Flower of Spring, Mrs. Mappin and Foster's Seedling. Pediodically a variegated-leaved geranium will sport a branch with just a narrow variegated band on the leaf margin; these are called the silver-leaved sports. Because they have less variegation and more green, they are usually more vigorous than their fully variegated form. This "sequential mutation" is noticed in other variegated foliaged plants genera also. These variegated-leaf geraniums will periodically sport completely white, albinotic shoots. These shoots survive on the parent plant for an amazing time period, sometimes even flowering, giving a striking display. Of course, these shoots won't survive if removed from the parent plant in an attempt to propagate an "all white plant."

Most tricolor geraniums are at least somewhat difficult to propagate, considerably slower growing and aren't outstanding in cutting production. For these reasons they don't lend themselves readily to commercial production. When they are grown for retail sales, it's usually best to allow additional time to produce specimen plants to be marketed at premium prices.

Fancy-leaved geraniums' outdoor performance varies with some giving a good flower and foliage display, while some of the most highly variegated ones have a tendency to "sunburn." A slightly shaded location is usually best. The popular varieties, Madam Languth, Happy Thought, Petals, Flower of Spring and Mrs. Parker, are usually very good performers. Some newer varieties of gold leaf types from England also provide unusual garden accents.

Ivy leaf geranium

White Mesh (see Color Fig. C20) and Crocodile are two very unusual ivy leaf geranium varieties. Their leaf variegation is caused by mesh-vein virus. This virus is quite difficult to transmit, though hobbyists have transmitted the virus by grafting into other ivy geraniums, giving the unusual foliage pattern to better blooming varieties. The virus appears to have no pathogenic effects. In hot weather the variegated foliage pattern is lost, apparently through virus suppression by heat; the variegation returns when weather moderates.

Stellar geranium

Stellar geraniums are so-called due to their star-shaped flowers and leaves. (See Color Figs. C21 and C22.) Some say the leaves look like gingko tree leaves. At one time people thought that the stellar form evolved from crosses between zonal geraniums and *Pelargonium formosum*. Through controlled breeding experiments, it has been determined that *P. formosum* isn't a distinct species, but rather a zonal geranium with a different leaf and flower form caused by a single gene mutation. This mutated gene produces two forms: stellar, with star-shaped leaf and flowers, and fingered flowers, with deeply cut leaves and the usually very narrow flower petals.

The original stellar geraniums came from Australia from the late Ted Both's breeding work. They were introduced to Europe and the Americas in the mid-1960s. Most stellars are low growing, being a little more compact than most zonal garden varieties. They bloom prolifically and have medium to large flower heads. Stellars have both single and double flowering forms; however, the single forms exhibit the star-shaped flowers more graphically. Further, most double forms often have considerably smaller flowers. Truly, this is one of the most overlooked novelty geraniums, since its garden performance is usually outstanding.

Scented geranium

Undoubtedly, the most popular novelty geranium category is the scented foliage type. Display and marketing is most critical for gardeners to become more familiar with them. It's important to explain to gardeners experiencing scented geraniums for the first time that these novel plants are grown for their scented leaves and not for their flowers. Though virtually all scented geraniums do flower, their flowers are usually small and rather insignificant. Like Regal geraniums, their close relatives, they bloom much better when they are grown under cool or cold night temperatures. Signs explaining scented geranium blooming are very important to good merchandising.

Why are these leaves scented? Undoubtedly, this scented foliage is a "defense" so that grazing animals won't browse on them, thus perpetuating these plants. How are these leaves scented? At the leaf surface, little hairs are visible (under strong magnification); between the hairs are small beads; rubbing the leaf breaks the beads, thus releasing fragrant oil. Signage asking gardeners to rub the leaves and enjoy the fragrances and flavors greatly increases awareness and purchases.

Many scented geraniums are pure Pelargonium species. Others are naturally occurring hybrids of these species. Still others are man-made hybrids. Hobbyists have made many crosses, so certain ancestry isn't known. Further, many seedling selections have been made over the years. Often these selections have different names, but very similar scents and appearances. This has led to some duplication.

Scented geraniums have many different scents and flavors. Four major categories exist: citrus types, including Lemon, Lime, Ginger, Orange and Strawberry; rose types, including Old-fashioned Rose, Attar of Rose, Rober's Lemon Rose, Cinnamon, Mint-scented Rose, Red-flowered Rose, Eucalyptus and a few others; and pungent types, including Nutmeg, Pine, Apple Cider, Fringed Apple, Peppermint, Old Spice and many others. The fourth category includes all scented geraniums that don't fall into the above three classes and includes the very popular Apple-scented geranium, plus Coconut, Balsam and many more. The Apple-scented geranium grows almost from a crown and doesn't really produce a good quantity or quality of cuttings; for this reason, it's usually grown from seed. The Apple-scented geranium makes an excellent hanging basket plant with large, silky leaves. One of the best blooming of all of the scented geraniums, it's frequently covered with hundreds of tiny white flowers.

Scented geraniums can be used in the garden as accent plants due to their striking foliage textures and colors, ranging from dark greens to silvery gray. Some varieties' vigorous growth habit and large, ultimate size make them useful as background plants in flower beds.

Scented geranium leaves can be used as aromatic herbs in sachets and potpourris; frequently, dry leaves are more fragrant than green leaves. The green or dry leaves can also be used in teas as well as in flavoring salads, breads, cakes and jellies.

Dwarf and miniature geranium

Though dwarf and miniature geraniums have existed for over 100 years, they're popular mostly with plant collectors. Because of their size, they are frequently called "windowsill" geraniums, a location they tolerate during the winter in areas where outdoor growing is impossible.

Miniatures are smaller than dwarfs, but the small growing characteristic is both inherited and environmental. If given luxurious growing conditions, miniature varieties outgrow dwarf varieties grown in small pots with poor growing conditions.

All unique zonal geraniums occur as dwarfs, including the rosebud, phlox, bird's egg, fancy leaf and stellar. A full color range exists, as well as singles and doubles.

New introductions *(Editor's note)*

There are more than 10,000 different geranium varieties. Shady Hill Gardens cultivates more than 1,100 different varieties. Their catalog has color photographs of many varieties.

In 1991, 1990 and 1989 more new geranium varieties were introduced to U.S. gardeners than ever before. Most notable in 1990 were the Florastar winners, Marilyn, Judy and Grace, bred by Bodger Seeds Ltd. and distributed by Oglevee Ltd. (See Color Fig. C32). These three floribunda geraniums are ideal in hanging baskets, window boxes and patio planters. Shady Hill Gardens introduced the new Delights, Cherry Delight and Flame Delight, also superb for hanging baskets.

The following were listed as new in 1991 by Shady Hill Gardens:

Stellar geranium: Dot Heath

Scented geraniums: Fringed Apple, Peacock, Golden Nutmeg, Roger's Delight, Ocean Wave

Ivy leaf geraniums: Plum Tart

References

[1]   Australian Geranium Society, The. 1978. *A Check List and Register of Pelargonium Cultivar Names.* Australia: Surrey Beatty & Sons Pty. Ltd.

[2]   Bagust, H. 1968. *Miniature Geraniums.* London: John Gifford Ltd.

[3]   Bennett, M. 1972. *Geraniums, The Successful Growers Guide.* New York: Hippocrene Books Inc.

[4]   Clark, D. 1988. *Pelargoniums.* London: The Hamblyn Publishing Group Limited.

[5]   Delamain, B. and D. Kendall. 1987. *Geraniums.* London: Christopher Helm (Publishers) Ltd.

[6]   Fogg, H.G. Witham. 1975. *Geraniums and Pelargoniums.* London: John Gifford Ltd.

[7]   Mastalerz, J.W. and E.J. Holcomb, Eds. 1982. *Geraniums,* 3d ed. University Park, Penn.: Pennsylvania Flower Growers Association.

[8]   Shellard, A. 1981. *Geraniums For Home and Garden.* North Pomfret, Vermont: David & Charles.

[9]   Taylor, J. 1988. *Geraniums and Pelargoniums.* Great Britain: The Crowood Press.

[10] Wilson, H.V.P. 1972. *The Joy of Geraniums.* New York: William Morrow & Company Inc..

[11] Wood, H.J. 1983. *International Pelargoniums or . . . Geraniums of the World.* Great Britain: Exeter, Devon: A. Wheaton & Co. Ltd.

[12] —— 1983. *Pelargoniums-Geraniums and Their Societies.* Great Britain: Exeter, Devon: A. Wheaton & Co. Ltd.

[13] —— 1987. *Pelargoniums The Growers' Guide to "Geraniums" and Pelargoniums Especially the Golden-Leaved Varieties.* Great Britain: Exeter, Devon: A. Wheaton & Co. Ltd.

CHAPTER 23

New Cultivars

Catherine Anne Whealy

During the last five years color trends in vegetative geraniums have shifted away from the traditional red to nonred colors. Previously a 70% red and 30% nonred market, the early 1990s' market is now 50% red and 50% nonred cultivars. It's predicted that nonred colors will be increasingly important. Salmon is the top nonred color; however, pink, coral and lavender are increasing in sales as better performing cultivars are introduced to the market.

Popular cultivars

Following are descriptions of the most popular cultivars currently available in North America. (See Color Fig. C23 to C42 for many of these varieties.)

Dark red

Crimson Fire. Oglevee. Dark red, semidouble flowers on medium green, zoned foliage. Late flowering cultivar, later than Sincerity and Yours Truly. Very vigorous and upright; may require multiple growth retardant applications. Excellent garden performance.

Fame. Oglevee. Intense dark scarlet, semidouble flowers on dark green, zoned foliage. Medium vigor. Good garden performance, but not high temperature-tolerant; good for cooler climates. U.S. Plant Patent #4785.

Ritz. Oglevee. True crimson red, semidouble flowers on medium to dark green foliage. Better branching and form than Crimson Fire. Later flowering than Yours Truly, but earlier than Crimson Fire. Vigorous growing cultivar; may require multiple growth retardant applications. U.S. Plant Patent pending.

Sassy Dark Red. Oglevee. Dark burgundy, semidouble flowers on dark green foliage. Earlier flowering than Ritz but later than Yours Truly.

Upright habit with good branching and medium vigor. U.S. Plant Patent pending.

Sunbelt Dark Red. Oglevee. Dark red, semidouble flowers on medium green, zoned foliage. Very late flowering. Vigorous growing cultivar, upright and not well-branched; may require multiple growth retardant applications. Good garden performance and high temperature tolerant.

Scarlet

Americana Scarlet. Goldsmith. Light orange scarlet flowers with medium green foliage. Reverse on petals. Later flowering than Yours Truly. Medium vigor. U.S. Plant Patent pending.

Eclipse Red. Goldsmith. Dark scarlet flowers on dark green foliage. Good flower-to-foliage contrast. Compact plant habit, but most vigorous cultivar in Eclipse series. U.S. Plant Patent pending.

Elizabeth. Oglevee. Scarlet red, floribunda double flowers with zoned, dark green foliage. Early flowering. Spreading habit, good for hanging baskets. High temperature tolerant and good garden performance. Replaces Brigette. U.S. Plant Patent pending.

Kim. Oglevee. Scarlet, semidouble flowers on medium green foliage. Medium vigor with good branching. Good in high light and high temperatures. U.S. Plant Patent #5311.

Mars. Fischer. Bright red, semidouble flowers on medium green foliage. Medium to high vigor. Good garden performance and rain tolerant. U.S. Plant Patent #5372.

Red Satisfaction. Ball/Oglevee. True scarlet semidouble flowers with medium green foliage. Earlier flowering than Kim. Medium vigor with good branching. Good garden performance and high temperature tolerant. U.S. Plant Patent #7641.

Sincerity. Oglevee. Large, scarlet semidouble flowers with dark green zoned foliage. Late flowering. Vigorous with few branches; may require multiple growth retardant applications.

Showcase Light Scarlet. Ball. Light orange/scarlet semidouble flowers contrasted with dark green foliage. Compact to medium vigor. Good garden performance. U.S. Plant Patent pending.

Sunset. Oglevee. Large, scarlet, double flowers with medium green foliage. Moderate to high vigor. Good garden performance. U.S. Plant Patent #7609.

Tango. Fischer. Scarlet semidouble flowers with dark green foliage. Large flower heads. Later flowering than Kim. Compact habit. Good shipping performance. U.S. Plant Patent #5933.

Yours Truly. Oglevee. Large, scarlet, double flowers on medium to dark green, zoned foliage. Early flowering. Medium to high vigor. Good garden performance.

Cherry red

Americana Cherry Red. Goldsmith. Nice, cherry red, semidouble flowers with darker green foliage. Compact habit, like Glaciers, with good branching. Good garden and high temperature performance. U.S. Plant Patent pending.

Atlantis. Fischer. True, cherry red, semidouble flowers with dark green foliage. Compact habit. High temperature tolerant and good garden performance. U.S. Plant Patent pending.

Showcase Cherry Red. Ball. Cherry red, semidouble flowers contrasted with dark green foliage. Early flowering. Medium vigor with branching. Good garden performance. U.S. Plant Patent pending.

Rose red

Americana Cherry Rose. Goldsmith. Intense magenta flowers on medium green foliage with slight zone. Compact plant habit. U.S. Plant Patent pending.

Disco. Fischer. Rose red, semidouble flowers with medium green foliage. Earlier flowering than Veronica. Compact to medium plant size with good branching. U.S. Plant Patent #5930.

Jazz. Fischer. Rose red, semidouble on medium to dark green foliage. Earlier flowering than Veronica. Medium to high vigor. Good garden performance, rain tolerant. U.S. Plant Patent #7499.

Veronica. Oglevee. Magenta, double flowers with medium green foliage. Early flowering. Moderate vigor with good branching. Good high light and temperature performance. U.S. Plant Patent #5054.

Coral

Americana Coral. Goldsmith. Bright coral flowers with medium green foliage. Good vigor, upright habit. Good garden performance, high temperature tolerant. U.S. Plant Patent pending.

Charleston. Fischer. Orange coral, semidouble flowers with dark green foliage. Good foliage to flower color contrast. Medium to high vigor. Good high temperature tolerance. U.S. Plant Patent #7393.

Eclipse Salmon Orange. Goldsmith. Coral, semidouble flowers on dark green foliage. Later flowering than Charleston. Upright in form, medium vigor. Good garden performance. U.S. Plant Patent pending.

Gloria. Fischer. Orange coral, semidouble flowers on medium green, zoned foliage. Medium to high vigor. High temperature tolerant. U.S. Plant Patent #7394.

Jubilee. Oglevee. Coral, semidouble flowers on medium green, zoned foliage. Reverse on the flower petals. Very vigorous with some branching; may require multiple growth retardant applications. High temperature tolerant and good garden performance. U.S. Plant Patent pending.

Showcase Bright Coral. Ball. Large, bright coral, semidouble flowers contrasted with dark green foliage. Medium vigor. Good garden performance. U.S. Plant Patent pending.

Sunbelt Coral. Oglevee. Coral, semidouble flowers with medium green foliage. Late flowering. Vigorous; may require multiple growth retardant applications. Good outdoor performance—Botrytis tolerant. U.S. Plant Patent #5757.

Dark salmon

Salmon Satisfaction. Ball/Oglevee. Dark salmon, semidouble flowers with medium green foliage. Early flowering. Medium vigor with good branching. Good garden performance. U.S. Plant Patent #7610.

Showcase Dark Salmon. Ball. Large, dark salmon, semidouble flowers contrasted with dark green foliage. Medium vigor with good basal branching. Good garden and high temperature performance. U.S. Plant Patent pending.

Medium salmon

Americana Salmon. Goldsmith. Medium salmon on medium green foliage with distinct zone. Medium vigor. U.S. Plant Patent pending.

Fidelio. Fischer. Salmon pink, semidouble flowers with light, medium green foliage. Compact habit, smaller and earlier than Schoene Helena. U.S. Plant Patent #5752.

Glacier Salmon. Oglevee. Medium salmon pink, double flowers with medium green foliage. Early flowering. Compact habit with branching. Not high temperature tolerant, good cultivar for Northwest. U.S. Plant Patent pending.

Judy. Oglevee. Floribunda type with double salmon flowers with white eye and zoned, dark green foliage. Early flowering. Spreading compact habit, good for hanging baskets. High temperature tolerant and good garden performance. U.S. Plant Patent #7322.

Pink Expectations. Oglevee. Large, salmon, double flowers on dark green, zoned foliage. Very early flowering, earlier than Wendy Ann and Schoene Helena. Medium habit and good branching. Good garden performance and high temperature tolerant. U.S. Plant Patent #5315.

Schoene Helena. Fischer. Large, salmon rose, semidouble flowers with medium green, zoned foliage. Medium to high plant vigor with good branching. High temperature tolerant and good garden performance. U.S. Plant Patent #5374.

Wendy Ann. Oglevee. Salmon pink, semidouble flowers on medium green zoned foliage. Vigorous growing cultivar with good branching; may require multiple growth retardant applications. Good garden performance.

Light salmon

Americana Light Salmon. Goldsmith. Light salmon, semidouble flowers on zoned, medium green foliage. Medium to high plant vigor. U.S. Plant Patent pending.

Eclipse Light Salmon. Goldsmith. Light salmon flowers with dark green, zoned foliage. Compact plant habit with good branching. U.S. Plant Patent pending.

Pink Camellia. Oglevee. Light salmon pink, semidouble flowers with medium green foliage. Early flowering. Compact plant habit.

Pink Champagne. Fischer. Light salmon pink, semidouble flowers with medium green, zoned foliage. Vigorous and upright. Good garden performance. U.S. Plant Patent pending.

Pink Satisfaction. Ball/Oglevee. Soft salmon pink, semidouble flowers with medium green foliage. Early flowering, flowers with Pink Expectations. Medium vigor with good branching. U.S. Plant Patent #7599.

Hot pink

Hot Pink Satisfaction. Ball Seed/Oglevee. Large, bright, hot pink, semidouble flowers with white eye and medium green foliage. Early flowering. Moderate to high vigor with good branching. Good garden performance and high temperature tolerant. U.S. Plant Patent pending.

Misty. Oglevee. Semidouble hot pink flowers with white eye and medium green, zoned foliage. Flowering is early. Moderate vigor and well-branched plants. Best for cooler climates. U.S. Plant Patent #7350.

Medium pink

Americana Pink. Goldsmith. Medium pink with white eye on medium green foliage. Good branching and medium vigor. U.S. Plant Patent pending.

Katie. Oglevee. Large, double, soft rose pink flowers with medium green, zoned foliage. Medium plant habit with good branching. Tolerates high humidity conditions. U.S. Plant Patent #5759.

Marilyn. Oglevee. Floribunda type with bright pink, double flowers with white eye and zoned, dark green foliage. Earliest flowering floribunda type. Spreading habit, good for hanging baskets. High temperature tolerant and good garden performance. U.S. Plant Patent #7247.

Melody. Oglevee. Semidouble, medium pink flowers with red and white flares and medium green foliage. Early flowering. Small but numerous flowers. Good branching and medium habit. U.S. Plant Patent pending.

Rio. Fischer. Single, pink and red bicolor flowers with dark green foliage. Better color contrast between flowers and foliage than Carnival; however, Carnival's flower color is more intense. Compact habit and slow growing under low temperature conditions. Not entirely resistant to shattering. U.S. Plant Patent #7422.

Risque. Oglevee. Medium pink, semidouble flowers with dark pink flares and a white eye with medium green foliage. Late flowering. Good basal branching and moderate vigor. High temperature tolerant. U.S. Plant Patent #6654.

Showcase Pink. Ball. True medium pink, semidouble flowers with white eye. Excellent contrast with dark green foliage. Compact to medium vigor and good plant form. Good garden and high temperature performance. U.S. Plant Patent pending.

Light pink

Americana Light Pink. Goldsmith. Very light pink flowers with medium pink flares and medium green foliage. Moderate to high vigor. U.S. Plant Patent pending.

Blues. Fischer. Large flowered, light pink, semidouble flowers with white eye and darker pink flares on medium green foliage. Later flowering than Pink Expectations. Medium vigor and good branching. Good outdoor performance. U.S. Plant Patent #5373.

Grace. Oglevee. Floribunda type with light pink, double flowers and zoned, dark green foliage. Early flowering. Spreading habit, good for hanging baskets. High temperature tolerant and good garden performance. U.S. Plant Patent #7321.

Lavender

Americana Rose. Goldsmith. Dark lavender flowers with dark flares on medium green foliage. Medium vigor. U.S. Plant Patent pending.

Brazil. Fischer. Fluorescent lilac, single flowers with medium to dark foliage. Medium vigor. Good outdoor performance. U.S. Plant Patent pending.

Danielle. Oglevee. Semidouble, lavender pink flowers with a darker eye with medium green foliage. Sport of Veronica. Medium to high vigor. Good garden performance and high temperature tolerant. U.S. Plant Patent #6055.

Laura. Oglevee. Semidouble, lavender flowers with a white eye and medium green foliage. Moderate vigor with good branching. Good in cooler climates. U.S. Plant Patent #7087.

Precious. Oglevee. True lavender, semidouble flowers with a small white eye on medium green foliage. Floriferous and early flowering. Medium vigor. U.S. Plant Patent pending.

Purple

Aurora. Oglevee. White-eyed, purple, double flowers with medium green, zoned foliage. Flowers later than Fox and Kardinal. Medium to high vigor with little branching.

Fox. Oglevee. Darkest purple geranium cultivar available. Fluorescent, double flowers with medium green foliage. Medium vigor. Earlier, better plant and flower form than Aurora. Good garden performance. U.S. Plant Patent #7083.

Kardinal. Fischer. Single-flowered purple with orange flares in center and medium green foliage. Earlier than Aurora. Vigorous plant habit with few branches; may require multiple growth regulator treatments. Good garden performance.

White

Alba. Fischer. White, semidouble flowers with medium green foliage. Early flowering, blooms with Snowhite. Medium to high vigor. U.S. Plant Patent #7392.

Americana White. Goldsmith. Pure white, semidouble flowers with medium green foliage. Most vigorous plant habit of all Americanas. Good garden performance. U.S. Plant Patent pending.

Showcase White. Ball. White, semidouble flowers with dark green foliage. Good flower to foliage color contrast. No pinking with low temperatures. Medium vigor. Good garden performance. U.S. Plant Patent pending.

Snowhite. Oglevee. Large, white, semidouble flowers with medium green foliage. Earlier flowering, darker foliage and larger flowers than Snowmass. Moderately vigorous and well-branched. Not high temperature tolerant. U.S. Plant Patent #5312.

Brocade

Ben Franklin. Oglevee. Melon pink, semidouble flowers with variegated foliage. Late flowering. Medium plant habit. U.S. Plant Patent #6218.

Mrs. Parker. Oglevee. Light pink, semidouble flowers with white eye and variegated foliage. Late flowering. Compact plant habit.

Wilhelm Langguth. Oglevee. Brick red, semidouble flowers with variegated foliage. Late flowering. Vigorous; may require multiple growth retardant applications. Not high temperature tolerant.

CHAPTER 24

Seedling Diseases

Charles C. Powell

Seedling geraniums may well have surpassed cutting geraniums in the U.S. greenhouse industry in the late '80s and early '90s. This hasn't presented any unique disease problems to growers, but neither has it provided relief from disease threat.

Seedling geranium pathology closely parallels cutting cultivar pathology that has been in use for many years. Literature contains few, if any, specific references to seedling geranium diseases, so most of this chapter is based on my observations of this crop's pathology, and observations and information from people working with zonal geraniums.

Cultivar susceptibility varies greatly with most seedling geranium diseases. This cultivar resistance hasn't been researched, documented or published adequately, but it's not uncommon to go into a greenhouse and see a disease widespread on one cultivar, yet barely affecting an adjacent cultivar. Each grower can, no doubt, make valuable observation about cultivars that seem to resist any disease in the greenhouse.

Another underlying principle of seedling geranium culture that impacts their pathology is that annual crops generally begin as seeds. Breaking the continued vegetative (cutting) propagation cycle has advantages from the pathogen epidemiology viewpoint. Viruses, systemic bacteria and fungi are commonly spread via symptomless cutting material. They are rarely, if ever, found on or in seeds. For instance, most seedling geraniums are as susceptible as zonal cultivars to bacterial wilt; however, the disease is rarely seen in greenhouses on seedling geraniums because the pathogen isn't seed transmitted. On the other hand, several fungal pathogens, such as Alternaria and Botrytis, can be transmitted into a crop or seed. So seedling geranium producers have a slightly different disease picture to use to plan health management programs.

Greenhouse production diseases

Damping-off

Damping-off, caused by *Pythium ultimum* and *Rhizoctonia* species, occasionally appears in seedling geranium crops. In general, I would suggest that the seedling geranium isn't particularly susceptible to the damping-off organisms. Damping-off occurs when growers don't germinate seedling geraniums under the proper temperature regime, although certain cultivars may be particularly susceptible to damping-off.

Keep the temperature high to get rapid and even seedling geranium germination. This also keeps seedlings generally free of damping-off. Some growers use fungicides to counter damping-off problems. Seedling geraniums can be somewhat sensitive to fungicides applied as soil drenches. It's better to control damping-off with cultural rather than with chemical methods.

Pythium black leg

Pythium black leg (Color Fig. C77) is a common disease on cutting geraniums. It's also quite common on certain seedling geranium cultivars. Pythium black leg is generally a problem resulting from too cool temperature and too much moisture surrounding geranium roots. It occurs on seedling geranium crops where condensed water drops from inside the greenhouse onto crop flats. The resulting wet spots are often sites of Pythium black leg infection. The disease then spreads from these spots and causes considerable loss when the crop matures.

Preventing Pythium black leg on seedling geraniums is much the same as it is on cutting cultivars. Prevention programs must be integrated. Improve drainage and don't allow water to stand around roots. Also, keep the soil and root temperature above 60 F or 65 F (16 C or 18 C). Chemicals such as Truban, Banrot, Subdue or Aliette combat Pythium black leg on. Again, watch for some cultivars' sensitivity to these fungicides applied as soil drenches. General sanitation programs, especially those relating to growing media infestation, are basic to black leg management.

Rhizoctonia root and crown rot

Several seedling geranium cultivars seem to be quite susceptible to Rhizoctonia root and crown rot. Rhizoctonia is often worse in greenhouses when the crop is subject to drought or salt stress. Such a condition can be common. Many growers limit water to control many cultivars' stretching and rank growth. When they do this, they invite Rhizoctonia attacks, especially when fertilizer levels are too high.

On seedling geraniums, Rhizoctonia moves up the stem and collapses the plant from the crown upward, similar to the effect of Pythium black leg. The difference is that the lesions and rotted tissue from the Rhizoctonia fungus grow on the leaf and lower stem tissue as it rots. This growing mycelium allows the fungus to move from plant to plant. Rhizoctonia can also move farther distances in dust.

Prevent Rhizoctonia crown and root rot by making sure the crop isn't subject to excessive drought or salt stress. In addition, apply chemical drenches of Cleary's 3336, Comain FL, Chipco 26019 or Terraclor to control the disease.

Sclerotinia crown rot

Sclerotinia crown rot (Color Fig. C93 and C94) can cause problems on several bedding plants, and it's been a problem in recent years on seedling geranium crops. This crown rot is a rapidly moving disease that causes significant damage quickly. It's favored by hot, humid temperatures and a closely spaced crop, which allows the fungus to move from plant to plant over the soil and lower plant parts.

The disease is easy to diagnose. The fungus is quite evident as a cottony growth over lower plant tissue. As the disease progresses and the fungus matures, black resting bodies (sclerotia) form on the leaves or stem. These vary in size. Most of them will be less than ⅛ inch (3 mm) in diameter and look like rat or mouse droppings!

No available fungicides have been shown to stop Sclerotinia infestation. Terraclor has been reported as being effective in some cases. If the disease appears in a crop, immediately space the crop, water more carefully to avoid excessive splashing and discard all infested plants as soon as possible. Between diseased crops where noted, make an extra effort towards sanitation by applying a good germicidal agent such as Greenshield or Prevent or thoroughly steaming all benches, pots and underlying soil. If germicidal agents are used, soak for 10 minutes. Never use these agents to sanitize soil or growing media. Sclerotinia sclerotia are impossible to eradicate from soil or growing media with anything but heat.

Botrytis leaf blight, crown rot and flower blight

The diseases caused by Botrytis (Color Fig. C70 to C76) are most common on seedling geraniums, perhaps because their germplasm is particularly susceptible, especially when compared with zonal varieties. No research on this possibility has been done. Another problem of seedling geraniums and Botrytis diseases concerns the flowers and their unique aging (sometimes "shattering") syndromes. I have seen Botrytis infections on leaves when seedling geraniums are growing under a hanging basket crop that is shedding plant debris onto them. The falling petals provide good infection spots for Botrytis fungus. The fungus begins to grow on petals and then invades leaf tissue, causing an ugly lesion or a blighted leaf. A third problem with seedling geraniums and Botrytis is seedling production. Botrytis can cause a crown rot on young seedling geraniums in greenhouses, especially under abnormally cool and damp growing conditions. As a crown rot, Botrytis usually kills the developing seedling. For control procedures, see Chapter 25, Foliar Diseases.

Viruses

Like zonal geraniums, seedling geraniums are susceptible to many plant viruses (Color Fig. C87 to C90). Tomato ringspot virus (Color Fig. C86) is seed-borne in many cultivars. Infected seed lots commonly yield plants that are 100% symptomatic. The plants show yellow or light green ringed spots on lower leaves that are quite evident in cool weather. As the days lengthen and warm up, symptoms disappear, and the plants emerge unaffected. Good crops are seemingly produced despite the infestation.

For the most part, other viruses found in seedling geraniums behave similarly, without major plant damage. It's particularly noteworthy that even a virus as potentially damaging as tomato spotted wilt virus seems to behave in a rather benign manner in seedling geraniums.

Outdoor bed diseases

Botrytis flower blight

Outdoors, when rainy, cool weather persists for several days, Botrytis may appear in a seedling geranium bed (Color Fig. C76). Botrytis attacks blossoms as they age and begin to form seed pods (Color Fig. C70). Picking off old flowers and seed heads is important to maintain the beauty of a seedling geranium bed. This procedure will also go a long way toward preventing unsightly Botrytis infections (Color Fig. C71). Additionally, spacing plants to allow good air circulation is important.

Bacterial leaf spot and blight

About the most serious situation with seedling geraniums grown outdoors is bacterial leaf spot (Pseudomonas) and blight (Xanthomonas) (Color Fig. C78 and C79). Many seedling geranium cultivars are quite susceptible to bacterial leaf spot and blight organisms. When infection occurs in a bed outdoors, it can quickly move throughout the bed via splashing water, insects or plant handling. Plants decline and die quickly. Disease spread and symptoms generally are worse under hot, rainy weather.

Fixed copper fungicides, especially Phyton-27, can slow bacterial blight organism spread. These fungicides leave an unsightly residue and can be somewhat phytotoxic on seedling geraniums. The best control would be to space plants and try to avoid splashing water during irrigations. If bacterial leaf spot and blight have been noticed in a seedling geranium bed, the disease may occur the following year because the bacterial inoculum may winter on crop residue left in that bed. Survival depends upon the decay rate of infected plant parts. Six months is about the survival limit for the causal pathogen. Turning under crop debris so that it decays more quickly lessens this survival time. In warmer climates where replanting might occur in less than six months, plant a different plant in locations where the disease had been a problem.

Alternaria leaf spot

This fungus disease has been noted occasionally on seedling geraniums grown outdoors. The leaf spot (Color Fig. C80) is almost identical to that caused by the bacteria mentioned above! Differences are that Alternaria leaf spot will occur in cooler weather. Also, leaf spots are somewhat darker and often exhibit a series of concentric rings, like those on a "bull's-eye" target.

I have rarely seen this disease bad enough to warrant control. Chipco 26019 is the only control specifically registered for this disease. Broad spectrum fungicides registered for other geranium fungal diseases would probably satisfactorily prevent Alternaria leaf spot. FORE, Phyton-27 or Daconil 2787 would probably be as effective as Chipco 26019.

Summary

Seedling geranium diseases closely parallel cutting cultivar diseases. Keep in mind important differences: a susceptibility to Rhizoctonia root and crown rot and the occasional presence of Sclerotinia crown rot and seed-borne viruses. Gardeners need to be aware of problems that bacterial leaf spot and blight disease can cause. All in all, a production program that optimizes an environment favoring seedling geraniums' vigorous growth and performance will go a long way toward ensuring freedom from most diseases.

CHAPTER 25

Foliar Diseases

Edema

Gary W. Moorman

Edema is a disorder that affects geraniums, causing leaves to yellow and die. It's thought to be due to adverse environmental conditions and, therefore, doesn't spread from plant to plant. Ivy geraniums, Irene cultivars and plants with a large root system as compared to shoot size are particularly sensitive to this disorder.

Small yellow spots in between veins are often the first symptom observed on the top of the leaf. Small, translucent, watery pustules are visible on the leaf underside below yellowed areas. These blisters and yellowing usually occur first on older leaf margins but then develop over the entire leaf (Fig. 25-1). The blisters enlarge and become brown with corky or scabby texture (see Fig. 32-1 in Chapter 32, Physiological and Environmental Disorders). The entire leaf may yellow, die and drop off in a pattern somewhat similar to that of bacterial blight.

Fig. 25-1. Early edema showing small, translucent blisters or pustules developing on geranium leaf underside.

Cause

Edema is thought to be caused by an imbalance of the plant's water uptake and water loss. It's most severe when soil temperature is warm and there's a high level of water absorption by the roots while, at the same time, water loss is low. This occurs when air is cool, relative humidity is high, light intensity is low (cloudy weather) and ventilation is poor. Water retention in the cells is thought to cause some cells to burst. As the broken tissue heals, it becomes dry and corky.

Mite (two-spotted, *Tetranychus urticae* Koch) feeding may also play a role in edema development. Potter and Anderson [4] screened ivy geranium cultivars for resistance to two-spotted spider mite attack and evaluated edema severity on mite-infested plants. There is a relationship in which mite-resistant cultivars develop less severe edema than mite-susceptible cultivars. The authors theorized that the plant's physiology or nutritional status that results in edema development may also favor mite survival. Edema- and mite-resistant cultivars were Double Lilac White, Sunset, Madame Margot, Amethyst and Salmon Queen. The most mite- and edema-susceptible cultivars included Sybil Holmes, Yale and Pascal.

Management

Edema can be lessened by using a well-drained potting mix and watering less frequently during cool, cloudy weather. Under such conditions, heat and ventilate the greenhouse to reduce humidity. Space plants to provide good air circulation. Water in the morning so that the soil isn't excessively wet overnight. Avoid wetting the leaves since wet leaves lose less water. Some growers report that removing saucers from hanging basket plants in which ivy geraniums are planted helps avoid edema. Finally, maintain good mite control.

References

[1] Balge, R.J., G.E. Beck and B.E. Struckmeyer. 1966. Characteristics of oedema and its occurrence in *Pelargonium hortorum. Proc. 17th Int. Hort. Congr. Md.* 1:482 (Abstr.).

[2] Digat, B. and J. Albouy 1976. *Donnees actuelles sur le probleme de l'oedeme du pelargonium. Pepinieristes Horticulteurs Maraichers* 168:51-55.

[3] Forsberg, J.L. 1979. *Diseases of Ornamental Plants.* Univ. of Ill. Special Pub. No. 3.

[4] Potter, D.A. and R.G. Anderson. 1982. Resistance of ivy geraniums to the two-spotted spider mite. *J. Am. Soc. Hort. Sci.* 107:1089-1092.

Botrytis Blight

Mary K. Hausbeck

Stem, leaf and flower blights caused by *Botrytis cinerea* are limiting factors in geranium (*Pelargonium* x *hortorum* L.H. Bailey) production.

Disease epidemics caused by *B. cinerea* typically occur in wet, humid environments. *B. cinerea* reproduces in diseased tissue and forms powdery gray conidia (gray mold) that can be picked up on air currents and transported to uninfected tissue. Conidia are considered the primary inoculum source for infecting stems and flowers and an important inoculum source for latent infections of cuttings [7]. Conidia require a film of water to germinate and infection occurs primarily between 59 F and 77 F (15 C and 25 C). Conidia production maximizes in an environment of high relative humidity and a temperature of 59 F (15 C) [4].

Controlling disease caused by *B. cinerea* offers challenges in each geranium production phase. Reliance on fungicides alone to control Botrytis blight increases the potential for *B. cinerea* resistance to the fungicide and could lead to control failure. For instance, *B. cinerea* strains resistant to benzimidazole fungicides have been isolated from many greenhouses in Pennsylvania. Strains that showed multiple resistance to benzimidazoles and dicarboximides were also observed but occurred less frequently [5].

Seed-propagated geraniums

Damping-off and leaf blight

B. cinerea is just one of several fungi that can cause seedling damping-off. Leaf blight may also be a problem, especially when leaves are wounded or damaged. Healthy leaves may become infected when petals from infected blooms fall onto leaves and initiate infection (Fig. 25-2).

Control
- Use only clean planting media and growing containers.
- Rogue diseased plants.
- Grow plants in low relative humidity (<85%).
- Water in the morning to allow the foliage to dry by evening.
- Use a protectant fungicide when necessary.
- Remove flowers from plants.

Fig. 25-2. Botrytis blight on leaf initiated by infected flower petals.

Stock plants

Stem blight

Stem blight typically begins in the stock plant's broken or cut stem surface and progresses downward, causing the entire stem to die and preventing development of shoot meristems that could be removed as cuttings (Color Fig. C72). In severe cases, stem blight extends into the plant base, resulting in plant death.

Disease caused by *B. cinerea* can be difficult to manage on stock plants because the conventional method of growing geraniums for cutting production is conducive to serious stem blight outbreaks. Stock plants' terminal buds are pinched at regular intervals or treated with the growth regulator ethephon (Florel) [8] to increase plant branching and the number of shoot meristems that can be removed as cuttings. This practice produces short, compact plants with dense canopies that limit light and air penetration and promote lower leaf senescence (Color Fig. C73) [6]. Closely spacing stock plants to maximize cutting production greatly enhances these conditions. Under specific environmental conditions, *B. cinerea* readily colonizes these senescent leaves and sporulates, providing ample conidia to infect stems wounded during cutting harvest (Color Fig. C74).

Airborne conidia

Airborne conidia of *B. cinerea* may be present among stock plants throughout the growing season [3]. Geraniums are intensively managed and require frequent grower involvement. Airborne conidia "showers" may occur whenever growers irrigate and fertilize (even when using a plastic tube drip system with emitters that prohibit splashing), spray pesticides, plant or harvest cuttings and fill or clean benches (Fig. 25-3, Fig. 25-4). As the crop matures, frequently the magnitude of the "showers" increases, apparently related to the number of blighted stems and necrotic leaves with sporulating *B. cinerea*.

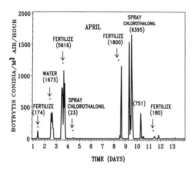 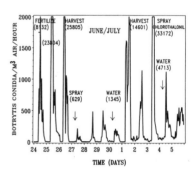

Fig. 25-3 and Fig. 25-4. Airborne conidia "showers" may occur among stock plants in association with growers' activities.

The occurrence of conidia "showers" during and immediately after cuttings harvest is important in disease management because cut stems on stock plants are susceptible to *B. cinerea* infection.

Control

- Apply a protectant fungicide immediately after the cutting harvest.
- Increase space between stock plants. A less dense plant canopy allows air circulation and results in a less suitable environment for *B. cinerea* germination, infection and sporulation [9]. Also, a less dense plant canopy allows fungicide coverage of wounded stems and sporulating *B. cinerea* on stems and senescent lower leaves.
- Remove plant debris on and underneath plant benches that could serve as a reservoir for sporulating *B. cinerea*. In an epidemic, you may need to remove senescent leaves with sporulating *B. cinerea* from individual stock plants.

- Limit grower activity immediately prior to and for 24 hours after harvest to minimize conidia "showers," thereby reducing the airborne conidia that could land on stems wounded during the cutting harvest.
- Reduce relative humidity for a minimum 24 hours immediately following the cutting harvest to limit stem blight [3].

Cuttings

Cuttings may develop cutting rot (Color Fig. C75) or leaf blight during propagation especially if they have been harvested from stock plants with sporulating *B. cinerea*. Conidia may lodge on the cutting leaf or broken stem surface at harvest. These conidia may subsequently germinate when placed in a propagation bench for rooting, where the wet and humid conditions established for optimum propagation promote conidia germination and cutting rot and leaf blight.

Conidia "showers"

Conidia "showers" may occur whenever growers perform an activity within the propagation area [3] (Fig. 25-5). Protecting newly planted cuttings from conidia "showers" is critical because the frequent and extended misting periods required at the onset of the propagation cycle favors Botrytis blight development.

It's important to control Botrytis blight in the propagation area because conidia "showers" occurring during cutting removal from the greenhouse for shipping may play a role in postharvest diseases. Disease

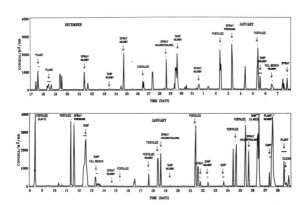

Fig. 25-5. Airborne conidia "showers" may occur among stock plants in association with growers' activities.

caused by *B. cinerea* has been implicated as a limiting factor in storage and shipment of unrooted and rooted cuttings [1]. Conidia deposited onto the plant surface may remain dormant or penetrate the plant surface and remain inactive. One or both resting stages may be activated during the postharvest period by physiological or biochemical changes in the cuttings and environmental conditions during shipping and storage.

Conidia impacted onto cutting leaves during shipping may germinate and infect under environmental conditions that occur if shipments are delayed or if plant material isn't removed from shipping boxes promptly. If infection begins on cuttings in the box, under favorable conditions it can continue after their removal, and during potting and placement on greenhouse benches. When symptomless plant parts from geraniums were evaluated for *B. cinerea* on the same day the shipments were received, *B. cinerea* was nearly always recovered [1]. *B. cinerea* sporulation was observed on the geraniums within five days after cuttings were received.

Control

- Treat newly planted cuttings with a protectant fungicide.
- Minimize the hours that foliage is wet and the relative humidity is high (>85%) by:
  - a) discontinuing misting and/or overhead watering early enough to allow foliage to dry by evening
  - b) heating or venting
  - c) increasing air circulation through fan utilization and increased plant spacing
  - d) providing adequate drainage to remove excess water from floors
- Rogue diseased cuttings immediately to minimize the material available for infection and subsequent sporulation of *B. cinerea* and reduce the occurrence and magnitude of conidia "showers."
- Minimize activities within the propagation area. Planting cuttings and removing diseased tissue from cuttings prior to shipping shouldn't occur within the propagation area because such activities often result in conidia "showers" that could infect nearby cuttings.
- Group plant material according to maturity. Newly planted cuttings should be propagated in a restricted activity area and should be physically separated from more established cuttings that could maintain sporulating *B. cinerea* on senescent leaves.
- Don't hold well-rooted cuttings in the propagation area because they don't require the wet and humid environment required for optimum nonrooted cutting propagation. Such mature cuttings typically exhibit senescent lower leaves with sporulating *B. cinerea*. Grower activity among these mature cuttings could provide conidia "showers" for nearby, newly planted cuttings.

- Remove cuttings from the shipping container immediately upon receipt, place them in a low relative humidity environment, apply a protectant fungicide and plant as soon as possible.

References

[1] Cline, M.N. 1987. Prevent Botrytis blight on geraniums. *Greenhouse Grower* Feb. 88-91.

[2] Hausbeck, M.K. 1990. "The epidemiology of *Botrytis cinerea* Pers. on the geranium (*Pelargonium* x *hortorum* L.H. Bailey)." Ph.D. dissertation, The Pennsylvania State Univ.

[3] Hausbeck, M.K. and S.P. Pennypacker. 1991. Influence of grower activity and disease incidence on concentrations of airborne conidia of *Botrytis cinerea* among geranium stock plants. *Plant Disease* 75:798-803.

[4] Jarvis, W.R. 1989. Managing diseases in greenhouse crops. *Plant Disease* 73:190-194.

[5] Moorman, G.W., and R.J. Lease. 1989. Fungicide resistance in *Botrytis cinerea* isolates from Pennsylvania greenhouses. (Abstr.) *Phytopathology* 79:1207.

[6] Rogers, M.N. 1982. Stock plants. *Geraniums,* 3d ed. Ed. J.W. Mastalerz and E.J. Holcomb. University Park, Penn.: Pennsylvania Flower Growers.

[7] Smith, P.M. 1967. The chemical control of grey mould, *Botrytis cinerea* Pers. ex Fr., in mist-propagated chrysanthemum cuttings. *Plant Pathol.* 16:157-159.

[8] Tjia, B. and H. Kim. 1975. Granulated ethrel and ARD-126 as potential branch inducing agents on geraniums. *Flor. Rev.* 157(4072):37, 74-75.

[9] Trolinger, J.C., and D.L. Strider. 1984. Botrytis blight of *Exacum affine* and its control. *Phytopathology* 74:1181-1188.

Fungal Leaf Spots

Arthur W. Engelhard

Alternaria leaf spot *(Alternaria alternata)*

Symptoms. Initial symptoms develop on lower leaf surfaces as small (1 to 2 mm in diameter), raised, water soaked-appearing spots that resemble blisters. Some have white centers. Leaf lesions may develop within 20 hours after the start of a continuous rain. Spots on the upper leaf surface usually aren't distinct at this time, but with time they become visible. Within a day the centers of the raised, blister-like spots start to collapse, resulting in sunken centers. These "new" lesions, 1 to 3 mm in diameter, may be numerous on a given leaf. Some spots develop a light yellow halo. They may increase in size up to 10 mm in diameter. Occasionally they develop concentric rings or coalesce with others to form necrotic areas on leaves (Color Fig. C80). Masses of dark brown to black spores are produced, especially on the older leaves. Chlorosis of infected leaf parts occurs and if infection is extensive, leaf abscission results. Conidiophores on older lesions on leaves still attached to the plants produce conidia and more extensive sporulation occurs on leaves on the ground (Fig. 25-6). The conidia are wind-disseminated. Lower or older leaves are infected more than younger leaves and defoliation progresses upward from older to younger leaves. All leaves on mature plants eventually develop symptoms.

Fig. 25-6. Old, established lesions with abundant sporulation occurring.

Occurrence. In Florida, the disease commonly occurs in warm weather when the temperature is too high for good geranium culture (April or later). In California [1], the disease appears during the winter when fog and rain are common along the west coast and plants aren't growing actively due to low night temperatures. It appears that in both Florida and California plants are likely to become infected when they are under stress (from heat or cold) and/or in a physiologically senescent stage.

Control. Removing infected leaves from plants and the ground to reduce airborne spores is very important in controlling this disease.

Cercospora leaf spot *(Cercospora brunkii)*

Symptoms. Leaf spots are initially 1 to 3 mm in diameter, sunken and pale green; they turn gray with age. Spots have the same appearance on both upper and lower leaf surfaces. Conidiophores and conidia develop on both leaf surfaces in the center of the gray spots, making them look darker and the centers appear raised, although they aren't. The conidia are wind-disseminated. Lesion margins may be slightly raised. At this stage a spot appears to have a thin, light-colored margin that is actually the sloped, marginal surface of the concave spot. Spots may coalesce, or leaf sections may be affected so that wedge-shaped necrotic areas extend in from leaf margins (Color Fig. C79). When many spots are present, the infection results in large necrotic areas and eventual leaf blight. Circular areas ("eyes") may be present in large necrotic areas. Tissues adjacent to lesions become chlorotic. In time an entire leaf may become chlorotic. Flowers and peduncles also are infected. The seedling cultivars Cherie, Ringo, Sooner Red and Sprinter Salmon are most susceptible, whereas Showgirl is more disease tolerant.

Leaf spots caused by *C. brunkii* resemble, in some aspects, those caused by *Pseudomonas cichorii*. Cercospora leaf spots generally are smaller than Pseudomonas leaf spots, have a more general chlorosis, cause leaves to abscise earlier, and are tan to brown in color, while Pseudomonas leaf spots are brown, dark brown or black. Both pathogens infect the inflorescences, causing them not to open. Pseudomonas leaf spots with progressive symptoms usually have a narrow, wet-appearing margin that Cercospora leaf spots do not.

Control. Use benomyl 50W fungicide at 0.25 pounds per 100 gallons (112 g per 378 l) tank-mixed with either chlorothanonil 75W or mancozeb 80W at 0.75 pounds per 100 gallons (337 g per 378 l) or Captan 50W at 1.0 pound per 100 gallons (454 g per 378 l) for good disease control.

Reference

[1]   Munnecke, D.E. 1956. *Pl. Dis. Reptr.* 40:452-453.

Rust

Joann Lee Rytter

Geranium rust, caused by the fungus *Puccinia pelargonii-zonalis,* is a pathogen of the cultivated garden geranium, *Pelargonium* x *hortorum.* The disease was first discovered in South Africa [3] where the majority of Pelargonium species are indigenous. Since then, the rust has spread throughout Europe, New Zealand, Australia and Hawaii. In 1967, the first reports of geranium rust in the United States occurred almost simultaneously in New York [2] and California [1]. In Pennsylvania, the disease was reported for the first time in 1971 [7]. Since that time, rust has been reported in several other states.

Etiology

Puccinia pelargonii-zonalis is in the family Puccinaceae and is an autoecious, microcyclic rust, completing its life cycle on one host. The pathogen produces rust pustules (uredia) that are embedded in leaf tissue. As the disease progresses, secondary pustules encircle the primary pustules creating a concentric circle effect. Urediniospores are the asexual spore stage and are broadly ovate. Teliospores, the sexual stage, are rarely produced [3, 4].

The pathogen continuously produces spores that can reinfect the same plant or spread to other plants. Free water is necessary for spore germination. The optimum temperature for germination is 61 F to 70 F (16 C to 21 C) for disease development [4, 5]. The fungus penetrates the plant through stomata on upper and lower leaf surfaces and eventually establishes itself within the plant.

Symptoms

Symptoms can occur within seven days after infection and first appear as small, yellow spots on the leaf underside. In 13 days, brown spore pustules develop in the center of each spot (Color Fig. C91). The color is rust, hence the disease name. Shortly after, a ring of secondary pustules surrounds the primary pustule, which then looks like two concentric rings (Color Fig. C92). Pustules eventually form on the upper surface as well; however, the concentric ring is usually lacking. Severely infected leaves become yellow and drop prematurely. Infected plants that still retain their leaves are severely blemished by the disease and aren't marketable. Eventually, overall plant appearance is diminished and occasionally flower quality and quantity are reduced.

Spread

In approximately 19 days from the original infection, the host epidermis is ruptured and the spores are released. These spores are spread by splashing water, air currents, workers handling the plants and moving infected stock plants. These spores can reinfect new leaf surfaces of the same plant or infect neighboring plants. Due to the repeating asexual cycle, this disease can spread very rapidly, especially under optimum temperatures.

Control

Geranium growers currently control geranium rust with combined cultural and chemical practices. Several cultural methods are available for prevention. Because free water is a prerequisite for spore germination, careful irrigation and adequate air circulation to maintain dry leaf surfaces diminishes rust infection. Use drip irrigation rather than overhead watering. Purchase certified, culture-indexed cuttings. Diligently inspect stock plants for rust, especially if cuttings are to be taken from these. Don't take cuttings from field-grown plants because rust may be present but not yet obvious.

Laughner [6] explored incorporating genetic resistance or immunity into the cultivated geranium. Rust resistance has been found in the Regal (*P.* x *domesticum*), the scented (*Pelargonium* spp.) and the ivy leaf geranium (*P. peltatum*). Commercial interspecific hybrids and primary hybrids (*P.* x *hortorum* x *P. peltatum*) produced in The Pennsylvania State University breeding program weren't hosts to rust, indicating that a genetic solution is possible. Biological rust fungus control by a bacterial antagonist has also been investigated [8, 9].

If geranium rust is present in your greenhouse, immediately rogue all infected plants. In the greenhouse apply a fungicide plus a spreader-sticker every seven to 10 days with one of the following: Dithane (mancozeb), Strike or Bayleton 25 (triadimefon), Daconil (chlorothalonil), PlantVax 75W (oxycarboxin) and Zyban (thiophanate-methyl plus mancozeb). You can use Carbamate WDG (ferbam) for rust infections outdoors.

References

[1]   California Department of Agriculture. 1967. Report of new or unusual plant pathogen. *Pl. Path.* A-67-8.

[2]   Dimock, A.W., R.E. McCoy and J.F. Knauss. 1968. Pelargonium rust. A new geranium disease in New York State. *NY State Flower Grower's Inc. Bul.* No. 268:1-3.

[3]   Doidge, E.M. 1926. A preliminary study of the South African rust fungi. *Bothalia* 2:98-99.

[4]   Harwood, C.A. 1974. "Factors influencing the development of geranium rust." Masters thesis, Univ. of California, Berkeley.

[5]   Harwood, C.A. and R.D. Raabe. 1979. The disease cycle and control of geranium rust. *Phytopathology* 69:923-927.

[6]   Laughner, J.L. 1985. Breeding and evaluation of Pelargonium for resistance to geranium rust, *Puccinia pelargonii-zonalis* Doidge. Masters thesis, Penn. State Univ.

[7]   Nichols, L.P. and L.B. Forer. 1972. Geranium rust discovered in Pennsylvania. *Plant Disease Reptr.* 56:759.

[8] Rytter, J.L. 1987. "The biological control of geranium rust, *Puccinia pelargoni-zonalis* Doidge, by a bacterial antagonist." Masters thesis, Penn. State Univ.

[9] Rytter, J.L., F.L. Lukezic, R. Craig and G.W. Moorman. 1989. Biological control of geranium rust by *Bacillus subtilis*. *Phytopathology* 79:367-370.

Bacterial Leaf Spots

Robert L. Wick

Xanthomonas campestris pv. *pelargonii, Pseudomonas cichorii, Pseudomonas syringae* pv. *syringae* and *Pseudomonas erodii* have been reported to cause leaf spots on Pelargonium. Leaf spots caused by Xanthomonas are covered in Chapter 26, Vascular Wilt Diseases, and won't be discussed here. *Pseudomonas erodii* has been reported in several countries including the United States; however, since no reference cultures exist, the bacterium is no longer considered a valid species. This chapter describes bacterial leaf spots caused by *Pseudomonas cichorii* and *Pseudomonas syringae* pv. *syringae* (Color Fig. C78).

Pseudomonas cichorii leaf spot

Pseudomonas cichorii was first reported found in Florida by Arthur Engelhard. The following descriptions are his as printed in *Geraniums III*.

"The appearance of leaf spots caused by the bacterium *Pseudomonas cichorii* varies depending on weather conditions. When plants are outside in the rain, dark brown to black, irregularly shaped necrotic areas 5 to 10 mm or larger develop (Fig. 25-7, 25-8). For a few hours after a rain, the spot margins appear water-soaked. Spots may extend or enlarge along the veins. They may coalesce and encompass large leaf sections. In some instances, the entire leaf may become necrotic and curl up. Leaves also curl and bulge when extensive infection and necrosis develops on leaf margins.

"Under environmental conditions less favorable to disease spread and development, such as when plants are exposed to dews or occasional overhead wetting, lesions develop that are sunken on both upper and lower leaf surfaces. The spots may have tan centers that may be raised slightly. A spot may have a dark margin and a yellow halo that may be as wide as the lesion or quite small to absent. During rains or excessively wet periods, existing lesions may develop water-soaked margins as they enlarge. Continued symptom development and disease spread stops in

Fig. 25-7. Wet leaf spots and necrotic areas after a rain.

Fig. 25-8 Leaf spots, necrotic areas on leaves and leaf margins, and a dead leaf.

dry weather or if the plant foliage is kept dry.

"Flower clusters are susceptible. Individual flowers fail to open when infection occurs in the bud stage. Occasionally, the flower stalk (peduncle) is infected (Fig. 25-9). A brown to dark brown lesion may extend to 10 mm long. Natural stem infection hasn't been observed and stem inoculations made in the greenhouse resulted in only superficial epidermal tissue discoloration. Infection in seed flats causes a rapid, soft, dark, wet decay of developing leaves. The necrotic area varies in size from a small spot to the entire leaf. Infected seedlings should be discarded."

P. cichorii should be of concern to growers because it causes disease on several ornamental plants and vegetables. Aglaonema, anthurium,

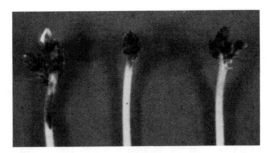

Fig. 25-9. Infection of flower buds and flower stem (peduncle) on the cultivar Ringo.

brassaia, caladium, dieffenbachia, dizygotheca, fatshedera, fatsia, hedera, monstera and philodendron are a few of the foliage plants reported to be hosts. Flowering plant hosts include chrysanthemum, gerbera, hibiscus and jasminum.

P. cichorii geranium leaf spot isn't well known outside Florida; however, *P. cichorii* commonly causes chrysanthemum bacterial blight in other states. Rane and Wick (unpublished data) cultured *P. cichorii* from chrysanthemum in Massachusetts and found it to cause geranium leaf spots.

Pseudomonas syringae leaf spot

In Massachusetts, *Pseudomonas syringae* pv *syringae* was identified as the cause of leaf spot on both seed and zonal geraniums (Fig. 25-10). *P. syringae* pv. *syringae* cultured from lilac was also found to cause geranium leaf spot. The disease has been noted in Pennsylvania and potentially may occur wherever geraniums are grown. Leaf spots are similar in appearance to those caused by *Pseudomonas cichorii*. The two bacterial leaf spots can't be distinguished without laboratory techniques. *P. syringae* pv. *syringae* has a much broader host range than *Pseudomonas cichorii*. In addition to bacterial blight of lilac, it causes diseases on several vegetable, ornamental and fruit crops, as well as grasses. *Celosia cristata*, *Chrysanthemum indicum*, *Dahlia pinnata*, *Delphinium* sp., *Forsythia* spp., *Hibiscus* sp., *Impatiens balsamina* and New Guinea impatiens, *Jasminum primulinum* and *Lupinus* sp. are some ornamental hosts.

Pseudomonas leaf spot control

The life histories of both bacterial pathogens are similar; thus, management practices for each will be considered together. Inspect incoming geranium crops to detect not only bacterial leaf spots but also other diseases and insect infestations. Don't allow plants with disease symp-

Fig. 25-10. Pseudomonas syringae *leaf spot on seed geraniums.*

toms in the greenhouse. Sanitation is an important disease management principle and is particularly pertinent to bacterial problems. Discard plants that develop symptoms in the greenhouse. Workers should wash hands after handling diseased plants or soil. Remove diseased plant debris from the growing area as much as possible. Bacteria are easily splashed from plant to plant by irrigation water. Consider practices that minimize splashing and reduce the leaf wetness period. Avoid handling plants when they are wet. Copper sprays are only marginally effective in controlling bacterial diseases and may be phytotoxic to geranium.

References

[1]   Engelhard, A.W. 1982. Bacterial leaf spot. *Geraniums,* 3d ed. Ed. J.W. Mastalerz and E.J. Holcomb, University Park, Penn.: Pennsylvania Flower Growers.
[2]   Wick, R.L. and K.K. Rane. 1987. *Pseudomonas syringae* leaf spot of *Pelargonium* x *hortorum.* (Abstr.) *Phytopathology* 77:1620.

CHAPTER 26

Vascular Wilt Diseases

Bacterial Blight

Margery Daughtrey and Robert L. Wick

The most destructive geranium disease is bacterial blight, a vascular wilt caused by *Xanthomonas campestris* pv. *pelargonii* (Brown) Dye. The disease has also been called bacterial stem rot, bacterial wilt or bacterial leaf spot, but the disease has so many different symptoms that the more comprehensive name "blight" is preferable. Even though specialist propagators use culture-indexing techniques, bacterial blight remains a serious and frequent disease problem for geranium growers, occasionally causing complete crop losses.

Host range and potential reservoirs

The commercial cultivars of both the zonal geranium, *Pelargonium* x *hortorum*, (cuttings and seedlings) and the ivy geranium, *P. peltatum*, are susceptible to bacterial blight. Some differences exist in symptom expression between cultivars, but no cultivars are significantly resistant. Hybrid geranium seedlings become infected from contact with zonal or ivy geraniums; seed transmission hasn't been a problem.

The Regal or Martha Washington geranium, *P. domesticum*, isn't susceptible to bacterial blight. *P. acerifolium, Pelargonium* Toronto, *P. tomentosum* and *P. scarboroviae* [4] are also resistant to *X. pelargonii* which may be useful to plant breeders seeking to improve disease resistance in geraniums for the trade.

Symptoms

Bacterial blight symptoms (Color Fig. C81 to C85) vary, depending on the cultivar and species affected and greenhouse environmental and cultural conditions. Symptoms alone aren't sufficient for making a diagnosis of bacterial blight: A laboratory diagnosis is necessary.

Leaf spots. When infection spreads through splashing or dripping water, small, water-soaked spots develop, first visible on the leaf underside. After a few days, these leaf spots become obvious on the leaf top (Fig. 26-1). The spots are tan to brown, round and generally $\frac{1}{16}$- to $\frac{1}{8}$-inch (2 to 3 mm) in diameter. They are slightly sunken with a well-defined margin. Ivy geraniums may develop leaf spots similar to those on zonal geraniums. The small spots caused by *X. pelargonii* help to distinguish them from common fungus leaf spots, but other bacteria may cause similar spots.

Fig. 26-1. Zonal geranium leaf infected with Xanthomonas pelargonii *showing slightly sunken, well-defined spots.*

Spots may be followed by yellow patches or brown wedges (Color Fig. C81, C83 and C85). (Similar, wedge-shaped brown areas in leaves may also be caused by Botrytis.) Bacteria spread from the leaf spots through the xylem (water-conducting tissue) of the leaf petiole into the stem, causing a systemic infection that leads to leaf wilt and plant death.

When plants are systemically infected through roots or via latently infected cuttings, the first symptom is lower leaf wilt (Color Fig. C82). These lower leaves often become chlorotic, beginning at the margins.

Water stress, root rot, Southern wilt (*Pseudomonas solanacearum*) or Verticillium wilt also cause geranium wilt. Roots are obviously discolored and decayed by Pythium, whereas wilting from bacterial blight, Southern blight or Verticillium infection may be visible on plants with healthy-looking roots. In stock plants, wilt and stem rot are the most common symptoms. Leaf spotting doesn't occur unless the bacterium is air-borne by water splash, which is often the case during propagation or after plants are set outdoors.

Systemically infected ivy geraniums don't wilt. Symptoms suggest nutrient deficiency or even mite infestations. Leaves may turn off color, or become necrotic with or without wedge-shaped sections (Color Fig. C85 and Fig. 26-2).

Fig. 26-2. Ivy geranium with bacterial blight, showing dry, brown leaves.

Stem rot. Stem rot is a later bacterial blight stage. The bacterial infection may first cause internal vascular discoloration; later, bacteria spread from the xylem inward to the pith and outward to the outer stem tissue, causing a visible brown to black discoloration of some stem portion. Leaf yellowing and wilting precede stem rot symptoms (Fig. 26-3 and 26-4).

Cutting rot. Cuttings infected with bacterial blight *X. pelargonii* may fail to root and slowly rot from the base upward. The leaves may wilt and show yellow or brown patches. In some cases, however, cuttings may have latent bacteria infection, so symptoms won't be visible during propagation. When infected cuttings are introduced to a greenhouse operation, symptom development is sometimes delayed for several months, particularly during cooler growing conditions. Plants started from healthy-looking cuttings may suddenly develop noticeable symptoms when warmer temperatures stimulate bacteria growth within the tissues.

Temperature effects

Warm temperatures speed the appearance of bacterial blight symptoms. Kivilaan and Scheffer [3] found that plants held at 81 F (27 C) showed symptoms seven days after inoculation, while three weeks elapsed before the first symptoms appeared on inoculated plants held at 60 F (16 C); symptom expression was even slower at 50 F (10 C). Disease development and symptom expression increased with rising temperatures from 50 F to 81 F (10 C to 27 C) during the first five weeks after inoculation. In general, symptoms are suppressed at low temperatures (50 F to 60 F; 10 C to 16 C), symptom development is enhanced at warm temperatures (70 F to 85 F; 21 C to 29 C), and symptom development is again suppressed at very high temperatures (90 F to 100 F; 32 C to 38 C). This symptom suppression at cooler temperatures explains the symptoms' sudden appearance in spring.

Fig. 26-3. Geranium plant infected with Xanthomonas pelargonii *showing early stem rot symptoms on cultivar Irene. Notice leaves wilting on plant's left side.*

Bacterium survival and spread

In cuttings. Infected stock plants may not show symptoms when cuttings are taken. Moisture and warmth supplied during cutting rooting often trigger symptom development.

On other plant species. *X. pelargonii* is known to attack only *Pelargonium* species, but there is some evidence that other bedding plant crops may support the bacterial inoculum for a limited time [2]. It's unlikely that plant species other than geranium are important in disseminating

X. pelargonii within the greenhouse industry: The bacteria are primarily distributed on or in infested geranium stock. When attempting to eradicate the bacteria from a greenhouse following a disease outbreak, however, other plant species are potential contamination reservoirs.

Fig. 26-4. Geranium plant infected with Xanthomonas pelargonii *showing blackened, shriveled areas on stems.*

On infested cutting knives. The most common means of disease transmission from one stock plant to another is the cutting knife. Because of this, the preferred method of taking cuttings is to break them from stock plants. If knives are used, they should be disinfested by dipping in 70% alcohol and flaming or soaking in 10% chlorine bleach (one part bleach plus nine parts water) for five minutes. Change knives between stock plants.

In soil. *X. pelargonii* doesn't survive in soil, but it does persist for a long time within geranium tissue debris. This bacteria survival depends on the infected plant parts' decay rate; therefore, removing geranium debris from the greenhouse following a disease outbreak is most important. Disinfest any greenhouse surfaces contaminated with *X. pelargonii* before reuse; discard, fumigate or steam-pasteurize contaminated media. Disinfest container surfaces and then reuse them for a crop other than geraniums.

By overhead water and plant contact. Rapid spread of bacterial blight disease occurs when overhead watering is used. The bacteria that cause the disease can also spread through foliage contact between a healthy and an infected plant, as well as through workers handling plants while taking cuttings or removing spent blossoms.

By insects. The greenhouse whitefly *Trialeurodes vaporariorum* can transmit *X. pelargonii* to healthy geraniums.

Control

No chemical treatments control *X. pelargonii*. Strict adherence to good sanitation practices, as well as a carefully administered culture-indexing program are the propagator's sole tools for supplying clean cutting stock. Individual geranium growers must rely on disease exclusion. In the event of a bacterial blight outbreak, they must depend on prompt symptom recognition and roguing out diseased plants.

Crop arrangement may also minimize bacterial blight losses: Don't place hanging ivy geranium baskets directly above bench- or floor-level geranium crops. Grow geraniums from seed in houses separate from geraniums from cuttings. Losses may be much more extensive in greenhouse ranges where geraniums of different types and from different sources are in close contact during production.

References

[1] Bugbee, W.M. and N.A. Anderson. 1963. Whitefly transmission of *Xanthomonas pelargonii* and histological examination of leafspots of *Pelargonium hortorum*. *Phytopathology* 53:177-178.
[2] Kennedy, B.W., F.L. Pfleger and R. Denny. 1987. Bacterial leaf and stem rot of geranium in Minnesota. *Plant Disease* 71:821-823.
[3] Kivilaan, A. and R.P. Scheffer. 1958. Factors affecting development of bacterial stem rot of Pelargonium. *Phytopathology* 48:185-191.
[4] Knauss, J.F. and J. Tammen. 1967. Resistance of Pelargonium to *Xanthomonas pelargonii*. *Phytopathology* 57:1178-1181.
[5] Munnecke, D.E. 1954. Bacterial stem rot and leaf spot of Pelargonium. *Phytopathology* 44:627-632.
[6] ——— 1956. Survival of *Xanthomonas pelargonii* in soil. *Phytopathology* 46:297-298.

Southern Bacterial Wilt

Ronald K. Jones

Southern bacterial wilt is caused by the common, soilborne bacterium *Pseudomonas solanacearum*. The disease was first described on geranium in 1979, in North Carolina by Strider, et al. Though not one of the major geranium diseases, it can cause serious losses to geraniums in the landscape, especially in southern regions where the bacterium is wide spread. *Pseudomonas solanacearum* has an extremely wide host range in field crops, vegetables and bedding plants (Table 26-1).

TABLE 26-1

Pseudomonas solanacearum common hosts

Family	Genus, species	Common names
Apocyanaceae	*Catharanthus*	Madagascar periwinkle
Balsaminaceae	*Impatiens balsaminu*	Garden impatiens
Compositae	*Ageratum* sp.	Ageratum
	Chrysanthemum morifollum	Chrysanthemum
	Gerbera sp.	Gerbera
	Tagetes erecta	Marigold
	Zinnia elegans	Zinnia
Labiatae	*Salvia* sp.	Salvia
Musaceae	*Musa* spp.	Banana
Solanaceae	*Capsicum* spp.	Pepper
	Lycopersicon esculentum	Tomato
	Nicotiana spp.	Tobacco
	Petunia hybrida	Petunia
	Solanum melongena	Eggplant
	Solanum tubersum	Potato
Tropaeolaceae	*Tropaeolum* spp.	Nasturtium
Verbenaceae	*Verbena hybrida*	Verbena

Symptoms

Southern bacterial wilt's initial symptoms appear as wilting on lower leaves followed in a few days by chlorosis and finally necrosis of affected leaves. Affected leaves can be necrotic within two weeks after infection with the terminal growth remaining green until the entire plant dies. Flowers on infected plants fail to open normally. As the disease progresses, vascular necrosis develops in the lower stem with the stem portions at the soil line turning brown, then black.

Symptoms of geranium Southern bacterial wilt are very similar to those caused by the bacterial blight pathogen, *Xanthamonas campestris* pv. *pelargonii*. With bacterial blight, leaf spotting is a common symptom but not with southern bacterial wilt.

Pseudomonas solanacearum is a soilborne pathogen and, therefore, infection usually occurs through wounds or natural openings in the roots. Roots and lower stems become necrotic, turn brown and then black; symptoms progress up the plant. The bacterial blight pathogen may enter the plant directly through leaves or through wounds on any plant part and stem necrosis may progress up or down.

Geranium susceptibility

Strider, et al. in 1982 screened 20 geranium cultivars for resistance to *Pseudomonas solanacearum*. Sixteen cultivars were normally propagated by cuttings and four cultivars were grown from seed. Three weeks after inoculation with *P. solanacearum,* all plants of all cultivars were dead. Based on this one study, no geranium cultivars are known to resist Southern bacterial wilt.

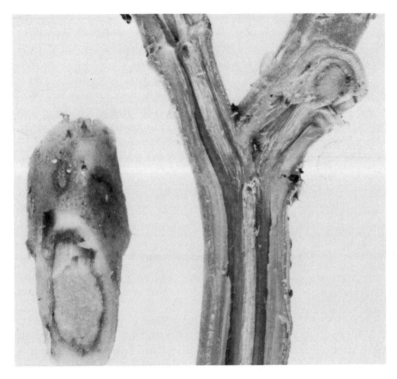

Fig. 26-5. Vascular necrosis of geranium infected with Pseudomonas solanacearum.

Control

Pseudomonas solanacearum is a common, soilborne bacterium in warmer climates around the world; the bacterium can survive in the soil for several years even without a susceptible host. The bacterium is very rare in greenhouse-grown pot crops such as geranium produced in peatlite or other soilless mixes on benches. This disease is, therefore, primarily a problem after the plants are set in soil in the landscape. To ensure freedom from Southern bacterial wilt, keep all greenhouse tools, benches, pots, media and plants free of infested soil.

Once plants become infected with *P. solanacearum,* they usually don't recover or survive. There's no preventive or curative chemical control of the disease. In ground beds infested with the bacterium, avoid growing susceptible species or remove soil and replace it with pathogen-free soil. Immediately remove infected plants from beds and destroy them.

References

[1]   Kelman, A. 1953. The bacterial wilt caused by *Pseudomonas solanacearum*— A literature review and bibliography. *NC Agri. Exp. Station Tech. Bul.* No. 99.
[2]   Strider, D.L. 1982. Susceptibility of geranium to *Pseudomonas solanacearum* and *Xanthamonas campestris* pv. *pelargonii. Plant Disease Reporter* 66: 59-60.
[3]   Strider, D.L., R.K. Jones and R.A. Haygood. 1981. Southern bacterial wilt of geranium (*Pelargonium hortorum*) caused by *Pseudomonas solanacearum. Plant Disease Reporter* 65:52-53.

Verticillium Wilt

Gary W. Moorman and Paul E. Nelson

Verticillium wilt, caused by the fungus *Verticillium albo-atrum,* attacks many greenhouse crops including chrysanthemum, rose and snapdragon. This pathogen is known to attack both *P.* x *hortorum* and *P.* x *domesticum* (Martha Washington or Regal geraniums) and the disease has been reported from New York [3], California [1] and Oregon [6, 13]. In the past few years the disease has been observed primarily on *P.* x *domesticum.*

Symptoms

The first symptoms of *P.* x *hortorum* consist of the collapse of a few leaves on the upper or middle main or side branches (Fig. 26-6). At first the affected leaves' petioles remain normal, but after a day or two, wedge-shaped areas on affected leaves turn yellow, or the entire leaf may turn yellow, and the petioles wilt. This is followed by a gradual drying of affected leaves and petioles, resulting in leaf drop. As the disease progresses, more leaves collapse, yellow, dry up and drop, resulting in affected plant area defoliation. Plant growth may also be stunted. The disease progress varies from plant to plant. In cases where progress is rapid the plant soon loses most of its leaves and the remaining, unwilted leaves are pale green and may show interveinal yellowing, which looks very much like a nutrient deficiency or a viral disease symptom. In the final stages an affected plant may show dieback or drying of branch tips and brown to black discoloration of the main stem and branches (Fig. 26-7). These stem areas are completely discolored in a cross section but above the externally discolored area the browning is limited to the vascular tissue.

Fig. 26-6. Springfield White geranium leaf infected with Verticillium albo-atrum *showing wilting and yellowing, early disease symptoms.*

Fig. 26-7. Springfield White geranium showing severe Verticillium wilt symptoms. Note stem discoloration and defoliation.

In some cases an infected plant may not show symptoms for weeks or even months. After this time, however, the entire plant may suddenly collapse as shown in Fig. 26-8. Shortly thereafter the plant will die.

Symptoms of Verticillium wilt on *P.* x *domesticum* are much the same as those on *P.* x *hortorum* except that the leaf wilting and collapse doesn't occur as readily. Affected leaves turn yellow but often don't wilt; however, the stunting of infected plants is often pronounced (Fig. 26-9).

Fig. 26-8. Verticillium wilt-infected plant (left) and healthy Irene geranium (right).

Fig. 26-9. Verticillium wilt-infected (left) healthy (right) and Pelargonium domesticum plants. Note the infected plant's stunting.

Species and cultivar susceptibility

When *P.* x *hortorum* cultivars Better Times, Olympic Red, Ricard, Wendy Ann, Diddon's Improved Picardy, Springfield White, Radio Red, Penny, Genie, Irene and Dark Irene were inoculated with *V. albo-atrum*, all cultivars were susceptible. The *P.* x *domesticum* cultivars The Princess, Marie Vogel, Graf Zeppelin and Mrs. Layal were also tested and all found to be susceptible to attack by the fungus [10].

Methods of spread

The Verticillium fungus can spread in infected soil and in symptomless but infected cuttings. The pathogen can survive in soil for long time periods in the absence of a suitable host. Therefore, steam sterilization or chemical treatment of the propagating medium and potting soil is necessary for disease control.

Similarity of symptoms to bacterial blight

The symptoms of Verticillium wilt and the stem rot phase of bacterial blight are similar in many respects [7]. Early symptoms of both diseases are leaf wilt, followed by defoliation, branch dieback, branch tip dieback and brown to black discoloration of the main stem and branches. It's possible that plants infected with *V. albo-atrum* have been mistakenly diagnosed as cases of bacterial blight and vice versa. It's also possible that one plant could be infected by both organisms at the same time. The only way to make a positive diagnosis is by making a laboratory culture from the plant in question.

Control

The incidence of Verticillium wilt in geraniums has been greatly reduced through using soilless potting mixes that are pathogen-free.

If soil is used in the potting mix, it should first be steam pasteurized to kill Verticillium. Chemical fumigants reduce but don't eliminate Verticillium in soil. No fungicides effectively protect plants against this fungus. Infected plants must be discarded.

Don't put Verticillium-infected plants in compost that will be used around plants unless the organic matter is thoroughly broken down and the temperatures in the pile are at least 150 F (66 C), high enough to kill Verticillium.

References

[1] Baker, K.F., W.C. Snyder and H.N. Hansen. 1940. Some hosts of Verticillium in California. *Plant Dis. Reptr.* 24:424-425.
[2] Bugbee, W.M. and N.A. Anderson. 1963. Whitefly transmission of *Xanthomonas pelargonii* and histological examination of leafspots of *Pelargonium hortorum. Phytopathology* 53:177-178.
[3] Dimock, A.W. 1940. Importance of Verticillium as a pathogen of ornamental plants. *Phytopathology* 30:1054-1055.
[4] Kelman, Arthur. 1953. The bacterial wilt caused by *Pseudomonas solanacearum* – A literature review and bibliography. *NC Agric. Exp. Stn. Tech. Bul.* 99.
[5] Kivilaan, A. and R.P. Scheffer. 1958. Factors affecting development of bacterial stem rot of Pelargonium. *Phytopathology* 57:1178-1181.
[6] McWhorter, F.P. 1962a. Diverse symptoms in Pelargonium infected with Verticillium. *Plant Dis. Reptr.* 46:349-353.
[7] ——— 1962b. Verticillium control must be considered when indexing geraniums. *Flor. Rev.* 130 (3363):11-12, 21.
[8] Munnecke, D.E. 1954. Bacterial stem rot and leaf spot of Pelargonium. *Phytopathology* 44:627-632.
[9] ——— 1956. Survival of *Xanthomonas pelargonii* in soil. *Phytopathology* 46:297-298.

[10] Nelson, P.E. 1962. Diseases of geranium. *NY State Flower Grower's Bul.* 201:1-10.

[11] Strider, D.L. 1982. Susceptibility of geranium to *Pseudomonas solanacearum* and to *Xanthomonas campestris* pv. *pelargonii. Plant Disease* 66:59-60.

[12] Strider, D.L., R.K. Jones and R.A. Haygood. 1981. Southern bacterial wilt of geranium caused by *Pseudomonas solanacearum. Plant Disease* 65:52-53.

[13] Torgeson, C.D. 1952. Observations of Verticillium wilt of geranium in Oregon. *Plant Dis. Reptr.* 36:51.

[14] Vaughan, E.K. 1944. Bacterial wilt of tomato caused by *Phytomonas solanacearum. Phytopathology* 34:443-458.

CHAPTER 27

Stem and Root Diseases

Pythium Root Rot and Blackleg

Gary W. Moorman

Seed geraniums are very susceptible to Pythium root rot and geranium cuttings to blackleg, diseases caused by several different Pythium species [13, 4]. Pythium, a fungus known since the 1800s [14], is common in field soil, sand or surface water sediment and dead roots of previous crops. It's easily introduced into pasteurized soil or soilless mixes by using dirty tools, dirty pots or flats, walking on or allowing pets to walk on mixes and by dumping mixes on benches or potting shed floors that haven't been thoroughly cleaned. If the sediment of surface water supplies enters an irrigation system, Pythium often causes major crop losses. The fungus can survive in a dormant state in dry soil for 12 years [12]. The use of Pythium-free soilless potting mixes has greatly reduced the incidence of Pythium root rot. When introduced into pasteurized soil or soilless mixes, however, Pythium can cause severe root rot because it has few competitors to check its activity. Pythium can probably be carried into potting mixes by fungus gnats, *Bradysia impatiens* [5], and shore flies, *Scatella stagnalis* [8]. This fungus poses a particular threat to seed geraniums grown in ebb and flow systems [15, 16]. If the reservoir is contaminated with debris or soil harboring Pythium, the fungus can spread to many plants quickly. If the fungus infests a cutting bed or if contaminated water is used in propagation, large losses due to blackleg can occur.

Although geranium seed cultivars differ in susceptibility to Pythium root rot [9, 2], overfertilized seed geraniums, regardless of cultivar, are highly susceptible [6, 7]. Silver thiosulfate (STS) sprayed on seed geraniums to prevent flower shatter renders plants more susceptible to this disease also [10].

Symptoms

In Pythium root rot, root tips that are very important in taking up nutrients and water are attacked and killed. Severe stunting can result. Infected root tips are brown and dead. The brown tissue on the outer root easily pulls off exposing a bare strand of vascular tissue. The plants wilt at midday and may recover at night, but eventually yellow and die. Infected root cells contain many microscopic, thick-walled spores. The fungus progresses up the stem causing discoloration and rot above the soil level (Fig. 27-1 and Color Fig. C77). In blackleg, a brown, shiny, wet-appearing area develops at the base of cuttings (Fig. 27-2). The rot turns coal black and quickly progresses up the cutting stem. A different cutting rot caused by the fungus Botrytis is usually dark brown to black but not shiny.

Management

Pythium root rot and blackleg are difficult to control once they have begun. Direct your efforts toward preventing these diseases before they

Fig. 27-1. Pythium root rot progress up a seed geranium stem.

begin. Pasteurize soil and sand with heat or chemical fumigant treatments to eliminate the fungus. Or, purchase a soilless potting mix since they are generally Pythium free. Potting mixes are now commercially available that contain Pythium-antagonistic microbes and that have physical properties that inhibit the fungus [11]. Cover treated soil or soilless mix and store it in an area that won't be contaminated by contact with nontreated soil. Also, keep ebb and flow system reservoirs covered.

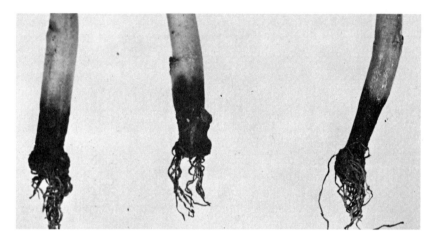

Fig. 27-2. Pythium blackleg causes a brown to black, shiny, water-soaked rot at the base of geranium cuttings.

Fertilize seed geraniums moderately and don't use silver thiosulfate on a crop in which Pythium root rot has been found.

Disinfest all surfaces, tools and equipment that contact the potting mix. If your greenhouse has been susceptible to Pythium root rot, apply a fungicide at planting and repeat the application at appropriate intervals. Etridiazole (Truban, Terrazole, Banrot) and metalaxyl (Subdue) are generally effective in protecting geraniums against Pythium root rot.

To prevent blackleg in cutting propagation, steam pasteurize the propagation medium and bed structure between crops. It's best to root cuttings in separate pots, cubes or peat pellets rather than in a common bed so that the fungus has less opportunity to spread from cutting to cutting. Don't rely on fungicides for Pythium control in cutting production since there's a risk of stunting or delayed rooting, depending upon growing conditions and cultivars treated.

References

[1]   Braun, H. 1924. Geranium stem rot caused by *Pythium complectens* N. Sp. *Journal of Agri. Research* 29:399-419.

[2]   Chagnon, M.C. and R.R. Belanger. 1991. Tolerance in greenhouse geraniums to *Pythium ultimum*. *Plant Disease* 75:820-823.

[3]   Dimock, A.W. 1959. Effective control of geranium diseases. *Florist's Exchange & Hort. Trade World* 132:15.

[4]   Farr, D.F., G.F. Bills, G.P. Chamuris and A.Y. Rossman. 1989. *Fungi on Plants and Plant Products in the United States*. St. Paul: American Phytopathological Society Press.

[5]   Gardiner, R.B., W.R. Jarvis and J.L. Shipp. 1990. Ingestion of *Pythium* spp. by larvae of the fungus gnat *Bradysia impatiens* (Diptera:Sciaridae).

[6]   Gladstone, L.A. and G.W. Moorman. 1989. Pythium root rot of seedling geraniums associated with various concentrations of nitrogen, phosphorus and sodium chloride. *Plant Disease* 73:733-736.

[7]   ———— 1990. Pythium root rot of seedling geraniums associated with high levels of nutrients. *HortScience* 25:982.

[8]   Goldberg, N.P. and M.E. Stanghellini. 1990. Ingestion-egestion and aerial transmission of *Pythium aphanidermatum* by shore flies (Ephydrinae: *Scatella stagnalis*). *Phytopathology* 80:1244-1246.

[9]   Hausbeck, M.K. 1985. "The effect of silver thiosulfate and fungicides on Pythium root rot of the seed propagated geranium." Masters thesis. Michigan State University.

[10]   ————, C.T. Stephens and R.D. Heins. 1989. Relationship between silver thiosulfate and premature plant death of seed-propagated geraniums caused by *Pythium ultimum*. *Plant Disease* 73:627-630.

[11]   Hoitink, H.A.J., Y. Inbar and M.J. Boehm. 1991. Status of compost-amended potting mixes naturally suppressive to soilborne diseases of floricultural crops. *Plant Disease* 75:869-873.

[12]   Hoppe, P.E. 1966. Pythium species still viable after 12 years in air-dried muck soil. *Phytopathology* 56:1411.

[13]   Miller, H.N. and R.J. Sauve. 1975. Etiology and control of Pythium stem rot of geranium. *Plant Disease Reporter* 59:122-126.

[14]   Plaats-Niterink, A.J. vander. 1981. Monograph of the genus Pythium. Baarn. The Netherlands: Centraal burau voor Schimmelscultures. *Studies in Mycology* No. 21.

[15]   Thinggaard, K. 1988. "Pythium and Phytophthora in greenhouse crops with recirculation of the nutrient solution." Fifth International Congress of Plant Pathology Papers. Kyoto, Japan.

[16]   ———— and A.L. Middelboe. 1989. Phytophthora and Pythium in pot plant cultures grown on ebb and flow benches with recirculating nutrient solution. *Journal of Phytopathology* 125:343-352.

Cottony Stem Rot

Arthur W. Engelhard

Symptoms

Infection by *Sclerotinia sclerotiorum* causes a brown, moist decay of the stem and attached petioles. Advancing margins of the affected area may have a dark green border 2 to 3 mm wide. Decay advances up and down the stem and to adjacent leaves. Leaves attached to affected areas wilt, have chlorotic veins and develop a diffuse marginal chlorosis.

Cottony tufts of the pathogen mycelium develop on affected stem parts, peduncles and inflorescences (Color Fig. C93 and C94). Eventually, typical sclerotia (dark brown to black oval-shaped fungous structures 4 to 7 mm long and 2 to 4 mm wide) that help the pathogen overseason develop on infected tissues.

A soft decay occurs when the flower heads are infected. The pedicels and inflorescences droop from the receptacle and hang down around the peduncles. Soon, white tufts of the pathogen appear among the pedicels followed by the dark colored sclerotia.

Control

Roguing diseased plants and removing all sclerotia from the area is important in controlling the disease in outdoor flower beds. Spores are produced on the sclerotia the following season if unchecked. Chipco iprodione 26019 50W and Ornalin vinclozalin 50W control cottony stem rot. Consult labels for clearance and directions.

Thielaviopsis/Black Root Rot

Robert G. Linderman

Black root rot, a disease of many plants, is caused by the soilborne fungus *Thielaviopsis basicola* (Berk. & Br.) Ferraris. It commonly occurs and can cause severe economic loss on poinsettias. Although it's relatively unknown on geraniums, it occurs in several Eastern states where

Fig. 27-3. Young geranium plants photographed two weeks after receipt from propagator. A is cultivar Ricard and B is cultivar Fiat. In both cultivars, the plant on the right is severely infected with Thielaviopsis basicola; the plant on the left only mildly infected.

geranium cuttings are propagated. Meyer [4] described infected plants in Connecticut as slow growing and stunted, with lower leaves that yellow and drop (Fig. 27-3) and with early and rather continuous flowering. Roots of plants showing these foliar symptoms, and from which *T. basicola* can be isolated, usually show typical Thielaviopsis root lesions, which are dark to black because of tissue discoloration and abundant production of dark, thick-walled chlamydospore chains and pigmented hyphae. (Fig. 27-4 and 27-5).

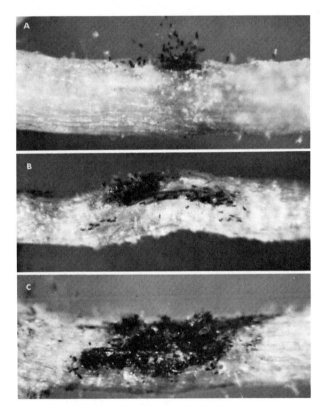

Fig. 27-4. Thielaviopsis root rot lesions. A small lesion with erupting mass of chlamydospores on an otherwise white, apparently healthy geranium root. B *and* C *show increasingly more severe black lesions.*

All the above symptoms are generally typical of *T. basicola* root rot on other hosts such as poinsettia, cyclamen, begonia, tobacco, cotton and bean. The etiology and epidemiology of *T. basicola* on geraniums is different in many respects, however, and these differences warrant further discussion.

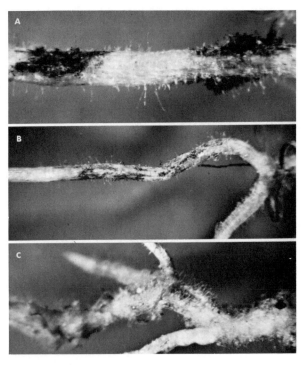

Fig. 27-5. Thielaviopsis root rot lesions. All three photographs are of black Thielaviopsis lesions on young roots that emerged as sidebranches from older, infected roots. Note that lesions in B and C are quite close to the old root. By the time the lesions are distinct, the healthy, white root portion had grown a considerable distance from its origin.

Pathogenicity

Even though *T. basicola* is reported to cause geranium black rot [4], its pathogenicity under experimental conditions hasn't been demonstrated. This was the primary objective of J.H. Haas [1, 2]. I have subsequently confirmed some of his findings. Nearly all attempts to introduce *T. basicola* into healthy geraniums by several artificial inoculation methods failed to give root or foliar symptoms that occur on infected commercial plants. Haas [1] investigated the effects of soil temperatures (50 F, 61 F, 70 F, 81 F and 90 F; 10 C, 16 C, 21 C, 27 C and 32 C), soil moisture and other fungal infections (Pythium, Rhizoctonia and *Fusarium solani*) on *T. basicola* pathogenicity on geraniums. Although *T. basicola* was reisolated from artificially infested soils at the end of all his experiments, he observed typical *T. basicola* root rot symptoms on inoculated geraniums only once on a plant that also exhibited foliar symptoms, apparently due to some leaf chlorosis virus(es). In some of Haas' experiments, however, inoculated plants were significantly stunted, as compared to uninoculated control plants, even though they exhibited no black lesions on their roots. I have made similar inoculations on geranium seedlings (Nittany Lion) with several clones of *T. basicola* known to be very virulent on cotton, cowpea and Pinto bean [3]. Although no significant foliar symptoms occurred, I observed root parasitism on young, inoculated plants. In these experiments, pure conidial inoculum was poured around the plant base. Microscopic examination of inoculated roots, two weeks later, indicated no obvious black root lesions; however, abundant chlamydospore production was evident on white, apparently healthy roots. These experiments corroborate Haas' observations that no black roots occurred on inoculated roots, but they also shed additional light on his suggestion that *T. basicola* is pathogenic on geraniums without being parasitic. I observed parasitism of at least root epidermis and outer cortex but without much visible damage at the cellular level. Commercial plant roots infected with *T. basicola*, do show visible cellular damage and typical black lesions. I presume that some physiological changes, environmental stresses or concomitant infections by other pathogens might predispose geraniums to *T. basicola* infections resulting in more severe root and foliar symptoms.

Predisposing factors

Several environmental factors might influence the incidence and severity of root infections by *T. basicola*. Haas [1] considered several of these in his inoculation studies. He observed no increase in disease incidence when soil temperatures were varied, or when roots were inoculated simultaneously with other root-infecting fungi isolated from geranium. I observed no significant increase in root infection severity when inoculated plants were watered more frequently than normal. R.H. Lawson and I made dual inoculations on geranium seedlings (Nittany Lion) with tobacco ringspot virus (TRSV) and *T. basicola* both isolated from geranium. We observed no increase in root rot on plants inoculated with both fungus and virus. *T. basicola* root infections were present on plants inoculated with the fungus, but root symptoms still weren't comparable to those on infected, commercial plants.

Pathogenicity variations

T. basicola incidence on geraniums is often high when geraniums are grown in greenhouses after poinsettias, since the fungus occurs on both hosts and is carried over in the soil as chlamydospores. Haas [1] attempted inoculations with geraniums isolates of *T. basicola* on beans and poinsettias. He observed root rot symptoms on beans inoculated with all isolates, but only some geranium isolates were pathogenic to poinsettias. Geranium inoculations with isolates from tobacco and poinsettias, as well as isolates from geranium, were all negative. I have inoculated geraniums (Nittany Lion seedlings) with isolates of *T. basicola* from cotton, bean, soybean and three from geranium. Only one isolate from geranium consistently gave significant root infections, but without severe, dark lesions. That isolate also caused severe root symptoms on poinsettia, and was moderately virulent on sweet peas.

Inoculation vs. natural infection

When young geranium seedling roots (Nittany Lion), inoculated by pouring conidia around the plant base, were examined two weeks later, the root system appeared white and healthy. Chlamydospores were visible and obvious, however, at higher magnifications of the dissecting microscope (64x). These infections' severity increased with time, but never resulted in typical black root lesions, even though the fungus was obviously living and proliferating on the geranium roots.

In contrast, commercial rooted cuttings, observed at Beltsville one to two weeks after they were received from the propagator, were stunted with chlorotic leaves (or chlorotic spots on the leaves) (Fig. 27-3), and showed severe black root rot (Fig. 27-4 and 27-5). The most severely infected roots were those formed in the heavy soil rooting medium. Those plants apparently lacked an adequate root system, and as a result the plants appeared stunted and chlorotic. Other apparently healthy plants of the same group with symptomless leaves were, in most cases, also infected with *T. basicola*. Roots on cuttings from several geranium cultivars were carefully examined. Ten plants that showed some leaf chlorosis or viruslike mottle symptoms were selected from each cultivar and compared with 10 plants with symptomless foliage. The soil was washed from the plant's roots, and the roots were examined microscopically. *T. basicola* root rot severity was estimated visually. Nearly all plants, whether they showed chlorotic leaf symptoms or not, were infected with *T. basicola*.

In general, root rot severity was greater on plants that also exhibited some foliar symptoms. The purpose of this examination was to see if *T. basicola* root rot was present only on plants showing some foliar chlorosis. This would have suggested a positive association of root rot and virus infection. Since root rot was present on all plants, I concluded that virus infection, as indicated by foliar chlorosis, wasn't a necessary predisposing factor to *T. basicola* root rot.

It's possible, however, that all the plants were virus infected, but only some expressed symptoms. Further, the viruslike symptoms visible on some leaves may have been present and visible on the source plant when cuttings were taken.

Many plants showed only slight to moderate root infections of *T. basicola* when received from the propagator. The bulk of the root system wasn't infected. When those plants were repotted in uninfested soil mix, new root growth was so rapid the developing root system outgrew the fungus for a time. When the roots of such plants were examined two years later, however, the entire root system was infected. Chlamydospores and severe black lesions were obvious and abundant. New roots being formed were white, but covered with hyphae and chlamydospores. Presumably, those roots would soon develop severe lesions and would eventually become necrotic.

Thielaviopsis cutting rot

Haas [1] studied the more noticeable disease symptoms of black basal stem rot. When uprooted cuttings were planted in soil infested with *T. basicola*, rooting was prevented, delayed or reduced. He found that the base of a new cutting was susceptible for only one to two weeks and highly resistant or immune following this initial period. Resistance preceded corking by one to two weeks, but other fresh stem wounds were also susceptible to invasion by *T. basicola*. He concluded that the importance of *T. basicola* on commercial geranium production seemed to be in the cutting rot phase rather than the reported root disease. Further, because only one- to two-week-old cuttings were susceptible to cutting rot, he considered its economic significance minimal.

Discussion and conclusions

Losses from Thielaviopsis root rot on geranium are probably greater than most growers realize. Plants may be nearly symptomless if infections occur on established root systems; however, light infections become more severe with time, and eventually the whole root system is involved. Cuttings that become infected soon after rooting (especially if rooted in heavy, poorly aerated, infested soil) probably will show foliar symptoms indicating a weakened, less efficient root system. Roots on these cuttings exhibit typical black root rot lesions on the first developing roots and symptomless root infections on newly formed roots. It seems likely that *T. basicola* root infections in geraniums occur only on new roots, and that black lesions don't appear until sometime later. Further, this infection pattern development may occur only on virus-infected cuttings.

One must conclude that *T. basicola* is probably a facultative parasite on geraniums, and that it apparently requires some predisposing factor to aid its geranium root invasion and its inducting of typical root and foliar symptoms. The fungus isolates from geranium apparently may be virulent on other hosts like poinsettia, but isolates from other hosts are apparently not virulent on geranium.

A chemical factor(s) may be present in geranium tissue related to its increased resistance or immunity. These chemicals may build up with time and thus limit *T. basicola* infections to early rooting stages. Since the geranium plant produces roots quite profusely, new root production may easily outrun the fungus and sustain the plant. Further, *T. basicola* geranium isolates aren't virulent enough to impair the root-producing potential, even after the whole root system becomes infected.

To control this disease, use only cuttings rooted in pathogen-free soil and soilless media. Since *T. basicola* can survive for extended periods as chlamydospores, take care to use only steam-sterilized or pasteurized soil or rooting medium. Future research may show that losses during or soon after rooting may be eliminated if cuttings are taken only from virus-free plants (see Chapter 29, Clean Stock Production).

References

[1]   Haas, J.H. 1962. The pathogenicity of *Thielaviopsis basicola* (Berk. & Br.) Ferraris to *Pelargonium* x *hortorum* B. Ph. D. thesis. The Penn. State Univ.
[2]   Haas, J.H. and J. Tammen. 1962. Pathogenicity of *Thielaviopsis basicola* to *Pelargonium hortorum. Phytopathology* 52:12.
[3]   Linderman, R.G. and T.A. Toussoun. 1968. Pathogenesis of *Thielaviopsis basicola* in nonsterile soil. *Phytopathology* 58:1578-1583.
[4]   Meyer, F.W. 1956. Black root rot of geraniums. Grower's Supplement, *Connecticut Florists Association Bul.* No. 261.

Bacterial Fasciation

Robert S. Dickey and Gary W. Moorman

Bacterial fasciation, also called leafy gall, is caused by the bacterial pathogen *Rhodococcus fascians*. The disease and pathogen were first described in 1936 by Lacey and Tilford. A general discussion of the disease on geranium was prepared by Nelson and published in 1962. The disease occurs in the United States and has been reported to occur in many other countries. The pathogen's host range includes plants in 24 families [2, 5, 13]. Hosts include begonia, chrysanthemum, dahlia, forsythia, gladiolus, kalanchoe, lily, nasturtium, petunia, phlox, poinsettia, shasta daisy and sweet pea. Although bacterial fasciation's economic importance on geranium production generally is considered minor, potential losses due to this disease shouldn't be ignored.

Symptoms

Bacterial fasciation is recognizable by a number of short shoots or stems at or near soil level. The shoots or stems are short, swollen, fleshy, twisted or aborted and produce misshapen leaves (Fig. 27-6). When clusters of the thinner, aborted stems occur, they may resemble a witch's broom. The pathogen also causes axillary buds on the stem below the soil level to produce several very short, hypertrophied shoots that eventually resemble galls (Fig. 27-7). These gall-like masses, often called leafy galls, sometimes develop so that they are visible just above the soil surface.

Fig. 27-6. Typical bacterial fasciation symptoms, caused by Rhodococcus fascians, *occurring at soil level. Note short, aborted stems and misshapen leaves.*

The plants aren't killed and the main stem and roots appear to be unaffected, although the plant growth may be somewhat stunted. The older parts of the gall-like tissue sometimes decay and are replaced by new growth, but usually there's very little death of the fasciated tissues. The aborted shoots' and leaves' above-ground color is normal while the underground portions are pale green or yellow. In the Netherlands, it has been demonstrated indirectly that *R. fascians* may be involved in basal rot of *Pelargonium zonale* mother plants and cuttings [11]. The evidence suggests that abnormal tissue of fasciations below the soil surface are more easily invaded by fungi that cause basal rot.

Etiology

Rhodococcus fascians is the bacterial pathogen that causes fasciation disease. The bacterium was originally described as *Phytomonas fascians* [18], was later designated as *Bacterium fascians* [9], and most recently renamed *Rhodococcus fascians* [6]. The bacterium is a Gram positive, slightly curved rod (0.5 to 0.9 by 1.5 to 4.0 mm), that is nonmotile,

Fig. 27-7. Typical bacterial fasciation symptoms occurring below soil level. Note development of short, hypertrophied shoots and gall-like growths (leafy galls) from stem axillary buds.

nonacid-fast and produces yellow to yellow orange colonies on media containing yeast extract [19]. Strain specialization of the pathogen for particular hosts hasn't been demonstrated although variation in virulence and artificial inoculation has been noted [13, 16].

All evidence indicates that the pathogen doesn't require wounds to produce its effect; however, only bud tissue can be stimulated to produce the typical fasciated growth. *R. fascians* causes rapid meristematic tissue division and hypertrophied shoot formation, which consist of small-celled, actively dividing meristematic tissue and large parenchyma cells [8]. The bacteria are mainly confined to the plant tissue's exterior surface and may develop as masses on the outside of abnormal tissue. The bacteria occasionally enter epidermal cells and intercellular spaces and cause some epidermal and subepidermal cell necrosis.

Lacey [10] suggested that *R. fascians* changes the normal auxin content or its distribution within the plant, thus causing abnormal growth. The application of kinetin to nodes of pea, *Pisum sativum*, has produced the same effect as inoculating the nodes with *R. fascians* [17]. Infected geranium stem tissues contain increased cytokinin activity and decreased indoleacetic acid as compared with healthy tissues [3]. In addition, cytokinin is produced by *R. fascians* [1]. This suggests that geranium leafy gall development may result from an altered auxin-cytokinin ratio caused by the pathogen producing cytokinin and/or the pathogen's ability to stimulate the host tissue to produce cytokinin.

The bacteria can survive in the soil [4, 15] and serve as an inoculum source for infecting succeeding crops. Soil can become infested by the bacteria through pieces of infected host tissue or by growing infected plants. The pathogen can be spread by soil water. *R. fascians* can be transmitted by nasturtium, schizanthus and sweet pea seed, but this occurrence on geranium seed hasn't been reported.

Control

Steam pasteurize the soil and propagating medium prior to planting disease-free cuttings and plants. Discard all doubtful or suspicious plants. Carefully inspect dwarf and miniature cultivars [12]. Remove and discard all infected stock plants and the soil in which they are growing to prevent pathogen spread in infested soil.

References

[1]   Armstrong, D.J., E. Scarbrough, F. Skoog, D.L. Cole and N.J. Leonard. 1976. Cytokinins in *Corynebacterium fascians* cultures. Isolation and identification of 6-(4-hydroxy-3-methyl-cis-2-butenylamino)-2-methylthiopurine. *Plant Physiol.* 58:749-752.

[2]   Baker, K.F. 1950. Bacterial fasciation disease of ornamentals in California. *Plant Dis. Rep.* 34:121-126.

[3]   Balazs, E. and I. Sziraki. 1974. Altered levels of indoleacetic acid and cytokinin in geranium stems infected with *Corynebacterium fascians. Acta Phytopathol. Acad. Sci. Hung.* 9:287-292.

[4]   Chekunova, L.N. and M.A. Chumaevskaya. 1967. *K. biologii vox buditelya fastsiatsii Zemylaniki Corynebacterium fascians* (Tilf.) *Dow. Biol. Nauki* 10:104-107. (Abst. in *Rev. Appl. Mycol.* 47:160. 1968).

[5]   Fairvre-Amiot, A. 1967. *Quelques observations sur la presence de Coryne-bacterium fascians* (Tilford) Dowson *dans les cultures maraicheres et florales en France. Phytiatrie-Phyto pharmacie* 16:165-176.

[6]   Goodfellow, M. 1984. Reclassification of *Corynebacterium fascians* (Tilford) Dowson in the genus Rhodococcus as *Rhodococcus fascians* comb. nov. *System. Appl. Microbiol.* 5:225-229.

[7]   Lacey, M.S. 1936a. Studies in bacteriosis. XXII. 1. The isolation of a bacterium associated with "fasciation" of sweet peas, "cauliflower" straw-berry plants and "leafy gall" of various plants. *Ann. Appl. Biol.* 23:302-310.

[8]   ———— 1936b. Studies in bacteriosis. XXIII. Further studies on a bacterium causing fasciation of sweet peas. *Ann. Appl. Biol.* 23:743-751.

[9]   ———— 1939. Studies in bacteriosis. XXIV. Studies on a bacterium associ-ated with leafy galls, fasciations and "cauliflower" disease of various plants. Part III. Further isolations, inoculation experiments and cultural studies. *Ann. Appl. Biol.* 26:262-278.

[10] ———— 1948. Studies on *Bacterium fascians.* V. Further observations on the pathological reactions of *Bact. fascians. Ann. Appl. Biol.* 35:572-581.

[11] Maas Geesteranus, H.P., P.C. Koek and Th. G.B.M. Wegman. 1966. *Corynebacterium fascians* and *Botrytis cinerea* in *Pelargonium zonale.* An

aspect from the many factors causing the wilting of pelargonium. *Neth. J. Plant Pathol.* 72:285-298.

[12] McWhorter, F.P. 1964. Miniature geraniums introduced the fasciation disease of pelargonium into Oregon. *Plant Dis. Reptr.* 48:913.

[13] Miller, H.J., J.D. Janse, W. Kamerman and P.J. Muller. 1980. Recent observations on leafy gall in Liliaceae and some other families. *Neth. J. Plant Pathol.* 86:55-68.

[14] Nelson, P.E. 1962. Diseases of geranium. Bacterial fasciation. *NY State Flower Growers Bul.* 201:6-8.

[15] Oduro, K.A. 1975. Factors affecting epidemiology of bacterial fasciation of *Chrysanthemum maximum. Phytopathology* 65:719-721.

[16] Sule, S. 1976. Bacterial fasciation of *Pelargonium hortorum* in Hungary. *Acta Phytopathol. Acad. Sci. Hung.* 11:223-230.

[17] Thimann, K.V. and T. Sachs. 1966. The role of cytokinins in the "fasciation" disease caused by *Corynebacterium fascians. Amer. J. Bot.* 53:731-739.

[18] Tilford, P.E. 1936. Fasciation of sweet peas caused by *Phytomonas fascians* n. sp. *J. Agr. Res.* 53:383-394.

[19] Vidaver, A.K. 1980. Corynebacterium. Ed. N.W. Schaad. *Laboratory Guide for Identification of Plant Pathogen Bacteria.* St. Paul:American Phytopathological Society.

CHAPTER 28

Viral Diseases

Stephen T. Nameth and Scott T. Adkins

Although in recent years more geraniums have been grown from seed, most are still propagated vegetatively from cuttings [32]. Vegetative propagation makes plant virus transmission from stock plants to cuttings a serious problem because infected plants are often symptomless [31]. While estimating the geranium virus importance is difficult [32], a commercial grower could suffer devastating losses if cuttings taken from virus-infected, asymptomatic stock plants were later exposed to environmental conditions that induced symptom expression in young plants. Reduced rooting and flowering as well as cosmetic damage have been attributed to virus infection [32]. Stone [30] reported that most viruses occur in low concentrations in Pelargonium and are difficult to isolate and identify. Problems in virus isolation coupled with virus-induced symptom variability complicate Pelargonium viral disease diagnosis.

Much research has focused on geranium viruses in European geranium-producing countries. A survey of geraniums in Denmark found cucumber mosaic virus (CMV), Pelargonium flower break virus (PFBV), several, little-studied viruses and yellow net vein, a viruslike disease for which no causal agent has yet been determined. The researchers concluded that these viruses had no practical importance to Danish geranium growers [21]. Similar surveys in Belgian nurseries found that about 70% of the *P. zonale* plants tested were virus infected. The following virus incidents were recorded in infected plants: Pelargonium ring pattern virus (PFBV), 80%; cucumber mosaic virus (CMV), 10%; tobacco mosaic virus (TMV), 10%; and tobacco ringspot virus (TRSV), 1.2% [33]. A survey in the United Kingdom found nine different isometric viruses infecting geraniums including tomato ringspot virus (ToMRSV), TRSV, PFBV and Pelargonium line pattern virus (PLPV) [30]. In 1959 [14] researchers observed that "Many brief reports in the literature describe various virus Pelargonium diseases, but little work has gone beyond description." Remarkably little has changed in the ensuing 30 years. The stagnation in this area is probably due largely to the difficulties of working with geranium viruses and lack of new technologies applied to this research area.

This chapter reviews the current status of geranium viruses and discusses new geranium virus identification and characterization methods/techniques that have developed since the last review.

Viral or viral-like agents

Pelargonium ringspot (tomato and tobacco ringspot virus)

Once commonly referred to as "Pelargonium ringspot virus," it's now known that these are two distinct but related viruses that cause similar geranium symptoms. While more than a dozen viruses have been reported in geraniums, tomato ringspot virus (ToMRSV) and tobacco ringspot virus (TRSV) are frequently considered to be the only economically important ones. Both a Canadian and Swedish study found that ToMRSV caused chlorotic spots and large, chlorotic ringspots on older geranium leaves (Color Fig. C86 and C87) [13, 25]. TRSV induced chlorotic rings and line patterns when inoculated into geranium seedlings [1]. Although symptoms disappeared as plants aged, TRSV was still recoverable [1]. Older seedlings and flowering plants weren't susceptible to TRSV when mechanically inoculated.

Though ringspot symptoms are characteristic of ToMRSV and TRSV infection, accurate diagnosis remains difficult because the virus is maintained at low concentration in the infected plant. This makes detection via indicator hosts and serology questionable. Other, more nondescriptive symptoms associated with ToMRSV and TRSV infection include: deformed foliage [13] and increased floret bud abortion [26].

Pelargonium flower break (Pelargonium flower break virus)

Once thought to be of little economic importance, [21, 30, 33] in recent years Pelargonium flower break virus (PFBV) has become a major problem in the geranium virus arena. A recent study in Denmark cited frequent incidence of ToMRSV and PFBV in their geraniums [22]. In the United States PFBV incidence lacks documentation, but some of my research indicates that PFBV incidence is increasing. Of the 77 (1990) geranium samples (some had viruslike symptoms, most didn't) submitted for disease analysis to The Ohio State University Plant and Pest Diagnostic Clinic, 38 plants (49%) tested positive for PFBV. Symptoms associated with PFBV infection include: flower color breaking, mild foliar tissue mottling and, in some cases, retarded plant growth, with small, rugosed flowers [20] (Color Fig. C90). PFBV is known only to infect *P. zonale* [30].

Pelargonium line pattern (Pelargonium line pattern virus)

First described in 1976 [29] and later characterized [23], Pelargonium line pattern virus (PLPV) has relatively low economic importance. Foliage symptoms associated with this virus include: yellow spots, rings and line patterns, some running parallel to leaf veins (Fig. 28-1). Though early attempts to characterize this virus appeared successful [23], recent attempts to more fully characterize this disease-causing agent have been less successful [9, 2]. Difficulties in isolating a distinct virus particle from plants showing Pelargonium line pattern symptoms have led some to believe that the Pelargonium line pattern agent may be a new type of virus or that it may be defective virus particle [2].

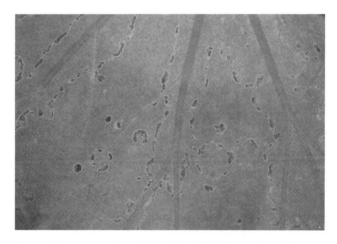

Fig. 28-1. Leaves from geranium infected with Pelargonium line pattern virus show symptoms like those seen above. Note the necrotic line patterns parallel to leaf veins.

Pelargonium leaf curl (Pelargonium leaf curl virus)

Symptoms associated with this disease are distinct from those associated with other geranium viruses. Pelargonium leaf curl or leaf crinkle [16] was first described in the United States in 1940 [11]. The causal agent, Pelargonium leaf curl virus (PLCV), is an isometric particle that is serologically distinct from other Pelargonium viruses [9]. As the name implies, the virus causes severe leaf curling or crinkling in affected plants (Fig. 28-2). This virus has little economic importance.

Fig. 28-2. Crinkle or leaf curl—yellow spots become brown and expanding leaves distort. (Photo from L.K. Jones.)

Pelargonium ring pattern (Pelargonium ring pattern virus)

Not to be confused with Pelargonium ringspot, Pelargonium ring pattern (PRPV) was first described in 1974 [28]. This virus induces trace ring patterns in *P. zonale* and is said to be the most common of the true Pelargonium viruses [28].

Pelargonium zonate spot (Pelargonium zonate spot virus)

Symptoms of this disease originally described in 1969 and 1979 [7] are concentric, chrome yellow bands on *P. zonale* foliage. The virus consists of quasi-isometric particles approximately 25 to 35 nanometers in diameter [6]. At this time there is no known vector and the disease is limited geographically to Southern Italy.

Leaf breaking and foliar mosaic (cucumber mosaic virus)

Cucumber mosaic virus (CMV) can cause severe cosmetic damage but usually isn't economically important [30, 31]. This virus was first reported in the United States in 1957 [16]. CMV induced severe mosaic, malformation, leaf breaking and stunting when inoculated into geranium seedlings [1]. As with other geranium viruses, CMV symptoms gradually became masked with age although the virus was still recoverable, and older seedling and flowering plants weren't susceptible to mechanical inoculation [1]. Older plants infected with CMV typically exhibit a mosaic pattern on leaves (Fig. 28-3 and Color Fig. C89).

Fig. 28-3. Leaf breaking and mosaic—disappearance of leaf zonation and irregular distribution of anthocyanin pigment without leaf distortion. (Photo from L.K. Jones.)

Yellow net vein (causal agent unknown)

No viral or viral-like agent has been found associated with this disease although it was originally described in 1961 [24] and later shown to be graft transmissible [12]. Symptoms associated with this disease are very striking (Color Fig. C88). Yellow net vein symptoms are most prevalent during short, cool winter days and tend to disappear during long, warm summer days. This disease has little economic importance.

Other viruses

Other viruses are known to infect geranium. The most notable are tobacco mosaic virus [20, 2], Pelargonium vein clearing virus [5] and tomato spotted wilt virus [4]. Tobacco rattle virus [20] has also been identified in geranium. These viruses have little or no economic importance and it's unclear what specific symptoms can be associated with their infections.

Virus diagnosis and identification

Prior to molecular biology-based techniques developed in the mid-1970s, much geranium virus research, particularly research associated with virus identification and characterization, was conducted using electron microscopy, indicator hosts and physical properties of the virus. Except for tobacco mosaic virus and tomato spotted wilt virus, most commonly encountered geranium viruses are isometric particles, about 30 nanometers in diameter [28]) (Fig. 28-4).

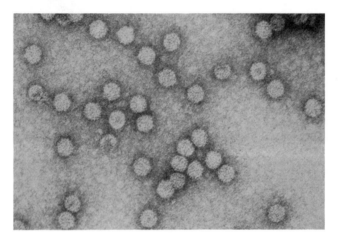

Fig. 28-4. Electronmicrograph of Pelargonium flower-break virus (magnified 90,500x). (Photo courtesy M.J. Anderson.)

This limits electron microscopy's usefulness in identifying these viruses. Coupled with this are similar symptoms on indicator hosts. In some cases, this has led to a possible misidentification of some geranium viruses. Within the last ten years, several researchers have applied polyclonal, antibody-based serology to geranium virus studies, most often using it to examine for relationships between various isometric viruses [23, 27, 30]. Serology is an extremely powerful technique, but its results are only meaningful if all viruses being used are correctly identified and characterized. With the advent of monoclonal antibody technology for detecting and characterizing plant viruses, newer, more highly sensitive and specific assays will be developed [8]. These assays, coupled with those based on nucleic acid homology [17], will dramatically increase geranium virus researchers' detection and characterization capabilities. Research is now underway using double-stranded ribonucleic acid analysis [3] and nucleic acid probes to identify and characterize geranium viruses (Fig. 28-5). These techniques will clarify the literature even further.

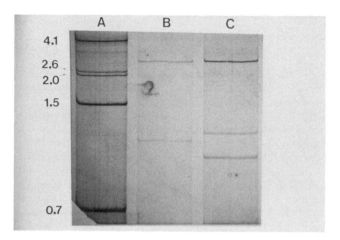

Fig. 28-5. Polyacrylamide gel showing double-stranded RNA profile of tobacco mosaic virus and cucumber mosaic virus (A); Pelargonium line pattern virus (B); and Pelargonium flower break virus (C). Numbers on left indicate approximate molecular weights.

Virus transmission

The primary virus transmission method in geraniums is through vegetative propagation via virus-infected stock material. Obviously if the stock plant from which cuttings are taken is virus infected, the subsequent cuttings will also be infected. Because of this, it's imperative that stock plants be subjected to virus-indexing (see Chapter 29, Clean Stock Production) to guarantee virus-free stock plants.

Tobacco (TRSV) and tomato ring spot virus (ToMRSV) belong to a virus group termed nepoviruses. One group property is that they are nematode transmitted. Both viruses are also seed transmitted in geraniums, though ToMRSV was the only one shown to be pollen transmitted [26].

Other viruses such as PFBV, PLPV and PLCV belong to a group of viruses that are readily transmitted by mechanical means, aren't insect transmitted and are quite commonly soil transmitted [15, 18], although not necessarily by soilborne organisms. Due to the easy mechanical transmission of these viruses, proper sanitation is vital during cutting or pruning of infected geraniums.

Virus control

Geranium virus control is primarily accomplished by producing virus-free stock via meristem culture [19]. This will be discussed in great detail in the following chapter.

References

[1] Abo El-Nil, E., A.C. Hildebrandt and R.F. Evert. 1976. Symptoms induced on virus-free geranium seedlings by tobacco ringspot and cucumber mosaic viruses. *Phyton.* 34:61-64.

[2] Adkins, S.T. 1991. "Improvement of geranium virus detection using viral-associated double-stranded RNA analysis and further characterization of Pelargonium line pattern virus." M.S. thesis, Ohio State Univ.

[3] —— and S.T. Nameth. 1989. Detection of virus in florist geranium using a simplified method of dsRNA analysis. *Phytopathology* 79:1197.

[4] Davis, R.F., V. DeHerrera, L. Gonzales and C. Sutula. 1990. Use of string-specific monoclonal and polyclonal antibodies to detect two serotypes of tomato spotted wilt virus. *Virus-Thrips-Plant Interactions of Tomato Spotted Wilt Virus.* Ed. H.T. Hsu. Proceedings of a USDA Workshop. USDA publication #ARS-87.

[5] DiFranco, A., M. Russo and G.P. Martelli. 1979. Isolation and some properties of pelargonium vein clearing virus. *Phytopathol. Mediterr.* 18:41.

[6] Gallitelli, D. 1982. Properties of a tomato isolate of Pelargonium zonate spot. *Ann. Appl. Biol.* 100:457.

[7] ——, A. Quacquarelli and G.P. Martelli. 1983. Pelargonium zonate spot virus. No. 272. *Commonwealth Mycological Institute / Association of Applied Biologists Descriptions of Plant Viruses.*

[8] Halk, E.L. and S.H. DeBoer. 1985. Monoclonal antibodies in plant disease research. *Annu. Rev. Phytopathol.* 23:231-350.

[9] Hollings, M. 1962. Studies of Pelargonium leaf curl virus. *Ann. Appl. Biol.* 50: 189-202.

[10] Hurtt, S.S. and R.L. Jordan. 1986. Partial characterization of an isometric virus from geranium. *Phytopathology* 76:1121.

[11] Jones, L.K. 1940. Leaf curl and mosaic geraniums. *Washington Agr. Exp. Sta. Bul.* 390: 3-19.

[12] Kemp, W.G. 1966. The occurrence of yellow-net virus in geraniums in Ontario. *Can. Plant Dis. Surv.* 46:81-82.

[13] —— 1969. Detection of tomato ringspot virus in Pelargonium in Ontario. *Can. Plant Dis. Surv.* 49:1-4.

[14] Kivilaan, A. and R.P. Scheffer. 1959. Detection, Prevalence and Significance of Latent Viruses in Pelargonium. *Phytopathology* 49:282-286.

[15] Martelli, G.P., D. Gallitelli and M. Russo. 1988. Tombusviruses. *The Plant Viruses-3.* Ed. R. Koenig. New York: Plenum.

[16] McWhorter, E.P. 1957. Virus diseases of geranium in the Pacific Northwest. *Plant Dis. Rep.* 41: 83-88.

[17] Miller, S.A. and R.R. Martin. 1988. Molecular diagnosis of plant disease. *Annu. Rev. Phytopathol.* 26:409-432.

[18] Morris, T.J. adn J.C. Carrington. 1988. Carnation mottle virus and viruses with similar properties. *The Plant Viruses-3.* Ed. R. Koenig. New York: Plenum.

[19] Oglevee-O'Donovan, W. 1986. Production of culture virus-indexed geraniums. *Tissue Culture as a Plant Protection for Horticultural Crops.* Ed. R.H. Zimmerman. Dordrecht: Martinus Nijhoff.

[20] Paludan, N. 1968. Virus diseases in Danish geranium cultures—Pelargonium leaf curl virus (PLCV). *Tidsskr. for Planteavl.* 72:211-216.

[21] ——— 1976. Virus diseases in *Pelargonium hortorum* specially concerning tomato ringspot virus. *Acta Hort.* 59:119-130.

[22] ——— and J. Begtrup. 1987. *Pelargonium Flower Break Virus and Tomato Ringspot Virus: Infection Trials, Symptomatology and Diagnosis.* Report no. 1895, Danish Research Service for Plant and Soil Science.

[23] Plese, N. and Z. Stefanac. 1980. Some properties of a distinctive isometric virus from Pelargonium. *Acta Hort.* 110:183-190.

[24] Reinert, R.A., A.C. Hildebrandt and G.E. Beck. 1963. Differentiation of viruses transmitted from *Pelargonium hortorum. Phytopathology* 53:1292-1298.

[25] Ryden, K. 1972. Pelargonium ringspot—a virus disease caused by tomato ringspot virus in Sweden. *Phytopath. Z.* 73:178-182.

[26] Scarborough, B.A. and S.H. Smith. 1977. Effects of tobacco and tomato ringspot viruses on The reproductive tissues of *Pelargonium* x *hortorum*. Phytopathology 67: 292-297.

[27] Stefanac, Z., N. Plese and M. Wrischer. 1982. Intracellular changes provoked by pelargonium line pattern virus. *Phytopath. Z.* 105:228-292.

[28] Stone, O.M. 1974. Some properties of two spirical viruses isolated from geraniums in Britain. *Acta Horticultura* 36: 113-119.

[29] ——— and M. Hollings. 1976. *Ann Rep. Glasshouse Crops Res. Inst.* for 1975: 119-120.

[30] ——— 1980. Nine viruses isolated from Pelargonium in the United Kingdom. *Acta Hort.* 110:177-182.

[31] Strider, D.L. 1985. Geranium. *Diseases of Floral Crops.* Ed. David L. Strider. New York: Praeger.

[32] Tayama, H.K., ed. 1988. *Tips on Growing Zonal Geraniums.* Ohio Cooperative Extension Service.

[33] Welvaert, W., G. Samyn and P. de Simpelaere. 1982. A virus survey on Pelargonium in Belgian nurseries. *Meded. Fac. Landbouwwet. Rijksuniv.* 47:1033-1038.

CHAPTER 29

Clean Stock Production

Culture-Indexing for Vascular Wilts and Viruses

Wendy Oglevee-O'Donovan

Vascular diseases are caused by pathogens that colonize a plant's water conductive tissues or vessels. These pathogens usually invade the plant through roots, stems or leaves and then move into the vessels, interfering with water and nutrient movement throughout the plant, eventually causing the plant to wilt. Major geranium vascular diseases are bacterial blight caused by *Xanthomonas campestris* pv. *pelargonii* and Verticillium wilt, caused by *Verticillium albo-atrum*. These pathogens often grow throughout the entire plant before the plant exhibits symptoms. For this reason, even with the most rigorous selection program based on observation, it's impossible to eliminate all infected cuttings. Under low light and cool temperatures, infected plants may not even be recognized.

Propagating infected cuttings carries pathogens to the propagation bench and then to the production areas where routine cultural practices further spread the pathogen. Again, this can and does take place without any obvious symptoms. In high light and warm temperature conditions, infected plant leaves begin to yellow. During the heat of the day, infected plants wilt but "recover" in the evening when the temperature is lower and the plant is under less stress. Eventually, infected plants wilt and never recover, resulting in plant death.

No chemical sprays or dips are available to cure or protect plants from vascular diseases, although some chemicals and soaps may slow down their spread. Because these diseases limit plant growth, growing plants with vascular diseases necessitates cultural practices to manage the disease and not the plant. This makes it impossible to fully utilize the crop's growth potential, adding time to the growing cycle and producing poor quality plants.

Vascular diseases are an economic problem directly due to plant death, and indirectly due to longer finishing time required and poorer plant quality. The only practical method for controlling vascular diseases in cutting geraniums is using pathogen-tested or "culture-indexed" propagating stock.

What is culture-indexing?

Culture-indexed geraniums implies an entire system designed to completely eliminate vascular pathogens. A specialist propagator of culture-indexed geraniums must have a zero tolerance level for *X. pelargonii* and *V. albo-atrum*. Culture-indexing geraniums is a total system of redundant testing for vascular pathogens coupled with intense selection and sanitation procedures. The pathway through this system is a unidirectional flow from tested material through controlled cutting buildup. All plants in the system must be renewed annually.

Geranium plant culturing is a laboratory technique to determine if a vascular pathogen is present in a cutting. A 2- to 3-inch (5- to 7½-cm) stem portion of the cutting to be tested is removed and surface sterilized. Under aseptic conditions paper thin cross-sectional slices are removed from within this entire stem section and placed into test tubes containing a clear, sterile liquid growth medium rich in sugar and vitamins.

If the stem section tested contains any of the pathogens in question, they are introduced into the medium where the bacteria and fungi grow quickly. After a two-week incubation period at room temperature, the tubes are observed for growth medium clarity. If any bacteria or fungi had been introduced into the culture tube, the medium becomes obviously "clouded." The cutting from which the stem section originated is then discarded without bacteria or fungi identification. Often times cuttings are discarded due to bacteria or fungi that aren't pathogenic; any indication of bacteria or fungi requires cutting be discarded immediately. This laboratory technique assumes that the pathogens are distributed in detectable quantities continuously throughout the plant's entire vascular system. This may not be true, and the pathogen may not be "sampled" with the chosen plant tissue [12].

For this reason, culture-indexing for geraniums *requires* that any cuttings testing negative for the first culturing are rooted and grown in isolation under warm conditions to encourage any systemic pathogen's growth. Cuttings are then removed from these first-generation cultured plants and the testing procedure repeated to yield second-generation cultured cuttings. These negative second-generation cuttings are rooted and grown in isolation under warm conditions to encourage any vascular pathogen multiplication. Then, for a third time, cuttings are taken and stem sections are tested for pathogens. Only negative, third-generation cuttings can enter the system and be labeled culture-indexed. This cleanup procedure takes at least one year. These culture-indexed cuttings are then rooted under strict sanitation as individual units in an isolation block called the "Nucleus Block."

Because it's economically impossible to culture index every plant for sale, cuttings can be taken from the Nucleus Block to increase the number of clean plants under strict sanitation in what is called the "Increase Block." Rooted cuttings from the Increase Block then serve as stock plants for production plantings, which provide cuttings for propagation and sale. The pathway from the Nucleus Block to the Production (or stock) Block is one way, and material never reverses through the system. All plant blocks must be renewed annually beginning in the Nucleus Block with new, culture-indexed geraniums. All phases of this system require regular inspection for any unusual symptoms or off-type plants. After introducing a three-generation cultured plant, only an additional one-generation culturing is required for its annual renewal.

Other methods for pathogen detection, such as ELISA and dot blot, are commercially available. Seong H. Kim of the Pennsylvania Department of Agriculture reported by personal communication that an indiscernible visual color change has been experienced from both the Agdia dot blot and AgriDiagnostics multiwell ELISA for *X. pelargonii* when the bacterial concentration was less than 10,000 cells per well or test. A bacterial concentration of less than 100 colony-forming units was difficult to recover on culture media; hence, the development of the three-generation culture-indexing system. These newer pathogen detection methods are much faster than culturing and are useful in rapid determination of suspect plants for commercial growers, if the infection is advanced. But due to the 100-fold decrease in test sensitivity, these methods haven't offered any advantages over the three-generation system for clean stock development. The techniques used in this system to produce clean plants are so rigorous and exacting that it's generally best that specialist propagators carry them out.

The major benefit of culture-indexed stock is its freedom from vascular wilt organisms previously described. The plants are also free of Pythium and root-rot pathogens. Because the Nucleus Block contains relatively few indexed plants, it's possible to implement an intensive program to select plants that produce superior quality and quantity, flowers and cuttings which then continue through the system. As a result, cuttings obtained from selected culture-indexed stock have increased growth potential, making a shorter production and finishing schedule possible. Using these selected culture-indexed cuttings with a proper sanitation program to produce high quality plants has eliminated losses due to vascular pathogens. (See "Sanitation" section). Please remember, the plants haven't been changed genetically and can be reinfected with vascular pathogens. Don't mix culture-indexed and non-culture-indexed plants and don't hold plants more than one year.

Further investigations in geraniums showed viruses were causing stunting, reduced vigor, fewer and small blooms with aborted florets, slower rooting and fewer breaks [6]. For this reason, culture indexing was further developed to eliminate economically limiting viruses.

Viruses and virus-indexing

Viruses are very small entities of ribonucleic acid (RNA) or deoxyribonucleic acid (DNA) genetic code material covered with a protective protein coat. These plant virus particles must be physically placed into a living plant cell for an infection to occur. This can be done by insects or nematodes feeding on infected plants and carrying these particles to healthy plants. Man, of course, working first with an infected plant and then a healthy one, spreads viral diseases throughout plants. Once in the plant cells, the viral particle uncoats and redirects the plant's energies into making products for viral replication instead of plant growth. This takeover of the plant's system causes nutritional imbalances, stunt, loss of vigor and other problems associated with viral infections. If a crop is selected for high performance in the presence of viruses, fewer obvious viral symptoms will be obvious in the crop. Such was the case with the selected culture-indexed geraniums when virus work was initiated.

Because selection based on symptoms isn't reliable, it's necessary to test for viruses. Because viruses can't be grown in the laboratory on culture media, culture-indexing won't detect them. Another testing method must be used. One "virus-indexing" method used extensively is the bioassay, which utilizes living plants. To virus-index geraniums using a bioassay, first make a test inoculum by grinding young leaves, florets or root tips of a suspect geranium in a mortar and pestle with a buffer solution. Lightly rub leaves of indicator plants, plants sensitive to geranium viruses, with this inoculum. If virus particles are present in the geranium plant part tested and, therefore, in the inoculum, the indicator plants become infected, evidenced by visual symptoms such as brown or yellow local lesions or spots on inoculated leaves, new growth dieback or even plant death.

This bioassay is designed to detect a general range of viruses. A more sensitive test at a different time of year is required for specific, known, distinctive viruses. This more sensitive, specific test is called ELISA. ELISA utilizes antibodies to specific geranium viruses to detect their presence in the form of a color reaction. Using these indexing methods, virus-infected plants can be detected and eliminated from a plant population. Often an entire variety or cultivar can be infected, however, and the plants must be put through an intensive program designed to eliminate viruses from a plant. This culture-virus-index program is an

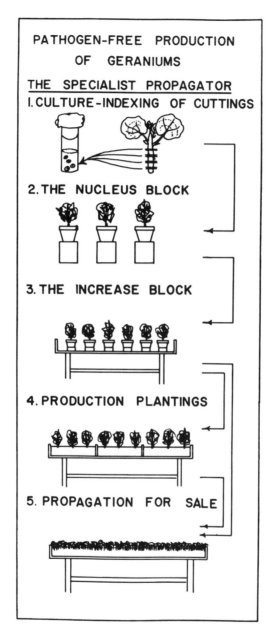

Fig. 29-1. This diagram illustrates the steps in culture-indexing geranium cuttings, establishing an increase block from culture-indexed cuttings and producing cuttings for sale.

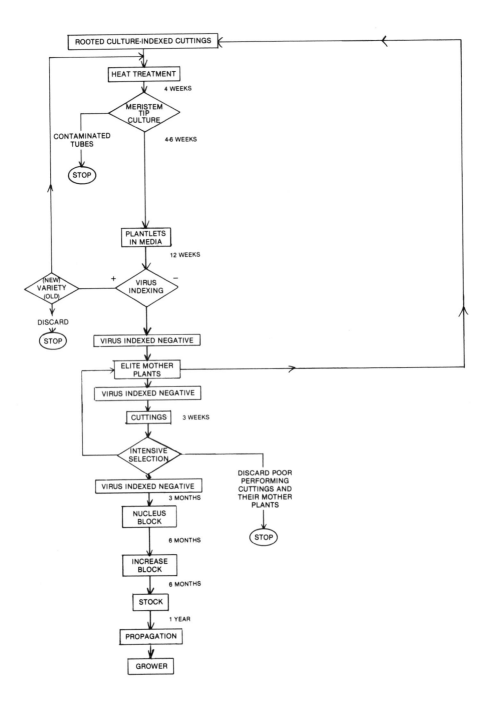

Fig. 29-2. Flow chart for production of CVI geraniums from CI geraniums.

extension of the original culture-index program previously described. Six-week-old, rooted, culture-indexed cuttings are placed in chambers where the temperature is gradually raised to 100 F (38 C) during the 16-hour light period and 95 F (35 C) during the 8-hour dark period. Plants are held in this environment for four weeks. Then, 0.5 mm growing tips or "meristem-tips" are removed from these plants under aseptic conditions and placed in test tubes of sterile medium that supports their growth and growth of the previously described vascular pathogens.

If any bacteria or fungi grow on the medium, the tube and its meristem are immediately discarded. After four to six weeks, the resulting plants are transferred to pasteurized soil as isolated units. These heat-treated, meristem-tip cultured plants, which at this point can only be stated as being culture-indexed for vascular pathogens, must be tested to determine if the viruses have been eliminated. The indexing should use the bioassay, ELISA methods.

The system used to produce virus-indexed plants is a three-step procedure: heat treatment, meristem-tip culture and virus-indexing, and must begin with a culture-indexed cutting. To heat treat and meristem-tip culture without virus-indexing is academic and worthless. The heat treatment and meristem-tip culture can help eliminate viruses and can produce more plants that test negative for the viruses. Virus-indexing to eliminate those plants still carrying virus is the key to the entire system. The resulting culture-virus-indexed plants are only as good as the systems used for indexing bacteria, fungi and viruses. Plants that have tested negative for vascular pathogens and for known economic viruses will then make up an Elite Nucleus Block. These culture-virus-indexed plants will provide cuttings for the Nucleus Block and from this point on, the selection program and build-up system is the same as previously described for culture-indexed plants.

Using culture-virus-indexed plants

Selected culture-virus indexed (CVI) geraniums are very vigorous growing plants. This vigor offers tremendous performance potential that can be harnessed by the grower and turned into profit. This means that growing methods that could never be used on nonculture-indexed material can now be used. In order to capitalize on the added performance potential offered by culture-virus-indexed geraniums, a systems approach to growing, using modern horticultural technology, must be implemented. In general, by taking advantage of the added growth potential offered in the CVI geranium, a June stock planting will give a harvest potential of 70 to 100 cuttings per square foot (6.5 to 9.3 per square m) of bench area, and a September planting could easily offer a harvest potential of 50 to 70 cuttings per square foot (4.6 to 6.5 per square meter) of bench area. The finisher will note additional advantages offered by the vigorous, selected culture-virus-indexed geraniums. The crop finishes with more uniformity and color and with less loss than the nonculture-indexed geraniums.

Because CVI geraniums are self-branching, a finished, 4-inch (10-cm) crop should never need pinching if grown properly. This removes two to four weeks from the finishing schedule. The most important factor to note when growing CVI geraniums is that the crop must be kept growing at all times. Never withhold water or fertilizer to limit the plant's growth; this damages the plant. Once damaged, the crop never recovers the original performance potential.

The ultimate consumer also realizes the benefits of clean geraniums in their outdoor performance. CVI geraniums flower more freely with larger blooms that last longer in summer heat. Depending on the cultivar selected, virus indexing alone has shown an increase in geranium blooms from 5% to 200%! Now, that is significant! After a rain the plants regain their blooms and provide a colorful show much more quickly than before the virus-indexing process.

Through the combined technology of science and horticulture, the floriculture industry and the consumer can enjoy the benefits of vigorous showy, high performance culture-virus-indexed geraniums.

Sanitation

Selected culture-virus-indexed geraniums have tremendous growth potential, but are just as susceptible to vascular, root rot and viral diseases as before the indexings. If any of these diseases are introduced during production, the plants become infected, and all the advantages gained by using culture-virus-indexed plants are lost. Therefore, never mix indexed stock with nonindexed stock and don't hold plants over from year to year. The best practice is to grow only indexed stock using proper sanitation and to completely replace plants annually.

Implement the following rules in a sanitation program for all culture- and culture-virus-indexed plants, not just geraniums.

Avoid using knives in taking cuttings. When you must use knives, keep several soaking in a disinfectant; after using one, exchange it for one that was soaking. Alternate knives in this way frequently; in Mother Blocks, use a fresh knife for every plant.

All soil, pots, benches, watering systems, wheel barrows, shovels, hoes—everything that comes into contact with the plants directly or indirectly—must be clean. Use steam sterilization as priority treatment. On items that can't be steamed, use a disinfectant such as bleach or hospital disinfectant for soaking. When possible, steam soil directly in containers. Avoid handling soil unnecessarily, and don't splash infested soil or expose treated soil to blowing dust. Steam benches or beds after each crop is removed even if it was in pots or flats.

Don't dip any plant material into any solution for any reason. One infected cutting will leave pathogens in the solution to be spread to all dipped cuttings. Dust hormone powders onto cuttings.

Discard and dispose of any suspect material. Keep employees from going to dump sites and then returning to the greenhouses.

Keep propagation in a separate area away from crop production and commercial areas. Place flats onto timber treated with copper naphthenate (never use creosote) for hardening plants outdoors. Never place flats directly onto the ground.

Spray for weeds around the greenhouse, under benches and in aisles; weeds can harbor diseases and insects. Always hang the hose nozzle on a hook near the bench.

If a hose nozzle is dropped on the ground, soak the hose in a disinfectant before using. Don't drag hoses on the ground and then across the bench.

Always wash hands between blocks of plants, benches and cultivars. Wear clean coats or coveralls when entering an isolated area, and remove and leave them in the area when finished.

To avoid introducing viruses into any crop, never use tobacco products in the greenhouse. After using tobacco products, wash hands before handling plant material.

In general, use clean stock in treated soil and containers and use sanitation measures to keep them clean. Knowing that diseases are spread by infected plants and soil, always think, "Could this action introduce a pathogen?" before you act. Don't just fight diseases, eliminate them.

References

[1]   Baker, K.F. et al. 1957. The U.C. system for producing healthy container grown plants. *Univ. Calif. Agr. Ext. Serv. Manual* 23.

[2]   Daughtrey, M. and J. Tammen. 1988. A new geranium threat from an old enemy. *The Tenn. Flower Growers Newsletter* Vol. 2, No. 5. The Univ. of Tennessee Institute of Agriculture.

[3]   Dimock, A.W. 1962. Obtaining pathogen-free stock by cultured cutting techniques. *Phytopathology* 52:1239-1241.

[4]   Hollings, M. 1965. Disease control through virus-free stock. *Ann. Rev. Phytopath.* 3:367-96.

[5]   Horst, R.K., S.H. Smith, H.T. Horst and W.A. Oglevee. 1976. In vitro regeneration of shoot and root growth from meristematic tip of *Pelargonium* x *hortorum* Bailey. *Acta. Hort.* 59:131-142.

[6]   Murdock, D.J., P.E. Nelson and S.H. Smith. 1976. A histopathological examination of Pelargonium infected with tomato ringspot virus. *Phytopathology* 66: 844-850.

[7]   Nelson, P.E., J. Tammen and R.R. Baker. 1960. Control of vascular wilt diseases of carnation by means of culture-indexing. *Phytopathology* 50: 356-360.

[8]   Oglevee, W.A., S.H. Smith and R.K. Horst. 1975. Isolation of tobacco ringspot virus from symptomless geraniums of the cv. Didden's Improved Picardy. *Proc. Amer. Phytopath. Soc.* 2:103. (abstr).

[9]   Smith, S.H. and W.A. Oglevee-O'Donovan. 1977. Meristem-tip culture from virus-infected plant material and commercial applications. *Plant Cell and Tissue Culture.* Ed. W.R. Sharp et al. Columbus: Ohio State Univ. Press.

[10] Strider, D.L., R.K. Jones and R.A. Haygood. 1981. Southern bacterial wilt of geranium caused by *Pseudomonas solanacearum. Plant Disease* 65:52-53.

[11] Tammen, J. 1960. Disease-free geranium from cultured cuttings. *Penn. Flower Growers Bul.* 117:6-9.

[12] Tammen, J. 1961. The production of geraniums free of bacterial stem rot and leaf spot. *Geraniums.* Ed. John W. Mastalerz. The Pennsylvania Flower Growers.

[13] Walsh, D.M., R.K. Horst and S.H. Smith. 1974. Factors influencing symptom expression and recovery of tobacco ring spot virus from geranium. *Phytopathology* 64:588. (abstr).

Controlling Viral Diseases

R. Kenneth Horst and
Michael J. Klopmeyer

Intensive cultural practices used in vegetative geranium production today are conducive to virus multiplication and spread. Most geranium cuttings on the market are obtained from the grower's own stock plants (self-propagators) or from propagators specializing in disease-free certification and delivery of rooted or unrooted forms. Since many geranium viruses are nonsymptomatic during most of the year, it's difficult to prevent further virus spread during normal cultural procedures by depending on symptoms alone.

Unlike fungal diseases, no chemicals or treatments are available to rid the plant of a viral pathogen. Thus, it's important to receive and maintain virus-free propagating material. There are approximately nine to 13 different viruses known to infect geraniums (see Chapter 28, Viral Diseases). Since geraniums are affected by many viruses, sensitive detection and eradication methods are important to establish virus-free propagating material. This chapter reviews methods available to detect viruses (virus-indexing), eradicate viruses from infected plants and maintain virus-free plant material in a greenhouse environment.

Virus-indexing

Virus-indexing or testing is an important element in a geranium certification program. Indexing can accurately verify the cleanliness of a given variety and determine if one or more viruses are present. Unfortunately, in many cases, one single testing method may not assure virus-free plants. Plant sampling methods, seasonality and testing method sensitivity and specificity all affect virus-indexing results.

Two major methods are currently used to detect geranium viruses: serology and bioassays (biological indicator plant inoculation and plant grafting). Alternate methods could include using sensitive nucleic probes specific for certain viruses [26] and visualizing the replicative form of the viral RNA (ribonucleic acid) isolated from infected plants [50].

Two common serological tests used to detect geranium viruses are enzyme-linked immunosorbent assay (ELISA) [43] and immunosorbent electron microscopy (ISEM) [35]. ELISA has proved a reliable and relatively inexpensive detection method for geranium virus detection. It can detect viruses in the host plant (geranium) with little loss in sensitivity and reliability compared to bioassays (indicator plants).

ELISA is accomplished by administering purified virus, i.e., tobacco ringspot virus (TRSV) [29] to rabbits to induce the formation of antibodies to TRSV. Six weeks later the antibodies are isolated from the rabbit's blood serum and prepared for the ELISA test [43]. ELISA is performed in commercially available, 96-well, polystyrene plastic plates. First, viral antibodies are added to wells in the plate to coat well surfaces. Next, a leaf extract or leaf disks are incubated in the plate well. This is followed by adding more antibodies (antibody conjugate) attached to the enzyme alkaline phosphatase; antibodies also bind to viral particles in the well, forming a double antibody sandwich. Finally, the colorless solution p-nitrophenylphosphate, which turns yellow when digested by the enzyme, is added. A yellow color indicates virus in the plant extract or disk, and the more intense the color, the more virus is present.

ELISA has been used to successfully and efficiently test for TRSV [30, 44], tomato ringspot virus (ToMRSV) [39, 38, 2], Pelargonium ring pattern virus (PRPV) [1] and Pelargonium flowerbreak virus (PFBV) (M. Klopmeyer, personal observation). ELISA successfully detected TRSV in crude leaf extracts and geranium seed extracts; it indicated ToMRSV in both crude leaf extracts and intact leaf disks. ELISA appears to be more sensitive and reliable in detecting TRSV than bioassay.

ISEM is a valuable detection technique if an electron microscope (EM) is available. Geranium plant sap is placed on an electron microscope grid and treated with the appropriate virus antiserum. If the plant virus is present in the extract, antibodies surround it. When this grid is viewed through an EM, the viral particles appear decorated and are easily seen. ISEM has detected ToMRSV, PFBV, PRPV and Pelargonium line pattern virus (PLPV), [35]. Since antisera are usually very specific, little or no cross reactivity is visible.

Bioassays are accomplished in three ways: by mechanically inoculating an indicator plant with sap extracted from the test plant; transversely cutting a tightly rolled geranium leaf and rubbing the cut end over the indicator plant leaf surface [35]; or by grafting a reliable indicator variety onto the test plant [21].

Virus symptoms may develop on the indicator plant leaves anywhere from three to 21 days following inoculation. Typically, two major symptoms are observed. The first, a local lesion response, is typified by small, necrotic or chlorotic lesions (1 to 2 mm in diameter) on inoculated leaves. This local lesion response may occur soon after inoculation. The second, a systemic response, is typified by mosaic or vein clearing symptoms on newly emerging leaves. This systemic response usually appears seven to 21 days after inoculation.

Geranium viruses can infect many plant species from several plant families. Plants found to exhibit symptoms when infected with crinkle virus were *Chenopodium amaranticolor, Antirrhinum majus* and *Nicotiana clevelandii* [12, 16]. Reinert et al. [41] tested a large range of host plant species for geranium virus symptoms. Chlorotic spot, mosaic and crinkle viruses produced local lesion reactions on *Cucumis sativus, Nicotiana tabacum, Salpiglossis sinuata, Pisum sativum, Phaseolus vulgaris* and *C. amaranticolor*. The nine viruses reported by Stone [48] to affect geranium can all be cultured in *Nicotiana clevelandii* and bioassayed in *Chenopodium quinoa*. These viral symptoms on *C. quinoa* are described in Table 29-1 (N. Paludan, personal communication). All can also be identified by ELISA in *N. clevelandii* sap, but rarely in geranium sap since virus concentration is low. *Chenopodium quinoa* is the most reliable indicator plant for most viruses infecting geranium (Fig. 29-3). This has been demonstrated for ToMRSV and PFBV [35], PRPV, cucumber mosaic virus (CMV) [51] and tobacco mosaic virus TMV, and a nonidentified isometric virus (PLPV) [37]. The lettuce strain of tomato spotted wilt virus (TSWV-L) has been reported to infect ivy geraniums (*P. peltatum*) [6]. This virus can be detected by ELISA or by inoculation onto *Nicotiana benthamiana*. Horst [21] also tested various Pelargonium species as possible indicators. These wild species were grafted to geranium varieties known to be infected with specific geranium viruses and some exhibited symptoms after grafting.

In some instances the presence of viruslike agents in a geranium may be beneficial [3]. Viruslike agents produced attractive streaks in ivy geraniums petals (Pelargonium petal streak agent) and in zonal gera-

Fig. 29-3 and 29-4. Left, *severe local lesion response on an inoculated leaf of indicator plant,* Chenopodium quinoa *(16 days after inoculation).* Right, *systemic response on* Chenopodium quinoa *12 days after inoculation with a purified isolate of tomato ringspot virus (T₀MRSV).*

nium foliage (Pelargonium net vein agent). Outside of these unique instances where a virus(es) may provide desirable phenotypic traits, it's still important to develop and maintain virus-free geranium stock. Thus, reliable virus-indexing procedures are essential.

Techniques for eliminating viruses

The virus-indexing procedures described above may reveal complete viral infection in many cultivars. Several techniques are used to eliminate viruses if visual selection of virus-free clones isn't possible. These include: (1) exploiting erratic virus distribution within plants, (2) chemotherapy, (3) heat therapy and (4) meristem tip culture [14]. Sometimes a combination may achieve greater success.

Some viruses are found erratically located within plants and the advantages of this phenomenon are obvious. Shoot tips from rapidly growing stems may be propagated and found to be virus-free. Holmes [17] has reported using this procedure in freeing chrysanthemums of aspermy virus. This technique has also been used by Hollings [13] to free geraniums of crinkle virus. Symptoms were found to be masked under greenhouse environmental conditions between March and September,

TABLE 29-1

Symptoms of nine known viruses that infect geranium after *Chenopodium quinoa* inoculation

Virus	Abbreviation	Symptoms
Pelargonium leaf curl	PLCV	Small necrotic white spots (pinhead size) five days after inoculation
Cucumber mosaic virus	CMV	Orange-colored local lesions after 5 to 8 days
Tomato ringspot virus	ToMRSV	Necrotic local lesions after
Tobacco ringspot virus	TRSV	4 to 6 days, followed by
Tomato black ring virus	TBRV	systemic, apical necrosis (especially for ToMRSV)
Pelargonium flowerbreak virus	PFBV	Yellow orange spots on inoculated leaves 15 to 20 days after inoculation; occasionally a systemic reaction
Pelargonium line pattern virus	PLPV	Yellow local lesions on inoculated leaves after 10 to 12 days
Pelargonium ring pattern virus	PRPV	Yellow local lesions on inoculated leaves after 10 to 12 days
Tobacco mosaic virus	TMV	Necrotic local lesions within 4 to 6 days

so small cuttings taken during this period were found to be free of crinkle virus in many instances.

Actual chemical cure of virus-infected plants hasn't been established as a reliable means of controlling plant viral diseases. Some success with chemical treatment of aspermy-affected chrysanthemums [22] and mosaic-affected carnations [45] has been reported. Incorporating antiviral substances into meristem tip culture media may yield rewarding results [25, 46]. Amantadine incorporated into a tissue culture medium for chrysanthemums resulted in chrysanthemum plantlets free of chrysanthemum stunt viroid (CSV) [18]. About 10% of these plants were CSV-free while no plants transplanted from nonsupplemented amantadine medium were found CSV-free. These reports haven't yet been confirmed. In some instances treated plants again show symptoms after treatment is stopped. Griffiths and Slack [8] reported that ribavirin, a potent viricide, eliminated potato viruses *X*, *S* and *M* from in vitro-treated plantlets with or without heat therapy. Using antiviral agents to free a whole plant of virus is difficult, presumably because virus metabolism is so closely integrated with plant metabolism. There have been no reports of attempts to control geranium virus diseases by using chemicals.

Heat therapy has been widely used to free plant materials of viruses. Two treatments have been used—hot water and hot air. Hot water treatments have been used most successfully in treating dormant tissues and depends on plant tissue withstanding higher temperatures than the virus [14]. Temperatures may range from 95 F to 130 F (35 C to 54 C). Hot air treatment has been more widely used and has generally given better results in both plant material survival and virus elimination. Hot air treatment's success depends on the viruses' inability to multiply and move readily within plants being exposed to air temperatures of 95 F to 104 F (35 C to 40 C). Successful treatment times may vary from one to 12 months depending on the plant cultivar and the specified virus. This technique has been successfully used with chrysanthemums and carnations [14].

Ornamental plant species may vary somewhat in their heat tolerance [9, 13, 47] and also in the ease with which they can be freed of virus [14]. In addition, all viruses haven't been eliminated by heat therapy. Hollings [11] and Paludan [32] failed to free chrysanthemums infected with chrysanthemum stunt viroid (CSV) and chrysanthemums infected with chrysanthemum chlorotic mottle viroid (ChCMV) [32]. Plants were freed of CSV by using very small meristematic tips. Mori and Hosokawa [28] achieved similar results using very small meristematic tips to eliminate virus from ornamental plants including carnation, chrysanthemum, dahlia, lily and iris. Low temperature therapy kept chrysanthemums free of CSV or ChCMV [33, 31]. Prolonged low temperature treatments at 40 F (5 C) combined with meristem tip culture yielded inactivation percentages ranging from 50% to 70%.

Heat treating geraniums infected with crinkle virus has been successful [13, 16]. Small shoots propagated from plants grown at 98.6 F (37 C) for four weeks were crinkle-free. Heat therapy has also been used to successfully free plants of TRSV and ToMRSV by Horst, et al. [19, 20] and PFBV and ToMRSV [35]. Geraniums, in general, aren't heat tolerant. Plants succumb after five to six weeks at 98.6 F (37 C). To overcome this problem plants were placed in chambers with 16-hour photoperiod and temperatures of 75 F (24 C) day and 70 F (21 C) night. The day and night temperatures were raised approximately 6 degrees F (3 degrees C) each 24 hours until reaching a final setting of 100 F (38 C) and 95 F (35 C) night. Heat therapy was considered initiated when final temperatures were reached. Plants were maintained at these conditions for four weeks after which meristem tips were removed for tissue culture.

Fluctuating temperatures successfully freed some plants of viruses and encouraged plant survival. Temperatures of 104 F (40 C) for four hours alternating with 60 F to 68 F (16 C to 20 C) for 20 hours daily for as long as eight weeks were used on potatoes [10]. Virus inactivation was complete after six weeks of treatment. An alternate heat treatment regimen effective in eliminating geranium viruses was daytime temperatures of 93 F (34 C) and nighttime temperatures of 68 F (20 C) for two months [34]. It's possible that fluctuating temperatures would assure greater success in virus elimination from geraniums that aren't heat tolerant. Plants must also be hardened somewhat prior to exposure to heat therapy [7]. Watering should be minimal, fertilization shouldn't encourage succulent growth and plastic pots shouldn't be used since they restrict soil aeration.

Meristem tip culture is extremely useful in virus elimination [14]. Meristematic tissues are known to be relatively virus-free. Differentiated leaves are excised from around the terminal growing point to expose this tissue dome. The meristematic tip includes not only the meristem dome but also the leaf primordia (Fig. 29-5). An excision is made just below the leaf primordia attachment and this piece of tissue is placed on a specific culture medium in a test tube. Survival and growth is greatly improved if the tissue taken includes slightly expanded leaf primordia. Morel and Martin [27] demonstrated that this technique could free ornamentals of viruses.

Successfully eliminating viruses by meristem tip culture depends on viruses' inability to invade or to persist in terminal meristems. The theory that the meristem is virus-free isn't always true; however, virus concentration diminishes in terminal meristematic tissue [23]. Hollings and Stone [15] consistently recovered virus from carnation meristem tips but were also able to obtain some that were virus-free. This procedure has been used to establish virus-free chrysanthemum, carnation and geranium plants [14, 19, 20]. Geraniums have been cultured successfully from meristem tips in several laboratories [4, 36, 19, 20, 42]. Standardized media is available for obtaining shoot and root regeneration from geranium meristem tips [19, 20, 42].

Although virus removal can sometimes be obtained by meristem tip culture alone, combining procedures may be better. Meristem tip culture from heat-treated plants has been used successfully [15, 40, 49, 19, 20]. Paludan [34] reported that combining heat treatment and meristem tip culture was effective in eliminating PFBV in 100% treated plantlets while ToMRSV was eliminated in 92% treated plantlets. Meristem tip culture alone eliminated PFBV in 75% treated plantlets while ToMRSV was eliminated in only 24% treated plantlets. Plants free of TRSV and ToMRSV significantly increase in commercial value because cuttings root faster, produce more flower buds, flower earlier and are more vigorous.

Maintaining virus-free plants

Plant material that has been freed of viruses is not immune. Therefore, good organization and plant health management are required to prevent reinfection. Many viruses are transmitted by insects, and some viruses such as TRSV and ToMRSV are transmitted by nematodes. Thus, insect and nematode control are extremely important. This is accomplished by adequate spray programs and soil sterilization. PFBV spreads easily among geranium plants by infected sap and contaminated knives [35]. Handling plants must be held to a minimum since viruses may also spread by handling first a virus-infected plant and then a clean plant. Good sanitation is essential since plant debris, unsterilized soil, contaminated knives and hands may be contamination sources. Therefore, the best method for geranium virus control and plant maintenance in a virus-free condition is to (1) burn virus-infested plants, (2) harvest only from nonsymptomatic plants, (3) buy virus-indexed plants from reputable sources and (4) use a program that includes sanitation and effective insect and nematode control.

Fig. 29-5. A, B and C. Geranium meristem tip culture. Left top, a carnation meristem tip showing apical leaf (see arrow) and two leaf initials (primordia). Left bottom, a geranium meristem tip showing two leaf initials surrounding hidden apical dome. Above, a young geranium plantlet four weeks after meristem tipping.

Conclusion

Many geranium varieties are probably infected with one or more viruses. Although these viruses may not cause significant economic losses, they will impact horticultural quality for the finisher and improved production for the propagator. Virus eradication studies [36, 19, 20, 5] indicate possible improvement in foliage, inflorescence and earliness to flower. Other characteristics that virus-free plants may improve include rooting, yield and vigor (Fig. 29-6). In the future the geranium market will become more competitive among specialist propagators. Product line differentiation won't be based only on color and pot performance, but also on whether plants are pathogen-free. Growers and specialist propagators must recognize the importance of disease-free, virus-free plants and take precautions to prevent the introduction and/or spread of these pathogens.

Fig. 29-6. Eight-week-old Wendy Ann geraniums (from an unrooted cutting) in a 4¾ -inch (12-cm) pot. The smaller plant is infected with Pelargonium flower break virus (PFBV) while the larger plant was heat treated, meristem tipped and tested free of PFBV by ELISA. Note the larger and more numerous inflorescences on the PFBV-free plant.

References

[1]   Balesdent, M.H., J. Albouy, J.P. Narcy, P. Robin, M. Tronchet and M. Lemattre. 1989. Enzyme-linked immunosorbent assay for the selection and certification of pathogen tested pelargonium. *Acta Hort.* 246:291-294.

[2] Bitterlin, M.W. and D. Gonsalves. 1988. Serological grouping of tomato ringspot virus isolates: implications for diagnosis and cross-protection. *Phytopathology* 78:278-285.

[3] Cassells, A.C. 1986. Pelargonium peltatum (ivy-leaf Pelargonium) Harlequin type: The use of beneficial infective agents and implications for genetic engineering of plants by non-integrating vectors. *Acta Hort.* 182:229-236.

[4] Chen, H.R. and A.W. Galston. 1967. Growth and development of Pelargonium pith cells in vitro. II. Initiation of organized development. *Physiol. Plantarum.* 20:533-539.

[5] Christensen, O.V. and N. Paludan. 1978. Growth and flowering of pelargonium infected with tomato ringspot virus. *J. Hort. Science* 53:209-213.

[6] Davis. R.F., V. DeHerrera, L. Gonzales and C. Sutula. 1990. Use of strain-specific monoclonal and polyclonal antibodies to detect two serotypes of tomato spotted wilt virus. "Virus-Thrips-Plant Interactions of Tomato Spotted Wilt Virus," proceedings of a USDA Workshop. *USDA-ARS.* 153-165.

[7] Fulton, J.P. 1954. Heat treatments of virus-infected strawberry plants. *Plant Dis. Reptr.* 38:147-149.

[8] Griffiths, H.M. and S.A. Slack. 1988. Potato virus elimination by heat and ribavarin. *Phytopathology* 78:838 (abstr).

[9] Hakkaart, F.A. and F. Quak. 1964. Effect of heat treatment of young plants on freeing chrysanthemums from virus B by means of meristem culture. *Neth. J. Plant Pathol.* 70:154-157.

[10] Hamid, A. and S.B. Locke. 1961. Heat inactivation of leaf roll virus in potato tuber tissues. *Amer. Potato J.* 38:304-310.

[11] Hollings, M. 1961. Virology. *Glasshouse Crops Res. Inst. Rept.* 72-78.

[12] ——— 1962a. Studies of Pelargonium leaf curl virus. I. Host range, transmission and properties in vitro. *Ann. Appl. Biol.* 50:189-202.

[13] ——— 1962b. Heat treatment in the production of virus-free ornamental plants. *Natl. Agr. Advis. Serv. Quart. Rev.* 57:31-34.

[14] ——— 1965. Disease control through virus-free stock. *Ann. Rev. Phytopathol.* 3:367-396.

[15] Hollings, M. and O.M. Stone. 1964. Investigations of carnation viruses. I. Carnation mottle. *Ann. Appl. Biol.* 53:103-118.

[16] ——— 1965. Studies of Pelargonium leaf curl virus. II. Relationships to tomato bushy stunt and other viruses. *Ann. Appl. Biol.* 56:87-98.

[17] Holmes, F.O. 1956. Elimination of aspermy virus from the nightingale chrysanthemum. *Phytopathology* 46:599-600.

[18] Horst, R.K. and D. Cohen. 1980. Amantadine supplemented tissue culture medium: A method for obtaining chrysanthemums free of chrysanthemum stunt viroid. *Acta Hort.* 110:315-319.

[19] Horst, R.K., S.H. Smith, H.T. Horst and W.A. Oglevee. 1976. In vitro regeneration of shoot and root growth from meristematic tips of *Pelargonium* x *hortorum* Bailey. *Acta Hort.* 59:131-142.

[20] Horst, H.T., R.K. Horst, S.H. Smith and W.A. Oglevee. 1977. A virus-indexing tissue culture system for geraniums. *Florists Rev.* 160:28, 29, 72-74.

[21] Horst, R.K. 1982. Controlling virus diseases. *Geraniums.* 3d ed. Ed. J.W. Mastalerz and E.J. Holcomb: University Park, Penn.: Pennsylvania Flower Growers.

[22] Howles, R. 1957. Attempts in the chemotherapy of virus-infected glasshouse plants. *Plant Pathol.* 6:46-48.

[23] Kassanis, B. 1957. The use of tissue culture to produce virus-free clones from infected potato varieties. *Ann. Appl. Biol.* 45:422-427.

[24] Kivilaan, A. and R.P. Scheffer. 1959. Detection, prevalence and significance of latent viruses in Pelargonium. *Phytopathology* 49:282-286.

[25] Loebenstein, G. and A.F. Ross. 1963. An extractable agent, induced in uninfected tissues by localized virus infections, that interferes with infection by tobacco mosaic virus. *Virology* 20:507-517.

[26] Miller, S.A. and R.R. Martin. 1988. Molecular diagnosis of plant disease. *Ann. Rev. Phytopathol.* 26:409-432.

[27] Morel, G. and C. Martin, 1952. *Guerison de dahlias atteints d'une maladie a virus. Compt. Rend. Acad. Sci. Paris* 235:1324-1325.

[28] Mori, K. and D. Hosokawa. 1977. Localization of viruses in apical meristem and production of virus-free plants by means of meristem and tissue culture. *Acta Hort.* 78:389-396.

[29] Newhart, S.R., C.P. Romaine and R. Craig. 1980. A rapid method for virus-indexing the florist's geranium. *HortScience* 15:811-813.

[30] —— 1982. Enzyme-linked immunosorbent assay for the detection of tobacco ringspot virus in *Pelargonium* x *hortorum*. *Penn. Agr. Exp. Sta. J. Series* No. 1200.

[31] Paduch-Chical, E. and S. Kryczynski. 1987. A low temperature therapy and meristem tip culture for eliminating four viroids from infected plants. *J. Phytopathology* 118:341-346.

[32] Paludan, N. 1980. Chrysanthemum stunt and chlorotic mottle. Establishment of healthy chrysanthemum plants and storage at low temperature of chrysanthemum, carnation, campanula and pelargonium in tubes. *Acta Hort.* 110:303-313.

[33] —— 1985. Inactivation of viroids in chrysanthemums by low-temperature treatment and meristem-tip culture. *Acta Hort.* 164:181-186.

[34] —— 1991. Elimination of viruses from Pelargonium. *COST 87 Pelargonium Micropropagation and Pathogen Elimination.* Ed. Applegren et al.

[35] Paludan, N. and J. Begtrup. 1987. Pelargonium flowerbreak virus and tomato ringspot virus: Infection trials, symptomatology and diagnosis. *Dan. J. Plant and Soil Science.* 91:183-193.

[36] Pillai, S.K. and A.C. Hildebrandt. 1968. Geranium plants differentiated in vitro from stem tip and callus cultures. *Plant Dis. Reptr.* 52:600-601.

[37] Plese, N. and Z. Stefanac. 1980. Some properties of a distinctive isometric virus from pelargonium. *Acta Hort.* 110:183-190.

[38] Powell, C.A. 1984. Comparison of enzyme-linked immunosorbent assay procedures for detection of tomato ringspot virus in woody and herbaceous hosts. *Plant Disease* 68:908-909.

[39] —— and M.A. Derr. 1983. An enzyme-linked immunosorbent blocking assay for comparing closely related virus isolates. *Phytopathology* 73: 660-664.

[40] Quak, F. 1957. Meristem culture, *gecombineered met warmtebehandeling, vor het verkrijgen van virus-vrije anjerplanten. Tijdschr. Plantenziekten* 63:13-14.

[41] Reinert, R.A., A.C. Hildebrandt and G.E. Beck. 1963. Differentiation of viruses transmitted from Pelargonium hortorum. *Phytopathology* 53:1292-1298.

[42] Reuther, G. 1983. Propagation of disease-free pelargonium cultivars by tissue culture. *Acta Hort.* 131:311-319.

[43] Romaine, C.P., S.R. Newhart and D. Anzola. 1981. Enzyme-linked immunosorbent assay for plant viruses in intact leaf tissue disks. *Phytopathology* 71:308-312.

[44] Romaine, C.P. and S.R. Newhart. 1982. ELISA: A rapid method for virus-indexing geraniums. *Pennsylvania Flower Growers Bul.* No. 337:1-3.

[45] Rumley, G.E. and W.D. Thomas. 1951. The inactivation of the carnation mosaic virus. *Phytopathology* 41:301-303.

[46] Sela, I., I. Harpaz and Y. Birk. 1964. Separation of a highly active antiviral factor from virus-infected plants. *Virology* 22:446-451.

[47] Stone, O.M. 1963. Factors affecting the growth of carnation plants from shoot apices. *Ann. Appl. Biol.* 52:199-209.

[48] ———— 1980. Nine viruses isolated from pelargonium in the United Kingdom. *Acta Hort.* 110:177-182.

[49] Thompson, A.D. 1956. Heat treatment and tissue culture as a means of freeing potatoes from virus Y. *Nature* 177:709.

[50] Valverde, R.A., S.T. Nameth and R.L. Jordan. 1990. Analysis of double stranded RNA for plant virus diagnosis. *Plant Disease* 74:255-258.

[51] Welvaert, W. and G. Samyn. 1985. Relative importance of pelargonium viruses in cutting nurseries. *Acta Hort.* 164:341-346.

CHAPTER 30

Host Plant Resistance

Richard A. Grazzini and Ralph O. Mumma

Host plant resistance is the natural way in which plants—the hosts that provide food and homes—defend themselves against pests and pathogens. Without host plant resistance, no plant would survive in nature for very long. It's obvious, however, that plants do survive and even thrive in most areas. Plants are able to defend themselves against insects and other arthropod pests of all sizes, as well as pathogenic organisms (fungi, bacteria, viruses and viruslike particles). Host plant resistance is active at all growth stages; in the seed, as the seed germinates, as the seedling grows and all subsequent stages. We take host plant resistance so much for granted that it's only when host plant resistance breaks down—and the pest or pathogen is able to successfully attack the plant—that we notice that host plant resistance is missing.

Plants are rich resources for pests, providing proteins, fats, carbohydrates and minerals. Storage structures, such as seeds, are favorite targets for many pests and pathogens. But in nature, seeds are also a very effective means of propagation. To be effective, the seed must have some mechanisms to keep pests from eating it, or eating all the seeds that a plant may produce. Some seeds have hard seed coats that pathogenic fungi can't penetrate. Some seeds taste bad, so many pests won't eat them. Some seeds are poisonous, so pests that eat them die. These mechanisms—hard seed coats, bad taste and toxic compounds—represent only a fraction of the many mechanisms by which plants resist attack by pests and pathogens.

How do plants resist pests and pathogens?

Plants resist pests and pathogens in many ways. A plant's outer surface may have structures that interfere with normal pest predation or alter surface ecology to limit fungal pathogen growth (Fig. 30-1). Internally, a plant may produce structures or biochemicals that limit a pest's ability to feed, or a pathogen's ability to infect and grow. For a detailed description, refer to the following books and review articles: [15, 8, 16].

Fig. 30-1. Magnified freeze fracture cross section ($^{10}/_{16}$ inch = 10 mm) of a rose petal infected with Botrytis cinerea shows a spore that has penetrated the cuticle and formed subsurface mycelium. (Photo courtesy Phil Hammer.)

Morphological mechanisms

Many host plant resistance mechanisms investigated involve particular morphological or anatomical structures that somehow interfere with a pest's normal activities. For instance, many Pelargonium species are covered with tiny hairs (trichomes) that impede small pests' movement (Color Fig. C68). These pests are less effective in feeding on plants with pubescent surfaces and avoid these plants if given a choice. If there is no choice—if the pest is provided only with a densely pubescent plant on which to feed—the pest feeds less, grows more slowly and is generally less effective in mating and reproducing. If the trichomes are removed from a pubescent plant, the pest feeds, grows and reproduces normally.

Many plant species' surfaces are covered with waxy coatings (cuticle) that can decrease susceptibility to pest and pathogen attack. These coatings form a mechanical barrier through which a pest or pathogen can't penetrate. In some cases, the waxy layer keeps the surface from becoming wet, and many pathogenic fungi can't grow under such conditions. There are many examples of pest responses to variations in the cuticular wax thickness and composition [35].

In some species, glandular trichomes secrete many substances that interfere with a pest's ability to survive or feed or reproduce. These mechanisms are generally considered to be biochemical in action.

Biochemical mechanisms

Plants produce both active and passive biochemical barriers to pests and pathogens. A passive defense is one in which the plant produces the substance regardless of an attacking pest's presence. Alternatively, an active defense is one in which the biochemical barrier is produced as a direct response to pest attack. A passive defensive compound can also be actively produced in higher quantities when the plant is under attack.

Biochemical resistance mechanisms may or may not be associated with a specific anatomical or morphological structure. It is, however, common for glandular trichomes to be associated with external exudates and for internal phytotoxic or cytotoxic biochemicals to be isolated within intracellular vacuoles.

Deterrents are mechanisms whereby a plant interferes with a pest or pathogen's normal activity. Similar to the mechanical barriers above, the secretion of sticky substances provides a very effective pest deterrent: the pest contacts the sticky substance and becomes entrapped. This entrapment need not be complete to be effective. If the sticky substance simply slows down a pest's activity, it will feed less, grow more slowly, mate less frequently and lay fewer eggs. Furthermore, a partially entrapped pest is more likely to succumb to predation. Over a few generations, pest populations increase much more slowly on those plants having sticky secretions. A wild potato species (*Solanum berthaultli*) appears to take this process one step further [23]. In this species, one type of glandular trichome produces a sticky substance, while a second produces an enzyme that causes this sticky substance to polymerize and become hard. Both trichomes are required for effective resistance; the net effect is that a pest becomes coated with the sticky secretion from one trichome and then becomes entrapped and immobilized on the sticky substance left by the enzyme action from the second trichome.

Many plants produce compounds that are apparently distasteful to pests. The plant may also contain chemicals that decrease the plant food quality when ingested by a pest. Thus, the resistant plant limits pest growth by producing chemicals that interfere with the pest's digestive metabolism, or that alter food usability. This mechanism works by decreasing the pest's survival and reproductive fitness.

A resistance mechanism that may be effective against one pest may actually attract a different pest species. Many members of the squash family, Cucurbitaceae, produce a bitter substance known as cucurbitacin, which provides an effective defense against insect pests such as the squash bug. The same substances, however, are strong attractants to cucumber beetles [3]. If a breeder were to select for high resistance against a pest repelled by cucurbitacins, the selection pressure would be *for* susceptibility to cucumber beetles. This is clearly not an effective strategy in areas where cucumber beetles are major pests.

Many plants produce compounds toxic to pests. Some pests, however, have developed an ability to consume plants that may be toxic to many other mite or insect species; they have developed mechanisms to detoxify toxins that may be lethal to closely related species.

Other plant-produced compounds repel insect pests before the pest attacks the plant. Many of these compounds belong to a class of biochemicals called terpenes. The characteristic citrus odor of lemon rind results from a terpene called limonene. Similarly, geraniol is a terpene with a sweet, geraniumlike odor. As with cucurbitacins, aromatic terpenes attract some pests while repelling others. Japanese beetles are strongly attracted by the geraniol scent. Commercial Japanese beetle traps are actually "baited" with geraniol.

Plant-pathogen defenses often appear to be much more active than those associated with arthropod pests. When a growing fungal hypha pushes into a cell, the cell reacts by producing a chemical (phytoalexin) that interferes with the fungus' ability to grow. Sometimes this response is so severe that the plant cell actually dies, producing an area of dead cells wherever the pathogen attacked. This suicidal hypersensitive response can actually be an effective resistance mechanism because it effectively limits fungus spread.

Pathogen resistance phenomena often display a continuous gradient from total immunity to marginal tolerance. A plant can be so completely immune to a pathogen that the pathogen won't grow or survive on the immune plant. A resistant plant may permit the pathogen spore to germinate and even penetrate the plant, but subsequent infection symptoms don't occur. The hypersensitivity reaction noted above is an extreme example. Tolerance is a phenomenon where the pathogen effectively infects the host plant, but the host doesn't die or suffer significantly. Manipulating tolerance is difficult; an infected but tolerant host plant may remain apparently healthy until the plant is stressed. Under severe water or nutrient stress, tolerant plants may succumb and exhibit disease symptoms.

Breeding resistant geraniums

Geraniums resist a wide variety of pests and pathogens; however, detailed investigations of resistance have been limited. One of the best investigated mechanisms is that with which geraniums resist small arthropod pests, such as aphids and mites.

Mite and small insect resistance

The surface of a zonal geranium (*P.* x *hortorum*) is covered with trichomes. Some of these trichomes are spiny and make it difficult for small pests to move. Other trichomes secrete a sticky substance on their surface and effectively trap small pests. Penn State researchers have intensively investigated geranium trichomes for the last 25 years.

In 1964, Richard Craig noticed that certain breeding lines were selectively attacked by spider mites. Upon examining the pedigrees of the plants involved, he realized that all susceptible lines were related, indicating the strong possibility of a genetic basis for mite-susceptibility. Chang et al. [4] examined leaf anatomy and morphology of resistant and susceptible genotypes and determined that leaves of resistant lines had more glandular trichomes and a thicker epidermis and leaf cuticle.

Stark [27] determined that there are four trichome types: short and tall spines and short and tall glandular hairs. The exudate on the tall glandular trichomes from resistant lines was sticky. Trichomes from susceptible lines had an exudate that was waxy, hard and not sticky.

Winner [34] used leaf disk bioassays to examine true-breeding resistant and susceptible inbreds, F_1 hybrids and F_2 and backcross generations. His observations of more than 9,000 leaf disks indicate that two-spotted spider mite resistance is genetically controlled and appears to be conditioned by a pair of complementary dominant epistatic loci.

Gerhold et al. [10] developed a simple method to collect the trichome exudate, determined by bioassay that the exudate was indeed active and determined that the active compound was a chemical closely related to aspirin: anacardic acid. He determined that the exudate from resistant inbreds consisted of two closely related anacardic acids.

Walters and Grossman examined the geranium leaf surface by scanning electron microscopy (Fig. 30-1) and determined the only consistent morphological difference between resistant and susceptible lines was in tall glandular trichome density. Resistant lines had high trichome densities, while susceptible lines had low trichome densities [30].

Walters et al. [33] examined both susceptible and resistant genotypes and determined that there was a simple chemical difference in the exudate composition. Tall glandular trichome exudate from resistant inbreds consisted mostly of *unsaturated* anacardic acids, while that from susceptible genotypes is mostly *saturated* (Fig. 30-2). A simple analogy is to consider liquid and solid cooking oils. Liquid oils (like corn oil) are mostly unsaturated, while solid cooking fats (such as lard or Crisco) are mostly saturated. Similarly, resistant trichome exudate is unsaturated, liquid and sticky. Susceptible trichome exudate is saturated, solid, nonadhesive—and ineffective as a sticky trap. It's interesting to note that the only chemical difference between the saturated and unsaturated forms, and therefore between resistant and susceptible genotypes, is the presence or absence of a single double bond in the acyl chain of the anacardic acid molecule.

Walters et al. [31] demonstrated that the anacardic acid resistance mechanism was effective against foxglove aphid as well as two-spotted spider mites and proposed that the sticky trap mechanism would be effective against pests small enough to be entrapped or substantially hindered by the sticky exudate. Subsequent investigations indicate that the exudate from resistant plants also interferes with the activity of enzymes proposed to be involved with arthropod reproduction [11] and with the feeding activities of large lepidopteran caterpillars [E. Yerger, personal communication].

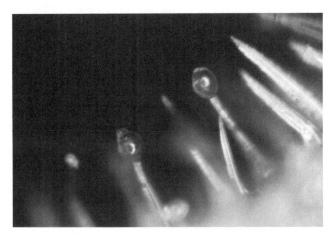

Fig. 30-2. Trichome exudate.

Walters et al. [32] also determined that resistant inbreds consistently produced resistant progeny, while susceptible inbreds produced susceptible progeny. Moreover, the trichome exudate composition of resistant inbreds' progeny was similar to that of the resistant parent, and the composition of the exudate from susceptible progeny closely resembled that of the susceptible parent. These observations indicated that the inbred lines were true breeding. When true-breeding resistant and susceptible inbreds were hybridized, the F_1 was observed to be resistant by bioassay and to have a biochemical composition similar to that of the resistant parent, indicating a strong genetic dominance.

Grazzini et al. [12] determined that inheriting the resistance phenomenon's biochemical basis was relatively simple. A single, dominant gene controlled the presence or absence of unsaturated anacardic acids in the trichome exudate and, therefore, determined whether or not a plant was resistant. Two epistatic loci were responsible for the relative composition of various anacardic acids in the exudate. A small number of codominant genes appeared to control tall glandular trichome density.

The interaction of trichome density and chemical composition was also examined. Chemical composition appeared to be significantly more important than trichome density. F_2 progeny with a tall glandular trichome density like that of the resistant parent (high) but an exudate chemistry like that of the susceptible parent (saturated) were susceptible. Similarly, F_2 progeny with a susceptible-like trichome density (low) and a resistant-like exudate chemistry (unsaturated) were clearly resistant.

We have begun to examine the evolutionary origins of this glandular trichome-mediated small pest resistance. The zonal geranium (P. x hortorum) is believed to have originated from the interspecific hybridization of P. inquinans and P. zonale and possibly other species from the Pelargonium section Ciconium (a taxonomic section is a group of closely related species). We recently examined the anacardic acid composition of common members of the section Ciconium. These observations lead us to propose that P. inquinans contributed both tall glandular trichomes and uniquely unsaturated anacardic acids to P. x hortorum. P. zonale has neither tall glandular trichomes nor anacardic acids. Other anacardic acid-containing species don't have the uniquely unsaturated anacardic acids of P. x hortorum and P. inquinans, suggesting that the capacity to produce anacardic acids may have evolved before the unique desaturation phenomenon. In addition, the capacity to produce anacardic acids appears to be tightly associated with tall glandular trichomes development. All Pelargonium species that produce anacardic acids do so only in tissues that possess tall glandular trichomes.

These detailed genetic and biochemical studies indicate that small pest resistance in geraniums is under relatively simple genetic control. Incorporating a screening technique for such resistance into a geranium breeding program can be readily accomplished. The important factor is the chemical composition of the exudate from one of the two types of glandular trichomes on the surface of P. x hortorum.

Differential degrees of two-spotted spider mite resistance were also observed among ivy geranium (P. peltatum) cultivars by Potter and Anderson [25]. There was no apparent relationship between mite resistance and the densities of either simple or glandular trichomes; however, the glandular trichomes they observed appear in the published scanning electron micrographs to be similar to those classified as *small* glandular trichomes by Walters et al. [31]. Such a correspondence would support the conclusion of Walters et al. that there is no relationship between small glandular trichomes and mite resistance, and the observations of Laughner [19] who noted that the leaves of the P. peltatum cultivar Salmon Queen (rated as relatively mite-resistant by Potter and Anderson) possessed only short glandular trichomes. The clear presence of a mite-resistance phenomenon in ivy geraniums suggests multiple resistance mechanisms in operation, since we have observed neither tall glandular trichomes nor anacardic acids in preliminary investigations of *P. peltatum*.

It's also possible to effectively incorporate mite resistance into ivy geraniums. Laughner [19] produced primary and secondary hybrids between the zonal and ivy geranium and backcrossed the hybrids to the zonal parents. She observed the presence of tall glandular trichomes on leaves and stems of the segregating backcross progeny. Recent observations of some of these backcross progeny between *P. x hortorum* and *P. peltatum* indicate that both tall glandular trichomes and unsaturated anacardic acid biosynthesis can effectively be transferred from zonals to ivies [D. Hesk, personal communication].

Whitefly resistance in Regal Pelargonium

Simple observations of whitefly populations on our small, pest-resistant geranium breeding lines indicates that the whitefly resistance mechanism may be different from the anacardic acid resistance mechanism noted above. Generally, both mite- and aphid-resistant and susceptible breeding lines are also whitefly resistant. In contrast, most zonal geranium (*P. x hortorum*) cultivars and one of the proposed parental species (*P. inquinans*) appear to be whitefly resistant. The other proposed primary parent species (*P. zonale*), however, appears to be whitefly susceptible, as well as mite and aphid susceptible.

In the Penn State greenhouses, there appears to be a slight variation for whitefly susceptibility in Richard Craig's Regal breeding populations. With routine pesticide applications, we see no differences. If we extend the intervals between pesticide applications, however, we can observe differences among breeding lines. Some lines are much more attractive to whiteflies than others. At this point, we can't distinguish between nonpreference and resistance. A nonpreference phenomenon—that is, one line is less attractive to whiteflies than another—may break down under either high pest pressure or a monoculture. If the only plant available to the whitefly is a nonattractive type, the pest may learn to feed on the unattractive plant. We don't consider nonpreference to be a very effective resistance mechanism, at least in intensively cultivated greenhouse crops. The possible appearance of apparent whitefly resistance in the Regal breeding populations, however, leads us to propose that whitefly resistance is indeed possible in *P. x domesticum*, and that it will be possible to incorporate whitefly resistance into commercial Regal breeding programs in a reasonable time frame.

Diseases

Bacterial blight resistance

In the 1950s, bacterial blight, incited by *Xanthomonas pelargonii*, was the most important disease affecting commercial geranium production in the United States. The devastating impact of this pathogen led to the development of culture-indexed geraniums [22, 28] and to the research that ultimately resulted in seed-produced geraniums [6].

Both Hellmers [17] and Munnecke [21] observed degrees of resistance among several species and cultivars of Pelargonium to the bacterial blight organism; however, the scope of their research was limited. In 1967, Knauss and Tammen studied the relative resistance of 38 species and cultivars to bacterial blight under standardized pathological conditions. They didn't observe any substantial resistance among cultivars of *P.* x *hortorum*; however, several species, including cultivars of *P.* x *domesticum* were significantly more resistant than the control *P.* x *hortorum* cultivar Better Times. Wainwright and Nelson [29] studied the histopathology of Pelargonium infected with bacterial blight and suggested that the apparent resistance of *P.* x *domesticum* may be related to tanninlike phenolic substances.

Shifeng Pan and Lowell Ewart of Michigan State [24] successfully hybridized *P.* x *hortorum* and *P. grandiflorum* (a proposed parent of *P.* x *domesticum*) to produce a Xanthomonas-resistant, interspecific hybrid. Whether or not it will be possible to transfer this resistance into a fertile zonal background isn't yet known. Such a transfer would be required to produce Xanthomonas resistance in both vegetative and seed-produced zonals. *P.* x *domesticum* is known to asymptomatically carry the Xanthomonas organism. For the "resistant" interspecific hybrid to be useful in breeding vegetatively propagated zonals, the resistance mechanism would need to produce complete immunity. Such immunity could produce "naturally clean stock," at least so far as Xanthomonas is concerned.

If the interspecific hybrid is fertile and can be used in a backcrossing program to *P.* x *domesticum*, it may also prove to be useful in transferring the whitefly resistance of *P.* x *hortorum* into *P.* x *domesticum*.

Botrytis resistance

There appears to be variation for Botrytis susceptibility in *P.* x *hortorum*. Metzler and Craig at Penn State [unpublished data] have observed differential resistance to *Botrytis cinerea* among inbred lines and their F_1 hybrids. Plants selected under the high humidity/high temperature stresses of the breeding program of the late Griffith Buck (Iowa State) are reported to have flowers that are somewhat Botrytis-tolerant. Furthermore, resistance to Botrytis flower blight may not indicate Botrytis resistance in leaves or stems. Hausbeck [unpublished data] indicated that differences in susceptibility are evident among zonal geranium cultivars; however, a high resistance level wasn't detected.

The slight degrees of Botrytis tolerance that exist in *P. x hortorum* may indicate that the resistance mechanism is under the control of multiple minor genes. Alternatively, observations of strong Botrytis resistance in other Pelargonium species may indicate that Botrytis resistance was not transferred into *P. x hortorum* during the interspecific hybridization process. In this case, the potential exists for hybridization of a strongly Botrytis-resistant species with *P. x hortorum* and recovering fertile Botrytis-resistant progeny to be used in a breeding program with the zonal parent.

Geranium rust resistance

The zonal geranium (*P. x hortorum*) is highly susceptible to geranium rust (*Puccinia pelargonii-zonalis* Doidge). The ivy geranium (*P. peltatum*) is immune to rust [2, 1, 20, 13]. The mechanism of rust resistance in *P. peltatum* may involve cuticular ridges surrounding the stomata on the leaf surface [19]. Puccinia spore germination occurs in droplets of free water on the leaf surface. The germ tube elongates until it "recognizes" a stoma and then expands into a matlike appressorium over the stomatal opening. An infection hypha arises from the appressorium and penetrates the stoma. The cuticular ridges around the leaf stoma of a resistant ivy geranium cultivar may deflect the germ tube from the stoma, thus effectively inhibiting infection.

The breeding program of Richard Craig (Penn State) effectively transferred this rust-resistance mechanism from the ivy geranium to the zonal geranium. Linda Laughner [19] hybridized *P. x hortorum* with *P. peltatum* in an attempt to transfer rust resistance from the ivy into the zonal. Despite very low fertilities, 22 rust-immune, interspecific progeny were recovered. Rust-immune, interspecific hybrids were backcrossed to zonal parents. From these backcrosses, progeny with immunity and varying degrees of rust resistance were recovered.

It's interesting that only rust-susceptible progeny had high densities of simple and glandular trichomes. Progeny scored by Laughner [19] as rust-resistant or rust-immune had few, if any, trichomes. At least three possible explanations for these observations exist. Trichomes may hold water droplets near the leaf surface, enhancing the rust spore's ability to germinate. Trichome exudate may also contain compounds that stimulate geranium rust spore germination [9]. The production of trichomes by the epidermis—or trichome exudate—may occur at the expense of cuticular wax formation. If the patterned deposition of cuticular wax is strongly involved in rust resistance, the development of trichomes and/or trichome exudate may minimize cuticular wax formation, thus rendering trichome-laden plants rust-susceptible.

Pythium resistance

The mechanism and inheritance of Pythium resistance hasn't been studied. Cline and Neely [5], working with vegetatively propagated cultivars, and Hausbeck et al. [14], examining seed-propagated hybrid cultivars, report cultivar differences in *Pythium ultimum* susceptibility, indicating that the phenomena may have a genetic basis.

Virus resistance

Many researchers have investigated the effects of viruses on geraniums; however, host plant resistance studies are virtually nonexistent. DeArmond [7] attempted to study resistance to cucumber mosaic virus in geranium. She postulated that if she could control the vector (green peach aphid, *Myzus persicae*) by using aphid resistant plants, the virus would be excluded and the plants would be functionally resistant. Although the plants were resistant to aphid feeding, the aphids still probed the plant and thus were able to transmit the virus.

Scarborough and Smith [26] studied the effects of tobacco ringspot (TomRSV) and tomato ringspot virus (TSRV) viruses on geraniums. TomRSV was both seed and pollen transmitted, while TRSV was only seed transmitted. No observations on resistance were conducted.

References

[1] Bode, F.A. 1972. Zonal geranium rust. *Flor. Rev.* 150(3875):30-31, 46.
[2] Bolag, A. 1966. *La rouilledu Pelargonium zonale. L'Her. Revue Hort Suisse* 39(1):2-5.
[3] Carroll, C.R. and C.A. Hoffman. 1980. Chemical feeding deterrent mobilized in response to insect herbivory and counteradaptation by *Epilachna tredecimnotata. Science* 209:414-416.
[4] Chang, K.P., R. Snetsinger and R. Craig. 1972. Leaf characteristics of spider mite resistant and susceptible cultivars of *Pelargonium* x *hortorum. Entomol. News* 83:191-197.
[5] Cline, M.N. and D. Neely. 1983. Wound-healing process in geranium cuttings in relationship to basal stem rot caused by *Pythium ultimum. Plant Disease* 67(6):636-638.
[6] Craig, R. and D.E. Walker. 1959. Geranium seed germination techniques. *Geraniums Around the World* 7(2):4-7.

[7]   DeArmond, V. 1976. "A virus-vector relationship in *Pelargonium* x *hortorum* Bailey." MS thesis, Penn. State Univ.

[8]   Duffey, S. 1986. Plant glandular trichomes: their partial role in defence against insects. *Insects and the Plant Surface,* ed. Barrie Juniper and Richard Southwood. London: Edward Arnold Publ. Ltd.

[9]   French. R.C., A.W. Gale, C.L. Graham, F.M. Latterell, C.G. Schmitt, M.A. Marchetti and H.W. Rines. 1975. Differences in germination response of spores of several species of rust and smut fungi to nonanal, 6-methyl-5-hepten-2-one and related compounds. *J. Agric. Food Chem.* 23:766-770.

[10] Gerhold, D.L., R. Craig and R.O. Mumma. 1984. Analysis of trichome exudate from mite-resistant geraniums. *J. Chem Ecol.* 10(5):713-722.

[11] Grazzini, R.A., D. Hesk, E. Heininger, G. Hildenbrandt, C.C. Reddy, D. Cox-Foster, J. Medford, R. Craig and R.O. Mumma. 1991a. Inhibition of lipoxygenase and prostaglandin endoperoxide synthase by anacardic acids. *Biochem. Biophys. Res. Comm.* 176(2):775-780.

[12] Grazzini, R.A., D.S. Walters, J. Harman, R. Craig and R.O. Mumma. 1991b. Inheritance of biochemical and morphological traits associated with small arthropod resistance in *Pelargonium* x *hortorum*. Submitted, J. Hered.

[13] Harwood, C.A. and R.D. Raabe. 1979. Disease cycle and control of geranium rust. *Phytopathology* 69:923-927.

[14] Hausbeck, M.K., C.T. Stephens and R.D. Heins. 1987. Variation in resistance of geranium to *Pythium ultimum* in the presence or absence of silver thiosulphate. *HortScience* 22(5):940-944.

[15] Hedin, P.A. 1983. *Plant Resistance to Insects,* ACS Symp. series #208, ed. Paul Hedin. Washington: Amer. Chemical Soc.

[16] ——— 1991. *Naturally Occurring Pest Bioregulators,* ACS Symp. series #449, ed. Paul Hedin. Washington: Amer. Chemical Soc.

[17] Hellmers, E. 1956. Bacterial leafspot of *Pelargonium (Xanthomonas pelargonii* [Brown]   Starr and Burkholder). Contr. Dep. Plant Path., Royal Vet. and Agr. College, Copenhagen 38:1-35.

[18] Knauss, J.F. and J.F. Tammen. 1967. Resistance of Pelargonium to *Xanthomonas pelargonii. Phytopathology* 57(11):1178-1181.

[19] Laughner, L.J. 1985. "Breeding and evaluation of Pelargonium for resistance to geranium rust, *Puccinia pelargonii-zonalis* Doidge." M.S. thesis, Penn. State Univ.

[20] McCoy, R.E. 1975. Susceptibility of *Pelargonium* species to geranium rust. *Plant Dis. Reporter* 59(7):618-620.

[21] Munnecke, D.E. 1954. Bacterial stemrot and leaf spot of Pelargonium. *Phytopathology* 44:627-632.

[22] ——— 1956. Development and production of pathogen-free geranium propagation material. *Plant Dis. Reporter Suppl.* 238:93-95.

[23] Neal, J.J., J.C. Steffens and W.M. Tingey. 1989. Glandular trichomes of *Solanum berthaultii* and resistance to the Colorado potato beetle. *Entomol. Exp. Appl.* 51:133-140.

[24] Pan, S., U.S. Gupta and L.C. Ewart. 1991. Bacterial wilt *(Xanthomonas pelargonii)* resistance in Pelargoniums. *HortScience* 26(6):101. Abstr. 256.

[25] Potter, D.A. and R.G. Anderson. 1982. Resistance of ivy geraniums to the two-spotted spider mite. *J. Amer. Soc. Hort. Sci.* 107(6):1089-1092.

[26] Scarborough, B. and S.H. Smith. 1977. Effects of tobacco and tomato ringspot viruses on the reproductive tissue of *Pelargonium* x *hortorum*. *Phytopathology* 67:292-297.

[27] Stark, R.S. 1975. "Morphological and biochemical factors relating to spider mite resistance in the geranium." Ph.D. thesis, Penn. State Univ.

[28] Tammen, J.F. 1960. Disease-free geraniums from cultured cuttings. *Penn. Flower Growers' Bul.* 117:1, 6-9.

[29] Wainwright, S.H. and P.E. Nelson. 1972. Histopathology of *Pelargonium* species infected with *Xanthomonas pelargonii*. *Phytopathology* 62:1337-1347.

[30] Walters, D.S., H.H. Grossman, R. Craig and R.O. Mumma. 1989a. Geranium defensive agents. IV. Chemical and morphological bases of resistance. *J. Chem Ecol.* 15(1):357-372.

[31] ———, R. Craig and R.O. Mumma. 1989b. Glandular trichome exudate is the critical factor in geranium resistance to foxglove aphid. *Entomol. Exp. Appl.* 53:105-109.

[32] ——— 1990. Heritable trichome exudate differences of resistant and susceptible geraniums. *Pesticides and Alternatives: Innovative Chemical and Biological Approaches to Pest Control,* ed. John E. Casida. Amsterdam:Elsevier Science Publ. BV.

[33] Walters, D.S., R. Minard, R. Craig and R.O. Mumma. 1988. Geranium defensive agents. III. Structural determination and biosynthetic considerations of anacardic acids of geranium. *J. Chem. Ecol.* 14(3):743-751.

[34] Winner, B.L. 1975. "Inheritance of resistance to the two-spotted spider mite, *Tetranychus urticae* Koch, in the geranium. *Pelargonium* x *hortorum* Bailey." MS thesis, Penn. State Univ.

[35] Woodhead, S. and R.F. Chapman. 1986. Insect behaviour and the chemistry of plant surface waxes. *Insects and the Plant Surface,* ed. Barrie Juniper and Richard Southwood. London: Edward Arnold Publ. Ltd.

CHAPTER 31

Insects and Other Pests

Integrated Insect and Mite Management

Richard K. Lindquist

The primary reasons for some growers experiencing more problems with insects and mites than others have not changed much in 75 years:

- Failure to detect insects and mites soon enough and lack of biological knowledge (not being familiar with the different developmental stages and approximate time to complete a generation).
- No sanitation or cultural control program (no weed control; vegetable gardens located just outside the greenhouse).
- Applying incorrect pesticide at the wrong rate, incorrect interval and/ or using poor spray techniques.

Although pesticide resistance is an important factor contributing to insect and mite problems, enough pesticides are available that if you pay attention to one or more of the above three items and begin a control program soon enough, you should achieve control. This sounds pretty complicated, especially for busy growers. After all, pests and diseases are usually listed way down on the list of direct costs associated with crop production (around 5%). When other costs, such as labor used to apply the pesticides and losses in production due to pests and diseases are included, however, the actual total may be 20% or higher. Therefore, not paying attention to the above items may be quite costly. Keep this in mind when calculating costs connected with pest and disease management.

Whether you are producing 100 or 1,000,000 geraniums, beginning a pest management program is the most important thing you can do. Forget about using biological controls until you have enough experience with conventional pest management. If you cannot control insects and mites with pesticides, biological control will probably be a failure as well.

Your goal should be to avoid problems on a mature crop by using correct procedures when plants are young. Some states have organized pest and disease management programs, or private consultants may be in your area. Take advantage of these programs if possible. The costs, if any, are minimal compared with plant value. Most of you, however, probably won't have access to an organized pest and disease management program and you'll have to go it alone.

Pest management program

Pest scouting, trapping and recordkeeping

Use yellow traps to help detect insects. Place traps vertically, at or just above plant tops. Use a system that allows you to raise traps as plants grow. Horizontal traps, placed within the plant canopy are effective in trapping insects, such as sweet potato whiteflies, fungus gnats and shore flies, especially when plants are small. Inspect traps at least weekly, and record or estimate the number of insects on each trap. Change traps weekly, if possible. Have a map of the greenhouse and mark trap locations on the map so you'll know where trouble spots are. For best results use about one to four traps per 1,000 square feet (93 square m). Many growers find this number too high to be practical and use at most one trap per 5,000 or 10,000 square feet (465 to 930 square m). Numbers of insects caught on sticky traps at these densities may or may not reflect population size, so trapping alone isn't enough.

Inspect plants in all greenhouse areas each week. Begin as soon as plants have been potted and placed in the greenhouse. Look at plants on bench ends as well as in centers. Infestations often begin as isolated pockets of insects, spreading out from there. Brush plant tops to see if any adults are present. Lift up some plants and look for eggs and immature stages on leaf undersides. Use a hand lens or other magnifying device to observe leaf undersides. Learn to recognize major pests and their development stages. Inspect whatever number of plants you can handle, because the chances of detecting an infestation in its early stages is directly related to number of plants inspected. Mark infested areas. Many growers use small, colored flags. Reinspect plants after pesticide applications. If you don't have time to do these things yourself, assign a dedicated employee to do them.

Physical/cultural controls

Eliminate weeds inside and immediately outside the greenhouse. Don't have your vegetable garden adjacent to your production area. If there are insects on plants outdoors, consider placing screens on vents and doors. Several types of screens are available that will restrict insect movement, but their use may not be practical for most commercial growers. Restricted air movement that may increase plant disease problems and keeping screens clean are two major problems with screening.

Proper pesticide use

No pesticide will be very effective if insects or mites are at outbreak levels before applications begin. This is the main reason for using sticky traps and inspecting plants: to make your pesticide applications more effective. Rotate chemical classes. Don't depend on and use the same product for every application. Make at least two applications in sequence before changing to another chemical class. Most insects and mites attacking geraniums have several developmental stages (egg, nymph, pupa, adult). A given pesticide isn't usually effective against all stages. By paying attention to sticky trap counts and plant inspections, you can adjust pesticide applications for maximum efficiency. For example, if there are many leaves containing whitefly eggs, delay a pesticide application until most or all eggs have hatched. Few pesticides are effective against eggs, but several kill nymphs.

Pesticide application method is very important. If you use an incorrect pesticide/pest/equipment combination, poor control will result. See the pesticide application section in this chapter for further information.

Keep records of pesticide applications, including environmental conditions when the applications were made. These data, plus records of pests on traps and plants provide much useful information.

After you have mastered pest detection, identification, monitoring and recordkeeping, you'll be able to relate the number of insects caught on sticky traps over a certain time period to the number on plants. You'll have a good idea of pesticide performance. You'll be able to establish a threshold level for the number of insects that can be tolerated and still produce a salable crop. Forget about zero insects and mites. It's doubtful whether most growers can achieve this without overusing pesticides. Choose a low pest number (five adults per trap per week) as a goal. Once this tolerable insect level has been achieved, you'll be able to experiment with different pesticide application schedules and techniques. For more information on the identification, biology and control of insects and mite pests see *Identification of Insects and Related Pests of Horticultural Plants* published by the Ohio Florists' Association, and the *Ball Pest and Disease Manual,* published by Ball Publishing.

Whitefly Biology and Management

John P. Sanderson and Gerard W. Ferrentino

Whiteflies can be common but serious pests of greenhouse ornamentals, including geraniums. They are primarily pests because their mere presence on an ornamental plant detracts from its aesthetic value. Their numbers are rarely allowed to increase to levels that actually affect a plant's vigor. At high numbers, however, their feeding can cause some plants to become chlorotic and reduce plant vigor. "Honeydew" excretions can cause leaves to become sticky and shiny, and serve as a substrate for grayish black sooty fungus growth, which detracts from a plant's aesthetic value and can interfere with photosynthesis. Some whitefly species, including those currently found in the greenhouse industry, are capable of transmitting certain plant viruses, although to our knowledge no viruses vectored by whiteflies have yet been reported on North American greenhouse ornamentals. There is one report [1] of whitefly mechanical transmission of *Xanthomonas pelargonii* (Brown), although this doesn't seem to be a common occurrence.

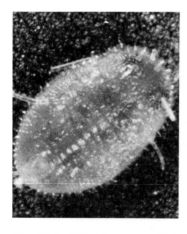

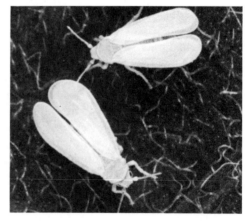

Fig. 31-1. Whitefly pupa. Note the eyespots in the upper right.

Fig. 31-2. Adult whiteflies.

Species

Two whitefly species are currently major pests of numerous greenhouse ornamentals. The greenhouse whitefly (GHWF), *Trialeurodes vaporariorum* (Westwood) has been a serious greenhouse pest for more than 100 years. GHWF infests a broad range of plants, including numerous plant genera in 82 families. This species has been collected from plants throughout the world. Since 1986, the sweet potato whitefly (SPWF), *Bemisia tabaci* (Gennadius), also known as the tobacco or cassava whitefly, has invaded and become established on certain greenhouse-grown ornamentals throughout North America and Europe. This species also has been recorded on a wide range of plants, including numerous plant genera in 63 families. SPWF is generally distributed throughout the world's more tropical regions. Initial SPWF collections from plants in the United States were recorded in California and Florida in 1904 and 1900, respectively. The sudden occurrence of this pest in North American greenhouse crops may be due to the development and transport of a different strain or biotype of the species. More than 60 plant viruses have been recorded as vectored by SPWF, although to date no incidence of virus transmission by SPWF on greenhouse ornamentals has been reported in North America.

GHWF is a common geranium pest, particularly on Regal geraniums (*Pelargonium* x *domesticum*). It also can infest zonal geraniums (*Pelargonium* x *hortorum*), although anecdotal reports indicate that certain cultivars are more susceptible to whiteflies than others. GHWF is currently more frequently encountered on geranium than SPWF.

This discussion will center on the identification, biology and management of whiteflies.

Identification

The ability to properly identify whitefly species and life stages is important for several reasons. Growers who fail to notice whether immature (nymphal) stages are present on leaf undersides may bring plants infested with significant numbers of whiteflies into their greenhouse or may fail to notice a serious infestation until clouds of adults emerge from the crop. Also, because some life stages are more susceptible to most insecticides than others, growers who can properly identify various life stages may be able to achieve better control with fewer sprays by timing insecticide applications to coincide with susceptible stages. Lastly, the two whitefly species may have biological differences, and it may be important to know which species is present for optimal control. GHWF is currently the most common whitefly pest of geranium, but SPWF has pestiferous potential.

Almost all whitefly life stages are confined to lower leaf surfaces. Except for adults, very rarely will any life stages be found on upper leaf surfaces. Adult whiteflies are small, white, flylike insects, from which the pest gets its name. Most growers need a hand lens capable of at least 10x magnification to see the immature life stages. Eggs are very tiny, spindle shaped, stand vertically on the leaf surface, and are attached to the leaf by a tiny pedicel or "stalk" at the base of the egg (Fig. 31-3). Eggs are often deposited on the leaf in a crescent-shaped pattern; otherwise they are scattered over the leaf either singly or in clusters. The eggs are pearly white when first laid, turning dark gray (GHWF) or amber brown to gray (SPWF) with time. The egg "shell" remains visible (with a hand lens) on the leaf surface after the crawler hatches, although it may be collapsed. The crawler and other nymphal stages are oval, greatly flattened and somewhat translucent with a white, light green or light yellow cast. The four nymphal stages are identified by their relative sizes; length and width increase with each successive molt (Fig. 31-3). The last nymphal stage is divided into three substages: the early fourth substage, transitional substage and pharate adult (adult enclosed within the nymphal cuticle) substage. The transitional and pharate adult substages are more plump than the early fourth substage, and with time the red eyes of the adult developing inside can be seen at one end of the body (Fig. 31-3). For this discussion we refer to the last two substages combined as the "pupal" stage, or pupa, although technically whiteflies don't have a pupal stage.

Use a hand lens or other magnifying device to distinguish between the two whitefly pest species. Concentrate on the pharate adult (pupal) stage to reliably differentiate between these two pests. The "pupal case" or "skin" left behind after the adult emerges can also be used for identification. No other life stages, except perhaps the adult, provides reliable identification.

The GHWF pupa has parallel sides that are perpendicular to the leaf surface, giving the nymph a disk-shaped or cake-shaped appearance. In side view, a SPWF pupa appears more rounded, or even pointed, with no parallel sides. GHWF has a tiny fringe of setae around the pupa periphery (or "rim") in a top view. SPWF pupa has no such fringe of setae around the edges. Both species may have several pairs of longer filaments (or "hairs") arising from the top of the pupa. Usually these filaments are larger and more obvious on GHWF than on SPWF, but this characteristic can vary depending on the host plant on which the insect developed. Therefore, these filaments' size isn't a reliable characteristic and shouldn't be used for identification. The SPWF also tends to be more yellow than GHWF, but color alone isn't a reliable characteristic.

The GHWF adult is larger than SPWF and holds its wings fairly flat over its abdomen in a plane that is rather parallel to the leaf surface. The SPWF adult is often slightly more yellow and holds its wings rooflike against its abdomen, with its wings at approximately a 45-degree angle with the leaf surface. Wings are held tightly against the body. Although the adults' appearance can be used to differentiate between these two species with confidence, use the pupal stage for confirmation. The characteristics of the pupal stage mentioned above aren't difficult to see with a hand lens of at least 10x magnification.

Biology

Knowing whitefly biology and life cycle is important for efficient, economical whitefly control. Whitefly eggs tolerate almost all currently registered whitefly insecticides. A crawler hatches from the egg, crawls a very short distance (usually within millimeters), settles on a leaf surface and begins feeding. Once settled, it doesn't move from this spot until immature development is complete and the individual emerges as an adult. Insecticide residues left on leaf surfaces after a spray can kill only those life stages that can crawl through the residue. Therefore, residual control is a consideration only for crawlers and adults. The other life stages can be killed by foliar insecticides only if they are contacted directly. Therefore, thorough lower leaf surface coverage and canopy penetration is crucial for effective chemical control. Better coverage and better control may be attained by initiating whitefly control on young plants while the plant canopy is open, respacing pots to open the canopy, modifying spray patterns and using spray equipment that provides good lower leaf surface coverage.

In addition to eggs, older nymphal stages and particularly pupae tolerate many currently registered whitefly insecticides. Thus, life stages most susceptible to most insecticides are the first, second and third nymphal stages and adults. This knowledge can lead to more efficient control by timing sprays against these susceptible stages. Sprays against a whitefly infestation at a time when the bulk of the population is in an insecticide-tolerant life stage may be largely wasted with most presently registered insecticides. Although it's possible that all life stages may be present on a crop at a given time, the *bulk* of an infestation is often comprised of two or three stages (adults laying eggs, or a combination of first, second and third instar nymphs; or pupae and emerging adults). This is possibly the result of pesticide applications that selectively kill susceptible stages, leaving behind more tolerant stages and thus developing an infestation made up primarily of only a few stages. Each life stage profile may require a different chemical control strategy, based on the life stages present, for the most efficient control. For example, if an infestation is comprised primarily of adults and eggs, smoke or aerosol may be effective against adults, followed by sprays against crawlers after the eggs hatch.

Data in Fig. 31-3 were generated from studies of GHWF and SPWF on poinsettias grown at temperatures that fluctuated between 65 F and 75 F (18 C and 23 C) and provide relative information about the duration of all immature stages. Developmental times for geraniums may differ slightly and may also vary by cultivar. This information can be used as a guideline to time spray applications at these temperatures. Developmental times are shorter at warmer temperatures. Note that for both species the time spent in insecticide-tolerant stages (eggs and pupae) amounted to about half the total developmental time. This, again, shows the need for timing sprays against the most susceptible stages for efficient whitefly control. If all life stages are present in roughly equal abundance, however, or timing sprays against susceptible life stages isn't possible for some reason, then multiple insecticide applications will be needed for control.

Knowing adult biology is also important for effective control. First, most adults of both species emerge from the pupal stage during the morning, between 6 and 11 a.m., although a few may emerge at various times throughout the day. Newly emerged whiteflies are probably more susceptible to sprays because their integument hasn't fully hardened and they haven't yet coated their body with wax. Second, the time

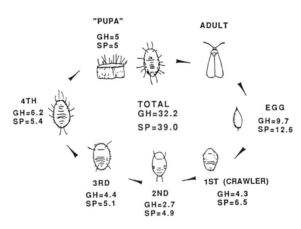

Fig. 31-3. Whitefly life cycle. Generalized life cycle and average duration (in days) of each life stage of greenhouse whitefly (GH) and sweet potato whitefly (SP) on poinsettias grown at temperatures fluctuating between 65 F and 75 F (18 C and 23 C).

between a female's emergence from the last nymphal stage until she begins laying eggs is called the preoviposition period, and this preoviposition period dictates maximum spray interval allowable before newly emerged females begin laying eggs. At temperatures between 65 F and 75 F (18 C and 23 C), apply sprays against adult SPWF and GHWF every three and four days, respectively, to kill adults prior to egg production. Female whiteflies of either species don't need to mate to reproduce. Unfertilized eggs develop into males; fertilized eggs develop into females. As many as 150 eggs per female may be laid on some crops over a 25-day period.

Detection and monitoring

Using a pest detection and monitoring (scouting) program allows one to determine which life stages are present in an infestation, how many and where, in order to make spray decisions. The following is the procedure used for whitefly control in New York state greenhouses.

The first step in the program involves analyzing the entire greenhouse site prior to introducing new plant material, so that current and future whitefly sources can be identified and hopefully eliminated. Whiteflies frequently appear on weeds, on plants from previous crops that haven't been discarded and on "pet plants" (plants that are not for sale but are kept in the greenhouse). Growers have a tendency to ignore pests on these plants, yet they can harbor serious pest numbers. Crops growing in adjacent greenhouses may also be infested, and if so, should be identified as a potential or real source of whitefly problems. Whiteflies in the pupal stage can survive and emerge as adults from leaves that have withered and dried, such as discarded plant material and weeds. Thus, discarded plant material must be properly disposed of.

Second, a thorough inspection of all incoming plant material is necessary. Don't assume that incoming plants are uninfested just because adult whiteflies aren't visible; eggs and other immature stages may be present. Pest management scouts or other adequately trained greenhouse employees should examine a proportion of all incoming plant material for whiteflies. Inspect all leaf undersides for any life stage. The species, number and life stage(s) of whiteflies on the incoming crop determines which control measures are appropriate.

Third, establish a weekly scouting/monitoring routine for the crop's duration. Any adequately trained employee whose vision is sufficient to see immature whiteflies with a 10x hand lens can do the scouting. The scouting routine involves three components: foliage inspection, "indicator plants" and yellow sticky cards.

Foliage inspection. Each week, choose plants arbitrarily on each bench throughout the crop, and inspect several leaves' lower surfaces from the lower, middle and upper plant portion for any whitefly life stage. Record location, plant variety, whitefly species, life stage and number of pests on each plant. Foliage inspection can help reveal infested areas, time insecticide applications and assess control achieved. A regular plant inspection routine can provide peace of mind because pest status is carefully monitored for crop duration.

Indicator plants. When infested plants are found during the foliage inspection, mark some plants with flagging tape or flags, and use them as indicator plants. Each week record the life stage and number of pests on these plants. Use the whitefly status on these plants to monitor the whitefly population's developmental status and time sprays against susceptible life stages and to assess spray efficacy. For example, pre- and postspray counts on these plants can reveal the control achieved by spray applications. Growers can evaluate whether coverage was adequate or whether an insecticide was effective.

Yellow sticky cards. Use weekly whitefly counts on yellow sticky cards to detect infestations that might be missed during foliage inspection and to monitor adult whitefly activity. Whitefly counts on sticky cards mirror increases and decreases in adult activity. Space 3- by 5-inch (8- by 13-cm) yellow sticky cards about every 50 feet (15 m) throughout the crop, positioning on stakes just above the crop canopy. Although some studies indicate that some SPWF adults should be monitored with cards in a horizontal position within or below the canopy and GHWF with cards in a vertical position above the canopy, we currently position cards vertically for both species. If time permits, check the cards twice weekly because of the 3- to 4-day period between adult emergence and egg laying.

Summarize weekly scouting records in a notebook, along with whatever control action (if any) is taken, including the date, to document the control action's effectiveness. In this way the grower can learn to interpret scouting reports and make optimal control decisions.

Crop status

Keep in mind the crop status for optimal whitefly control. Because thorough coverage and good canopy penetration is so important for effective chemical control, it's critical to control an infestation early in the crop while the canopy is still small and coverage isn't as difficult to achieve. Also, if scouting reports reveal few or no whiteflies toward the end of the crop, a grower may not need to spray thereafter if there's insufficient time for a detectable infestation to develop before the crop is sold. This can save time, money and insecticides, eliminate the significant risk of phytotoxicity when the bracts are showing color and free a grower to concentrate on other aspects of crop production and sales.

Spray application technique

Whitefly chemical control efforts are often most efficient (best kill with fewest sprays) when directed against the young, immature stages, particularly before the crop canopy becomes dense and precludes adequate spray coverage. Application equipment and technique that provide the best lower leaf surface coverage (where immature stages are located) is usually most efficient.

Biological control

Whitefly fungal pathogens are currently under commercial development and look promising for biological control. At least three fungal species are under investigation: *Verticillium lecanii, Aschersonia aleyrodis* and *Paecilomyces fumosoroseus*. United States and European companies are pursuing commercial use of these fungi. Biological control of whiteflies, particularly GHWF, with the commercially available whitefly parasitoid *Encarsia formosa* is another promising control option on some crops and in some situations; however, little research has been done on geranium. Predaceous beetles, including *Delphastus pusillus,* also are being evaluated for whitefly biological control.

Reference

[1]   Bugbee, W.M. 1962. Whitefly transmission of *Xantromonas pelargonii*. *Phytopathology* 52: 5.

Aphids

James R. Baker

About 4,000 aphid species (Fig. 31-4) have been described for the world (including the woolly aphids and gall aphids, which don't attack geraniums). Fortunately, most aphid species are restricted to a few host plants, usually in the same genus, and most of those genera aren't in the geranium family. At least 25 aphid species have been reported from plants in the Geraniaceae (Table 31-1). Westcott [24] listed four aphid species that infest geranium and Pelargonium. Of these she considered the geranium aphid, the green peach aphid and the potato aphid to be most important; the wild geranium aphid could be a pest to the cranes bills' fanciers (*Geranium* spp., particularly *Geranium richardsonii* Fisch. & Trautv.). Scented varieties, Martha Washington (Regal) and ivy geraniums are most susceptible to aphids [8]. The geranium aphid attacks celery, doves foot (*Geranium molle*), calceolaria, calla lily, chrysanthemum, cineraria, *Geranium pusillum, Pelargonium hortorum,* storks bill (*Erodium cicutarium*), verbena and viola.

Aphids are small, fragile insects and the only insects with a pair of cornicles on the abdomen (Fig. 31-5). Aphids seem to become resistant to

Fig. 31-4. Adult aphid (center) and nymphs.

insecticides rapidly, appear overnight and reproduce with frightening speed. On the other hand, aphid populations tend to collapse suddenly. Aphids, which are genetically identical, may have wings or be wingless [25]. Winged aphids readily disperse by air. Dickson [6] found that 75 miles of desert wasn't a sufficient barrier to prevent a massive aphid migration, but simply screening the vents can prevent aphids from entering the greenhouse [1].

TABLE 31-1

Aphids reported from Geraniaceae

Common name	Scientific name and discoverer	Reference and taxa
Banana aphid	*Pentalonia nigronervosa* Coquerel	[16]: from Pelargonium
Bean aphid	*Aphis fabae*	[2]: from Pelargonium
Black bean aphid	*Aphis fabae*	[2]: from Pelargonium
Carrot-willow aphid	*Cavariella aegopodii* (Scopoli)	[14]: from cultivated geranium, probably Pelargonium
Foxglove aphid	*Aulacorthum solani* Kaltenbach	[16]: from Erodium; [2]: from Geranium
Glasshouse -potato aphid	*Aulacorthum solani* Kaltenbach	[16]: from Erodium; [2]: from Geranium
Geranium aphid	*Acyrthosiphon malvae* (Mosley)	[16]: from Pelargonium; [2] listed *Acyrthosiphon malvae geranii* as a distinct subspecies attacking Geranium— herb robert, crane's bill.
Green peach	*Myzus persicae* (Sulzer)	[16]: from Pelargonium)
Potato aphid	*Macrosiphum euphorbiae* (Thomas)	[14, 21]: from "wild geranium" in Utah
Shallot aphid	*Myzus ascalonicus* Doncaster	[2]: from Geranium
Tulip bulb aphid	*Dysaphis tulipae* (Boyer de Fonscolombe)	[14]: from cultivared geranium, probably Pelargonium
Tulip aphid	*Dysaphis tulipae* (Boyer de Fonscolombe)	[14]:from cultivared geranium, probably Pelargonium
Wild geranium aphid	*Amphorophora geranii* Gillette and Palmer	[24])
Wild geranium aphid	*Amphorophora coloutensis* Smith and Knowlton	[14] called this the wild geranium aphid from "wild geranium" in Utah; [21] collected it from *Geranium fremontii* Torr. and *G. richardsonii* Fisch. and Trantv. in Colorado, Idaho and Utah noting the aphid often caused distortion.
	Aphis extranea Walker	[16]: from Pelargonium

Common name	Scientific name and discoverer	Reference and taxa
	Aphis rumicis Linnaeus	[16]: from Geranium
	Aphis urticata Gmelin	[16]: from *Geranium divaricatum*
	Cryptaphis geranicola Shinji	[16]: from Geranium
	Hyadaphis foeniculi (Passerini)	[14]: from cultivated geranium, probably Pelargonium
	Indoidiopterus geranii (Chowdhuri, Basu, Chakrabarti, Raychaudhuri)	[5]: from *Geranium divaricatum*
	Kakimia cefsmithi Knowlton	[22]: from *Geranium pusillum* and G. *richardsonii*
	Kakimia crenicorna C.F. Smith and Knowlton	[22]: from "wild geranium" in Utah
	Macrosiphum bosqi Blanchard	[16]: from Pelargonium
	Macrosiphum geranii Oestilund	[22], 1978: from *Geranium maculatum* in California, the upper Midwest and Pennsylvania
	Macrosiphum salviae Bartholomew	[20]: from Geranium; this aphid ranges from southern United States into Argentina
	Microlophium carnosum (Buckton)	[16]: from Erodium and Pelargonium
	Myzus targionii Del Guercio	[16]: from Pelargonium

Aphids feed by inserting slender mouthparts into plant phloem tissue and sucking out sap. Guthrie et al. [10] and Bornman and Botha [3] demonstrated that aphids feed almost exclusively in the sieve elements of green plants' phloem tissue. As the aphids insert their tiny, threadlike mouth parts, they inject saliva into tissue. The saliva forms a sheath around the slender stylets inside the leaf. Aphids then suck out sap, which is rich in sugars but poor in other nutrients and excrete excess water and sugars as a sweet, sticky liquid called honeydew. Infested plants are often disfigured by honeydew and sooty molds (dark fungi) that grow in the honeydew. Aphids molt several times as they grow and the molted skins adhere to the honeydew, which also detracts from the geranium's appearance.

Aphids start flying at light intensities of 100 to 1,000 footcandles (1 to 11 klux) and by midday, tremendous numbers of aphids may be flying. Once aphids become airborne, their color preference changes from bright blue to yellow [13]. This is one reason to refrain from wearing yellow, yellow green or blue clothing, as aphids may be inadvertently carried into the greenhouse on such clothes.

Most aphids in North America exhibit cyclical reproduction in which no males are produced for most of the growing season (parthenogenetic reproduction). In response to fall's shorter days, colder temperatures and other factors, males are born. Males eventually mate with special females, which then lay overwintering eggs. In greenhouses aphids are never chilled and their food supply is unlimited. Under these conditions aphids usually stay in cyclical reproduction indefinitely [11]. A great advantage to parthenogenesis is that most offspring are females and therefore are capable of producing offspring without having to waste time searching for males or waiting for fertilization before eggs develop. Parthenogenetic aphid embryos begin developing just after ovulation so that an aphid may have embryos developing within embryos. Thus, these parthenogenetic nymphs are expectant grandmothers even before they are fully mature [7]!

The carrot-willow aphid *(Cavariella aegopodii)* is a medium-sized aphid with a projection of the last abdominal segment above the cauda. There is a short terminal process on the antenna and swollen cornicles [2]. This aphid is found principally on willow and Umbelliferae and occurs throughout America, Australia, Europe, Japan, Korea, the Middle East, New Zealand, Rhodesia and India [18].

The foxglove or glasshouse-potato aphid *(Aulacorthum solani)* is a medium-sized, greenish yellow, shiny aphid with cylindrical tapering cornicles (Fig. 31-6). A dark green patch of pigmented material usually shows through the cuticle at each cornicle base. It can produce sexual forms and lay eggs on many different host plants, although in Britain it often overwinters parthenogenetically [2]. This aphid is practically cosmopolitan in distribution [18].

The geranium aphid *(Macrosiphum malvae)* is medium-sized, pale green or yellowish with long, tapering cornicles and cauda (Fig. 31-7). It infests Pelargonium in greenhouses and homes [2]. This insect is distributed throughout America, Europe and West Bengal in India [18].

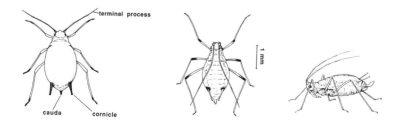

Fig. 31-5, left, *Wingless adult aphid. Fig. 31-6,* center, *Geranium aphid. Fig. 31-7,* right, *foxglove aphid.* (Fig. 31-5 and 31-6 drawings courtesy James R. Baker.)

The green peach aphid or peach-potato aphid (*Myzus persicae*) is a pale green to yellow green, smallish insect (2 to 2.4 m) with long, laid-back cornicles, which are slightly swollen and knobs inside the antennae base that slant inward (Fig. 31-8). The green peach aphid is shaped somewhat like a fat football. The winged forms have a black patch on the top of the abdomen. The green peach aphid resists many insecticides and persists on floral crops after pesticide applications. It feeds on various plants; although geraniums aren't a favored host, this aphid infests them readily. In Southern United States, the green peach aphid apparently maintains parthenogenetic reproduction throughout the year.

The potato aphid (*Macrosiphum euphorbiae*) is a fairly large, long-bodied, grayish green or pink aphid with long legs, antennae and cornicles. The cornicles have a reticulate pattern. The primary host (plant on which eggs are laid in the fall) is rose [2]. The potato aphid can transmit bean yellow mosaic virus, iris mild and iris severe mosaic viruses, tulip breaking virus, lily symptomless virus, cucumber mosaic virus and lily rosette, narcissus yellow stripe, narcissus white streak, narcissus latent, narcissus degeneration and narcissus late season yellows to susceptible plants. Migrating aphids may settle briefly and feed long enough to transmit viruses [15]. Thiodan seems to be the only pesticide for which the potato aphids have developed some resistance [James Walgenbach, personal communication].

The shallot aphid (*Myzys ascalonicus*) is similar to the green peach aphid, but is gray yellow and has shorter, more swollen cornicles [2].

The tulip bulb or tulip aphid (*Dysaphis tulipae*) is a whitish or tan aphid that lives all year parthenogenetically on tulip, iris, gladiolus and crocus bulbs in the ground or in storage (Fig. 31-9). This aphid is a much more important threat to bulb crops than to geraniums. The winged forms have black heads, thoraces and black markings on their abdomens [2]. Heavily attacked corms and bulbs grow poorly and emerging shoots are often distorted. This aphid is a vector of narcissus yellows streak virus [15, 17].

Control

Resistance to insecticides has been reported for 12 aphid species, particularly the green peach aphid. According to Hussey et al. [12], it's possible for aphids in one section of a large greenhouse range to resist an insecticide, and aphids in another section to be susceptible. Certainly green peach aphids from various parts of the country differ in their pesticide susceptibility. For example, Robb et al. [19], working in California, obtained greater than 90% control of green peach aphids on chrysanthemums with Orthene and Avid, 35% control with Talstar and

Fig. 31-8, left, *green peach aphid.*
Fig. 31-9, right *tulip bulb aphid.*
(Fig. 31-9 courtesy James R. Baker.)

0% control with Knox-Out 2FM. Yost et al. [26], also working in California, controlled virtually 100% green peach aphids on chrysanthemums using Talstar, Avid and Orthene. Growers in North Carolina report that Talstar, Avid and Orthene don't always give 100% control. These chemicals' inconsistent performance isn't surprising as the aphids in these situations undoubtedly had different exposure pesticide backgrounds.

Resistance to insecticides is acquired by a population by sexual recombination (for those aphids with a male stage), mutation, mitotic recombination or a suppressed genes mechanism in which characteristics such as resistance aren't expressed until some outside factor impinges on the aphid population (such as halfhearted insecticide application). In the suppressed gene mechanism theory, those genes that confer resistance are present in an individual but aren't expressed because such expression is detrimental to the aphid in the absence of an insecticide. When an insecticide doesn't quite kill that aphid, some of its offspring may display greater resistance to that pesticide than the parent displayed. This may explain the sudden resistance to Mavrik. Over the past few years, flower growers have reported that Mavrik worked well for aphid control for a short period after which, even at 18 times the recommended rate, aphids survived.

Because resistance to some pesticides is almost complete in the green peach aphid, don't use pyrethroids for green peach aphids since that may cause them to scatter. Pyrethroids also eliminate parasites, allowing the aphid population to increase faster! Of the organophosphates, Orthene still seems to work best. (With sprays, consider using a spreader sticker and a buffering agent to get the pH to around 6.) Lannate is the best nonsystemic spray, but it's not registered for greenhouse use. Growers report that Thiodan may be losing its effectiveness. Avid gives mediocre aphid control. Results with soap have been encouraging. Griswold in 1927 found that Ivory soap (¼ ounce per quart or 7.8 mm per l) plus nicotine sulfate (¼ teaspoon per quart or 2 mm per l) adequately controlled aphids and that the soap was absolutely necessary for control. It seems that resistance to soaps and oils isn't likely to develop. In our experience, Lannate, soap (M-Pede, Insecticidal Concentrate) and oil

(Ultra-Fine Spray) significantly outperformed the pyrethroids, Mavrik, Talstar and Tame.

In a nutshell, for best aphid control:

- Allow no one wearing blue, green or yellow to enter your greenhouse.
- Screen all vents to prevent aphids from flying in.
- When applying an insecticide, treat thoroughly using Orthene, soap or oil in an effort to eliminate the entire population on the first treatment.

References

[1] Baker, J.R. and R.K. Jones. 1990. An update on screening as part of insect and disease management in the greenhouse. *NC Flower Growers Bul.* 35(6):1-3.

[2] Blackman, R. 1974. *Aphids.* London and Aylesbury: Ginn & Co. Ltd.

[3] Bornman, C.J. and C.E.J. Botha. 1973. The role of aphids in phloem research. *Endeavor* 32:129-33.

[4] Breakey, E.P. 1957. Bulbous iris insects. *Handbook on Bulb Growing and Forcing.* Northwest Bulb Growers Assoc. Mt. Vernon, Wash.: Skagit Valley J.C.

[5] Chowdhuri, A.N., R.C. Basu, S. Chakrabarti, D.N. Raychaudhuri. 1969. Aphids (Homoptera) of Simal (Himachal Pradesh) India. *Oriental Insects* 3(1): 83-92.

[6] Dickson, R.C. 1959. Aphid dispersal over southern California deserts. *Ann. Entomol. Soc. Amer.* 522:368-72.

[7] Dixon, A.F.G. 1985. *Aphid Ecology.* New York: Chapman & Hall.

[8] Graf, A.B. 1976. *Exotica,* Series 3. East Rutherford, NJ: Roehrs Co. Inc.

[9] Griswold, G.H. 1927. The control of aphids on house plants. *New York State College Agr. Ext. Bul.* 162:1-15.

[10] Guthrie, F.E., E.V. Campbell and R.L. Baron. 1962. Feeding sites of the green peach aphid with respect to its adaptation to tobacco. *Ann. Entomol. Soc. Amer.* 55: 42-46.

[11] Huxley, T.H. 1858. On the agamic reproduction and morphology of Aphis.— Part 1. *Trans. Linn. Soc.* 22:193-219.

[12] Hussey, N.W., W.H. Read and J.J. Hesling. 1969. *The Pests of Protected Cultivation.* London: E. Arnold (Pub.) Ltd.

[13] Kennedy, J.S., C.O. Booth and W.J.S. Kershaw. 1961. Host finding by aphids in the field. III. Visual attraction. *Ann. Appl. Bio.* 49:1-21.

[14] Knowlton, G.F. 1983. Aphids of Utah. *Utah State Univ. Exp. Sta. Res. Bul.* 509:155.

[15] Lane, A. 1984. *Bulb Pests.* Ministry of Agr. Fisheries and Food. London: Her Majesty's Stationery Office.

[16] Patch, E.M. 1938. Food-plant catalogue of the aphids of the world including the Phylloxeridae. *Maine Agr. Exp. Sta. Bul.* 393:35-431.

[17] Penna, R.J., W.M. Morgan, M.S. Ledieu, D. Price and A. Lane. 1984. Pest and disease control of protected crops outdoor bulbs and corms. Reprinted from Scopes, 1983. Ed. N. and M. Ledieu. 1983. *Pest and Disease Control Handbook,* 2d ed. Croydon: BCPC Publ.

[18] Raychaudhuri, D.N. 1980. *Aphids of North-east India and Bhutan.* Calcutta: Zool. Soc.

[19] Robb, K.L., J.K. Virzi, H.A. Yoshida, J.T. Yost and M.P. Parrella. 1986. Control of aphids in chrysanthemums grown for cut flowers, summer, California. *Insecticide and Acaricide Tests* 11:379.

[20] Smith, C.F. and M.N. Cermeli. 1979. An annotated list of Aphididae (Homoptera) of the Caribbean Islands and South and Central America. *NC Agr. Res. Serv. Tech. Bul.* No. 259.

[21] Smith, C.F. and G.F. Knowlton. 1983. A key to the species of aphids (Homoptera: Aphididae) on wild Geranium spp. in the United States, with description of a new species. *Proc. Entomol. Soc. Washington* 85(4):686-90.

[22] Smith, C.F. and D.S. Parron. 1978. An annotated list of Aphididae (Homoptera) of North America. *NC Agr. Exp. Sta. Tech. Bul.* No. 255.

[23] Warkentin, D. 1988. Controlling aphids on chrysanthemums. *Proceedings of the Fourth Conference on Insect and Disease Management on Ornamentals*. In press.

[24] Westcott, C. 1973. *The Gardener's Bug Book*. Garden City, New York: Doubleday Co. Inc.

[25] White, W.S. 1946. The environmental conditions affecting the genetic mechanism of wing production in the chrysanthemum aphid. *Amer. Natur.* 80:245-70.

[26] Yost, J.T., G.W. Ferrentino and M.P. Parrella. 1985. Control of the green peach aphid on potted chrysanthemums. *Insecticide and Acaricide Tests* 11:382.

Two-Spotted Spider Mites

Dave Smitely

One pest frequently found on geraniums, particularly ivies, is the two-spotted spider mite. Although spider mites are found outdoors in most parts of the world, they rarely become numerous enough to damage plants because predaceous mites and insect predators feed on them. In food crops or horticultural crops where insecticides are used, however, spider mites may become a serious plant pest. Spider mites tolerate pesticides more than their natural enemies. Also, spider mites develop pesticide resistance rapidly. Unfortunately, the greenhouse environment is nearly ideal for spider mites. Broad spectrum insecticides are generally used on flower crops to guarantee top quality plants free of unsightly insect injury. Insecticide applications eliminate predators, and warm temperatures year-round allow spider mites to reproduce rapidly. Another reason that spider mites are serious greenhouse pests is that the enclosed structure effectively isolates that population and prevents breeding with individuals from more susceptible populations. It's well established by biologists that isolated populations, with all individuals being repeatedly treated with a pesticide, develop resistance rapidly. In fact, this is considered the ideal environment to produce a resistant population. Because spider mites resist most insecticides, chemicals with specific activity against mites, called miticides, are needed to achieve adequate control.

Damage and biology

Spider mites damage plants by destroying leaves' epidermal cells when they feed. Their saliva may also be somewhat toxic to some plants. In geraniums this is manifested as infested leaf yellowing. The major symptom of infected plants, however, is white or brown stippling in pockets around the infested leaf base, followed by entire leaf bronzing and necrotic leaf margins. Spider mites also produce webbing on heavily infested leaves and move towards the plant top seeking new leaves or opportunities to be blown or carried to other plants.

Spider mites are more closely related to spiders and ticks than to insects. They have eight legs instead of six, two main body parts instead of three and lack wings. The adult female is 0.5 mm long, just barely visible as a moving dot on a white piece of paper. Each female may produce up to 120 eggs and deposits them on the leaf surface. They are too small to see without a microscope or hand lens. The first of three

Fig. 31-10. Two-spotted spider mites and eggs.

larval stages emerges from the egg and begins feeding on leaf cells. All larval stages resemble the adult, but the first stage has only six legs. The adult male (0.35 mm) is smaller than the female and more triangular.

Spider mite populations may grow rapidly and seem to explode within a few days under warm conditions due to a short generation time at high temperatures. Like all arthropods, spider mites develop at a rate directly dependent on their environmental temperature. The time needed for spider mites to go from egg-larva to I-larva to II-larva to III-adult-egg (one generation) is 30 days at 60 F (16 C), 14½ days at 70 F (21 C) and 3½ days at 90 F (32 C). Also, because spider mites are very tiny, they live close to the leaf surface, which under sunny conditions may be 9 to 10 degrees F (5 C) warmer than air temperature. Outdoors, spider mite populations undergo diapause in temperate regions to survive the winter. Two-spotted spider mites diapause as adult females. When the photoperiod begins to shorten in fall and air temperatures decrease, females look for sheltered sites to overwinter. Spider mites may or may not diapause in greenhouses, depending on population genetics and environmental conditions. Sometimes only part of the population diapauses and the rest remain active. In some areas, diapausing mites present a serious problem when they break diapause in late winter or early spring because infestations appear suddenly. In southern England spider mite diapause is induced in fall if the photoperiod shortens to less than 13 hours and if greenhouse temperatures become cooler.

Management

Many European greenhouse vegetable growers have adopted biological control as their principal approach to mite management. Releasing predaceous mites for spider mite control has been very successful in these systems. Little, if any, research has been conducted to determine if geraniums can be produced with biological control instead of chemical control. Some growers believe that biocontrol won't work on geraniums because quality demands are too high. This is like saying something can't be done because it's never been done before. Geraniums are actually a good candidate for biological control because the major pests are whiteflies and spider mites, two arthropods that have been successfully managed with biological control on other plants. Some research is needed to pinpoint potential pitfalls in a biocontrol program for geranium production. It may be worthwhile investing in biocontrol research because whiteflies and mites have a great ability to develop pesticide resistance. The chemical control history of spider mites in greenhouses is a series of initial success stories followed by complete failure some two to 10 years after introducing each new product. There's no reason to believe new miticides will fare any better; there are no exceptions to the "resistance" story. The only thing spider mites haven't developed resistance to are predators. Therefore, the most promising future management strategies lie with biocontrol.

In addition to resistance problems, solutions to environmental issues also require reduced chemical use. Key components of environmental stewardship and pesticide safety are: (1) worker safety; (2) consumer safety; (3) groundwater contamination; and (4) surface runoff to streams and ponds. In some areas public pressure and government regulations have restricted pesticide products that can be used and have instituted monitoring programs for pesticide runoff. Although biocontrol and IPM (integrated pest management) are desirable, most growers rely on miticides to guarantee clean, damage-free geranium plants. The basic management strategies used are: (1) preventive chemical control; (2) IPM; (3) biological control; and (4) biological control with a chemical finish. I'll briefly describe each strategy.

Preventive chemical control. One approach is to assume that you will have spider mites at some point and apply miticides from the beginning of the growing period to prevent them from becoming a problem. The standard way to do this is to apply a broad spectrum insecticide-miticide once every three weeks, starting about four weeks after germination. Any miticide listed in Table 31-2 can be used as preventive applications, but the most popular ones are Vydate L as a soil drench (systemic), Oxamyl as a soil granular (systemic) or Avid as a foliar spray. Scout plants in every section once each week for spider mite damage. If spider injury is found, apply a miticide once every week until mites are no longer found.

TABLE 31-2

Miticides labeled for spider mite control on geraniums (U.S.)

Product name	Common chemical name	Formulation	Resistance problems	Phytotoxicity* Leaves	Flowers
Avid	Abamectin	0.15 EC	–	No	No
Kelthane	Dicofol	35 WP	X	No	No
Malathion	Malathion	25 WP, 50 EC	X	No	Yes
Mavrik	Fluvalinate	2 F	X	No	–
Oxamyl	Oxamyl	10 G	X	No	?
Pentac	Dienochlor	50 WP, 4 F	?	No	No
Talstar	Bifenthrin	10 WP	X	No	No
Vendex	Hexakis	50 WP, 4 F	–	No	–
Vydate L	Oxamyl	24 SL	X	No	Yes

*Potential leaf and flower injury problems caused by insecticide applications.

IPM. The IPM philosophy is to use the minimum pesticide necessary to achieve satisfactory control. Greenhouses should be mite free before starting a new crop. Eliminate any weeds that may harbor spider mites. If diapausing mites are a problem, start heating your greenhouse with full lighting (greater than 13 hours per day) one week before putting in live plant material. Diapausing mites and insects will emerge and starve if the greenhouse is vacant. Examine all new plant material brought into the greenhouse for mites and insects and treat them with an insecticide or miticide if you find any infestation signs.

Scouting is critical to an IPM approach. Either a greenhouse employee must be trained and given a specific scouting responsibility, or a scout should be hired to examine each section every one to two weeks. Investing some time in a scout usually saves time and money by reducing pesticide applications or improving plant quality. A good scout will find the first mite infestations, notify the grower when to start miticide applications and know when adequate control is achieved. The scout will also know from weekly records if a miticide isn't working or if spray coverage was adequate.

Another aspect of IPM is selecting geranium cultivars most resistant to spider mites. More research is needed to identify the most resistant and most susceptible cultivars. Resistant cultivars require fewer miticide applications (see Chapter 30, Host Plant Resistance).

Biological control. Biological control must be implemented as a complete pest management program. You can't use predator mites to control spider mites and spray insecticides for whitefly control. Assume that every insecticide is toxic to predators and parasites, unless specific information indicates otherwise. The most successful biocontrol programs use various predators and parasites and no insecticides or miticides. Some products unlikely to destroy predator mites or *Encarsia formosa* (for whitefly control) are kinoprene, insecticidal soap, horticultural spray oils and *Bacillus thuringiensis*. Take care when using oil or soap to avoid damaging geranium plants. Kinoprene may be effective for whitefly control without killing predaceous mites and *Bacillus thuringiensis* for caterpillar control without killing Encarsia or predaceous mites. Biological control success rides on early predator and parasite release and repeated releases to establish a predator-prey balance before insect and mite populations grow so large that it's impossible to release enough predators to have one predator for every 100 prey.

Biological control programs for geranium growers are experimental at this point. More work is needed to develop reliable programs.

Biological control with chemical finish. An alternative to a season-long biocontrol program is to start with biocontrol and avoid pesticides as long as predators are effective. If biocontrol doesn't work or if it breaks down at some point because of pesticide interference (such as a fungicide toxic to predators) or an unexpected pest problem (such as western flower thrips when flowering is initiated), then finish the crop with chemical control. It's not possible to start with chemicals and switch to biocontrol. A single insecticide or miticide application when the plants are small may preclude using predators for the entire growing season, but a grower can start with biocontrol and switch to chemical control for the remaining season.

Caterpillars and Slugs

Murdick McLeod

Many caterpillars can damage geraniums. Some common caterpillars and their typical damage symptoms are described below. Consult local extension specialists for specific control recommendations.

Geranium plume moth

The geranium plume moth, *Platyptilia pica* Walsingham, is generally a problem in California, but is occasionally transported on cuttings to greenhouses in the northeast United States. Adults are long, slender, brownish moths with bilobed forewings and a wingspan of 15 to 25 mm. Eggs are deposited anywhere on the host plant.

Damage is caused by larvae that mine leaves in addition to feeding externally on leaves, buds and flower parts. Larval development requires three to five weeks. Fully mature larvae are about 10 mm long, green to red colored and are covered with many setae swollen at the tip. Brown pupae hang from plant leaves' undersides. This pest may occur in greenhouses year-round.

Cabbage looper

Cabbage looper, *Trichoplusia ni* (Hubner), (Fig. 31-11) is a pest of many cultivated plants. Moths have brown forewings with a distinctive, silver spot near the wing's center. Adults are attracted to lights at night and occasionally find their way into greenhouses where they deposit eggs on

Fig. 31-11. Cabbage looper larva.

available host plants. Adults can produce 275 to 350 eggs.

Larvae feed on leaf undersides, removing a tissue layer to produce "windows." Larvae have three pairs of prolegs and move about with a characteristic looping motion. Early instars have several white stripes running the body's length, but later instars have less notable striping. Larvae grow 30 mm long before pupating in a loosely woven cocoon, usually located on the leaf underside. Larvae feed for two to four weeks.

Slugs

Slugs (Fig. 31-12) aren't insects at all but belong to a group of creatures known as mollusks. Because they are invertebrate animals that feed on agricultural crops, they are frequently included in discussions of insect pests. Slugs are soft-bodied creatures with two pairs of tentacles located on the head. They move about by gliding on a mucus path secreted from a gland located just below the mouth. Activity is generally restricted to nighttime and, therefore, slime trails are often the only evidence of slug presence. Many different species occur in the greenhouse.

Eggs are deposited in groups of one to several dozen in cracks or crevices of the soil surface. Immature slugs have the same body shape as adults but are smaller and lighter in color.

Fig. 31-12. Slugs.

Slug feeding damage is visible as holes chewed in leaf or stem tissue. Both adult and immature slugs feed on plant tissue. Because slugs are active at night and hide during the day, they may not be visible when damage is noticed, but their slime trails help identify the culprit.

Limited chemical compounds are effective for slug control. Baits that are lethal when fed upon are commonly used in slug control.

Fungus Gnats

Mark E. Ascerno

Fungus gnats (primarily *Bradysia* spp.) are common geranium pests. Fungus gnat adults are less than ⅛-inch (3 mm) long, gray to black flies with relatively long, delicate legs. The presence of a distinctive, Y-shaped vein in each wing (Fig. 31-13) distinguishes fungus gnats from other adult flies. This characteristic is best seen under magnification.

Not requiring magnification and easier to see are antennae that are noticeably longer than the head (Fig. 31-14). Fungus gnats, although easily agitated, are poor fliers and tend to stay on soil surfaces.

Fig. 31-13. Fungus gnat showing Y-*shaped vein in the wing.*

Adult fungus gnats don't directly damage plants. When abundant, however, they can become a nuisance. Customer complaints, particularly through retail operations and supermarket chains, create pressure to control fungus gnats during production.

Fig. 31-14. Fungus gnat adult antennae.

Fungus gnats undergo a complete change in their body form (metamorphosis) in which the immature stage doesn't look at all like the adult. Fungus gnat larvae are wormlike, translucent to opaque white in color and have a distinctive, shiny black head capsule (Fig. 31-15).

Fungus gnat larvae normally develop in fungi-infested soils, feeding on decaying plant material [1]. Once established, however, larvae can damage plants by feeding on small roots and root hairs. In some cases the larvae may extend their damage by tunneling in tender, succulent

Fig. 31-15. Fungus gnat larva.

tissue. Tunneling in geranium stems is usually associated with root rot.

Fungus gnats probably occur more frequently and create more problems in plug production than in conventional geranium production. This is related to high humidities associated with plug crops and abundant small roots and root hairs favored by fungus gnat larvae.

For many reasons growers tend to react to fungus gnat problems only after the adult stage becomes apparent. Unfortunately, this may place the grower in a difficult, catch-up situation. The relationship of beneath-the-bench fungus gnat populations to crop problems, the fact that fungus gnats thrive under the same conditions that favor soil fungi, that there is a free-flying, nonfeeding adult stage and that larvae are found in the soil all set the stage for a control program that integrates several different approaches.

Larvae control

Proper greenhouse sanitation is important to reduce fungus gnat populations. Growers who allow weeds to remain beneath benches are asking for problems. The weeds trap moisture and contribute decaying organic material that serve as food for fungus gnat larvae. In addition, plant parts from the crop can drop, be thrown or swept under the bench. Fungus gnats usually get their start under the bench.

Several approaches can make the underbench environment inhospitable for fungus gnats. Keeping the area free of weeds and plant debris reduces available food. Do this by handweeding and cleanup or applying herbicides.

Other techniques may eliminate unwanted weeds and prevent fungus gnat development. Growers using these approaches claim they are effective, but I haven't found research data to verify or refute claims.

One approach uses a slurry of hydrated lime at 1.5 pounds per gallon of water (180 g per l), which is distributed under the bench. Applying hydrated lime is messy, potentially irritating to the skin, and takes up to two days to dry. A month's residual is reported at this rate.

A second approach uses copper sulfate at 1 pound per gallon (120 g per l, .5 kg. per 3.8 l) of water. Spray the mix on all under-bench soil. A three-month residual is claimed at this rate. In either case, crop contact can result in damaged plant tissue.

Growers who have used both techniques feel that copper sulfate is easier to work with than hydrated lime. Overall, the easier application and longer residual action make copper sulfate less expensive than hydrated lime even though copper sulfate is more expensive per pound.

Even with good under-bench control, in-crop control is often needed. Diazinon (Knox-Out), oxamyl, Gnatrol and bendiocarb (Turcam, Dycarb) have been reasonably good at controlling fungus gnat larvae. Since most larvae feed close to the soil surface, light drenching is usually effective. (Because registrations are constantly changing, be sure to check with your local cooperative extension service before using any pesticide.)

Adult control

Adult fungus gnats may also need control, particularly if the population is out of hand. Resmethrin in aerosol form has given good adult control.

Monitoring fungus gnat population levels is also important in a good control program. The adult stage is attracted to the color yellow and can be trapped on yellow cards coated with a sticky substance. Hang the traps just above the crop surface, check regularly for fungus gnat adults and record the number found. Look for small flies that have antennae longer than the head and a Y vein on each of its two wings (Fig. 31-16). An increasing population requires greater control efforts, while one that is decreasing indicates control techniques are working. Traps are available commercially. Finally, since fungus gnat larvae do best when fungi are present in the soil, control soil fungi and root rots in particular. Using the appropriate fungicide helps reduce fungus gnat problems. Check your local extension service for recommendations.

It may also be helpful to understand some conditions that add to fungus gnat problems. Cloudy, cool days slow evaporation and increase condensation. Thus, potting media and soil beneath benches remain wet

Fig. 31-16. Fungus gnat adult on yellow sticky card. Note the antennae and Y vein.

for longer periods encouraging fungal growth. Increased fungi results in increased fungus gnat populations.

Although weather can't be controlled, we can influence how fast things dry out. In addition, weather can be a barometer of how much attention and time will have to be spent on fungus gnat control.

The growing media used may also influence fungus gnat populations [2]. Certain media, especially those containing composted bark or having a peat base, may produce more fungus gnats than other media types. Lindquist et al. [2] also found that some insecticides were more effective in some media than in others. Thus, using more artificial growing media could contribute to increased fungus gnats. This doesn't mean that growers should return to soil-based mixes. Rather, they should consider fungus gnats as an insect that probably requires greater attention.

References

[1] Kennedy, M.K. 1974. Survival and development of *Bradysia impatiens* (Diptera: Sciaridae) on fungal and nonfungal food sources. *Ann. Entomol. Soc. Amer.* 67:745-749.

[2] Lindquist, R.K., W.R. Faber and M.L. Casey. 1985. Effect of various soilless root media and insecticides on fungus gnats. *HortScience* 20:358-360.

Pesticide Application: High- and Low-Volume Methods

Richard K. Lindquist and Charles C. Powell

Pesticides (including fungicides, insecticides and miticides) are important to most geranium pest and disease management programs. Knowing major aspects of pesticide application helps you make informed decisions concerning what application methods and equipment are best for you.

Any pesticide application's most important objective is to deliver pesticide to a target in sufficient concentration to control the pest or pathogen involved. The target may be an entire plant, a specific plant part or the growing medium, as well as the pest or pathogen. Defining the target is important to proper pesticide application. Is the objective to hit airborne pests? Pests on or in leaf surfaces? Are the pests on all plant parts or only in certain areas? Are the pests in the growing medium or under benches?

Keep in mind that the pesticide applied and its mode of action are crucial to any application's success or failure. Does the pesticide have vapor or systemic activity? If so, it will *redistribute* from its deposition point and reach other areas. Many fungicides are preventive (they prevent a pathogen from becoming established). Most insecticides are eradicative (they eradicate a pest that happens to be present, but don't prevent future infestations).

Pesticide application is a two-step process: deposition and distribution. Deposition is applying the pesticide to the target area, and distribution is getting the material to the correct area in the amounts required to be effective. With any applicator, you must first distribute pesticide to the area, then deposit it on the target.

Spray drop size is important in determining any application's deposition and distribution. Table 31-3 shows the number of different sized spray drops that can be produced from 1 quart (about 1 liter) of liquid. It's easy to see that small drops potentially increase coverage. Research has shown that the large number of small drops produced by some applicators, if distributed properly, gives better pest and disease control than fewer, larger drops, if deposition proceeds properly. Small droplet deposition depends greatly on target characteristics (foliage canopy thickness, plant height, bed, floor or bench-grown plants) and equipment characteristics (air-assisted movement, flow rate).

TABLE 31-3

Theoretical spray coverage*

Drop diameter (microns)	No. drops per square centimeter
10	19,099
20	2,387
50	153
100	19
200	2.4
400	0.3
1000	0.02

*In drops per square cm, applying 1 liter per hectare (0.4 quart per acre) with different sizes of drops.

High-volume sprays

High-volume sprays (HV) are the most traditional way of applying pesticides to geraniums in greenhouses and outdoors. Equipment and methods haven't changed much over the years. These sprays involve mixing a certain quantity of pesticide with a large volume of water and spraying the plants or growing medium to some point of wetness. Water is used in two ways: to dilute the pesticide concentrate and as a carrier to deliver material to the target and deposit it on the target.

High-volume sprays can be inefficient. For instance, when the target is a small flying insect, only about 2% to 6% applied pesticide actually reaches its intended target, with the remaining material being lost through evaporation, drift or runoff. HV spray inefficiency in such cases is related to spray drop size. Most volume in HV sprays consists of large drops (greater than 100 to 400 microns diameter), but there are also significant numbers of very small drops. Neither very large nor small spray drops deposit well on target surfaces.

HV spray efficiency can be significantly increased with proper application techniques. With techniques such as using high pressure, getting close to plants and moving the spray nozzle in an arc, foliage canopy penetration and leaf surface coverage can be quite uniform.

High-volume spray equipment is widely available, relatively inexpensive and remains the only legal way to apply many pesticides. Every greenhouse should have an HV sprayer. This application method is, however, time consuming, often inefficient, wasteful, environmentally undesirable and may delay reentry into the greenhouse because of the time it takes the crop to dry.

Low-volume sprays

These sprays often utilize specialized equipment, including thermal pulse-jet foggers, mechanical aerosol generators and electrostatic mist sprayers. They are designed to eliminate many disadvantages of conventional HV applications. They often take less time, use less water or oil to dilute and carry pesticide (no runoff, faster reentry), may use less pesticide and produce most of the spray volume in small spray drops, which are supposed to be more efficient.

Low-volume applications are not new. Various types of LV equipment have been in use for decades. Advantages and limitations of using certain types of presently available low-volume equipment follow.

Thermal pulse-jet foggers

Thermal foggers have been used for more than 30 years. Originally, foggers were used to apply vapor-active fumigants. As might be expected, they are very effective at this. Thermal foggers also are effective at applying residual pesticides, often depositing as much pesticide as an HV application. Most thermal foggers disperse both liquid or wettable powder formulations. A carrier, or dispersal agent, to be mixed with or used in place of water, may be specified by the manufacturer to help ensure a persistent fog.

Pesticide deposition within the plant canopy and on leaf undersides is often poor, so using thermal pulse-jet applicators with certain pesticide/pest combinations won't be successful. Using a pesticide that redistributes after application through systemic or vapor action can be very effective.

Thermal foggers produce very small drops, mostly from 10 to 50 microns in diameter, that are able to move rather long distances from the applicator. With some of the larger units available, the drops will travel more than 200 feet (65 m). Liquid flow rates also vary with the unit size. For example, a small fogger will disperse 10 liters (2.6 gallons) in 30 minutes, and the largest model will disperse 20 liters (5.2 gallons) in 30 minutes. The area covered will depend on whether a wettable or liquid formulation is applied. Foggers aren't used to apply pesticides to small areas for spot treatments. They are designed to treat large areas quickly.

The primary advantage of using a thermal fogger is the short time required to apply a pesticide to a large greenhouse area. Thermal foggers can be very effective if the correct pesticide is used. Disadvantages include poor foliage penetration and leaf underside coverage. Unless the greenhouse is ventilated, the small drops produced by thermal foggers remain suspended for hours after application. This can be a potential safety problem.

Aerosol generators

Several different aerosol generators are now being sold. These sprayers use air pressure, supplied by a compressor, to break up spray liquid into small drops, sometimes less than 5 microns in diameter. Air is also the main method of moving the spray around the greenhouse and onto the foliage. The greenhouse air circulation system, horizontal air flow or overhead convection tubes complete much of the spray movement.

Aerosol generators are supposed to disperse both liquid and dry pesticide formulations. Some growers have experienced nozzle clogging with powder formulations. We have had the same problems when using some liquid formulations. The pesticide can be mixed with water in the tank, the application set to take place when no people are present and the area ventilated before any workers reenter the greenhouse. The flow rate from these sprayers is less than 2 fluid ounces (59 ml) per minute, so it will take about 2½ hours to apply 7 liters (1.5 gallons) of spray.

Our results in commercial greenhouses, using fluorescent tracer material as well as nonsystemic insecticides, have shown that deposition and distribution vary quite widely, and that deposition on upper leaf surfaces was much higher than on undersides. In smaller research greenhouse experiments, however, greenhouse whitefly larvae control on undersides of leaves with bifenthrin was excellent. Similar results have been obtained against two-spotted spider mites and melon aphids.

The obvious advantage of using this sprayer is that the application can be made without anyone being in the greenhouse. Questions that need answering are largely safety related, such as the quantity of pesticide remaining in the air after application and off-target deposition on benches, fans or walls.

Air-assisted electrostatic sprayers

Electrically charging spray drops to better cover leaf undersides and reduce spray drift is an idea that has been around for many years. We have evaluated three types of electrostatic, low-volume applicators. In general, our results showed that air assistance from a fan on the sprayer or an air compressor gave better pest and disease control than nonair-assisted sprayers.

Two air-assisted electrostatic sprayers are being sold in the United States. These sprayers produce drops with a volume median diameter about 30 microns. About four to 16 gallons of spray are applied per acre (37.5 to 150 l per ha), depending on pressure and walking speed. Both liquid and powder formulations can be applied. Sprays are applied by walking through the greenhouse, aiming the spray ahead and slightly downward toward the crop with a sweeping motion. Practice is necessary to obtain even coverage and deposition. The time required to treat an area is somewhere between that required to use a thermal fogger and high-volume spray. One acre can be treated in about one to two hours.

We have conducted several experiments in commercial and research greenhouses with the air-assisted sprayers. Our results showed a "plant position" effect: Pest control was better and fluorescent tracer deposition was higher on plants nearest the sprayer nozzle. This effect varied with pesticide and sprayer. Fig. 31-17 summarizes fluorescent particle deposition within a potted geranium plant canopy after electrostatic sprays using three spray volumes. Total deposition was related to spray volume, as might be expected. Furthermore, the electrostatic charging effect was most pronounced at lower spray volumes. Most deposition was on upper leaf surfaces; however, excellent greenhouse control of immature whitefly has been obtained using bifenthrin, indicating deposition of at least some pesticide on leaf undersides. Deposition and distribution of pesticides applied with this (and other) application equipment depends on plant type and spacing. No sprayer can effectively penetrate a closely spaced, mature crop. Experiments are still in progress, using fluorescent tracer and bioassays for both pests and diseases, attempting to obtain more information on these sprayers' characteristics.

Fig. 31-17. Thermal fogger.

Fig. 31-18. Aerosol generator.

Fig. 31-19. Electrostatic sprayer.

Conclusions

Any pesticide application method and equipment has advantages and disadvantages. Every greenhouse should have a good high-volume sprayer. After that, you need to determine which, if any, of the low-volume sprayers best fits your greenhouse size, design and crops.

The basic facts are that, in many cases, low-volume application technology is well ahead of legality. Any application method not prohibited on the label can be used to apply a pesticide; however, label directions for a specific dilution (1 pound per 100 gallons; 120 g per 100 l) effectively prevent low-volume applications. These sprays must be made at very high concentrations (10 to 25 times normal) to be effective. This makes the legality of using such equipment questionable when applying traditionally labeled pesticides. Some pesticide labels, particularly on recently registered products, mention concentrate sprays or specify pesticide amount to apply per acre. As new labels are registered, more pesticides should become legal for low-volume sprayers.

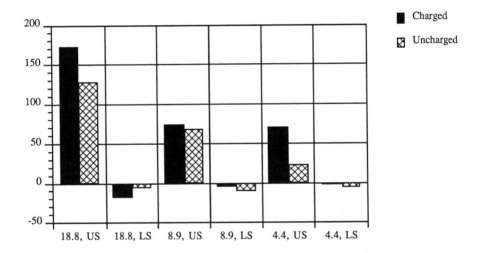

Fig. 31-20. This graph shows the fluorescent tracer deposition (FPS) per square centimeter. FPS on upper (US) and lower (LS) leaf surfaces using charged sprays at three spray volumes (18.8, 8.9 and 4.4 gallons per acre; 178, 84, 42 l per ha).

CHAPTER 32

Physiological and Environmental Disorders

Ralph N. Freeman

Growers producing geraniums can experience various difficulties at one time or another. When a difficulty or disorder occurs, it's important to determine what it is and its cause. Many times the grower is an excellent diagnostician and able to identify the disorder. From time to time, an unbiased diagnostician is necessary. In that case it's wise to engage an horticultural consultant or extension specialist.

Identifying disorders and correcting the situation isn't always easy, nor can it always be done quickly. The diagnostician—whether a grower, consultant or extension specialist—must be objective and systematic in thinking, have a discerning eye for accurate observations and the capacity to ask the right questions, determine and evaluate what has been done and decide if those procedures have any relationship to the problem or disorder. He must also be sensitive to growers (feelings and attitudes), have an intuitive insight into problems, be analytical, have a rational approach, obtain relevant facts, analyze facts and data collected and arrive at a reasonable solution.

Identifying disorders, evaluating the cause and rectifying the situation isn't always easy. There's no question, it takes a special person. Many times the problem is identified visually, but other times intensive testing, laboratory culturing, chemical analysis or other procedures must be employed. Some tests may be done quickly while others will require considerable time, requiring much patience from the grower.

A real resource for growers diagnosing many problems affecting geraniums (or any other crop) is literature. Numerous researchers and authors have contributed much information to help growers become diagnosticians. Being a diagnostician requires reading, studying and critical evaluation along with a good knowledge of plant structure and function.

The remainder of this chapter reviews many common symptoms observed on geraniums at one time or another. The areas noted involve physiological problems as well as pest problems such as diseases, insects and weeds. The areas to be considered are:

- Edema
- Petal shattering
- Gases and volatiles
- Plant wilting
- Misshapen plants or plant parts
- Leaf and plant color

Edema

Edema, sometimes called oedema, occurs on stems, petioles and leaf undersides as small, pimplelike swellings filled with plant fluids (Fig. 32-1). They frequently enlarge, turn brown and become corky. This condition has been well described previously in this book (see Chapter 25, Foliar Diseases).

This isn't a disease but a physiological problem that frequently occurs on geraniums. It's more prevalent on cutting-type geraniums and particularly on many ivy geraniums. Some varieties are more prone to edema than others.

It often appears during cool, cloudy weather when soil is moist and warm. During these conditions plants can take in luxurious amounts of water without transpiring it. Pimplelike blisters develop, and they after a period of time burst and leave behind corky scars (Fig. 32-1).

Edema is frequently confused with spider mite injury.

To help control this condition, select cultivars resistant to edema, space plants to allow good air circulation and don't overwater during

Fig. 32-1. Edema on an ivy geranium. Note the pimplelike blisters and corky areas that develop in an advanced stage.

cool, cloudy conditions. Irrigate/fertilize early in the day and use a light, well-drained growing mix.

Petal shatter

On some hybrid geranium cultivars flower petals fall off while the crop is still on the bench as well as during shipping. This condition, called "petal shattering," has been described by several authors.

Many seed geranium cultivars, more commonly called hybrid geraniums, are very susceptible to this condition. At Michigan State University, Miranda found that seed geraniums had an area of small cells between the petal and the receptical resembling an abcission zone. When petals shattered, they always broke off in this area. This attachment area was found to be narrower than that in the cutting types and the ratio of fresh weight to attachment width wasn't greater in seed geraniums than cutting geraniums. Thus, the breaking can occur more easily in this area.

Ethylene concentrations as low as 0.1 ppm in the air around the plant were found to be a causal factor. Ethylene is generated internally (within plants) and affects growth and development at low concentrations whether the plant is exposed to it in the air or through material sprayed on the plant.

The Dutch began using silver thiosulfate as a compound to help reduce blossom shatter. Workers at MSU experimented further and perfected its use for the North American grower. The concentration and usage suggested is as follows:

(1) Weigh 20 grams (¾ ounce) silver nitrate ($AgNO_3$) and dissolve in one pint distilled water.

(2) Now weigh 120 grams (4½ ounces) of sodium thiosulfate (prismatic) [$Na_2S_2O_3(5H_2O)$] and dissolve in another 1 pint (½ l) distilled water.

(3) Pour the silver nitrate solution (from Step 1) into the sodium thiosulfate solution (from Step 2) slowly, while stirring briskly. This will result in a total of 2 pints or 1 quart (.9 l) solution.

(4) For seed geraniums use 2 teaspoons (10 ml.) stock solution to 1 gallon (4 l) water for a direct spray solution. Spray ⅓ fluid ounce (10 ml) per plant when the buds begin to open and color is just visible. This stock solution will prepare 96 gallons (363 l) spray, which is enough to treat 36,000 plants.

(5) When measuring and mixing, use only glass or plastic equipment. Use plastic sprayers when mixing and making applications. Metal containers deactivate the STS solution.

(6) Treated plants must be Pythium-free. Infected plants have been found to die quickly following STS application. Use a fungicide to help control Pythium prior to spraying STS.

To help reduce blossom shatter, ship plants in bud stage, select cultivars more resistant to this problem, apply STS (silver thiosulfate) at the appropriate time, grow at suggested production temperatures, finish the crop cool and keep the greenhouse ethylene-free.

Gases and volatiles

A number of gases and volatiles can affect geranium production, some positively, others negatively. Growers may wish to refer to a very detailed discussion of air pollution effects on ornamental crops by Marlin Rogers in the *Ball RedBook* 14th edition. That discussion reviews air pollutant effects on many crops.

Flue gases

Incomplete combustion in the firebox of hot air furnaces or boilers; puff backs in windy, gusty weather; leaky heat exchangers; leaky flue pipes; lack of oxygen when burners ignite (especially in tight, polyethylene greenhouses) have been found to affect geraniums and many other crops. The effects, depending on concentration and length of exposure, include hardened growth, petal drop and necrotic spots of varying shapes in the older leaves' interveinal areas. Many of these problems result from sulfur dioxide (SO_2) injury (Fig. 32-2).

Ethylene

Geraniums are known to be poor crops to ship in closed containers. Frequently, when shipments are delayed and left in warm areas, the plants deteriorate, resulting in very yellow plants with some epinasty developing. Geraniums are excellent ethylene producers in dark, warm and closed conditions.

Florel (Ethephon or Ethrel)

Florel is frequently used on stock plants to encourage development of many cuttings at one time. This product sprayed on plants releases ethylene. Although there's a beneficial effect at specific concentrations (allows a cutting peak to develop), plants treated often appear poor in quality shortly after treatment (some yellowing, some epinasty or elongated petioles).

Fig. 32-2. Air pollution injury on geraniums as a result of sulfur dioxide (SO_2) in the greenhouse.

Wood preservatives

Green cuprinol and CCA treated wood is safe to use in greenhouses if the products are washed and adequate aeration follows installation. Other wood preservatives are very toxic to many plants, including geraniums, and shouldn't be used. The chemicals themselves and their volatiles can be detrimental to numerous crops.

Paints

The only paints and stains that should be used in greenhouses are those designed for "greenhouse use." Only one or two manufacturers supply these products. Use only "paints" available from horticultural suppliers. The volatiles in paints available from hardware stores, lumber yards, discount houses and building supply centers may be very troublesome to greenhouse crops.

Carbon dioxide

Introducing carbon dioxide into the greenhouse atmosphere at approximately 1,000 ppm each day from 9 a.m. to 4 p.m. during the fall, winter and spring production period has beneficial effects: increased production, higher quality plants and/or cuttings and shorter cropping periods. The use can be extended if HID lights are used during the night. An extensive discussion by Freeman on carbon dioxide usage appears in the 14th and 15th editions of the *Ball RedBook*.

Plant wilting

Occasionally, growers experience wilted plants. Wilting is nothing more than a water deficit occurring in the plant itself. In other words, the water loss is greater than the uptake; thus, the plants react by what is commonly known as wilting.

There are many causes of wilting:

• Soluble salts in the growing mix are too high, restricting water uptake (Fig. 32-3).
• Insufficient water in the growing mix for roots to take up. This may be due to excessive air movement in certain greenhouse areas causing localized wilting, lack of irrigation or poor aeration in the growing mix.
• Vascular systems affected by common diseases such as Pythium, Rhizoctonia, bacterial blight (*Xanthomonas pelargonii*), *Thielaviopsis basicola* and *Verticillium alboatrum*.
• Severe nematode activity on the root system. Due to modern cultural methods, this is rarely seen today.
• Extensive leaf (foliage) area in relation to the root system. This creates a water deficit.
• Bright and sunny weather following dark weather periods. Here the plants are acclimated to dark weather conditions and when the bright, sunny weather occurs in combination with higher temperatures in the greenhouse, plants become stressed and wilt.
• The substrate (growing mix, soil) may be poor quality and texture, providing inadequate root system aeration.
• Improper paint or herbicides may cause plants to wilt. Usually there are other symptoms affecting leaf shape, coloration or poor growth.
• Certain herbicides at low rates in the growing mix or water supply may induce wilting before other symptoms are expressed. At the present time only Surflan and/or Diquat are suggested for greenhouse use.

Misshapen plants or plant parts

Plants may appear to be misshapen at various times. Misshapen plants or plant parts may appear with the following symptoms:

Crinkled leaves. One or more viral diseases or a growth regulator overdose may cause this. (See Chapter 28, Viral Diseases and Chapter 7, Growth Regulating Chemicals.)

Spindly plants. Improperly spaced plants (too close together) and/or inadequate light sometimes in combination with high production temperatures may cause this.

Fig. 32-3. An overapplication of Osmocote and improper distribution in the mix (note the band of Osmocote approximately 2 inches (13 cm) above the bottom of the root ball) caused this plant to wilt, then develop necrotic areas at the leaf edges. Using the proper rate and distributing it in the mix alleviates this difficulty.

Stubby roots. Roots become flaccid then turn black within a few days and soon become stubby and thick, along with a dying meristem, perhaps from a calcium deficiency. Also, some leaves may become smaller and growth stunted.

Leaf cupping. This is normally a viral disease resulting in the upward cupping of the leaf laminate. The leaves become extremely distorted and misshapen. (See Chapter 31, Viral Diseases.)

Curled leaf edges. The omnivorous leafroller larvae "roll or tie" leaves together with webbing material. The rolled leaves are very obvious.

Yellow vein network on leaves. This condition is probably due to a viral disease causing a yellowing along the veins on both older and young leaves. These outstanding symptoms have caused affected plants to serve as stock plants. Cuttings taken were grown and plants displaying what is termed as net-vein virus are sold as a special item. (See Chapter 28, Viral Diseases.)

Flecks on petals. White or light-colored flecks of varying shapes may be due to pesticide damage, thrips feeding and/or Pelargonium flower break, a virus.

Leaf spots. Leaf spots occur in varying shapes, locations, sizes and colors, depending on the cause. These range from small pieces of rust dropping from overhead heating pipes; shading compound leaking through laps in the glass or finding entry through ventilators or other openings; spots from paint; pesticide residues; feeding from insects (slugs or thrips); and diseases such as Botrytis viral diseases, bacterial leaf spot, rust, Cercospora or Alternaria. Fig. 32-4 shows leaves treated with chemical overdose, thus causing injury (spots and necrotic areas at the leaf edge).

Fig. 32-4. Geranium leaf showing necrotic spots and leaf edges resulting from a chemical overdose.

Leaf holes. Various size holes in leaves are frequently caused by cabbage loopers, slugs and geranium plume moths. (See Chapter 31, Insects and Other Pests.)

Short, stubby growth at plant base. Occasionally bacterial fasciation, caused by *Corynebacterium fascians*, occurs at the plant base causing several short shoots or stems at or near the soil line. They are short, swollen, fleshy, twisted or aborted and produce misshapen leaves resembling a witches-broom. Plants aren't killed, but the plant growth may become stunted or retarded.

Reduced flowers. Plants exposed to water stress, overwatering or temperatures greater than 80 F (27 C) may have fewer flowers.

Crook neck. This occurs on certain cultivars and is more prevalent on older ones.

Root deterioration. The root system may deteriorate due to diseases (see Chapter 28, Viral Diseases), excess fertilizers, soilborne pests such as nematodes, fungus gnats, compact growing mix (not enough aeration and drainage) or overwatering.

Germination reduced. If the germination percentages are drastically reduced, possible causes may be irrigating with cold water below 60 F (16 C), media temperature during germination below 70 F (21 C), media temperature above 90 F (32 C), insufficient oxygen for germination or very compact growing media.

Leaf and plant color

To the skilled diagnostician, leaf color and condition indicate a plant's general health and nutrient condition. Although visual symptoms are an unreliable diagnostic tool, soil testing and foliar analysis provide more

TABLE 32-1

General diagnostic guidelines

Symptom	Probable cause
Leaves light green; some red pigments on major veins and petioles; leaf edges may curl up or down; reduced growth.	Nitrogen deficiency
Leaves yellow; unthrifty plants.	Excess ammonia nitrogen
Oldest leaves yellow.	Leaf on petioles originating at or below growing mix surface. Insufficient light for lower leaves due to crowding plants. Plants grown too dry.
Yellow brown leaf discoloration followed by brown necrotic spots.	Underwatering
Small leaves; necrotic blotches at older leaf margins; margins turn up; purple red color.	Phosphorus deficiency
Oldest leaves light green.	Potassium deficiency Magnesium deficiency
Leaves dull in color; some chlorosis; veins very green; some red blotches.	Potassium deficiency
Marginal necrosis.	Potassium deficiency
Younger leaves light green, chlorotic; zonation prominent then lost; reduced leaf size; chlorosis at leaf edge.	Iron deficiency

Symptom	Probable cause
Younger leaf chlorosis.	Iron deficiency
Steel green gray leaves.	High soluble salts
	High nutrient levels
	Moisture stress
Mid- to older leaves dull in color with slight red tinge and chlorotic.	Iron-Manganese Syndrome (When manganese is present in excessive quantities, iron solubility is decreased and some chlorosis results.)
Upper leaves and terminals white.	Very high temperatures
Leaves have general red tinge and yellow color.	Low temperatures and/or nutrition
Meristem dies; roots flaccid; root tips black; roots thickened.	Calcium deficiency
Necrotic spots on older leaves between veins.	Sulfur dioxide (air pollution injury)
Blotchy light/dark green areas on leaves.	May be due to virus (see Chapter 28, Viral Diseases)
White cottony areas on leaves and stems.	Mealybugs may be the cause (see Chapter 31, Insects and Other Pests)
Black mold on leaves.	Black sooty mold (nonpathogenic, unsightly, caused by heavy populations of whiteflies, aphids, mealy bugs or scales excreting honeydew on which the black sooty mold grows and develops)
Veins in leaves more or less parallel, leaf color may be blotchy green.	Exposure to air pollution or weed control compounds; virus (see Chapter 28, Viral Diseases)
Beige (light brown or tan) mold on mix; surface may reduce percolation of water and fertilizer into mix.	Nonpathogenic fungus
White cottony areas on leaves and stems.	Mealybugs may be the cause (see Chapter 31, Insects and Other Pests)
Black stubs and dieback on plants.	Seen on internode from which cutting was taken. Wound is infected with Botrytis
Black or brown black lesions on stems near soil line and below.	Pythium, Rhizoctonia or other diseases (see Chapter 27, Stem and Root Diseases)

Detailed descriptions on nutrient deficiencies and toxicities are well described in Chapter 5, Fertilization and Post [32].

exact information about the nutrient status and nutrient trends in either the growing mix or dry plant matter. Table 32-1 gives some general guidelines—possible indications as to why plants may respond as they do. Keep in mind a visual diagnosis won't show nutrient imbalances, interactions and other problems that are exhibited when soil tests and foliar analyses are properly used on a regular basis. Visual inspections provide only a general guide to what may be a problem.

References

[1] Armitage, A.M. and W.H. Carlson. 1981. Hybrid geranium shatter. *BPI News* April, 1981.: 5.
[2] Ball, V. 1982. Important: Geranium petal shatter spray. *GrowerTalks* January.: 22-23.
[3] ——— 1985. Hybrid or seed geraniums. *Ball RedBook,* 14th ed. Ed. Vic Ball. Reston, Va.:Reston Publications.
[4] ——— 1991. Seed geraniums. *Ball RedBook,* 15th ed. Ed. Vic Ball. Geneva, Ill.:Ball Publishing.
[5] Carlson, W.H. 1981. Found: An end to petal shattering in seed geraniums. *Amer. Veg. Grower.* October.
[6] ——— 1991. Ivy geraniums. *Spartan Ornamental Network.* (computer database). Mich. St. Univ.
[7] Daughtrey, M. 1991. Diseases of geraniums. *Long Island Hort. News.* March.: 1-3.
[8] Dickey, R.S. 1982. Bacterial fasciation. *Geraniums,* 3d ed. Ed. J.W. Mastalerz and E.J. Holcomb. University Park, Penn.:Penn. Flower Growers.
[9] ——— Root rots, bacterial fasciation. *Geraniums,* 3d ed. Ed. J.W. Mastalerz and E.J. Holcomb. University Park, Penn.: Penn. Flower Growers.
[10] Engelhard, A.W. 1982. Alternaria leaf spot. *Geraniums,* 3d ed. Ed. J.W. Mastalerz and E.J. Holcomb. University Park, Penn.:Penn. Flower Growers.
[11] ——— Bacterial leaf spot. *Geraniums,* 3d ed. Ed. J.W. Mastalerz and E.J. Holcomb. University Park, Penn.:Penn. Flower Growers.
[12] ——— Cercospora leaf spot. *Geraniums,* 3d ed. Ed. J.W. Mastalerz and E.J. Holcomb. University Park, Penn.:Penn. Flower Growers.
[13] Forer, L.B. and L.P. Nichols. 1982. Rust. *Geraniums,* 3d ed. Ed. J.W. Mastalerz and E.J. Holcomb. University Park, Penn.:Penn. Flower Growers.
[14] Fosberg, J.L. 1975. *Diseases of Ornamental Plants.* Special Pub. No. 3. Univ. of Ill.
[15] Freeman, R. 1985. The importance of carbon dioxide. *Ball RedBook,* 14th ed. Ed. Vic Ball. Reston, Va.:Reston Publishing.
[16] ——— 1991. The importance of carbon dioxide. *Ball RedBook,* 15th ed. Ed. Vic Ball. Geneva, Ill.: Ball Publishing.
[17] Holcomb, E.J. and J.W. White. 1982. Fertilization. *Geraniums,* 3d ed. Ed. J.W. Mastalerz and E.J. Holcomb. University Park, Penn.:Penn. Flower Growers.
[18] Langhans, R.W. 1980. *Greenhouse Management,* 2d ed. Halcyon Press.
[19] ——— 1990. *Greenhouse Management,* 3d. ed. Halcyon Press.

[20] Langhans, R.W., J.C. Neal, T.C. Weiler, J.P. Sanderson, M.L. Daughtrey and R.K. Horst. 1990. 1990 Cornell greenhouse crop production guideline for New York State, integrated crop management for geraniums. Cornell Coop. Ext. Cornell Univ.

[21] Laurie, A., D.C. Kiplinger and K.S. Nelson. 1958. *Commercial Flower Forcing,* 6th ed. McGraw-Hill.

[22] ——— 1968. *Commercial Flower Forcing,* 7th ed. McGraw-Hill.

[23] ——— 1979. *Commercial Flower Forcing,* 8th ed. McGraw-Hill.

[24] Lawson, R.H. 1982. Viruses. *Geraniums,* 3d ed. Ed. J.W. Mastalerz and E.J. Holcomb. University Park, Penn.:Penn. Flower Growers.

[25] Lindquist, R.K. 1982. Insects and related pests. *Geraniums,* 3d ed. Ed. J.W. Mastalerz and E.J. Holcomb. University Park, Penn.:Penn. Flower Growers.

[26] Nelson, P.E. and L.P. Nichols. 1982. Bacterial blight. *Geraniums,* 3d ed. Ed. J.W. Mastalerz and E.J. Holcomb. University Park, Penn.:Penn. Flower Growers.

[27] Nichols, L. 1961. Fasciation. *Geraniums,* ed. J.W. Mastalerz. Penn. Flower Growers.

[28] ——— 1961. Oedema. *Geraniums,* ed. J.W. Mastalerz. Penn. Flower Growers.

[29] ——— and P.E. Nelson. 1971. Foliage diseases. *Geraniums,* 2d ed. Ed. J.W. Mastalerz. Penn. Flower Growers.

[30] ——— 1982. Botrytis blight. *Geraniums,* 3d ed. Ed. J.W. Mastalerz and E.J. Holcomb. University Park, Penn.:Penn. Flower Growers.

[31] ——— 1982. Edema. *Geraniums,* 3d ed. Ed. J.W. Mastalerz and E.J. Holcomb. University Park, Penn.:Penn. Flower Growers.

[32] Post, K. 1949. *Florist Crop Production and Marketing.* Orange Judd Publisher.

[33] Reinert, R.A. 1961. Geranium virus diseases. *Geraniums,* ed. J.W. Mastalerz. Penn. Flower Growers.

[34] Rogers, M.N. 1985. Pollution. *Ball RedBook,* 14th ed. Ed. Vic Ball. Reston, Va.: Reston Publ.

[35] Romaine, C.P. and R.H. Lawson. 1982. Viruses. *Geraniums,* 3d ed. Ed. J.W. Mastalerz and E.J. Holcomb. University Park, Penn.:Penn. Flower Growers.

[36] White, J.W. 1971. Fertilization. *Geraniums,* 2d ed. Ed. J.W. Mastalerz, Editor. Penn. Flower Growers.

CHAPTER 33

History

Linda J. Laughner

Our modern geranium, *Pelargonium* x *hortorum,* is a man-made species with a complex heritage. The species is probably a little more than 100 years old. Through these years, changes and improvements have contributed to the geranium's present popularity. Today, the geranium is one of our most valuable floricultural crops. In 1988 the U.S. Census of Horticulture Specialties reported total sales value as a bedding and potted plant at over $173 million.

Taxonomically, the geranium belongs to the genus Pelargonium, not Geranium; however, Pelargonium and Geranium are both genera found in the large plant family known as Geraniaceae. More than 90% of the *Pelargonium* species are native to South Africa.

In 1597, the great English herbalist, John Gerard, wrote of *G. columbinum,* "... the seed set togither like the head and bill of a birde, whereupon it was called Cranes bill or Storks bill, as are also all the others of his kind." Thus both genera got their names from the characteristic shape of the fruit that forms. The word "geranium" comes the Latin *geranos,* meaning crane; "Pelargonium" is derived from *pelargos,* Latin for stork.

It is believed that the first Pelargonium to arrive in England from South Africa was *P. triste,* brought from the Cape in 1632 by John Tradescant (Fig. 33-1) famous botanist, plant hunter and gardener to Charles I of England. This was the period when the Dutch East India Company operated a trade route from the Netherlands around the coast of Africa to India. The Cape of Good Hope became an important stopover on this voyage and a Dutch colony was established there.

By the early 1700s the appearance of *Pelargonium* species in English and European gardens was well documented. African geraniums appeared in *Hortus Elthamensis* in 1732. This was an account of the plants in the garden of a Dr. Sherard, written and illustrated by Dillenius (Fig. 33-2), Oxford University's first Professor of Botany. Phillip Miller's (Fig. 33-3) *Gardeners Dictionary,* published in 1733, listed 20 African Pelargoniums. About the same time a publication from Amsterdam, *Rariorum*

Fig. 33-1. Left, *John Tradescant; Fig. 33-2.* Right, *James Dillenius.*

Africanum Plantarum, by Johannes Burman also contained illustrations and descriptions of *Pelargonium* species.

Although differences had been noted between the native European geranium and the South African Pelargonium, until 1738 they were classified together as Geranium. It was the Dutch botanist Burman, who first insisted that the two should be given separate classifications. He designated the African species as *Pelargonium* and the native species as *Geranium.* This wasn't immediately accepted; 15 years later, when Linnaeus wrote *Species Plantarum,* he still lumped the two classifications together under Geranium. The main feature that distinguishes the Pelargonium is a nectar tube that runs from the uppermost sepal down the pedicel. The bump in the stalk marks the end of the nectar tube. The tube isn't always apparent, however, as it is in double-flowered cultivars.

Many species described by Linnaeus as *Geranium* were again given the name *Pelargonium* by the French botanist L'Heritier. Because of L'Heritier's political involvements and the intrusion of the French Revolution, his manuscript was never published. He did correspond frequently with William Aiton, however, and L'Heritier's descriptions of Pelargonium were eventually published by Aiton in *Hortus Kewensis,* a catalog of plants grown at the Royal Botanic Gardens at Kew, England. It was this publication that finally established Pelargonium as a separate genus. Though Aiton actually published the descriptions, he gave rightful credit to L'Heritier. Thus, when you see, for example, *P. capitatum* (L) L'Her ex Ait, credit for the name goes to three people: L'Heritier and Aiton's names are joined with the Latin ex, meaning "from" for establishing the genus as Pelargonium. The *L* stands for Linnaeus, who gave the species name.

During the late 1700s and early 1800s, Pelargonium importation from South Africa increased dramatically for several reasons. For one thing, Great Britain controlled the Cape of Good Hope at this time. More important, in England especially, this coincided with a growing interest in cultivating plants under glass, and exotic plant species were especially popular. Many Pelargonium species were cultivated and new forms began to appear. Unfortunately, systematic botany was still in its early stages at this time. As a result, a hybrid's parentage, if known, was rarely recorded.

Botanical publications were numerous at this time. One work is especially important in depicting the emergence of our modern geranium: Robert Sweet's *Geraniaceae,* published in five volumes from 1820 to 1830. This extensive work contained 5,000 illustrations of species and hybrids. In addition, hybrid parents were recorded when known.

The first record we have of Pelargonium in America dates to 1760. At this time, seed was sent to John Bartram in Philadelphia. Records indicate that potted Pelargoniums were sold in Philadelphia by 1789. Thomas Jefferson also played a role in bringing Pelargoniums across the Atlantic, sending them home after seeing them in France.

It's difficult to pinpoint when the Pelargonium we know today as the garden geranium appeared. Through history it has been known as the Fish geranium, Horseshoe geranium, Hybrid Perpetual, Zonate and Zonal. The name "zonal" was extremely unfortunate and confusing because it also refers to the native species, *P. zonale.* L.H. Bailey, in 1900, was the first to group the popular garden geraniums as *Pelargonium hortorum.* As of 1952, the International Code of Nomenclature deemed this name legitimate with one small change: the species should be written *Pelargonium* x *hortorum* to acknowledge the group's hybrid nature.

Since the time of Robert Sweet, botanists have tried to place Pelargonium species into groups. In his work *Flora Capensis* in 1860, Harvey divided Pelargoniums into 15 sections based on similarities of either plant habit or flower structure. In his monograph of 1912, Knuth (Fig. 33-4) retained Harvey's divisions. Two other authorities, Clifford and Moore, writing in the 1950s, changed this classification very little. Clifford divides the genus into 14 subgenera, rather than 15. Only three of these subgenera are important when considering horticulturally important Pelargoniums today; from Pelargium we have *P.* x *domesticum,* the Regal geranium; from Dibrachya comes *P. peltatum,* the ivy geranium; finally, from Ciconium we have *P. x hortorum.* Moore, in 1955, divided the cultivated taxa of the genus into 11 groups designated by Roman numerals. Of interest here is that Moore groups together Ciconium and Dibrachya, emphasizing the taxonomic connection between the ivy geranium and the cultivated garden geranium. More recently, Van der Walt [17] established 16 groups in the *Pelargonium* genus.

It's most often reported that the garden geranium, *P. x hortorum,* is a complex hybrid derived from hybridization among members of the

Fig. 33-3. (Left,) *Phillip Miller; Fig. 33-4.* Right, *Knuth.*

subgenus Ciconium. Since few records exist from the time when the garden geranium was developed, speculation on its probable parentage isn't a simple matter. Several authorities' speculations has yielded seven possible contributors: *Pelargonium inquinans, P. zonale, P. hybridum, P. frutetorum, P. scandens, P. acetosum* and *P. stenopetalum.* Historically, *P. inquinans* and *P. zonale* have been regarded as primary parents, with other species contributing traits to a lesser extent. (*Editor's note:* Current research indicates that *P. stenopetalum* (now known as *P. x burtonii*) is actually an interspecific hybrid, presumably between *P. acetosum* and *P. zonale.*)

Several studies have been designed to scientifically assess the relationships among *Pelargonium* species. Harney used paper chromatography to study biochemical relationships between *P. x hortorum* and its putative ancestors. She compared chromatographic leaf extract patterns from 14 hortorum cultivars and the seven species believed involved in their development. The two most likely species, *P. inquinans* and *P. zonale,* had remarkably similar profiles, differing only in one compound. Except in one case, each species or cultivar had a unique profile. The exception is *P. inquinans* and the hortorum cultivar Penny Irene, which match perfectly; although their biochemical profiles match, outwardly the two plants are not alike. The best evidence Harney could find to verify that a species was a parent is the presence of species specific compounds. Harney concluded that *P. inquinans, P. zonale, P. frutetorum* and *P. scandens* contributed to *P. x hortorum* development but couldn't make the same conclusion about *P. stenopetalum, P. hybridum* or *P. acetosum.*

Crossability studies, that is, testing the possibility of a cross being made between two species or cultivars, have also shown relationships within the genus. Harney and Chow studied crossability among four of the probable ancestors of *P. x hortorum—P. inquinans, P. zonale, P. stenopetalum,* and *P. scandens.* Since unsuccessful crosses can be due to (1) failure of the ovule to be fertilized, (2) lack of seed set or incomplete seed development, or (3) lack of seed germination, these criteria were used to determine total fertility of the cross. Interspecific crosses between *P. zonale* and *P. inquinans* were most successful, producing relatively high quantities of viable seed. Crosses using *P. stenopetalum* or *P. scandens,* for the most part, were relatively low. Harney believes that the relative ease of crossing between *P. inquinans* and *P. zonale* confirms that these two species probably contributed most to the development of *P. x hortorum.* This was further confirmed when Chow and Harney used the same four species to test crossability between the species and a diploid *P. x hortorum* cultivar. As was expected, crosses of *P. x hortorum* with *P. inquinans* or *P. zonale* were most successful; those with *P. stenopetalum* and *P. scandens* were relatively unsuccessful.

In the early days of garden geranium development, two descent lines were apparent. One emphasized characteristics of *P. inquinans*: These had well-rounded florets with broad petals. The other emphasized *P. zonale* heritage: Petals were narrower, but flower trusses were much larger. This last type was commonly called Nosegay geraniums.

In what Clifford refers to as "the hybridizing craze of the 1850s," the best of both types gradually merged. The result was larger trusses with broad, more rounded petals. Other favorable changes were evident, including double flowers. Although doubling had been noted before this time, Victor Lemoine in 1864 was the first to create a line of double cultivars. Also, the extensive hybridization and selection during these years led to increased flowering time. Finally, the color range increased greatly, including the first white, broad-petaled type and also petals with various color combinations such as white centers or margins.

In 1870, a change took place that would have important and longlasting effects on garden geranium characteristics. The details aren't clear, but apparently a large size sport (mutation) occurred on a normal size plant. Paul Bruant, a French propagator, recognized this unique change and from it created a new race of geraniums. This new strain has been called the Bruant Race, French Type or *Gros Bois*. These geraniums are characterized by thicker stems, heavier leaves, larger flower clusters, thicker and often shorter flower stalks and in general, a more vigorous appearance. It's generally believed, though difficult to verify, that the mutation was chromosome doubling. This was the origin of tetraploid geraniums. Because of their desirable characteristics, a strong trend began that emphasized tetraploid selection. Today, most asexually produced, commercially successful cultivars are tetraploid.

Bruant's tetraploids were the starting point for several famous plant lines. The Ricards, developed by Bruant in 1894, were the first of these lines. Also from Bruant comes the Fiat group, originating from a cultivar called Gorgeous. The Fiats consist of sports of this cultivar and sports of sports. The most popular of all geranium lines so far, however, has been the Irenes. Although the parentage of the first Irene isn't known for certain, it's believed by some to be a cross between a Fiat and a Ricard. Charles Behringer, Warren, Ohio, introduced the first Irene in 1942. Success wasn't immediate, but following the introduction of Irene to England a few years later, its popularity skyrocketed—both in England and the United States. From 1958 to 1965 the cultivar Irene accounted for one-fourth of all the geraniums grown in the United States. That is really only part of the story because Irene also produced a number of sports. In fact, in 1964 alone, 11 different new Irene sports were released. When Wood wrote his book on Pelargoniums in 1966, he reported that 75% of geraniums grown in the United States were Irenes. Their success is due largely to their reliability—growers and consumers alike could count on their rapid growth and abundant, early flowering.

The diverse heritage of *P.* x *hortorum* contributes to the great intrinsic variation within the species. After nearly 200 years of breeding Pelargoniums, man has tapped this potential to create a highly successful horticultural crop. Knowing the species' heritage can facilitate further improvement by helping us locate and incorporate desirable traits.

During the 1960s, intensive breeding programs in the United States and abroad yielded many superior cultivars to compete with the Irenes. Yoder Bros. introduced notable new cultivars, such as Hildegaard, Sincerity, Cherry Blossom and Snowmass. Improved cultivars, which were more heat tolerant, were released from the breeding program at Iowa State University. These include the Sunbelt Series and the cultivar

Toreador. In Europe, the work of Wilhelm Elsner in Dresden, Germany, resulted in several excellent cultivars such as Karminball, Bruni and Veronica (see Chapter 35, Intellectual Property Protection, Table 35-1).

Not all advances in geranium research during the 1960s and 1970s can be credited to the plant breeders; plant pathologists also played a role. The idea of culture-indexed geraniums originated at the Pennsylvania State University with James F. Tammen. This system provides for propagating geraniums by cuttings from stock plants free of pathogenic bacteria and fungi. A few years later, the process merged with virus-indexing through the efforts of Samuel Smith at Penn State and R. Kenneth Horst at Cornell. In addition to lowering fertility, viruses can slow growth and delay flowering. Thus, when culture-virus-indexing is applied to existing cultivars, their full genetic potential can be realized (see Chapter 29, Clean Stock Production).

Also in the 1960s a new trend in geraniums began to develop. Until then, all commercially successful geraniums were propagated asexually. During this decade, seed-propagated geraniums became a reality. Richard Craig and Darrell Walker of The Pennsylvania State University developed the first commercial, seed-produced inbred, Nittany Lion Red. The first F_1 hybrids were not far behind with the release of Carefree F_1 from PanAmerican Seed Company and New Era from Harris Seed Company. With the advent of seed-propagated geraniums, the diploid geranium saw a revival.

Seed geraniums have come a long way since Nittany Lion Red was introduced. Perhaps the most notable change has been crop time reduced by half. The first Carefree hybrids bloomed in 120 to 140 days. In the early 1980s, the East German Diamonds marked the first dramatically reduced crop time, leading to rapid earliness developments. Today, the Multibloom series can be flowered in only 70 days.

Color range has increased greatly, with series such as the Orbits boasting as many as 17 colors. Though red remains the traditional color, new shades of hot pink or magenta, rich salmon or bicolor hybrids such as Hollywood Star are enticing the consumer to try colors other than red.

Many changes have been inspired by plug technology and mass production. In the early seed cultivars, germination was erratic, often occurring in flushes over one to two weeks. Breeding and advances in seed technology have improved germination characteristics. Now, plug growers can depend on a uniform, high germination rate in three to five days. Since the introduction of Sprinter, geraniums have become more compact, with smaller foliage and shorter flower stalks. By selecting for more dwarf habits and Cycocel response, breeders have created hybrids that grow well in cell packs or pot tight, yet mature into vigorous, well-branched geraniums. The highly successful Red Elite has become the model for judging these traits crucial for pack performance.

For the past 25 years vegetative and seed geranium developments have followed separate paths. Today these paths are merging. Since the introduction of Denholm Seeds' Summer Showers in 1986, ivy geraniums have become a seed product. Likewise, tetraploid geraniums are no longer the domain of vegetative propagators. Two recent releases from PanAmerican Seed, Freckles and Tetra Scarlet, mark the advent of tetraploids from seed. These new classes of geraniums expand Pelargonium's usefulness, and set the stage for further development.

References

[1]   Bode, F.A. 1965. The Irene story. *Geraniums Around the World.* 12(4):76-78.
[2]   Chow, T.W. and P.M. Harney. 1970. Crossability between a diploid *Pelargonium* x *hortorum* Bailey cultivar and some of its putative ancestral species. *Euphytica* 19:338-348.
[3]   Clifford, D. 1956. The parents of the geranium. *J. Roy. Hort. Soc.* 81:20-22.
[4]   ——— 1970. *Pelargoniums including the popular Geranium.* London: Blandford Press.
[5]   Craig, R. 1965. "The future of geraniums." Presentation at the Penn. Flower Growers Conference, University Park.
[6]   ——— 1982. Chromosomes, genes and cultivar improvement. *Geraniums,* 3d ed. Ed. J.W. Mastalerz and E.J. Holcomb. University Park, Penn.: Pennsylvania Flower Growers.
[7]   ——— 1983. Geraniums for the 80s. *Flor. Rev.* 9:21-22.
[8]   Harney, P.M. 1966. A chromatographic study of species presumed ancestral to *Pelargonium* x *hortorum* Bailey. *Can. J. Genet. Cytol.* 8:780-787.
[9]   ——— 1976. The origin, cytogenetics and reproductive morphology of the zonal geranium: a review. *HortSci.* 11(3):189-194.
[10] Knicely, W.W. 1964. "Chromosome numbers and crossability studies in the genus Pelargonium." M.S. Thesis. The Penn. State Univ.
[11] Krauss, H. 1955. *Geraniums for Home and Garden.* New York: Macmillan Company.
[12] Moore, H.E. 1955. Pelargoniums in cultivation (II). *Baileya* 3:71-97.
[13] Moore, H.E. and P.A. Hyypio. 1982. Taxonomy of Pelargoniums in cultivation. *Geraniums,* 3d ed. Ed. J.W. Mastalerz and E.J. Holcomb. University Park, Penn.: Pennsylvania Flower Growers.
[14] Swinbourne, R.F.G. 1973. The unit Pelargonium, botanically and horticulturally classified. The Proc. of the First Intl. Sem. on Nomen. and Regis. of Pelargonium cultivars. *Science House.* Sydney.
[15] Tyler, H.A. 1969. The French type geranium (I) *Geraniums Around the World* 17(3):8-10.
[16] ——— 1970. The French type geranium (II) *Geraniums Around the World* 17(4):8-12.
[17] Van der Walt, J. and P. Vorster. 1988. "Pelargoniums of Southern Africa 3. National Botanic Gardens, Kirstenbosch, Republic of South Africa." (Note: The reader is also directed to the other writings of J.J.A. Van der Walt.)

[18] Voigt, A.O. 1982. Status of the industry. *Geraniums,* 3d ed. Ed. J.W. Mastalerz and E.J. Holcomb. University Park, Penn.: Pennsylvania Flower Growers Association.

[19] Wilson, Helen Van Pelt. 1965. *The Joy of Geraniums.* New York: M. Barrows and Co. Inc.

[20] Wood, H.J. 1966. *Pelargoniums.* London: Faber and Faber Ltd.

CHAPTER 34

Breeding Geraniums for 2000 and Beyond

Richard Craig

Since *Geraniums III* was published in 1982, many changes have occurred in geranium science and technology. Scientists have not only learned more about geranium cytology, genetics, biochemistry, taxonomy and morphology, but also have initiated important studies in biotechnology and molecular genetics. Fundamental studies have resulted in isolating and cloning geranium genes and within the year these genes will be inserted into other Pelargoniums and other species.

We are at a crossroads of science and technology. Breeding programs initiated in the late 1970s and early 1980s have resulted in dozens of new cultivars of both seed- and cutting-propagated zonals, ivies and Regals. As we view the 1991 catalogs, we're amazed that most asexually propagated cultivars are patented; it's not a coincidence that the geranium industry's growth is related to the increased protection of intellectual properties and the resulting use of royalty income to support research and development programs.

Breeding programs

Geranium breeding will be discussed in four parts:
(1) Selection criteria for breeding programs
(2) Breeding asexually propagated cultivars
(3) Breeding seed-propagated cultivars
(4) Information on specific breeding objectives
Geranium cultivars developed for these different purposes require diverse developmental programs and techniques. Those who are interested in producing their own seedlings should see a discussion of pollination techniques by Craig [16].

Selection criteria

Prior to reviewing actual breeding programs, consider what criteria are used to select superior progeny in breeding programs. More details on some of these criteria appear at the end of this section. These criteria include:

- Earliness and continuity of flowering
- General vigor
- Plant height and growth habit
- Response to environment and stresses
- Leaf color, size, shape and placement
- Tolerance to garden conditions
- Leaf zonation
- Seed germination
- Flower color, size, number and type
- Stock plant productivity/rooting
- Improved postproduction traits
- Pollen production
- Pest resistance
- Seed production

Breeding asexually propagated cultivars

Most asexually propagated cultivars of *P.* x *hortorum, P.* x *domesticum, P. peltatum* and other species have been developed by commercial and amateur plant breeders. It's not within the scope of this chapter to review geranium development history in detail, as this has been done by other authors [15, 43]. Further, no attempt will be made to present an extensive list of persons who have been involved in breeding cultivars within the genus Pelargonium. The names of these breeders are presented in literature, particularly in the publications of the International Geranium Society, the British Pelargonium and Geranium Society and the *Journal of the Australian Pelargonium and Geranium Society.* Extensive bibliographies have been presented by Hieke [26] and Clifford [15] and these are supplemented by literature cited in these references. Additional information on modern history is included in Chapter 35, Intellectual Property Protection.

The early history of geranium improvement in Europe has been detailed by Roberts [35], Hieke [26] and Clifford [15] and in the United States by Butterfield [14]. Prominent geranium breeders have been located in many countries especially Australia, Czechoslovakia, England, France, Germany and the United States.

Bode [5] presented an excellent review of *P.* x *domesticum* breeding, including its European and American history (see Chapter 33, History). Schmidt [39] reviewed *P. peltatum* breeding and presented many cultivars' parentage. Ross [37] has provided the history of dwarf geraniums (particularly *P.* x *hortorum*) and lists plant breeders associated with these cultivars. The tetraploid *P.* x *hortorum* cultivar history has been presented by Bode [4] for the Irenes and in general by Tyler [40, 41].

Public plant breeders have also been responsible for developing asexually propagated geranium cultivars. In the United States the most productive programs have been at Iowa State University, where many important tetraploid cultivars have been developed [42, 6, 7, 8, 9, 10, 11, 12]. The University of New Hampshire has also introduced several geranium cultivars. Currently, The Pennsylvania State University has an active breeding program on asexually propagated tetraploid geraniums and has introduced both zonal and Regal cultivars. More information on these programs' results is presented later in this chapter.

As a result of interspecific hybridizations in the 17th and 18th centuries, it's assumed that the geranium's germplasm was originally quite diverse. This diversity resulted from assembling a number of species into *P. x hortorum* (see Chapter 33, History). All early cultivars were diploid and the first tetraploid form (Bruant or French Strain) wasn't reported until 1880. It's reasonable to assume that the earliest cultivars had a limited color range and that these cultivars had single florets. Cultivars propagated by cuttings may be either diploid or tetraploid; however, most commercial types are tetraploid and recent plant breeding research has concentrated on them.

Three methods have been used to improve asexually propagated cultivars:

(1) Sports or mutant selection

(2) Strain selection

(3) Hybridization (in the future to include recombinant DNA protocols)

Sport selection. Mutant types selection has had major importance in improving geraniums. It requires the breeder or grower to recognize a new phenotype as distinct from the parent plant's phenotype. Asexual propagation of the sport can result in a new, distinct cultivar.

Mutations can be of three basic types: gene mutations, chromosome mutations including changes in chromosome number, and cytoplasmic mutations. In all types it's assumed that the mutation occurs in a single cell, which then gives rise to other cells, tissues and organs. Of particular importance are mutations in meristem (growing point) cells. Chimeras are special mutations that result when plants are made up of several chromosomally or genetically different tissue layers. In chimeral plants destruction of the epidermal layers and replacement by second layer cells result in organs that may be quite distinct from the parental phenotype. A more detailed discussion of plant sports and chimeras can be found in Dermen [21].

Most geranium sports are spontaneous mutations; however, irradiation can be used to hasten their occurrence. Richards [34] reported that he developed three diploid cultivars by irradiation: Orchid Princess, Royal Purple and King Midas (with bright, citrus orange flowers). Ross

[38] has reported using strontium 90 in geranium breeding. Further research with irradiation could be valuable in geranium improvement and such research might be useful in explaining major mutation developments in geranium evolution.

Wilson [43] discussed various *P. x hortorum,* "families," some derived mainly through the sporting process. Several examples are Better Times, Ricards, Fiats and Irenes. Note that both diploid and tetraploid geraniums are capable of mutation. The apparent greater number of tetraploid cultivars that have arisen from sports is probably due to their relative importance in the industry.

The following data indicates mutations' importance in modern geranium evolution.

TABLE 34-1

Mutation	Date
Double florets	1864
White flowers	1859 to 1860
Tetraploidy	1880
Colored-leaved types	As early as 1734
Dwarfness	As early as 1799

Source: [15].

These mutations have greatly improved the geranium. Of particular importance would be the mutation from diploid to tetraploid and the mutation from single to double florets. It's assumed that these major mutations were incorporated into cultivars by hybridization.

One major drawback to using sporting as a cultivar improvement method is the narrow germplasm base that is maintained. Most sports are single gene mutations, especially flower color sports, so the remainder of the genotype (all genes except the mutant gene) is identical to that of the parent plant. Thus, all parent genes, both beneficial and deleterious, express themselves in the mutant phenotype. A narrow germplasm can be a disadvantage because it limits the species' or cultivar's adaptability. Sports or mutations on patented cultivars are often governed by propagation licenses; such licenses usually don't affect inventorship with respect to a plant patent; however, this may not be the case with general patents.

Strain selection. Strain selection differs from sport selection only in the intensity of the observed differences and the time required for improvement. In strain selection, the objective is to improve the cultivar by selecting types that are either cultivar characteristics (as originally developed) or cultivar improvements small enough not to distinguish it as a new cultivar. Strain selection can be practical with any cultivar regardless of its origin.

Assuming that every cultivar originates as a single plant and that thousands or even millions of propagants are cloned from the original stock, there are a great number of chances for mutations to occur. In the previous section, we dealt with mutations affecting major characters such as flower color or floret type; in this discussion we are interested in all other plant genes that affect less visible characters. We are interested in dimensional or quantitative characters. These may include flowering time, flower size, plant size, vigor and plant quality.

A practical example of strain selection would be selecting your best plants for propagative stock for the following year's production. At one time, we at The Pennsylvania State University observed a grower who selected his propagative stock from plants left over after the sales season. He had followed this same procedure for several years. We were contacted when he realized that a great proportion of his plants were not going to produce flowers by Memorial Day. By selecting his propagative stock from leftover plants, he had actually selected for later and later flowering plants—a negative strain selection.

Careful strain selection assures improved cultivars. Certain cultivars are designated as improved selections, such as Diddens Improved Picardy. In strain selection, we're hoping to eliminate deleterious mutations and to increase beneficial mutations.

Strain selection has played a major role in developing culture-indexed and culture-virus-indexed geraniums. An integral part of these procedures is greenhouse and field evaluation of asexually propagated progeny from each mother plant—from each culture-indexed cutting or from each meristem designated as free of fungal, bacterial or viral pathogens. Mericlone differences have been observed in research at Penn State (unpublished data) and elsewhere, and have been used to improve commercial cultivars.

Intercultivar hybridization. This breeding technique has been used very successfully with geraniums to ensure germplasm by recombining genes from both parents. Generally, commercial cultivars or breeding parents are cross-pollinated to produce hybrid seedlings. Since most commercial cultivars are extremely heterozygous and usually tetraploid, inbred lines aren't easily produced. The parents' inherent heterozygosity causes great variation among hybrid seedlings. The breeder must grow and evaluate seedlings and select types that are promising. Normally, the selected seedlings are asexually propagated and evaluated as cutting-produced plants (clones). Potential cultivars are evaluated in several stages based on various plant characteristics. Initially, flower color and size and various foliage characteristics are noted. Rooting response must also be evaluated early. After the initial evaluation, more cuttings are propagated and tested for growth habit, time required for flower production, insect and disease resistance and response to various environmental conditions including temperature, carbon dioxide and fertilization. The cultivar's general overall quality, including flower keeping quality and tolerance to stress and nutritional disorders, must also be evaluated. Later, the most promising lines are tested on a wider scale in actual greenhouse tests and in the field, under garden conditions. When the breeder feels that a superior cultivar has been produced, it is patented and introduced into the commercial trade.

New cultivar introduction. Asexually produced cultivars can be patented and thus can be protected for a 17-year period. (see Hutton [29] for an excellent discussion of plant patents and Chapter 35, Intellectual Property Protection). Patented cultivars earn a royalty for the plant breeder. This royalty is money earned for each cutting sold and represents a return to the breeder to support research.

Most new commercial cultivars are sold as culture-virus-indexed plants or certified cuttings. Prior to a new cultivar's commercial sale, it's produced as clean stock, so all cuttings sold are propagated from stock plants that index free of known systemic diseases.

During the last two decades, excellent asexually propagated geraniums for greenhouse production have been bred and introduced by both university and commercial plant breeders. Table 35-1 in Chapter 35, Intellectual Property Protection, lists all patented cultivars that have been developed. The following summary tells an interesting story: The 1970s reflect the programs by Yoder Brothers from work by William Duffett, W.W. Knicely and Walter Jessel—Snowmass, Cherry Blossom, Yale and Harvard and by Griffith Buck at Iowa State University—Pearlie Mae Red, Hazel and Marian; additional Iowa State cultivars were patented in 1986. Both of these programs were discontinued.

TABLE 34-2

Dates	Number of patented Pelargoniums
1930 through 1939	0
1940 through 1949	8
1950 through 1959	0
1960 through 1969	2
1970 through 1979	4
1980 through 1989	103
1990 through October, 1991	43

The 1980s were a period of intensive breeding efforts by PAC in Dresden, Germany—breeders Wilhelm Elsner and Guenter Hofmann and later, Christa Hofmann. Perhaps the most outstanding cultivars from this program were Veronica (Purpurball), Pink Expectations (Palais) and Kim (Alex). Pelargonien Fischer in Hillschied, Germany, contributed many cultivars through the outstanding research of Ingeborg Schumann. Among the most noteworthy introductions are Blues, Mars, Tango and Schone Helena. Also in the late 1980s, Blair Winner of Denholm Seeds introduced such cultivars as Victoria, Carnival and Claret. Charles Heidgen of Shady Hill Gardens in Batavia, Illinois, patented Fireworks, Confetti and the Delight series. A major zonal program was initiated at Penn State University in 1980 and resulted in the introduction of Ben Franklin, the Classic series and Juliet (Risque); these were developed to expand zonal flower color range and to improve garden performance. Other programs that resulted in new cultivars in 1989 were those of Gunter Duemann and Sigfried Klemm, both in Germany.

With the development of significant research programs in the mid-1980s, introductions continued at a fast pace in the early 1990s. One significant event was the introduction of floribundas, developed by Douglas Holden at Bodger Seeds, Lompoc, California. Ivy geraniums were also receiving the attention of plant breeders (Jacque, Bernard and Maurice Gillou in France) and Ingeborg Schumann. During this period we see the first introductions (Satisfaction series) from PanAmerican Seed Co. and the Americana and Eclipse series from Goldsmith Seeds. Pelargonien Fischer and Penn State University continue to introduce both zonal and Regal cultivars.

As we look to the future, it's obvious that several active breeding programs should continue to introduce new and improved geraniums.

Breeding seed propagated cultivars

Breeding programs by commercial seed companies have resulted in a large number of F_1 hybrid seed propagated cultivars. The Joseph Harris Seed Company and PanAmerican Seed Co. introduced the first F_1 hybrid geraniums—New Era and Carefree F_1 hybrids, respectively. In 1971, Goldsmith Seeds Inc. and Sluis and Groot introduced the first pot plant cultivar—Sprinter Scarlet. From 1971 to 1991, over 150 commercial cultivars were introduced.

Major breeding efforts have been conducted by the seed companies mentioned above and by Denholm Seeds; Farmen in Italy; Flora Nova in England; and PAC in East Germany.

Early research at universities with seed-produced geraniums include a breeding progress report by Barnhart [3] at the University of California, the release of three seed-produced lines from the West Virginia University [22] and the release of a number of inbred lines from the University of Maryland [32].

An active research program with seed-produced geraniums has been conducted at The Pennsylvania State University since 1958. The Pennsylvania Agricultural Experiment Station released the cultivar Nittany Lion Red in 1963. The research at Penn State was aimed at the diploid and tetraploid inbred lines production and collection of research information, which has been shared with other plant breeders for commercial application.

Uniform progeny is the major factor required for commercial geranium production. In seedlings, uniformity results from genetic homozygosity of inbred lines and defined uniformity of F_1 hybrids. It should be noted that almost all commercially available seedling cultivars are diploid; thus, problems associated with tetraploidy aren't generally involved in current breeding programs. One notable breakthrough has been the development of Freckles and Tetra Scarlet by PanAmerican Seed Co.; this should lead the way to additional tetraploid seed propagated cultivars in the future.

Inbred line development. Breeders have used various sources in establishing the germplasm that is incorporated in seed-propagated cultivars. Some sources that have been used are asexually propagated diploid cultivars, both commercial and novelty types, and seedling mixtures. Regardless of the source, inbred lines are developed by self-pollination and selection of acceptable types. The number of generations required for inbred lines development depends on the number of heterozygous genes in the parental stock and the degree of uniformity desired in the inbred. Inbreeding depression hasn't been a serious problem in geranium breeding.

The following example illustrates Nittany Lion Red development, an inbred line also used as the first commercial, seed-propagated cultivar.

1958   Grew plants of Florist Mix—Selected and self-pollinated a red seedling.

1959   All progeny were red flowered. Selected and self-pollinated a red seedling.

1960   All progeny were red flowered. Selected and self-pollinated a red seedling.

1961   Enter line in Penn State Trials; bulked seed from several selections.

1962   Entered line in All-America Selections.

1963-64 Produced seed in commercial quantities.

1965   Seed made available to public as Nittany Lion Red.

F₁ hybrid development. All the criteria mentioned earlier aren't usually found in a single inbred; however, inbred hybridization from diverse parentage allows combination of acceptable characteristics into an F_1 hybrid.

When test hybrids are first evaluated, many are discarded as unacceptable for further trial. Of the hundreds of test hybrids evaluated, only one or two ultimately become a cultivar.

Promising hybrids must be evaluated for seed production efficiency since seed cost is directly related to seed production cost. Some inbreds are much more efficient seed producers than others. F_1 hybrid evaluation must be carried out both in the greenhouse and in the field. Initial testing is limited to the breeder's facilities, but ultimately all F_1 hybrids must be evaluated under commercial growing conditions. When the breeder feels that a superior cultivar has been developed, it's given a name and released for sale. The time involved for an F_1 hybrid geranium development is about five to seven years.

Since the breeder has a choice of releasing either inbreds or F_1 hybrids, why does he usually invest the extra time, effort and money into the development of F_1 hybrids? The first reason is heterosis or hybrid vigor, which is associated with F_1 hybrids. The second and more important reason is that development of F_1 hybrids provides cultivar protection for the breeder. As long as the breeder controls the inbred lines, they maintain exclusive control of the F_1 hybrid. It's especially important to control F_1 hybrid seed production since they can't be patented by plant patents; essentially, inbred lines are protected as trade secrets. Through this protection, the breeder is able to earn a return on the long-term investment in research that is necessary for cultivar development.

Specific objectives in breeding programs

Flower color. Flower color isn't a limiting factor in commercial development of new geranium cultivars. The production of inbred lines, which are true breeding for the various flower colors, is a relatively easy task for the plant breeder. Tetraploid parents for asexually propagated cultivars also display diverse flower colors. More important is the development of lines with a balanced horticultural phenotype; that is, those combining acceptable characteristics such as early flowering, compact plant habit and tolerance to extremes in cultural and environmental factors. To obtain a balanced inbred, it's necessary to select progeny in segregating generations produced from crosses of various inbreds or breeding parents.

Many flower colors are available in geraniums. These include orange, purple, magenta and various shades of red, salmon and pink. Bicolored flowers are also available within the germplasm of *P. x hortorum*; note the cultivars Rio, Carnival, Juliet and Freckles.

Fig. 34-1. Geranium seed production: (upper left) petal removal before pollen is ripe; (upper right) anther emasculation; (lower left) floret being manually pollinated; (lower right) mature seeds ready for harvest 30 days after pollination.

Floret type. Prior to 1970, all commercially available, seed-grown cultivars had single florets. The major problem with single-flowered cultivars is their tendency to shatter. In 1970, a limited series of semidouble-flowered cultivars were introduced. They were named Double Dip and a mixture of them was named Parfait. These have been followed by the Marathon series and Double Steady Red; none of these has attained any significant place in the market.

The first uniform, semidouble F_1 hybrid geraniums (diploid) were produced at Penn State in 1962 from crosses of single- and double-flowered inbreds. Although these semidoubles were acceptable horticultural quality, they weren't introduced as commercial cultivars. Diploid semidouble cultivars' commercial application has been limited mainly because their seed production has been inefficient. To produce semidouble

F_1 hybrids, it's most efficient to use single-flowered inbreds as seed parents since they are easier to emasculate than double-flowered inbreds and they normally produce more florets per inflorescence than doubles. If this procedure were used, then obviously the double-flowered inbred would be used as the pollen parent. The greatest problem with double-flowered inbreds is a paucity of pollen grains, since extra petals in double florets are produced mostly at the stamens' expense. Thus, the production of double-flowered inbreds, which have good pollen production, must precede the introduction of semidouble cultivars. Using double-flowered inbreds as seed parents hasn't been feasible since emasculation and removal of double florets' petal parts is extremely difficult. Even if emasculation were practiced, doubles' petals adhere tightly to florets, and this promotes fungal infection.

An interesting approach to semidouble-flowered cultivar development is to use tetraploid breeding lines. Both Badr and Horn [2] and Buswell [13] have shown that geraniums are autotetraploid and exhibit tetrasomic inheritance ratios. Badr and Horn, working only with commercial cultivars as parents, indicated that the locus s [17] segregated tetrasomically. These results are similar to those observed in Penn State's breeding program. In theory, one should be able to produce F_1 hybrids between semidouble parents that are predominantly, if not totally, semidouble-flowered. This technique would avoid the use of both fully single ($ssss$) and fully double ($SSSS$) parents. Further research is needed to establish if this procedure is valid.

Growth habit. Plant height and growth habit (in addition to earliness) are probably the major horticultural characteristics needed for commercial greenhouse production. All cultivars introduced prior to 1971 were the tall phenotype and developed for use as bedding plants. Since 1971, with the introduction of Sprinter Scarlet, geranium breeders have been successful in producing more compact F_1 hybrids—examples of these are Elite, Ringo, Orbit, Pinto and Diamond series. A major advancement was made by Goldsmith with their Multibloom cultivars. Most cultivars are still treated with growth retardants to produce more compact plants.

Henault and Craig [25] classified geraniums as tall, semidwarf (intermediate between tall and dwarf) and dwarf. They observed that semidwarf F_1 hybrids could be produced from crosses of tall and dwarf parents and that both parents affected F_1 hybrid height. The semidwarf F_1 hybrid has smaller leaves, more branches and more compact growth than the tall phenotype. Their dwarf and semidwarf plants had dark green foliage—a pleiotropic effect of the dw locus.

Commercial plant breeders have supposedly used other genetic loci (undefined) to produce more compact F_1 hybrids. It is postulated that these cultivars have the $dwdw$ genotype (tall phenotype and normal

Fig. 34-2. Pelargonium x *hortorum seeds.*

green foliage) and that additional loci act as modifiers. These compact F_1 hybrids are self branching and are usually early flowering, two additional reasons that may explain some of their compactness. The major reason for producing compact geraniums is that they are more space efficient in the greenhouse.

In the future, we will have dwarf and semidwarf tetraploid cultivars with semidouble flowers. Of particular interest would be the introduction of F_1 hybrid dwarf cultivars for house plants, border plants and novelty use.

Flowering earliness. From the commercial flower grower's viewpoint, flowering earliness is the most important character influencing the decision to grow geraniums from seed. Related to earliness is plant height at flowering. With most tall cultivars, there's a direct relationship between time to flower and plant height; thus, a plant flowering in the least number of days has the more acceptable height. This would be generally true with semidwarf F_1 hybrids except that height is usually in a more optimal range.

The number of days from sowing to flowering depends on many cultural and environmental factors [19, 20]; however, the hybrid's genetic constitution is of prime importance. Earliness to flower appears to be a dominant trait in tall x dwarf F_1 hybrids; thus, the F_1 hybrid is usually as early as the earlier parent. Hanniford [24], using dwarf inbreds, concluded that earliness to flower is dominant and controlled by two epistatic loci that control the amount of cumulative PPF required for floral initiation. Additional loci may be involved in other geranium lines. Plant breeders have been successful in developing inbreds that flower in the least number of days when grown under optimum cultural conditions and they have used these inbreds as parents for F_1 hybrids. The earliest commercial F_1 hybrids flowered in 80 to 100 days from sowing.

Earliness is also critical for asexually propagated geraniums; thus, it should be an important criteria in breeding programs. Selections should be made during the normal flowering season. Usually, clonal propagants from selected seedlings are trialed along with standard early flowering cultivars. Selection for earliness is critical in clean stock programs where clones are produced from individual meristems; we have noted significant differences among mericlones of the same cultivar.

Germination. Prior to 1958, it was generally accepted that geranium seed germination was a slow, erratic and inefficient process. Many authors noted that 40% to 51% germination over a several month period was normal. Craig and Walker [18] stated that poor germination resulted because geraniums' seed coat was impermeable to water, oxygen or both. They devised a scarification technique (removing a small part of the seed coat) that resulted in 100% germination two weeks after sowing. Control lots, not scarified, had a germination percentage of 40% in the same time span. McWhorter [30] observed that scarification could also be done by cutting a small piece from the seed's pointed end.

All commercially available seeds are scarified by either mechanical or acid scarification techniques. These processes are costly and not always dependable. Several researchers have hypothesized that genetic variability existed for this character. This hypothesis was favored for the following reasons: (1) within most seeds lots that were not scarified, a few seedlings germinated quickly; (2) positive selection pressure created by sowing unscarified seeds appeared to increase germination in certain lines' progeny in several commercial breeding programs; and (3) in the early scarification research 40% of the control (not scarified) seeds germinated within two weeks after sowing. To test the hypothesis Fries and Craig [23] sowed seed of many true breeding lines and germinated them under controlled environmental conditions. Only those seedlings that germinated quickly were grown further. Seeds were produced on these selections and germination was tested again. The results indicated that natural germination is a heritable trait and that improvement is possible through selection. Inbred lines are available that germinate 100% without scarification. Some of these have been selected for three to five generations.

Reed [33] studied the effects of pH and soluble salts on seed germination and concluded that geranium seeds germinate well under a wide pH range (4.4 to 6.9) and soluble salt levels (1,000 to 3,800 micromhos). The highest seedling fresh weights were generally observed at the highest soluble salt levels at each pH.

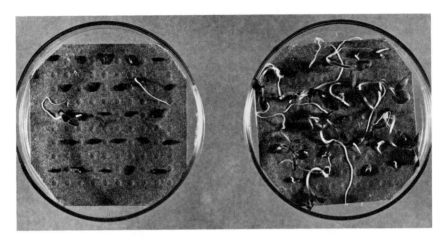

Fig. 34-3. Manual scarification effect on P. x hortorum seed germination 10 days after sowing. The treated seeds are on the right.

Nell et al. [31] observed no differences in water uptake, respiration rate and seed coat morphology between cultivar seeds that differed in germination rates. Differences between cultivars were based on their behavior under commercial production conditions. When tested under laboratory conditions, rates were quite similar and thus, their conclusions relate only to the two cultivars evaluated.

Rogers [36] reported Ethephon's positive effect on geranium seed germination, perhaps suggesting a mechanism that has not yet been investigated. Additional valuable research on seeds and germination has been conducted by Horn et al. [27, 28] and Bachthaler [1].

Several areas that still require investigation are the inheritance of high natural germination, the mechanism that differentiates easy-to-germinate and difficult-to-germinate seeds and the incorporation of this trait into commercial seed produced cultivars.

References

[1]   Bachthaler, E. 1985. Influence of scarification methods on the germination of seeds of *Pelargonium zonale. Gartenbauwissenschaft* 50(1):14-19.

[2]   Badr, M. and W. Horn. 1971a. *Genetische Untersuchungen an diploiden und tetraploiden Pelargonium zonale - Hybriden. Z. Pflanzenzuchtg* 66:203-220.

[3]   Barnhart, D. 1957. Preliminary observations on a geranium breeding project. *Geraniums Around the World* 5(2):15-16.

[4]   Bode, F.A. 1965. The Irene story. *Geraniums Around the World* 12(4):76-78.

[5]   —— 1966. Character study of a queen. *Geraniums Around the World* 14(2):28-35.

[6]   Buck, G.J. 1964. Geraniums make a comeback. *Geraniums Around the World* 11(4):76-77, 91.

[7]   —— 1973a. Waltztime geranium. *HortScience* 8(5):421.

[8]   —— 1973b. Hawkeye geranium. *HortScience* 8(5):422.

[9]   —— 1978a. Hazel geranium. *HortScience* 13(1):67.

[10] —— 1978b. Marian geranium. *HortScience* 13(1):68.

[11] —— 1978c. Pearl Bailey geranium. *HortScience* 13(1):69.

[12] —— and I.R. Lambert. 1969. Geraniums, popular summer bedding plants. *Geraniums Around the World* 17(1):4, 14.

[13] Buswell, G.E. 1978. "Flower color and anthocyanin inheritance of tetraploid *Pelargonium* x *hortorum* Bailey." Ph.D. thesis, The Pennsylvania State Univ.

[14] Butterfield, H.H. 1963. California growers take part in history. *Geraniums Around the World* 10(4):77, 92.

[15] Clifford, D. 1970. *Pelargoniums Including the Popular Geranium,* 2d ed. London: Blandford Press.

[16] Craig, R. 1962. Geranium pollination techniques. *Geraniums Around the World* 10(2):29-30, 47.

[17] —— 1963. "The inheritance of several characters in the geranium, *Pelargonium* x *hortorum* Bailey." Ph.D. thesis, The Pennsylvania State Univ.

[18] —— and D.E. Walker. 1959. Geranium seed germination techniques. *Geraniums Around the World* 7(2):4-7.

[19] —— 1960. Temperature and seedling geraniums. *Penn. Flower Growers Bull.* 114:3, 6-7.

[20] —— 1963. The flowering of *Pelargonium* x *hortorum* Bailey seedlings as affected by cumulative solar energy. *Proc. Am. Soc. Hort. Sci.* 83:772-776.

[21] Dermen, H. 1960. Nature of plant sports. *Am. Hort. Mag.* 39:123-173.

[22] Fortney, W.R. 1966. Colorful mountaineers. *W. Va. Ag. Exp. Sta. Bul.* 523:1.

[23] Fries, R. and Craig, R. 1966. "Selection for high, early germination in *Pelargonium* x *hortorum.*" Unpublished report. The Pennsylvania State Univ.

[24] Hanniford, G.G. 1980. "The inheritance of early flowering in the geranium, *Pelargonium* x *hortorum* L.H. Bailey." M.S. thesis, The Pennsylvania State Univ.

[25] Henault, R.E. and R. Craig. 1970. Inheritance of plant height in the geranium. *J. Hered.* 61(2):75-78.

[26] Hieke, K. 1962. Assortment of *Pelargonium zonale* ait. of Pruhonice 1958-1962. *Acta. Pruhonica* (Research Institute of Ornamental Gardening in Pruhonice, Czechoslovakia). English summary.

[27] Horn, W., E. Bachthaler and M. Badr. 1973. Germination and flowering time in F_1 cultivars and strains of *Pelargonium zonale* hybrids. *Gartenbauwissenschaft* 38(5):391-408.

[28] —— and E. Bachthaler. 1976. *Versuche an samlings und klongeneration bei Pelargonium zonale hybriden. Z. Pflanzenzuchtg* 77:181-186.

[29] Hutton, R.J. 1969. Patenting of new cultivars. *Roses.* Ed. J.W. Mastalerz and R.W. Langhans. University Park, Penn.: The Pennsylvania Flower Growers and New York Flower Growers.

[30] McWhorter, F.P. 1959. How to cut or chip geranium seeds to increase germination. *Geraniums Around the World* 7(4):6-7.

[31] Nell, T.A. and D.J. Cantliffe. 1981. Seed dormancy and germination of geranium *(Pelargonium hortorum* Bailey). *J. Am. Soc. Hort. Sci.* 106:509-513.

[32] Nugent, P.E. 1967b. Inbreds, hybrids and breeding lines. *Geraniums Around the World* 15(1):12-17.

[33] Reed, H. 1974. "Germination of geraniums *(Pelargonium* x *hortorum* Bailey) as influenced by temperature, media, soluble salts and pH." M.S. thesis, The Pennsylvania State Univ.

[34] Richards, E. 1958. Notes on colors in zonals. *Geraniums Around the World* 6(1):11.

[35] Roberts, D.M. 1954. The geranium family. *Gardeners Chronicle* 136:214-125, 244.

[36] Rogers, O.M. 1987. Ethephon overcomes seed scarification requirements of Pelargonium. *Acta Horticulturae* 201:165-170.

[37] Ross, M.E. 1969. History of miniatures. *Geraniums Around the World* 16(4):6-8.

[38] ——— 1970. The irresistibles. *Geraniums Around the World* 18(1):10.

[39] Schmidt, W.E. 1957. Collecting and hybridizing ivy-leaf. *Geraniums Around the World* 5(4):12, 13, 24.

[40] Tyler, H.A. 1969. The French type geranium. Part I. *Geraniums Around the World* 17(3):8-10

[41] ——— 1970. The French type geranium. Part II. *Geraniums Around the World* 17(4):8-12.

[42] Volz, E.C. 1951. Better geraniums for Iowa. *Trans. Iowa Hort. Soc.* 86:131-137.

[43] Wilson, H.v.P. 1965. *The Joy of Geraniums.* New York: M. Barrows and Co. Inc.

CHAPTER 35

Intellectual Property Protection

Richard Craig

New plant cultivars and novel processes by which plants are propagated or produced are considered to be intellectual properties. As intellectual properties they can be protected by:

- Plant patents—asexually reproduced plants
- Plant variety protection certificate—sexually reproduced inbreds
- General patents (utility patents)
- Trade secrets—sexually and asexually reproduced plants and plant processes

Some attorneys are even suggesting other novel approaches for protecting plant cultivars, such as design patents and copyrights.

Current status

A high activity level exists with respect to intellectual property protection of plants at both national and international levels.

Nationally, concerns include the choice of protection mechanisms—a current movement is toward using the general patent for plants instead of the traditional plant patents or plant variety protection legislation. While costs are greater and time for processing may be longer, most experts feel that the protection scope will be improved.

On an international level, several concerns are evident. One concern is the status of the International Union for the Protection of New Varieties of Plants (UPOV) with respect to its adoption by current and future member nations. Many legal and ethical questions need to be resolved regarding international transport of patented cultivars; these involve plants imported into a country where the patent has been allowed and also when patented cultivars are exported to any country that doesn't have equitable reciprocal protection laws and practices.

Rationale for breeder's rights

The term intellectual property best describes a new plant cultivar—intellectual referring to the creativity involved in plant breeding and property to indicate that it is the personal property of the plant breeder or the organization licensed to propagate it.

Breeder's rights is a concept (and a law) recognizing that a specific cultivar's breeder (intellectual property) has guaranteed rights under the law and that these rights are protected through legislative, regulatory and judicial processes.

Intellectual property protection dates back to 200 B.C. in Greece; however, the first official legislation related to the subject occurred in 1474 in Venice. One of the first people to benefit from this legislation was Galileo, who had invented a machine for irrigating land. More recently the "Universal Declaration of Human Rights" of the United Nations included the following statement, "Everyone has the right to protection of the moral and material interests resulting from any scientific, literary or artistic production of which he is the author."

In essence, intellectual property protection is a simple contract between an inventor and the federal government. If an inventor discloses details of his invention to the public so this information can be used to further scientific development, then the federal government grants the inventor a time-limited monopoly on the invention's use and sale.

A landmark legal decision was Chakrabarty vs. Diamond, which says, "Congress intended statutory subject matter to include anything under the sun that is made by man." This decision relates to plant and animal material as well as inert material. Also, in this age of genetic engineering, it can apply to patenting plants and/or animals, since it's possible to "make" new plants and animals using recombinant DNA techniques.

Plant patents

The Townsend-Purnell Act of 1930 specifically protects plant cultivars that are reproduced asexually but not if they are reproduced by seeds. The inventor of a new, asexually propagated plant is required only to describe the plant to the best of his or her ability. This is different than general patent procedures, which require rather detailed descriptions. About 6,200 patents have been granted for such plants in the past 61 years.

In the congressional hearings that occurred prior to the Plant Patent Act passage, those who argued for the act wanted to give plant breeders the same status that inventors of chemical and mechanical processes have through patent laws. It was stated that, "The purpose of the bill is to afford agriculture, so far as practicable, the same opportunity to participate in the benefits of the patent system as has been given industry and thus assist in placing agriculture on a basis of economic equality with industry. The bill will remove the existing discrimination between plant developers and industrial inventors." It was hoped that the bill would afford a sound basis for investing capital in plant breeding and consequently stimulate plant development through private funds.

During his lifetime, Luther Burbank had supported this viewpoint. "A man can patent a mouse trap or copyright a nasty song, but if he gives to the world a new fruit that will add millions to the value of earth's annual harvests, he will be fortunate if he is rewarded by so much as having his name connected with the result."

Thomas Edison is quoted at the congressional hearing, "Nothing that Congress could do to help farming would be of greater value and permanence than to give the plant breeder the same status as the mechanical and chemical inventor now have through the patent law. There are but few plant breeders. This (bill) will, I feel sure, give us many Burbanks."

Plant variety protection

Another congressional act of importance to plant breeders is the Plant Variety Protection Act of 1970, which provided protection for inbred plant varieties produced from seed. This act includes both a farmers exemption and a research exemption. A sample of protected seeds must be deposited in the National Seed Laboratory at Fort Collins, Colorado; 3,116 varieties have been protected as of September, 1991—68.6% in the agronomic group, 4.5% flowers and 26.9% vegetables. (Statistics were provided by the National Seed Laboratory and the Plant Variety Protection Office, based on the Progress Report, September 1991.) Certificates have been issued in 83 crops; no geranium cultivars have been protected under this act.

The definitions and rules of construction in the Plant Variety Protection Act indicate that the term novel variety may be represented by, without limitation, seed, transplants and plants, and is satisfied if there is:

(1) Distinctness in the sense that the variety clearly differs by one or more identifiable morphological, physiological or other characteristics (which may include those evidenced by processing or product characteristics, for example, milling and baking characteristics in the case of wheat) as to which a difference in genealogy may contribute evidence,

from all prior varieties of public knowledge at the date of determination within the provisions of sections 42; and

(2) uniformity in the sense that any variations are describable, predictable and commercially acceptable; and

(3) stability in the sense that the variety, when sexually reproduced or reconstituted, will remain unchanged with regard to its essential and distinctive characteristics with a reasonable degree of reliability commensurate with that of varieties of the same category in which the same breeding method is employed.

General patents

General patents (also called utility patents or Sec. 101 patents) are identical to those given for other inventions. The first general patent use for plants occurred recently. Not only does the general patent law protect against unlicensed sale, use or reproduction of the invention, but it also covers the parts and embodied idea. Implied is the inventor's right to license rights for use or sale of the patented product in return for royalties to be paid to the inventor and/or organization that employs the inventor.

To be patentable an invention must exhibit:

(1) Utility—usefulness

(2) Novelty—no prior use, publication or sale; there is a one year limit in the United States but none for foreign patents.

(3) Nonobviousness—would not be obvious to a person skilled in the art (of plant breeding).

Geranium cultivar protection

Virtually all asexually propagated cultivars available from clean stock programs are patented; these include zonals, ivies and Regals. The criteria for patenting—new and distinct, nonobviousness, prior asexual propagation—provide a broad base for geranium improvement. Similar criteria would apply to general patents.

Distinct infers that a new cultivar has characteristics that clearly distinguish it from existing cultivars. During the past several years, flower color and patterning, leaf color and greenhouse/garden performance have dominated the patent scene. In the future, such traits as host-plant resistance, interspecific hybrids, postproduction characteristics and stress-related tolerances will receive more attention. New implies that it hasn't been commercialized; there is a one-year limit.

Nonobviousness is not well defined in the case of plant patents but is that portion of patent law "that requires an inventive step"—an improvement over prior art. Cultivars resulting from a planned breeding program would seem to satisfy the requirement of nonobviousness, while plants developed from clonal selection or as spontaneous sports or mutations would require greater inspection.

A complete list of all patents allowed for Pelargoniums is included in Table 35-1. I believe that all improved cultivars will be patented in the future. Seed-reproduced cultivars usually aren't protected since they are F_1 hybrids; parental inbreds that are cross-pollinated to produce F_1 hybrid cultivars are protected as trade secrets although they can also be protected by all other available mechanisms.

TABLE 35-1

Patented Pelargonium cultivars 1930 to October 1991

Year	Patent number	Cultivar name	Class	Inventor(s)
1940	355	Fiat Queen	Zonal	Charles A. Brown II
1940	24			Felix L. Sturn
1942	558	Clara Marie		Charle Fiore
1946	683	Winter Cheer		Helen R. Bohannon
1946	684	Summer Charm		Helen R. Bohannon
1946	685	Spring Glory		Helen R. Bohannon
1946	686	Autumn Glow		Helen R. Bohannon
1949	875			William J. Haese
1966	2621	Patricia Andrea		Frank Andrea
1969	2868			Mark A. Miller
1970	2969	Ekdahls Poitezone		Paul H. Ekdahl
1972	3173	Snowmass	Zonal	William E. Duffet and Walter W. Knicely
1972	3174	Cherry Blossom	Zonal	William E. Duffet and Walter W. Knicely
1974	3498			Frank Andrea
1976	3874	Yale	Ivy	William E. Duffet and Walter W. Knicely
1976	3875	Harvard	Ivy	William E. Duffet and Walter W. Knicely
1976	3888	Cornell	Ivy	William E. Duffet and Walter W. Knicely
1976	3941	Happy Cherub		William E. Duffet and Walter W. Knicely
1976	3942	Coy Cherub		William E. Duffet and Walter W. Knicely
1976	3943	Spry Cherub		William E. Duffet and Walter W. Knicely
1977	4039	Pearlie Mae Red (Sunbelt Scarlet)	Zonal	Griffith J. Buck

Year	Patent number	Cultivar name	Class	Inventor(s)
1977	4040	Hazel (Sunbelt Rose)	Zonal	Griffith J. Buck
1977	4041	Marian (Sunbelt Hot Pink)	Zonal	Griffith J. Buck
1978	4215			Frank Andrea
1981	4653	Super Waltztime (Sunbelt Salmon)	Zonal	Griffith J. Buck
1981	4778			Walter H. Jessel Jr. and William E. Duffett
1981	4785	Fame	Zonal	Walter H. Jessel Jr. and William E. Duffett
1983	5054	Veronica (Purpurball)	Zonal	Wilhelm Elsner
1983	5057	Glacier Crimson (Bruni, Red Beauty)	Zonal	Wilhelm Elsner
1983	5059	Glacier Carmen (Karminball)	Zonal	Wilhelm Elsner
1984	5310	Fireworks		Charles F. Heidgen
1984	5311	Alex (Kim)	Zonal	Guenter Hofmann
1984	5312	Perlenkette (Snowhite)	Zonal	Guenter Hofmann
1984	5313	Lachsball (Glacier Salmon)	Zonal	Guenter Hofmann
1984	5314	Jubilaeum (Glacier Dark Red)	Zonal	Guenter Hofmann
1984	5315	Palais (Pink Expectations)	Zonal	Guenter Hofmann
1984	5368	Fledermaus	Zonal	Ingeborg Schumann
1984	5369	Flirtpel	Zonal	Ingeborg Schumann
1984	5370	Fortuna	Zonal	Ingeborg Schumann
1984	5371	Polka	Zonal	Ingeborg Schumann
1984	5372	Mars	Zonal	Ingeborg Schumann
1984	5373	Blues	Zonal	Ingeborg Schumann
1984	5374	Schone Helena	Zonal	Ingeborg Schumann
1984	5375	Olymp	Zonal	Ingeborg Schumann
1986	5752	Fidelio	Zonal	Ingeborg Schumann
1986	5757	Orangeade (Sunbelt Coral)	Zonal	Griffith J. Buck
1986	5758	Troubadour	Zonal	Griffith J. Buck
1986	5759	La Sevillana (Katie)	Zonal	Griffith J. Buck
1987	5928	Twist	Zonal	Ingeborg Schumann
1987	5929	Mercury	Zonal	Ingeborg Schumann
1987	5930	Disco	Zonal	Ingeborg Schumann
1987	5931	Waltz	Zonal	Ingeborg Schumann
1987	5932	Columbia	Zonal	Ingeborg Schumann
1987	5933	Tango	Zonal	Ingeborg Schumann
1987	5938	Casino	Zonal	Ingeborg Schumann
1987	5939	Bolero	Zonal	Ingeborg Schumann
1987	5940	Volcano	Zonal	Ingeborg Schumann
1987	5941	Satellite	Zonal	Ingeborg Schumann
1987	5942	Champagne	Zonal	Ingeborg Schumann
1987	5980	Gemini	Zonal	Ingeborg Schumann

Year	Patent number	Cultivar name	Class	Inventor(s)
1987	5992	Cabaret	Zonal	Ingeborg Schumann
1987	6012	Macy	Regal	Wolfgang Kirmann
1987	6014	Betty	Regal	Wolfgang Kirmann
1987	6015	Shirley	Regal	Wolfgang Kirmann
1987	6017	Gypsy	Regal	Wolfgang Kirmann
1987	6018	Rosy	Regal	Wolfgang Kirmann
1987	6019	Josy	Regal	Wolfgang Kirmann
1987	6020	Peggy	Regal	Wolfgang Kirmann
1987	6021	Sally	Regal	Wolfgang Kirmann
1987	6022	Lilly	Regal	Wolfgang Kirmann
1987	6023	Candy	Regal	Wolfgang Kirmann
1987	6027	Mary	Regal	Wolfgang Kirmann
1987	6028	Dolly	Regal	Wolfgang Kirmann
1987	6029	Vick	Regal	Wolfgang Kirmann
1987	6030	Honey	Regal	Wolfgang Kirmann
1987	6055	Danielle	Zonal	Daniel T. Busch
1987	6063	Claret (Glacier Claret)	Zonal	Blair L. Winner
1987	6064	Victoria	Zonal	Blair L. Winner
1987	6065	Valerie	Zonal	Blair L. Winner
1987	6066	Diplomat	Zonal	Blair L. Winner
1987	6072	Lucy	Regal	Wolfgang Kirmann
1987	6073	Tutti-frutti	Zonal	Ingeborg Schumann
1987	6080	Micky	Regal	Wolfgang Kirmann
1987	6218	Ben Franklin	Zonal	Richard Craig
1988	6219	Paris	Zonal	Richard Craig
1988	6220	Cassandra	Zonal	Richard Craig
1988	6247	Helen	Zonal	Richard Craig
1988	6286	Gugino's Nell		John L. Gugino
1988	6351	Confetti		Charles F. Heidgen
1988	6352	Fire Chief		Charles F. Heidgen
1988	6353	Night Watch		Charles F. Heidgen
1988	6354	Peach Delight		Charles F. Heidgen
1988	6355	Watermelon Delight		Charles F. Heidgen
1988	6356	Candy Delight		Charles F. Heidgen
1988	6366	Duster		Charles F. Heidgen
1988	6378	Calypso	Zonal	Richard Craig
1988	6379	Siren	Zonal	Richard Craig
1988	6416	Scarlet Red Veronica	Zonal	Cleveland Ott and D. Tostevin
1989	6593	Mandarin	Zonal	Blair L. Winner
1989	6602	Pink Silver Crown	Ivy	P. Jacobsen
1989	6605	Red Silver Crown	Ivy	P. Jacobsen
1989	6636	Carnival	Zonal	Blair L. Winner
1989	6652	Capri (67)	Zonal	Blair L. Winner
1989	6654	Juliet (Risque)	Zonal	Richard Craig and Leon Glicenstein
1989	6657	Sabrina	Zonal	Scott C. Trees
1989	6708	Plaisire		Gunter Duemenn

Year	Patent number	Cultivar name	Class	Inventor(s)
1989	6716	Praeludium		Gunter Duemenn
1989	6717	Feeling		Gunter Duemenn
1989	6761	Serenade		Gunter Duemenn
1989	6994	Klefisall		Siegfried Klemm
1989	6995	Klefice		Siegfried Klemm
1989	6996	Manepi		Siegfried Klemm
1989	6997	Mareli		Siegfried Klemm
1989	7014	Mapursit		Siegfried Klemm
1989	7015	Klerissa		Siegfried Klemm
1989	7016	Kledap		Siegfried Klemm
1989	7017	Kledaph		Siegfried Klemm
1989	7018	Markoni		Siegfried Klemm
1989	7059	Duerom		Gunter Duemenn
1989	7060	Dueleg		Gunter Duemenn
1989	7080	Isabell		Christa Hofman
1989	7083	Fox	Zonal	Christa Hofman
1989	7085	Cherry		Christa Hofman
1989	7086	Rebeca		Christa Hofman
1989	7087	Laura	Zonal	Christa Hofman
1989	7107	Kledol		Siegfried Klemm
1990	7122	Marlimon		Siegfried Klemm
1990	7149	Marpanet		Siegfried Klemm
1990	7153	Marktex		Siegfried Klemm
1990	7199	Manepert		Siegfried Klemm
1990	7208	Marlicat		Siegfried Klemm
1990	7246	Brigette	Floribunda	Douglas Holden
1990	7247	Marilyn	Floribunda	Douglas Holden
1990	7257	Marix		Siegfried Klemm
1990	7271	Simone (Guipolac)	Ivy	Jacques Guillou, Bernard Guillou and Maurice Guillou
1990	7273	Gigi	Ivy	Jacques Guillou, Bernard Guillou and Maurice Guillou
1990	7321	Grace	Floribunda	Douglas Holden
1990	7322	Judy	Floribunda	Douglas Holden
1990	7343	Crystal	Regal	Richard Craig and Glenn Hanniford
1990	7350	Misty	Zonal	Richard Craig and Leon Glicenstein
1990	7351	821-(82-116-13)	Zonal	Richard Craig and Leon Glicenstein
1990	7352	Mimi (Guidonal)	Ivy	Jacques Guillou, Bernard Guillou and Maurice Guillou
1990	7353	Renee (Guisuma)	Ivy	Jacques Guillou, Bernard Guillou and Maurice Guillou
1990	7356	Duege		Gunter Duemenn

Year	Patent number	Cultivar name	Class	Inventor(s)
1990	7358	Nicole (Guicordan)	Ivy	Jacques Guillou, Bernard Guillou and Maurice Guillou
1990	7359	Duelyr		Gunter Duemenn
1990	7360	Nanette (Guitaril)	Ivy	Jacques Guillou, Bernard Guillou and Maurice Guillou
1990	7373	Camille	Ivy	Jacques Guillou, Bernard Guillou and Maurice Guillou
1990	7374	Klefizon		Siegfried Klemm
1990	7385	Fiswig		Ingeborg Schumann
1990	7387	Majestic	Regal	Richard Craig and Glenn Hanniford
1990	7388	Fisfid		Ingeborg Schumann
1990	7391	Fismanon (Solidor)	Ivy	Ingeborg Schumann
1990	7392	Fisbal (Alba)	Zonal	Ingeborg Schumann
1990	7393	Fiscab (Charleston)	Zonal	Ingeborg Schumann
1990	7394	Fisglo (Gloria)	Zonal	Ingeborg Schumann
1991	7410	Klecerol		Siegfried Klemm
1991	7422	Fisrix		Ingeborg Schumann
1991	7467	Allure	Regal	Richard Craig and Glenn Hanniford
1991	7494	Breneso		Siegfried Klemm
1991	7499	Fispol		Ingeborg Schumann
1991	7538	Fantasy	Regal	Richard Craig and Leon Glicenstein
1991	7567	Fisrom		Ingeborg Schumann
1991	7576	Centennial	Zonal	Richard Craig and Leon Glicenstein
1991	7599	Pink Satisfaction	Zonal	Scott C. Trees
1991	7609	Sunset	Zonal	Blair Winner
1991	7610	Salmon Satisfaction	Zonal	Scott C. Trees
1991	7620	Flair	Regal	Richard Craig and Glenn Hanniford
1991	7627	208 (81-344-3)	Zonal	Richard Craig
1991	7641	Red Satisfaction	Zonal	Scott C. Trees
1991	7656	Splendor '315 - (8348-1)	Regal	Richard Craig and Glenn Hanniford

Most cultivars are protected on an international basis; this is generally accomplished through various breeder's rights legislation. International Union for the Protection of New Varieties of Plants (UPOV) is the international agreement responsible for protecting breeder's rights.

Plant breeding must be viewed as a business that operates like other business operations. There are creative employees, facilities, equipment and taxes. Unlike many other horticultural businesses, plant breeding requires a long time for product development.

A breeding program might not result in a commercial product for eight to 20 years; however, investment of funds and time must be made continually. The return on the breeder's investment is realized when the first cultivars are commercialized. Without royalty or license fees no breeder would risk this financial investment. In addition, legal costs for intellectual property protection must be considered. The royalty for floricultural plant cultivars is crop dependent; a current estimate for zonal geraniums is $0.02 per unit. A typical scenario for a program of investments and returns would be:

Assume: 8 years' investment in the breeding program without any financial returns

Assume: only $250,000 is invested per year for employees, facilities, equipment, interest

Assume: royalties are two cents per propagant or $20,000 per million units

Assume: 10 patented commercial cultivars are developed in the program.

Total investment: $2,000,000

The number of propagants to return investment costs without a profit: 100,000,000 units or 10 million of each cultivar over the cultivar's commercial life, which may be as short as five years.

One might argue that sports can be discovered and patented and their cost is minuscule when compared to development through a breeding program. An example of a cultivar that originated as a sport is Danielle (sport of Veronica). Patented cultivar sports (mutations) are usually controlled in licensing agreements for this reason. A sport, spontaneous or induced, usually results from a simple genetic change; note that the tens of thousands of other genes remain unaltered and were due to the plant breeder's creative genius.

One caveat—the discoverer of the sport is usually the inventor (plant patent) for legal purposes; however, ownership may be in question if the inventor of the original cultivar fails by license agreement to reserve an interest in the sport.

Documentation

A novel plant cultivar must have distinct, uniform and stable characteristics created and/or incorporated by the inventor. A full-color photograph is required. The photograph is usually made by a professional photographer since the color must match the descriptions, usually based on color charts, referenced in the patent document.

Plant breeders currently use visual descriptions; however, plant "fingerprinting" by way of chromosome counts, chemical composition, machine identification of color by reflectance-absorption methods, light and electron microscopy of morphological or anatomical features, chromosome structure and genetic descriptions are modern methods that can be used to document the invention's novel nature.

Fingerprinting will be more sophisticated in the future; we are now in the era of molecular genetics and the use of recombinant DNA and DNA-restriction enzymes. The DNA sequence determines protein structure and ultimately the biochemical profile (fingerprint) of living organisms. Techniques such as restriction fragment length polymorphism protocols provide exquisite detail of differences in plant cultivars. The data are obtained at the DNA level, thus are unaffected by the environment and aren't subject to the time of gene expression; plants can be identified at any development stage. Documentation might also include biochemical markers such as specific enzymes, isozyme patterns, plant pigments, amino acids and other specific biochemical compounds that might be present in detectable quantities or specific combinations characteristic of the novel plant. Sophisticated documentation can be used in the patent description; however, it's even more important to have this information in infringement cases.

What is infringement?

Propagation of a protected cultivar for any purpose without a license is an infringement of the inventor's rights. Although a research exemption is recognized in the Plant Variety Protection Act, it's not implicit in either the plant patent legislation or for plants protected by a general patent. Burden of proof that someone has infringed on a novel plant cultivar patent is with the patent holder. Plant patent infringement is simply stated to include three prohibitions: asexual reproduction, selling and/or using the plant so reproduced. Protected geraniums (patented) are the property of the inventor who often assigns propagation rights to a primary propagator. In a sense the inventor or his parent company is really renting or leasing mother plants to a propagator to produce cuttings or propagules. This act of renting/leasing speaks to the payment of royalties for each propagule produced. The primary propagator may also have the right to sublicense certain propagation rights to secondary propagators.

Licenses are legal contracts and include several important stipulations, including provisions for royalty enforcement and collection, a protocol for handling sports (mutations) and requirements for plant labeling.

The flower grower who purchases propagules from a licensed propagator has the responsibility not to propagate patented plants (unless a license has been granted) and to attach an identifying label to each plant marketed. Royalties are usually paid to the propagator when cuttings are purchased, so the producer has already satisfied this part of the law.

Failure to pay royalties is illegal and a financial affront to the inventor; it's really no different than stealing or illegally converting property.

Most propagators or producers know the legal boundaries with respect to propagation; some, however, are unaware that propagation for any reason, whether for their own use, for trials or for sale, is considered infringement.

The Plant Variety Protection Act of 1970, described as Public Law 91-577, lists specific infringements (performance without authority, prior to expiration of the right to plant variety protection) as follows:

(1) to sell the novel variety, or offer it or expose it for sale, deliver it, ship it, consign it, exchange it, or solicit an offer to buy it, or any other transfer of title or possession of it;

(2) to import the novel variety into or export it from the United States;

(3) to sexually multiply the novel variety as a step in marketing (for growing purposes) the variety; or

(4) to use the novel variety in producing (as distinguished from developing) a hybrid or different variety therefrom; or

(5) to use seed which had been marked propagation prohibited or progeny thereof to propagate the novel variety; or

(6) to dispense the novel variety to another, in a form which can be propagated, without notice as to being a protected variety under which it was received; or

(7) to perform any of the forgoing acts even in instances in which the novel variety is multiplied other than sexually, except in pursuance of a valid United States plant patent; or

(8) to instigate or actively induce performance of any of the foregoing acts.

Making the decision to patent a variety

I get dozens of questions each year from growers, avocational horticulturists and commercial corporations about protecting novel plants they have created or discovered. They often feel that the novel plant will be interesting to a commercial propagator or seed company. These are the questions that I normally ask:

(1) What are the plant genus and species?

(2) What is the origin of their invention or discovery?

(3) How is the plant uniformly reproduced: by seed or by asexual propagation such as cuttings, grafts or divisions?

(4) If a sport, what was the parent cultivar and was it patented?

(5) Why do they feel that the new cultivar has commercial potential? What competition exists in the industry?

(6) Are they willing to spend the money and time to pursue intellectual property protection?

I then give them the following information:

- A list of attorneys who specialize in the intellectual property protection of plants and/or plant related processes.
- Names of commercial companies involved in propagation and/or breeding the crop.
- Names of experts who may assist them in documenting characteristics of the novel plant or process.
- References to published information on intellectual property protection.

The decision to protect intellectual properties develops from your consultation with an attorney. However, you invalidate your rights if more than one year elapses between public disclosure of a cultivar and patent application. The plant inventor must:

(1) Keep excellent records of the invention; records should be dated and signed. Photodocument the invention.

(2) Maintain tight control over the invention; don't share it with anyone without a valid testing license (your attorney can develop a testing license for your use). Don't sell propagants or plants of the invention. If you are patenting a process, don't discuss your ideas with anyone without a signed confidentiality agreement.

(3) Prepare a disclosure document in consultation with an attorney at the earliest possible time. Use a prior (recently issued) patent or certificate for your crop as a model.

(4) Propagate the invention to ensure it's propagable and uniform; this is a requirement of the plant patent legislation.

Significant issues for the '90s

- Further movement toward the internationalization of intellectual property protection.
- New cultivar protection with general patents rather than continued plant patent or plant variety protection.
- Increased litigation both nationally and internationally, particularly with respect to "pirated" plants exported into the United States and European Economic Community.
- Plant part protection and research exemption redefinition.

- Legislative or judicial interpretations of "minimum distance" and "essentially derived" concepts with respect to plant intellectual properties.

For information concerning U.S. plant patents and federal trademarks on plants, contact:

The National Association of Plant Patent Owners
1250 *I* Street, N.W.
Suite 500
Washington, DC 20005
Telephone 202/789-2900.

Also available at the following address are: copies of specifications and drawings of all patents, sold at $1.50 each (order by patent number); copies of registered marks, sold at $1.50 each (identify the trademark by registration number or give the name of the registrant and approximate date of registration); *General Information Concerning Patents* (pamphlet, $2.00); *Basic Facts About Trademarks* (pamphlet, $1.75); *Patent Official Gazette* (by subscription); annual indexes (prices vary). Write to:

The Superintendent of Documents
U.S. Government Printing Office
Washington, DC 20402.

References

[1] American Association of Nurserymen. 1990. *Plant Patent Directory*. Washington, D.C.: AAN.

[2] American Society for Horticultural Science, Northeast Region. 1991. The horticultural dilemma—trademarks, patents, royalties and cultivars. *HortScience* 26(4): 359-365.

[3] Asgrow Seed Company. 1989. A chronicle of plant variety protection. Asgrow Seed Co., Kalamazoo.

[4] Bent, S.A., R.L. Schwaab, D.G. Conlin and D.D. Jeffrey. 1987. *Intellectual Property Rights in Biotechnology Worldwide*. New York: M. Stockton Press.

[5] Chu, M.A. 1990. Bringing intellectual property law down to earth: An introduction to the Plant Variety Protection Act. *J. Proprietary Rights* 2(5):2.

[6] CIOPORA. 1988. Fifth International Colloquium on the Protection of Plant Breeders Rights. September 10-11, 1987. CIOPORA, Geneva.

[7] Congress of the United States, Office of Technology Assessment. 1984. "Commercial biotechnology: An international analysis." Report OTA-BA-218. U.S. Government Printing Office, Washington.

[8] ——— 1989. "New developments in biotechnology: Patenting life." Report OTA-BA-370. U.S. Government Printing Office, Washington.

[9] Crop Science Society of America, American Society of Agronomy and Soil Science Society of America. 1989. "Intellectual property rights associated with plants." ASA Special Publication No. 52. CSSA, ASA, SSSA, Madison.

[10] Hutton, R.J. 1969. Patenting of new cultivars. *Roses*. Ed. J.W. Mastalerz and R. Langhans.

[11] Jeffrey, D.D. et al. 1984. Patent protection for plants. *Genetic Engineering News* 4(3).

[12] ——— 1990. The plant description requirements of United States patent law. *Diversity* 6(3) and 6(4).

[13] Hauser, A. 1989. Industrial property protection for advanced biotechnological processes and products. *Industrial Property* April:161.

[14] Karney, G.M. 1989. Intellectual property in the 1990s: Patenting higher forms of life. Special Report No. 4., "Special Report Series on Biotechnology." Washington, DC: Bureau of National Affairs, Washington.

[15] National Association of Plant Patent Owners. 1990. *Plant Patents and Federal Trademarks on Plants*. Washington, DC.

[16] Neagley, D.D. Jeffrey and Diepenbrock. 1984. Section 101 patents— panacea or pitfall? *American Patent Law Association, Selected Legal Papers* 1(2).

[17] Verity, C.W. and D.J. Quigg. 1988. *The Story of the U.S. Patent and Trademark Office*. Washington, D.C.: Department of Commerce.

[18] Whaite, R. et al. 1989. Biotechnological patents in Europe: the draft directive. *European Intellectual Property Rev.* 585:145.

[19] World Intellectual Property Organization. 1985. "Industrial property protection of biotechnological inventions." Publication BIG/281. WIPO, Geneva.

Additional information sources

Supplied by W.H. Willott Jr., J.A. Battcha and D.D. Jeffrey for the American Society for Horticultural Science Intellectual Property Rights Working Group, July 22, 1991.

United States statutes and legislative history:

35 United States Code, Section 161-164 *(The Plant Patent Act)* and cases reported thereunder.

Congress of the United States, House Committee on Patents. 1930. *Plant Patents.* Hearings before the Committee, April 9, 1930. U.S. Government Printing Office, Washington.

7 United States Code, Section 2321 *et seq (The Plant Variety Protection Act)* and cases reported thereunder.

Congress of the United States, House Committee on Agriculture. 1970. *Plant Variety Protection Act.* HR 91-1605. U.S. Government Printing Office, Washington.

International Union for the Protection of New Varieties of Plants (UPOV), Geneva:

UPOV. 1985. "International Convention for the Protection of New Varieties of Plants of December 2, 1961; additional act of November 10, 1972; and revised text of October 23, 1978." Document 293(E).

——— 1985. "Plant variety protection laws and treaties." Document 651(E) and supplements thereto.

———— 1986. *Plant Variety Protection* (UPOV newsletter). No. 51.

———— 1991. "Diplomatic conference for the revision of the International Convention for the Protection of New Varieties of Plants." Geneva, March 4, 1991. Final draft. DC/91/138.

———— "General information." Publication 408(E).

———— "Collection of the texts of the UPOV Convention and other important documents established by UPOV." Part 1. Text and documents. Document 644(E).

———— "Collection of the texts of the UPOV convention and other important documents established by UPOV." Part 2. Test guidelines. Volume 1-6. Document 645(E).

———— "The first 25 years of the International Convention for the Protection of New Varieties of Plants." Document 879(E).

———— "Nomenclature, UPOV Symposium Record, Geneva, October 12, 1983."

A

A-Rest 65, 69
Aerosol generators 345, 346
Air-assisted electrostatic sprayers 346
Alkalinity 52
Alternaria 215, 219, 228, Color Plate C80
Aphids 182, 324–331
  Control 328
  Resistance 329
Auxin 65, 66

B

B-Nine 69
Bacterial blight. *See* Xanthomonas
Bacterial fasciation 262–266
  Control 265
  Etiology 263
Bacterial leaf spot 218, 233, 237
Bacterial stem rot. *See* Xanthomonas
Bacterial wilt. *See* Xanthomonas
Ball 208–212, 214
Bemisia tabaci 317
Beneficial rhizobacteria 27
Bird's egg-flowered geranium 202
Black root rot. *See* Thielaviopsis
Blackleg. *See* Pythium
Blindness 6
Bonzi 65, 70, 109
Boron 44, 48, 49, Color Plates C52–C54, C60
Botrytis 96, 115, 182, 183, 185, 215, 217, 218, 223, 300, 307, Color Plates C70–C72, C74–C76
Botrytis cinerea. See Botrytis
Bradysia impatiens. *See* Fungus gnats
Breeder's rights 390
Breeding 373–388
  Cutting geraniums 374
  Flower color 381
  Hybrid seed geraniums 379
  Inbred lines 380
  Irradiation 375
  Regal geraniums 175
  Resistance 158
  Royalties 378
  Sport selection 375
  Strain selection 376

C

CEC 9–11
CVI 282, 283, 284
Cabbage looper 336
Cactus-flowered geranium 202, Color Plate C16
Calcium 41, 47, Color Plates C47, C51
Carnation-flowered geranium 202
Caterpillars 336–337
Cercospora brunkii. See Cercospora leaf spot
Cercospora leaf spot 229, Color Plate C79
Certified stock 75
Chimeras 375
Clean stock production 103, 277–297
CO_2 100, 109, 129, 130, 355
  Plant quality 142
Composting media 20
Containers
  Recycling 160
Copper 44
  Deficiency 49
Corynebacterium fascians 358
Cost accounting 145–155
Cost of production, 145–155
  Cutting geraniums 149–151
  Fixed costs 146
  Hybrid seed geraniums 152–155
  Labor 148
  Losses 149
  Overhead 146–147
  Variable costs 148
Cottony stem rot 255, Color Plate C93
Crown rot 216, 217
Cucumber mosaic virus 267, 270, Color Plates C89, C90
Cultivars 207
  Color mix 173
Culture indexing 181, 277–286
  CVI 282, 283, 284
  Nucleus block 279
  Sanitation 284
  Virus-indexing 280
  Viruses 267–275, 280–285
Culture virus indexed. *See* CVI
Cutting geraniums
  Breeding 374–379
  Cost of production 149
  Cuttings 97, 104

G

Gases 354
Geranium plume moth 336
Geranium rust 230, 232, 308, Color
  Plates C91, C92
  Resistance 232
Germicides 217
Germination 116–119, 125–126,
  153
  Bench germination 118
  Damping off 216
  Germination room 117
  Medium 119
  Plug production 113–123,
    125–134
  Production costs 153
  Sweat chamber 118
  Water quality 119
Gibberellic acid 65–66, 83, 110,
  135, 138
Goldsmith 208, 209, 210, 211,
  212, 213
Greenhouse whiteflies. *See* White-
  flies
Growing media 3–24, 106, 115,
  122
  Beneficial microbes 25–29
  Buffering ability 9
  CEC 9–11
  Compaction 21
  Composting 20
  Germination medium 119
  Mycorrhizal fungi 26
  Nutrient availability 10
  Nutrient balance 16
  Pasteurization 95
  pH 11–15, 19, 42, 80, 108
  Porosity 5
  Root rhizosphere 25, 28
  Saturated paste analysis 16, 18,
    45
  Soluble salts levels 17
  Storage 22
  Surfactants 8
  Wetting agents 8
Growth regulators 65–74, 109–
  110, 121
  A-Rest 65, 69
  Auxin 65
  Auxins 66
  B-Nine 69
  Bonzi 65, 70
  Cycocel 65, 68, 129
  Cycocel and light 129

Ethylene 65
Florel 68, 84
Gibberellic acid 66, 83
Sumagic 65, 70

H

Heat therapy 88
History 363–371
  Crossability 367
  Origin of term zonal 365
  Taxonomy 363
Hormodin 98
Host plant resistance 299–311
  Bacterial blight 307
  Biochemical mechanisms 301
  Botrytis 307
  Breeding 302
  Mites 302
  Morphological mechanisms 300
  Pythium 309
  Rust 308
  Viruses 309
  Whitefly 306
Hybrid seed geraniums 2, 125–
  134, 379
  Blindness 6, Color Plate C69
  CO_2 129
  Diseases 215–219
  Germination 116–119, 125, 385
  Growth regulators 65–74, 121
  Light 120–121, 127–129
  Marketing 157–163, 165–170
  Petal shattering 353–354
  Plug production 113–123, 125–
    129
  Production costs 152–155
  Scheduling 122, 131–133
  STS 72–73
  Temperature 130–131
  Timing 131–133
  Transplanting 127, Color Plate
    C65

I

IBA 183
In vitro. *See* Tissue culture
Inbred lines 380
Insect traps 314
Insects
  Aphids 324–331
  Cultural controls 314
  Fungus gnats 338–342
  Host plant resistance 299–311

Flowering 177–178, 183
Forcing 184
Growing media 182
Growth regulators 184
History 175
Light 179
Photoperiod 179, 185
Pinching 182, 183
Postharvest 186
Propagation 180–182
Rooting 182
Spacing 185
Stock plants 180
STS 186
Temperature 177–179, 182, 183
Whitefly resistance 306
Relative humidity 109
Resmethrin 341
Retailing 168–170
Display 169
Reverse day and night temperature
60, 111
Rhizobacteria 27–28
Rhizoctonia 20, 95, 216, 356
Rhizosphere 25–26, 28–29
Rhodococcus fascians. See Bacterial fasciation
Root rhizosphere 28
Root rot. *See* Pythium and Thielaviopsis
Rosebud geranium 201, Color Plates C13, C14
Royalties 378
Rust 230, 308
Resistance 232

S

STS 72–73, 172, 186, 251, 353
Saturated paste analysis 16, 18
Scatella stagnalis. See Shore flies
Scented geranium 204
Sclerotinia 217, 255, Color Plate C94
Control 255
Scouting 314, 335
Seed geranium varieties. *See* Hybrid seed geraniums
Seeders 114
Shattering 171
Shipping 171–172
Shore flies 251
Silver thiosulfate. *See* STS
Slugs 337–338
Southern bacterial wilt 242–245

Spider mites 196, 222, 331–336, Color Plate C68
Biological control 333, 335
Chemical control 334
Control 333
IPM 335
Management 333
Resistance 331
Scouting 335
Sprayers
High-volume 344
Low-volume 345
Standard geraniums. *See* Tree geraniums
Stellar geranium 203, Color Plates C21, C22
Stem blight 224
Stock plants 75–85, 95–97, 150, 181–182, 224
Botrytis 224
Certified stock 75
Containers 77
Cost 150
Cutting quality 84
Diseases 78
Fertilization 80
Framework 77, 81
Harvest 82
Insects 78
Light 79
Nutrition 97, 98
Regal geraniums 181–182
Substrate 78
Taking cuttings 82
Tree geraniums 83
Storage 171–172
Subirrigation 35
Sulfur 41, 49, Color Plate C62
Sulfur dioxide 354
Sumagic 65, 70
Surfactants 8
Sweat chamber 118
Sweetpotato whitefiles. *See* Whiteflies

T

Temperature 59–60, 108, 111, 130–133, 142, 172, 177–180, 184–185
DIF 60, 111
Ivy geraniums 192
Leaf unfolding rate 59
Plant development rate 59
Plant height 60

Plant quality 142
Plug production 116, 120
Regal geraniums 177–
    180, 184, 185
Shipping and storage 172
Thermal foggers 345
Thielaviopsis 20, 255–262, 356
    Control 262
    Symptoms 256
Thrips 182
Tissue analysis 46
Tissue culture 87–95, Color Plate
    C67
    Growth medium 91
    Limitations 89
Tobacco mosaic virus 267, 271
Tobacco rattle virus 271
Tobacco ringspot virus 267, 287,
    Color Plate C86
Tomato ringspot virus 90, 218, 267
Tomato spotted wilt virus 271
Topiary. *See* Tree geraniums
Transplanting 122, 127
Tree geraniums 83, 135–137
Trialeurodes vaporariorum 317
Tricolor geraniums 203
Tulip-flowered geranium 201,
    Color Plate C15
Two-spotted spider mites 196, 222,
    331–336

U

U.S. Department of Agriculture 1
UPOV 389, 397, 401

V

Varieties. *See* Cultivars
Vegetative propagation 87–102, 95–
    106, 226–228
    Botrytis 226–228
    Shipping cuttings 227
    Tissue culture 87–102
Verticillium wilt 238, 245–249,
    356
    Control 248
    Spread 247
    Symptoms 245
Vesicular-arbuscular mycorrhizae
    fungi 26
Viral diseases 218, 267–275, 280–
    285, 309
    Control 273, 286
    Culture-indexing 277–286
    Diagnosis 272

Eliminating viruses 289
Heat treatment 88, 291
Identification 272
Meristem tip culture 292
Pelargonium flower break virus
    268
Pelargonium leaf curl virus 269
Pelargonium line pattern virus
    269
Pelargonium ring pattern virus
    270
Pelargonium ringspot virus 268
Pelargonium vein clearing virus
    271
Pelargonium zonate spot virus
    270
Tobacco mosaic virus 271
Tobacco rattle virus 271
Tobacco ringspot virus 268, 287
Tomato ringspot virus 268
Tomato spotted wilt virus 271
Transmission 273, 293
Symptoms 290
Virus-indexing 287
Yellow net vein 271

W

Water quality 31–32, 45–
    46, 51, 119
    Alkalinity 52
    Analysis 45–46
Wetting agents 8
Whiteflies 182, 316–323
    Biological control 323
    Control 319
    Detection 321
    Identification 317
    Life stages 318
    Monitoring 321
    Timing spray applications 320

X

Xanthomonas 90, 96, 103, 107,
    181, 237–242, 307, 356, Color
    Plates C81–C85
    Control 242
    Cutting rot 239
    Hosts 237
    Leaf spots 238
    Stem rot 239
    Symptoms 237
    Temperature 239
    Transmission 241

Y

Yellow net vein 271, Color Plate
   C88
Yellow sticky cards
   Trapping 322

Z

Zinc 44, 49
Zonals. *See* Cutting geraniums

or **No** choice for proceeding with a particular operation, or informs you of the status of an operation:

```
Save document? Yes (No)                                   (Text was not modified)
```

A selection line is a one-line menu running across the bottom of the screen, from which you can select another option of a WordPerfect feature:

```
1 Figure; 2 Table Box; 3 Text Box; 4 User Box; 5 Line; 6 Equation: 0
```

To activate a prompt or selection-line option, place the mouse pointer on the option and click the left mouse button, or press the number or boldface letter for the option. You can press ↵ to accept the default option, which is represented by the number at the end of the selection line. (If the number 0 appears, pressing ↵ will return you to the editing screen or the previously displayed menu.) If you want to cancel a particular menu, press F7 or F1 (or ESC) to return to the editing screen or to step backward to the previously displayed menu.

A Closer Look at
■the Visual Approach

Unlike most computer books, the format for this book relies on graphics, not text, to present the material—no more wading through endless technical discussions to find out how to use a WordPerfect feature. Each chapter is broken up into practical exercises that present key graphics and written instructions, followed by pictures of what you actually see on your computer screen.

Here is an example of the kind of step you will follow in the exercises in this book:

■ *1*   Select **T**ools, **D**ate **T**ext or press SHIFT F5
1 or **T** to automatically place the current
date on the screen.

```
November 14, 1990_
```

Some screen illustrations have notes that provide additional information about the screen you are viewing, as in the following example:

■ *4*   Select **L**ayout, **A**lign, Hard **P**age or press
CTRL ↵ to create a hard page break, and a
second page.

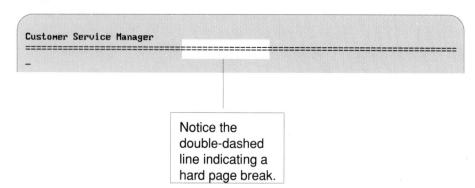

```
Customer Service Manager
===================================================================
_
```

Notice the
double-dashed
line indicating a
hard page break.

You will notice that most numbered steps have two parts: instructions, followed by a picture of what you should see on your screen. The instructions, which consist of text and key graphics, always offer you several methods for activating a WordPerfect feature. You can select an option using the main-menu bar and pull-down menus, or you can use the function keys. And you are always free to use just the mouse, just the keyboard, or a combination of the two.

A function-key instruction that reads "or press `SHIFT` `F5` `1` or `T`" tells you to press the Shift key, hold it down while pressing the F5 key, then release Shift-F5 and press the number 1 *or* the letter T to select the menu option.

Notice the step-by-step flow of the next two instructions:

■ 1   Select **F**ile, E**x**it or press `F7`.

```
was placed on the container. This will not happen again. My
apologies. Mr. Beck, our new CEO, and I look forward to seeing you
at the convention in October.

Best Regards,



Brian Knowles
Customer Service Manager




Save document? Yes (No)
```

■*2*   Select **Y**es or press [Y] or [↵].

```
Document to be saved: _
```

Step 1   "Select **F**ile, E**x**it" instructs you to click the right mouse button
           or press [ALT][-] to display the main-menu bar, place the mouse
           pointer on **F**ile and click the left mouse button, then do the
           same thing to select E**x**it from the pull-down menu. (You can
           also use the arrow keys to highlight the options and press [↵].)
           The instruction "or press [F7]" means you can press [F7] instead
           of selecting File and Exit on the menus. When you complete
           one of these instructions, your screen should look like the one
           shown below the step 1 instructions.

Step 2   "Select **Y**es" instructs you to place the mouse pointer on **Y**es
           and click the left mouse button. The instruction "or press [Y] or
           [↵]" means you can press [Y] or [↵] on the keyboard instead of
           using the mouse. (When you see [↵] in the instructions, press
           the Enter key.) When you complete one of these instructions,
           your screen should look like the one shown below step 2.

You will also see notes throughout the book like this one:

Note   *Deleting Codes* Use Reveal Codes to delete and replace codes that may
format your document incorrectly. For a listing of WordPerfect codes, see the inside
covers of this book.

Notes give you supplementary information about the feature you're learning. One of these might refer you to a related section in the book, give you a helpful hint or warning, or offer a useful or interesting observation. Notes that give special information on using the mouse begin with a graphic icon of a mouse.

There is another graphic element in this book that you should become familiar with before you start the exercises. When you see two keys that slightly overlap, like this:

■2　Select **E**dit, **B**lock or press ⎡ALT⎤⎡F4⎤

it means you should press and hold down the first key (⎡ALT⎤ in this case), press the second key (⎡F4⎤) and release it, then release the first key. This accomplishes the same thing as selecting **E**dit on the main-menu bar and selecting **B**lock on the pull-down menu.

■*Before You Begin*

Many people become intimidated when they have to read instructions of any sort. I want to give some quick advice about intimidation to those of you who are new to computers: Forget it! This book is designed to make getting started with WordPerfect as easy and as painless as possible.

So relax and begin at the beginning. By the time you complete the first chapter, you will have created and edited your first document, and will have saved it for future editing and printing. I think you will be surprised by how easy it is.

1

- *Getting*
- *Started*

■ If you are ready to begin learning Word-Perfect 5.1, you've turned to the right chapter. If you haven't read the Introduction, please do so, for it contains valuable information that you need to know before starting your computer and Word-Perfect. Also, take a look at the appendices at the back of the book and make sure you have installed WordPerfect on your hard or floppy disk and have told WordPerfect what printer you'll be using.

In this chapter, I will guide you through these basics:

Starting WordPerfect

Creating, editing, and printing your first document

Getting help

Starting WordPerfect
▪from a Hard Disk

Starting WordPerfect from a hard disk is very simple.

▪ *1*   Turn on your monitor and computer, and if prompted to do so, type in the current date. Press ⏎. If the date is already correct, just press ⏎.

```
Current date is Wed 11-14-90
Enter new date (MM-dd-yy):
Current time is 12:16:45.08
Enter new time:_
```

▪ *2*   Type in the current time if prompted to do so, and press ⏎. If the time is already correct, just press ⏎.

```
Current date is Wed 11-14-90
Enter new date (MM-dd-yy):
Current time is 12:16:45.08
Enter new time:

C>_
```

The C prompt appears.

1

> **Note**   Incorrect entries will result in an error message. Don't panic! You will be presented with the same prompt at which you made the incorrect entry. Proceed by simply entering the correct information at the prompt.

■*3*   Type **CD WP51** and press ⏎ If Word-
Perfect 5.1 is installed in a subdirectory with
another name, type **CD**, then the directory
name, and press ⏎

```
C>CD WP51

C>_
```

> The status line that you see at the bottom of the screen when you start WordPerfect can display measurements in various kinds of units. Don't be alarmed if you see some-thing other than inches.

■ *4*   Type **WP** and press ⏎. You'll be prompted to type in your registration number. Do so and press ⏎. (You only have to do this the first time you run WordPerfect.) You'll see the editing screen with the status line at the bottom.

Doc 1 Pg 1 Ln 1" Pos 1"

Doc 1 Pg 1 Ln 1" Pos 1"

| The number of the document you are working on. | The page you are currently viewing. | The line of the document where the cursor is currently positioned. | The position of the cursor. It is now at the left margin. |

1

Starting WordPerfect
■*from Floppy Disks*

Starting WordPerfect on a computer with dual floppy-disk drives requires you to do a little disk-shifting.

■*1*   Place your DOS system disk in drive A, and turn on your computer. If necessary, respond to the date and time prompts. (See the previous section, "Starting WordPerfect from a Hard Disk," steps 1 and 2, for more information on these prompts.)

```
Current date is Wed 11-14-90
Enter new date (mm-dd-yy):
Current time is 12:16:45.08
Enter new time:

A>_
```

■*2*   Take out the DOS disk, and place the Word-Perfect 1 disk in drive A and a blank, formatted disk in drive B.

■*3*   Type **B:** and press ⏎. This logs you onto drive B. Your document will automatically be saved to the disk in drive B.

```
A>B:

B>_
```

■4 Type **A:WP** and press ⏎. When prompted,
place the WordPerfect 2 disk in drive A and
press any key. You'll be prompted to type in
your registration number. Do so and press
⏎. You'll see the editing screen with the
status line at the bottom.

Doc 1 Pg 1 Ln 1" Pos 1"

Doc 1 Pg 1 Ln 1" Pos 1"

The number of the document you are working on.

The page you are currently viewing.

The line of the document where the cursor is currently positioned.

The position of the cursor. It is now at the left margin.

1

> **Note** The editing screen will be blank when you start WordPerfect, except for the *status line* in the lower-right corner. The lower-left corner will be blank until you retrieve a file for editing. Then you'll see the name of the document, and the disk drive and subdirectory where the file is located.

■ *Creating a Sample Document*

Let's construct a simple letter. Typing on a computer keyboard is like typing on a typewriter except you do not have to press ⏎ (Enter) at the end of a line. WordPerfect's *word wrap* feature automatically wraps the sentence to the next line for you. Don't worry about mistakes — you'll have a chance to correct any you make in a moment.

■ *1*   Select **T**ools, Date **T**ext or press [SHIFT][F5] [**1**] or [**T**] to automatically place the current date on the screen. (In Chapter 3, you'll learn more about this feature.)

```
November 14, 1990_
```

■ *2*   Press ⏎ four times, and type what you see here:

     Mr. Randall Jones ⏎

     2153 Ocean View Drive ⏎

     Burbank, CA 91505 ⏎

⏎

⏎

Dear Randall: ⏎

⏎

Your order of July 12 was traced to Nome, Alaska. It should arrive at your office on August 25. Apparently, an incorrect address label was placed on the container. This will not happen again. My apologies, and I look forward to seeing you at the convention in September. ⏎

⏎

Best Regards, ⏎

⏎

⏎

⏎

Brian Knowles ⏎

Customer Service Manager

Note Do not press ⏎ after the word *Manager*. If your screen shows a different number of lines, it may be because you have a different monitor. Don't worry about it, everything's fine.

Your letter should look like this:

```
November 14, 1990


Mr. Randall Jones
2153 Ocean View Drive
Burbank, CA 91505


Dear Randall:

Your order of July 12 was traced to Nome, Alaska. It should arrive
at your office on August 25. Apparently, an incorrect address label
was placed on the container. This will not happen again. My
apologies, and I look forward to seeing you at the convention in
September.

Best Regards,


Brian Knowles
Customer Service Manager_
                                        Doc 1 Pg 1 Ln 4.83" Pos 3.4"
```

Congratulations! You have just created your first document. See how easy it is?
Now let's make some changes to the letter.

■*Moving the Cursor*

The cursor is now located at the end of the letter. To edit your letter, you must
move the cursor to the place you want to make a change. Use the mouse or the
arrow keys located to the right of the main keyboard. On older keyboards, these
keys may be combined with the numeric keypad. Holding an arrow key down
moves the cursor continuously in the indicated direction.

■ *1*   Place the mouse pointer on the *C* in *Customer* and click the left button, or press ⬅ until the cursor is under the *C*.

> Brian Knowles
> <u>C</u>ustomer Service Manager
>
>   Doc 1 Pg 1 Ln 4.83" Pos 1"

■ *2*   Place the mouse pointer on the *Y* in *Your* and click the left button, or press ⬆ until the cursor is under the *Y*.

> <u>Y</u>our order of July 12 was traced to Nome, Alaska. It should arrive
> at your office on August 25. Apparently, an incorrect address label
> was placed on the container. This will not happen again. My
> apologies, and I look forward to seeing you at the convention in
> September.

■ *3*   Place the mouse pointer on the *u* in *July* and click the left button, or press ➡ until the cursor is under the *u*.

> Your order of J<u>u</u>ly 12 was traced to Nome, Alaska. It should arrive
> at your office on August 25. Apparently, an incorrect address label
> was placed on the container. This will not happen again. My
> apologies, and I look forward to seeing you at the convention in
> September.

■4 Place the mouse pointer on the *I* in the last
sentence and click the left button, or press
⬇ until the cursor is under the *I*.

```
Your order of July 12 was traced to Nome, Alaska. It should arrive
at your office on August 25. Apparently, an incorrect address label
was placed on the container. This will not happen again. My
apologies, and I look forward to seeing you at the convention in
September.
```

Now play with the mouse and/or arrow keys for a moment, just to get the feel of
moving the cursor. When you're ready, position the cursor back at the *I* in the
last sentence, and move on to the next exercise.

■*Inserting Text*

Let's modify the letter by inserting text at the *I* in the last sentence. If you see the
word *Typeover* in the lower-left corner of your screen, press **INS** (insert) to make
it disappear. (You'll learn about the Typeover feature in just a moment.)

■1 Type **Mr. Beck, our new CEO, and.** Press
the spacebar.

```
Your order of July 12 was traced to Nome, Alaska. It should arrive
at your office on August 25. Apparently, an incorrect address label
was placed on the container. This will not happen again. My
apologies, and Mr. Beck, our new CEO, and I look forward to seeing you at the co
September.
```

■*2* Press ⬇ to automatically adjust the text between margins.

> Your order of July 12 was traced to Nome, Alaska. It should arrive
> at your office on August 25. Apparently, an incorrect address label
> was placed on the container. This will not happen again. My
> apologies, and Mr. Beck, our new CEO, and I look forward to seeing
> you at the convention in September._

Using Typeover is another way you can place new text in a document. You type over the existing character, which is deleted and replaced with the new character. Be careful! If you forget you are in the Typeover mode, you may accidentally delete text you want to keep.

■*1* Move the cursor to the *J* in *July* in the first sentence.

> Your order of July 12 was traced to Nome, Alaska. It should arrive
> at your office on August 25. Apparently, an incorrect address label
> was placed on the container. This will not happen again. My
> apologies, and Mr. Beck, our new CEO, and I look forward to seeing
> you at the convention in September.

■*2* INS to turn the Typeover feature on. (You will see the word Typeover appear highlighted in the lower-left corner of the screen.)

> Brian Knowles
> Customer Service Manager
> Typeover
>   Doc 1 Pg 1 Ln 3" Pos 2.4"

1

■*3*   Type **June**.

> Your order of June_12 was traced to Nome, Alaska. It should arrive
> at your office on August 25. Apparently, an incorrect address label
> was placed on the container. This will not happen again. My
> apologies, and Mr. Beck, our new CEO, and I look forward to seeing
> you at the convention in September.

■*4*   [INS] to turn the Typeover feature off.

> Brian Knowles
> Customer Service Manager
>
>   Doc 1 Pg 1 Ln 3" Pos 2.8"

■*Deleting Text*

Now you're ready to correct any mistakes you might have made when typing the letter. In this section you'll first use [BKSP] (Backspace) to delete text, then you'll use [DEL] (Delete).

■*1*   Move the cursor to the *M* in *Mr. Beck.*

> Your order of June 12 was traced to Nome, Alaska. It should arrive
> at your office on August 25. Apparently, an incorrect address label
> was placed on the container. This will not happen again. My
> apologies, and Mr. Beck, our new CEO, and I look forward to seeing
> you at the convention in September.

■*2*   [BKSP ←] six times.

> Your order of June 12 was traced to Nome, Alaska. It should arrive
> at your office on August 25. Apparently, an incorrect address label
> was placed on the container. This will not happen again. My
> apologies<u>M</u>r. Beck, our new CEO, and I look forward to seeing
> you at the convention in September.

■*3*   [.] (period) and press the spacebar once.

> Your order of June 12 was traced to Nome, Alaska. It should arrive
> at your office on August 25. Apparently, an incorrect address label
> was placed on the container. This will not happen again. My
> apologies. <u>M</u>r. Beck, our new CEO, and I look forward to seeing
> you at the convention in September.

Practice by correcting any other mistakes until your letter looks like the one above.

[DEL] deletes the character at the cursor location. If you press and hold down [DEL], you will pull text to the left and can delete as many characters as you want. This happens quickly, so be careful.

■*1*   Move the cursor to the *S* in *September*.

> Your order of June 12 was traced to Nome, Alaska. It should arrive
> at your office on August 25. Apparently, an incorrect address label
> was placed on the container. This will not happen again. My
> apologies. Mr. Beck, our new CEO, and I look forward to seeing you
> at the convention in <u>S</u>eptember.

1

■*2*　　DEL nine times.

```
Your order of June 12 was traced to Nome, Alaska. It should arrive
at your office on August 25. Apparently, an incorrect address label
was placed on the container. This will not happen again. My
apologies. Mr. Beck, our new CEO, and I look forward to seeing you
at the convention in _
```

■*3*　　Type **October**.

```
Your order of June 12 was traced to Nome, Alaska. It should arrive
at your office on August 25. Apparently, an incorrect address label
was placed on the container. This will not happen again. My
apologies. Mr. Beck, our new CEO, and I look forward to seeing you
at the convention in October._
```

Viewing the
■*Document before Printing*

View Document is a fun and useful feature of WordPerfect. In fact, you'll find it a necessity when you work with graphics. Gone are the days of having to print a test document to see how it looks; now you can see how it looks right on your monitor.

Let's take a look at how the letter you just created will look when printed, before you print it.

■ *1*   Select **F**ile, **P**rint or press `SHIFT` `F7`.

```
Print

       1 - Full Document
       2 - Page
       3 - Document on Disk
       4 - Control Printer
       5 - Multiple Pages
       6 - View Document
       7 - Initialize Printer


Options

       S - Select Printer                HP LaserJet Series II
       B - Binding Offset                0"
       N - Number of Copies              1
       U - Multiple Copies Generated by  WordPerfect
       G - Graphics Quality              Medium
       T - Text Quality                  High




Selection: 0
```

The View Document feature comes in handy when working with a document that has lots of text effects and graphics. If you stay on the editing screen, you may not be able to get a good sense of what these elements will look like when printed, so you can pop in and out of View Document to get a preview.

1

■*2*   Select **V**iew Document or press [**6**] or [**V**].

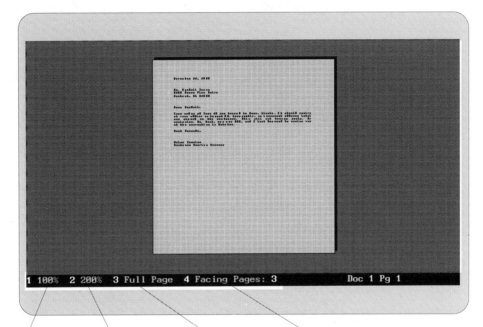

Select **100%** or press [**1**] to enlarge the display to show the page at its actual size.

Select **200%** or press [**2**] to see the page at twice its actual size.

Select **Full Page** or press [**3**] to see a full-page layout.

Select **Facing Pages** or press [**4**] to see two consecutive pages, back-to-back.

To resize the document with the mouse, place the pointer on the document size you want and click the left button.

■*3*   Click the right mouse button or press F7 to
return to the document screen.

> was placed on the container. This will not happen again. My
> apologies. Mr. Beck, our new CEO, and I look forward to seeing you
> at the convention in October.
>
> Best Regards,

Note   When in View Document, use any cursor key to view portions of a document not shown on the View Document screen. PG UP and PG DN will move you from the displayed page to the previous or following page.

■*Printing the Document*

Now it's time to see if what you have created on the screen will come out of the printer.

> The Print menu
> has many op-
> tions for control-
> ling the quantity
> and quality of
> your printouts.

1

■ *1* Select **F**ile, **P**rint or press SHIFT F7.

```
Print

      1 - Full Document
      2 - Page
      3 - Document on Disk
      4 - Control Printer
      5 - Multiple Pages
      6 - View Document
      7 - Initialize Printer


Options

      S - Select Printer              HP LaserJet Series II
      B - Binding Offset              0"
      N - Number of Copies            1
      U - Multiple Copies Generated by  WordPerfect
      G - Graphics Quality            Medium
      T - Text Quality                High




Selection: 0
```

■ *2* Select **F**ull Document or press 1 or F. If
you are using the continuous-feed or sheet-
feeder mode of printing, the letter now prints,
and you return to the document screen.

```
was placed on the container. This will not happen again. My
apologies. Mr. Beck, our new CEO, and I look forward to seeing you
at the convention in October.

Best Regards,
```

If you are using manual-feed mode, complete steps 3 –7. A beep will alert you to
insert your paper manually. In a multiple-page document, this beep sounds after
each page is printed.

■ *3* Select File, Print or press [SHIFT][F7]

```
Print

       1 - Full Document
       2 - Page
       3 - Document on Disk
       4 - Control Printer
       5 - Multiple Pages
       6 - View Document
       7 - Initialize Printer


Options

       S - Select Printer              HP LaserJet Series II
       B - Binding Offset              0"
       N - Number of Copies            1
       U - Multiple Copies Generated by   WordPerfect
       G - Graphics Quality            Medium
       T - Text Quality                High




Selection: 0
```

1

■*4*   Select **C**ontrol Printer or press **4** or **C**

```
Print: Control Printer

Current Job

Job Number:  2                           Page Number:  1
Status:      Printing                    Current Copy: 1 of 1
Message:     None
Paper:       Standard 8.5" x 11"
Location:    Manual feed
Action:      Insert paper
             Press "G" to continue

Job List

Job  Document              Destination        Print Options
 2   (Screen)              LPT 1



Additional Jobs Not Shown: 0



1 Cancel Job(s); 2 Rush Job; 3 Display Jobs; 4 Go (start printer); 5 Stop: 0
```

■*5*   Insert paper, then select **G**o or press 4 or
            G. The document is printed.

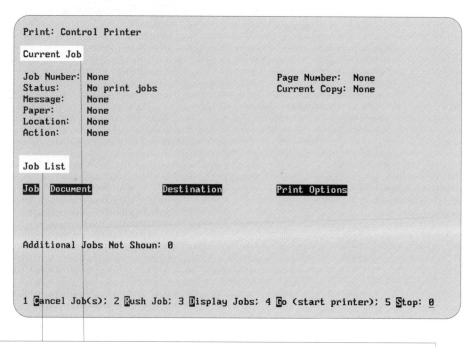

```
Print: Control Printer

Current Job

Job Number:  None                           Page Number:   None
Status:      No print jobs                  Current Copy:  None
Message:     None
Paper:       None
Location:    None
Action:      None


Job List

Job  Document                  Destination          Print Options




Additional Jobs Not Shown: 0



1 Cancel Job(s); 2 Rush Job; 3 Display Jobs; 4 Go (start printer); 5 Stop: 0
```

Notice there is no Current Job or Job List information displayed on the menu.

■*6*   F1 (Cancel) or click the center mouse button
            to return to the Print menu.

```
Print

        1 - Full Document
        2 - Page
        3 - Document on Disk
        4 - Control Printer
        5 - Multiple Pages
        6 - View Document
        7 - Initialize Printer
```

1

Pressing [F1](Cancel) returns you to the previous menu. Pressing [F7](Exit) or the right mouse button skips this menu and returns you directly to the document screen.

■ *7* Click the right mouse button or press [F7] to
return to the document screen.

```
was placed on the container. This will not happen again. My
apologies. Mr. Beck, our new CEO, and I look forward to seeing you
at the convention in October.

Best Regards,
```

Saving Your Document *■and Exiting WordPerfect*

If you want to keep this letter for future editing, you must save it on a floppy or hard disk. Remember, the letter you see on your monitor is only saved in temporary memory (RAM).

Try to remember to save your data often while you work. Losing a few minutes' worth of data is not so bad as losing an hour's worth if there is a sudden loss of power to your computer.

Read Chapter 14 to learn how to have files saved automatically.

■*1* Select **F**ile, **E**xit or press F7.

Save document? Yes (No)

If you select No, the document will disappear from the screen and will be gone for good. The default is Yes, which prevents you from inadvertently losing the document.

■*2* Select **Y**es or press Y or ↵.

Document to be saved: _

You are requested to type in the file name you want for this document.

■*3* If you are using a computer with a hard disk, type **MYLETTER** and press ↵. If you have floppy disks, go to step 5.

Exit WP? No (Yes)   (Cancel to return to document)

1

■*4*   Select **Y**es or press 〔**Y**〕. The MYLETTER
document is saved, and you are returned to
the C prompt in DOS. (Skip steps 5 and 6.)

```
C>_
```

■*5*   If you are using a computer with floppy-disk
drives only, type **B:MYLETTER** and press
〔⏎〕. (You type **B:** because your document
disk is in drive B.)

```
Exit WP? No (Yes)                    (Cancel to return to document)
```

■*6*   Select **Y**es or press 〔**Y**〕 to exit WordPerfect.

```
B>_
```

Note   Use uppercase or lowercase letters when you type the file name —
WordPerfect accepts it either way. File names can be from one to eight characters
long and can start with either a letter or a number. (You can modify WordPerfect
so that it accepts much larger file names, but they will be truncated to eight characters in any DOS directory.)

■*Getting Help*

WordPerfect's powerful word processing features are so numerous that you'll probably have trouble remembering many of the keyboard and menu commands. Well, you can rest easy because WordPerfect has an internal help feature. Let's see how it works.

Restart WordPerfect as you learned at the beginning of this chapter, and go to the blank editing screen. In this section you'll get some help setting the left and right margins.

■ *1*   If you have a dual floppy-disk computer, remove the document disk from drive B and replace it with the WordPerfect 1 program disk. Select **H**elp, **H**elp or press F3. (Selecting **H**elp, **I**ndex displays an alphabetical listing of WordPerfect features.)

```
Help                                           WP 5.1   11/06/89


        Press any letter to get an alphabetical list of features.

             The list will include the features that start with that letter,
             along with the name of the key where the feature is found.  You
             can then press that key to get a description of how the feature
             works.


        Press any function key to get information about the use of the key.

             Some keys may let you choose from a menu to get more information
             about various options.  Press HELP again to display the template.











Selection: 0                               (Press ENTER to exit Help)
```

1

Note If you are a floppy-disk user, you will see the following message if you forget to place the WordPerfect 1 disk in drive B before you select **Help**: "WPHELP.FIL not found. Insert the diskette and press drive letter:" Remove the document disk from drive B and replace it with your WordPerfect 1 disk. Press [B] and you will see the screen shown on the previous page.

■ 2 [M] to display an alphabetical index of those features that begin with the letter *m*.

```
Features [M]                    WordPerfect Key   Keystrokes

Macro Editor                    Macro Define      Ctrl-F10
Macro Commands                  Macro Commands    Ctrl-PgUp
Macro Commands, Help On         Macro Define      Ctrl-F10
Macros, Define                  Macro Define      Ctrl-F10
Macros, Execute                 Macro             Alt-F10
Macros, Keyboard Definition     Setup             Shft-F1,5
Mail Merge                      Merge/Sort        Ctrl-F9,1
Main Dictionary Location        Setup             Shft-F1,6,3
Manual Hyphenation              Format            Shft-F8,1,1
Map, Keyboard                   Setup             Shft-F1,5,8
Map Special Characters          Setup             Shft-F1,5
Margin Release                  Margin Release    Shft-Tab
Margins - Left and Right        Format            Shft-F8,1,7
Margins - Top and Bottom        Format            Shft-F8,2,5
Mark Text For Index (Block On)  Mark Text         Alt-F5,3
Mark Text For List (Block On)   Mark Text         Alt-F5,2
Mark Text For ToA (Block On)    Mark Text         Alt-F5,4
Mark Text For ToC (Block On)    Mark Text         Alt-F5,1
Master Document                 Mark Text         Alt-F5,2
Math                            Columns/Table     Alt-F7,3
More... Press m to continue.

Selection: 0                              (Press ENTER to exit Help)
```

The "Margins - Left and Right" feature
is activated by pressing [SHIFT] [F8] [1] [7].

■*3* [SHIFT] [F8]

Format

Contains features which change the current document's format. Options on
the Line, Page and Other menus change the setting from the cursor
position forward. Document Format options change a setting for the entire
document. To change the default settings permanently, use the Setup key.

If Block is On, press Format to protect a block. You can use Block
Protect to keep a block of text and codes together on a page (such as a
paragraph which may change in size during editing).

1 - Line

2 - Page

3 - Document

4 - Other

Note: In WordPerfect 5.1, you can enter measurements in fractions (e.g., 3/4")
as well as decimals (e.g., .75"). WordPerfect will convert fractions to
their decimal equivalent.

Selection: 0   (Press ENTER to exit Help)

More than 500
features are
described in
the Help index.

■4   1 or L

```
Format: Line

    1 - Hyphenation

    2 - Hyphenation Zone

    3 - Justification

    4 - Line Height

    5 - Line Numbering

    6 - Line Spacing

    7 - Margins - Left and Right

    8 - Tab Set

    9 - Widow/Orphan Protection



Selection: 0                              (Press ENTER to exit Help)
```

■5   7 or M

```
Margins - Left and Right

    Sets the left and right margins from the cursor forward.  If your cursor
    is not at the left margin when you set margins, a [HRt] code is inserted
    before the setting.
```

■ 6   F3 to display the WordPerfect features
template. Selecting **H**elp, **T**emplate or pressing F3 twice always displays the template.

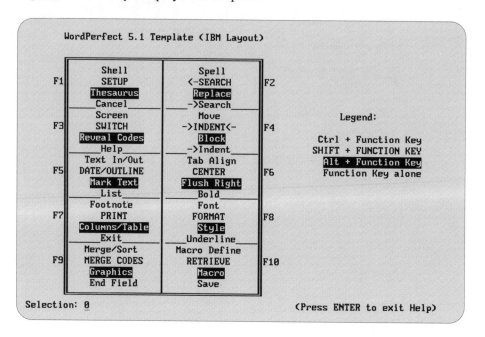

■ 7   ↵ You're back at the blank editing screen.

Doc 1 Pg 1 Ln 1" Pos 1"

1

Experiment for a few minutes by selecting **H**elp, **H**elp or by pressing `F3`, then press different function-key combinations and letter keys to see the help that is offered. I think you'll find it fun and interesting. Press `↵` when you're finished.

Note Activating Help from within a WordPerfect menu will display helpful information about that menu. For example, if you want to create new tabs and aren't sure how to do it, press `SHIFT` `F8` to display the Format menu, then press `F3` for help with formatting. Press `L`, followed by `T`, to display help with tabs.

Go on to Chapter 2, where you'll learn some more editing features to help speed up your work.

2

- Basic
- Editing
- Techniques

■ **N**ow that you have a little experience, let's move on to some more editing features of Word-Perfect 5.1. By the end of this chapter, you will begin to feel very comfortable with WordPerfect and should have enough skills to begin creating your own documents, if you aren't already doing so. Here's what you'll learn:

Retrieving a saved document

Creating soft and hard page breaks

Advanced cursor movement

Advanced deletion techniques

Restoring deleted text

Saving an edited document

Retrieving a ◾*Saved Document*

Once you have created a document and stored it as a file on a disk, you may want to recall it to the screen for further editing. When you retrieve a file, WordPerfect places a copy of the file on the Edit screen, leaving the original file intact on the disk. When you save the newly edited version of the file, it replaces, or writes over, the original version.

WordPerfect provides two easy methods for retrieving a file: entering the file name and selecting the file name from a list of files. Before you begin, start WordPerfect (refer to Chapter 1 if you need to); you'll see the Edit screen.

Retrieving by File Name

You can retrieve a specific document if you know the name of the file. Let's retrieve the file named MYLETTER, which you created in Chapter 1. If you have a dual floppy-disk system, be sure to place the document disk with the MYLETTER file stored on it in drive B.

◾*1* Select **F**ile, **R**etrieve or press SHIFT F10.

```
Document to be retrieved: _                                    (List Files)
```

■*2*   Type **MYLETTER** and press ⏎. (Type
B:MYLETTER if you have a dual floppy-
disk system so that WordPerfect will look on
the document disk in drive B, then press ⏎.)

```
November 14, 1990


Mr. Randall Jones
2153 Ocean View Drive
Burbank, CA 91505


Dear Randall:

Your order of June 12 was traced to Nome, Alaska. It should arrive
at your office on August 25. Apparently, an incorrect address label
was placed on the container. This will not happen again. My
apologies. Mr. Beck, our new CEO, and I look forward to seeing you
at the convention in October.

Best Regards,


Brian Knowles
Customer Service Manager
C:\WP51\MYLETTER                                Doc 1 Pg 1 Ln 1" Pos 1"
```

■*3*   F7 N N to exit without saving the document (remember, the file is still stored on the
disk). You will remain in WordPerfect with a
clear screen.

Retrieving a File with List Files

WordPerfect's List Files feature is a powerful file management tool. It provides
you with a list of all the files in a selected directory. It's like a master document
register of all the files in a conventional file cabinet, except this file cabinet is
electronic.

■ *1*   Select File, List Files or press F5.

Dir C:\WP51*.*   (Type = to change default Dir)

To display a list of files on another disk, type the drive letter where the disk is located, followed by a colon (for example, **B:**).

■ *2*   With the mouse pointer on the prompt, click the right button, double-click the left button or just press ↵. (Your list of files will be different from the one displayed here, so don't be alarmed if your screen looks different.)

```
11-14-90                    Directory C:\WP51\*.*
Document size:        0    Free:  5,005,312 Used:  5,071,473      Files:      230

.       Current    <Dir>                    ..     Parent    <Dir>
LEXICONS.         <Dir> 11-19-89 03:07p    {LF}     .DIR      5,500  06-30-89 01:35p
20CENT   .       17,594 11-26-89 05:09p    8514A    .VRS      4,866  11-06-89 12:00p
ALTA     .WPM        71 04-26-89 09:02p    ALTB     .WPM        155  05-23-88 09:40a
ALTC     .WPM        71 04-28-89 03:24p    ALTD     .WPM         71  05-03-89 04:31p
ALTE     .WPM        75 05-20-88 02:06p    ALTF     .WPM        122  09-09-89 12:52p
ALTG     .WPM        57 09-08-89 10:29a    ALTI     .WPM         76  05-21-89 05:20p
ALTJ     .WPM       100 05-20-88 01:54p    ALTL     .WPM        227  01-18-89 09:06a
ALTP     .WPM        67 04-27-89 11:14a    ALTQ     .WPM         71  05-12-89 04:40p
ALTR     .WPM        89 05-17-89 11:49a    ALTRNAT  .WPK        919  11-06-89 12:00p
ALTS     .WPM       182 02-02-89 09:01a    ALTT     .WPM         67  04-26-89 08:54p
ALTU     .WPM       178 01-31-89 08:44a    ALTU     .WPM        173  05-23-88 09:45a
ALTX     .WPM        69 04-26-89 08:55p    ALTY     .WPM         80  07-28-89 04:41p
ALTZ     .WPM        71 04-26-89 09:03p    APPA     .ADD      5,034  11-27-89 04:49p
APPA     .W51     5,550 11-27-89 05:08p    APPB     .W51     16,013  12-04-89 12:28p
APPC     .W51     7,781 12-02-89 05:43p    APPD     .W51      6,287  11-14-89 05:20p
APPNOTE  .TXT     3,695 06-01-88 12:00a    AREACODE.ARC     12,928  08-04-89 09:32p
ARROW-22.WPG       116 11-06-89 12:00p    ATI      .VRS      6,041  11-06-89 12:00p
BALLOONS.WPG     2,806 11-06-89 12:00p    BANNER-3.WPG        648  11-06-89 12:00p

1 Retrieve; 2 Delete; 3 Move/Rename; 4 Print; 5 Short/Long Display;
6 Look; 7 Other Directory; 8 Copy; 9 Find; N Name Search: 6
```

■*3* Using the mouse pointer or the arrow keys,
move the highlight to the MYLETTER file.

```
MACROS   .WPK   32,835  11-06-89 12:00p │ MAKESFX .COM      872  06-01-88 12:00a
MANUAL   .DOC   90,625  06-01-88 12:00a │ ME      .EXE   58,368  03-12-87 03:30p
MEHELP   .1      1,429  03-12-87 03:30p │ MEM     .MEX    6,451  03-12-87 03:30p
MPM      .MEX    5,875  03-12-87 03:30p ▼ MYLETTER.         765  11-14-90 10:58a

1 Retrieve; 2 Delete; 3 Move/Rename; 4 Print; 5 Short/Long Display;
6 Look; 7 Other Directory; 8 Copy; 9 Find; N Name Search: 6
```

To highlight a file not displayed on the screen, press the left button, hold it
down, and move the highlighter to the top or bottom of the screen. The files not
visible will scroll onto the screen. Release the button to stop scrolling.

The List Files screen displays three data about each file in a directory: the name of the file and its extension, the size of the file in bytes, and the date and time the file was last saved.

■*4* Double-click the left mouse button (with the pointer on MYLETTER) or press **6** or **L** to look at the contents of the file, and verify that this is the document you want to retrieve. You cannot edit in the Look mode, but you can use the up and down arrow keys to view the file's contents.

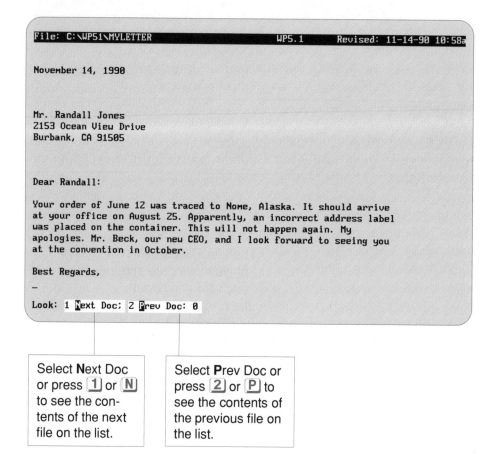

```
File: C:\WP51\MYLETTER                          WP5.1        Revised: 11-14-90 10:58a

November 14, 1990


Mr. Randall Jones
2153 Ocean View Drive
Burbank, CA 91505


Dear Randall:

Your order of June 12 was traced to Nome, Alaska. It should arrive
at your office on August 25. Apparently, an incorrect address label
was placed on the container. This will not happen again. My
apologies. Mr. Beck, our new CEO, and I look forward to seeing you
at the convention in October.

Best Regards,
—

Look: 1 Next Doc; 2 Prev Doc: 0
```

Select **N**ext Doc or press **1** or **N** to see the contents of the next file on the list.

Select **P**rev Doc or press **2** or **P** to see the contents of the previous file on the list.

2

■*5*   Click the right mouse button, press the
spacebar, or press [F7] to exit the Look mode
so you can retrieve the file for editing.

```
MACROS   .WPK    32,835  11-06-89 12:00p   MAKESFX .COM       872  06-01-88 12:00a
MANUAL   .DOC    90,625  06-01-88 12:00a   ME      .EXE    58,368  03-12-87 03:30p
MEHELP   .1       1,429  03-12-87 03:30p   MEM     .MEX     6,451  03-12-87 03:30p
MPM      .MEX     5,875  03-12-87 03:30p ▼ MYLETTER.          765  11-14-90 10:58a

1 Retrieve; 2 Delete; 3 Move/Rename; 4 Print; 5 Short/Long Display;
6 Look; 7 Other Directory; 8 Copy; 9 Find; N Name Search: 6
```

■*6*   Select **R**etrieve or press [1] or [R] to place
the file on the blank editing screen.

```
November 14, 1990



Mr. Randall Jones
2153 Ocean View Drive
Burbank, CA 91505


Dear Randall:

Your order of June 12 was traced to Nome, Alaska. It should arrive
at your office on August 25. Apparently, an incorrect address label
was placed on the container. This will not happen again. My
apologies. Mr. Beck, our new CEO, and I look forward to seeing you
at the convention in October.

Best Regards,



Brian Knowles
Customer Service Manager
C:\WP51\MYLETTER                              Doc 1 Pg 1 Ln 1" Pos 1"
```

The following list describes the List Files options displayed at the bottom of the List Files screen:

- **R**etrieve—Retrieves a file stored on a disk and copies it into temporary memory (RAM)
- **D**elete—Deletes a file from a disk
- **M**ove/Rename—Changes the name of a file; moves a file to another directory or disk
- **P**rint—Prints a file from a disk
- **S**hort/Long Display—Assigns a longer, more descriptive name (up to 30 characters) to a file (see Chapter 14)
- **L**ook—Lets you preview a document before retrieving it
- **O**ther Directory—Lists files located in another directory or on another disk; creates a new directory
- **C**opy—Records a copy of a file onto another disk or into another directory
- **F**ind—Finds a particular file or files, according to conditions you specify, when you can not remember the file name
- **N**ame Search—Locates a file when you type in the file name

Creating Soft
■and Hard Page Breaks

Your MYLETTER document is just one page long, but you will often work on larger documents with many pages. WordPerfect paginates in two ways—using soft and hard page breaks.

WordPerfect inserts a *soft page break* automatically every 54 lines of text unless you change the default top and bottom margins (you'll learn how to change the defaults later).

You enter a *hard page break* manually. This is useful when you have a small amount of text on a page and don't want to press ⏎ repeatedly to get to the next page.

Let's create a second page in your MYLETTER file using a soft page break, then a hard page break. Then we'll add some text to the second page.

■ 1 Using the mouse or the arrow keys, move the cursor to the space after *Manager* in the letter's closing.

```
Brian Knowles
Customer Service Manager█
C:\WP51\MYLETTER                                  Doc 1 Pg 1 Ln 4.83" Pos 3.4"
```

■ 2 ⏎ 31 times. Look for the dashed line across the screen that indicates a soft page break.

```
--------------------------------------------------------------------------------
C:\WP51\MYLETTER                                  Doc 1 Pg 2 Ln 1" Pos 1"
```

The page indicator automatically changes to page 2.

■ 3 ⟸ 31 times to delete the hard returns and return the cursor to the word *Manager.*

```
Customer Service Manager_
```

■*4* [CTRL][↵] to create a hard page break, and a
second page.

```
Customer Service Manager
========================================================================
_
```

Notice the double-dashed line indicating a hard page break.

Type this text on the second page:

OFFICE MEMO

TO: Sales Department

FROM: Charles Beck, CEO

SUBJECT: Monthly Sales

In the short time I've been with the company, our monthly sales have
increased over 50%. This is directly due to the outstanding performance of the Sales Department.

Thank you for your fine efforts. Keep up the good work through the
holiday season and everyone will see a nice year-end bonus.

Now move on to learn some of the advanced cursor movements in WordPerfect
5.1. If you need to stop here and want to save what you've done, go to the last
exercise in this chapter, "Saving an Edited Document," and follow the instructions. When you want to start again, retrieve MYLETTER and pick up where
you left off.

■*Advanced Cursor Movement*

2

You know the most basic methods of moving the cursor—placing the mouse pointer and clicking the left mouse button, or using the arrow keys to move one character or line at a time. Let's look at some ways to move the cursor more quickly and efficiently.

Moving One Word at a Time

■*1* CTRL ← four times to move the cursor four words to the left.

```
performance of the Sales Department.

Thank you for your fine efforts. Keep up the good work through the
holiday season and everyone will see █ nice year-end bonus.
```

■*2* CTRL → four times to move the cursor back.

```
performance of the Sales Department.

Thank you for your fine efforts. Keep up the good work through the
holiday season and everyone will see a nice year-end bonus.█
```

Moving to the Beginning and End of a Line

■ *1*   HOME ← to move the cursor to the beginning
of the line.

```
performance of the Sales Department.

Thank you for your fine efforts. Keep up the good work through the
holiday season and everyone will see a nice year-end bonus.
```

■ *2*   HOME → or END to move the cursor to the
end of the line.

```
performance of the Sales Department.

Thank you for your fine efforts. Keep up the good work through the
holiday season and everyone will see a nice year-end bonus.
```

Moving to the
Beginning and End of a Document

■ *1*   HOME HOME ↑ to move to the beginning of
the document.

```
November 14, 1990


Mr. Randall Jones
```

2

■*2* HOME HOME ↓ to move to the end of the document.

> performance of the Sales Department.
>
> Thank you for your fine efforts. Keep up the good work through the
> holiday season and everyone will see a nice year-end bonus.█
>
> C:\WP51\MYLETTER   Doc 1 Pg 2 Ln 3.17" Pos 6.9"

Moving One Page at a Time

■*1* PG UP to move to the top of the previous page.

> November 14, 1990
>
>
> Mr. Randall Jones

■*2* PG DN to move to the top of the following page.

> OFFICE MEMO
>
> TO: Sales Department
>
> FROM: Charles Beck, CEO

Moving to the Bottom and Top of a Page

■ *1*   [CTRL] [HOME] [↓] to move to the bottom of the current page.

> performance of the Sales Department.
>
> Thank you for your fine efforts. Keep up the good work through the holiday season and everyone will see a nice year-end bonus.█
>
> C:\WP51\MYLETTER   Doc 1 Pg 2 Ln 3.17" Pos 6.9"

■ *2*   [CTRL] [HOME] [↑] to move to the top of the current page.

> █FFICE MEMO
>
> TO: Sales Department
>
> FROM: Charles Beck, CEO

Moving One Screen at a Time

The [+] (plus) and [−] (minus) keys on the numeric keypad move the cursor one screen at a time. (Be sure Num Lock is off when you use these keys.) If the cursor is not at the top or bottom of the screen, you will have to press the [+] or [−] key twice—once to move to the top or bottom of the displayed text, and a second time to move up or down one full screen of undisplayed text.

■*1*   ⌷‾ twice to move up two screens.

```
November 14, 1990



Mr. Randall Jones
```

■*2*   ⌷₊ twice to move down two screens.

```
performance of the Sales Department.

Thank you for your fine efforts. Keep up the good work through the
holiday season and everyone will see a nice year-end bonus.
```

Scrolling through Text with the Mouse

■*1*   Place the mouse pointer in the center of the screen. Press the right mouse button, hold it down, and move the pointer to the top edge of the screen. The text scrolls down the screen. Release the mouse button to stop scrolling.

```
November 14, 1990_                    ■



Mr. Randall Jones
```

■*2*   Press the right mouse button, hold it down,
and move the pointer to the bottom edge of
the screen. The text scrolls up the screen.

```
performance of the Sales Department.

Thank you for your fine efforts. Keep up the good work through the
holiday season and everyone will See a nice year-end bonus.
C:\WP51\MYLETTER                                    Doc 1 Pg 2 Ln 3.17" Pos 4.3"
```

When you work with text that extends off the left or right edges of the screen,
you can press and hold the right mouse button and move the pointer to the left or
right edge of the screen to scroll left or right.

As you learn to use WordPerfect, you'll become familiar with the basic
keystrokes for moving the cursor. You can also refer to the table inside the cover
of this book for a complete list.

■*Advanced Deletion Techniques*

You've learned that DEL and BKSP delete characters one at a time. This can be time
consuming when you need to erase large amounts of text. WordPerfect provides
several methods for doing this more quickly.

Return the cursor to any character in the word *through* in the next-to-last line of
the memo. Let's experiment a little.

Deleting Words

■ *1* CTRL BKSP← four times to delete a word with
each stroke of the Backspace key.

```
performance of the Sales Department.

Thank you for your fine efforts. Keep up the good work and
everyone will see a nice year-end bonus.
C:\WP51\MYLETTER                              Doc 1 Pg 2 Ln 3" Pos 6.5"
```

■ *2* ↓ The line length changes to adjust for the
deleted text.

```
performance of the Sales Department.

Thank you for your fine efforts. Keep up the good work and everyone
will see a nice year-end bonus.
C:\WP51\MYLETTER                              Doc 1 Pg 2 Ln 3.17" Pos 4.1"
```

Deleting Lines

■ *1* Move the cursor to the beginning of the last
line of the memo.

```
performance of the Sales Department.

Thank you for your fine efforts. Keep up the good work and everyone
will see a nice year-end bonus.
C:\WP51\MYLETTER                              Doc 1 Pg 2 Ln 3.17" Pos 1"
```

■*2*   `CTRL` `END` to delete the entire line of text.

```
performance of the Sales Department.

Thank you for your fine efforts. Keep up the good work and everyone
█
C:\WP51\MYLETTER                              Doc 1 Pg 2 Ln 3.17" Pos 1"
```

No matter where you place the cursor, this command deletes all characters to the right of the cursor position.

Deleting to the End of a Page

■*1*   Move the cursor to the word *This* in the
second sentence of the memo.

```
SUBJECT: Monthly Sales

In the short time I've been with the company, our monthly sales
have increased over 50%. This is directly due to the outstanding
performance of the Sales Department.
```

■*2*   `CTRL` `PG DN` You are asked to confirm the
deletion.

```
performance of the Sales Department.

Thank you for your fine efforts. Keep up the good work and everyone

Delete Remainder of page? No (Yes)
```

2

■3 Select **Y**es or press **Y** to delete everything
after the cursor.

```
In the short time I've been with the company, our monthly sales
have increased over 50%. █




C:\WP51\MYLETTER                              Doc 1 Pg 2 Ln 2.5" Pos 3.5"
```

You'll learn more about deleting blocks of text in Chapter 8. Continue with the
next exercise to restore the text you just deleted.

■*Restoring Deleted Text*

What happens when you accidentally delete some text and then decide you want
to keep it? WordPerfect can restore up to the last three deletions you have made.
Let's restore the last two sections of text you deleted in MYLETTER.

■1 Select **E**dit, **U**ndelete or press **F1**. The Cancel key displays the Undelete prompt at the
bottom of the screen.

```
In the short time I've been with the company, our monthly sales
have increased over 50%. This is directly due to the outstanding
performance of the Sales Department.

Thank you for your fine efforts. Keep up the good work and everyone

Undelete: 1 Restore; 2 Previous Deletion: 0
```

WordPerfect highlights the last deletion you made when you press **F1**.

Restoring Deleted Text Click the center button or the left and right buttons at the same time on a three-button mouse. Click both buttons at the same time on a two-button mouse.

■*2*   Select **R**etrieve or press **1** or **R** to restore
the last thing you deleted.

```
In the short time I've been with the company, our monthly sales
have increased over 50%. This is directly due to the outstanding
performance of the Sales Department.

Thank you for your fine efforts. Keep up the good work and everyone
█
C:\WP51\MYLETTER                              Doc 1 Pg 2 Ln 3.17" Pos 1"
```

■*3*   Select **E**dit, **U**ndelete or press **F1** to display
the last thing you deleted.

```
Thank you for your fine efforts. Keep up the good work and everyone
This is directly due to the outstanding performance of the Sales
Department.

Thank you for your fine efforts. Keep up the good work and everyone

Undelete: 1 Restore; 2 Previous Deletion: 0
```

2

■*4*   Select **P**revious or press 2 or P to display
the text you deleted *before* the last deletion.

```
Thank you for your fine efforts. Keep up the good work and everyone
will see a nice year-end bonus.




Undelete: 1 Restore; 2 Previous Deletion: 0
```

■*5*   Select **R**estore or press 1 or R to restore
the text.

```
Thank you for your fine efforts. Keep up the good work and everyone
will see a nice year-end bonus.



C:\WP51\MYLETTER                                     Doc 1 Pg 2 Ln 3.17" Pos 4.1"
```

Pretty handy, don't you think? You'll probably find the Undelete feature to be a
lifesaver on more than one occasion.

■*Saving an Edited Document*

In Chapter 1 you learned how to save a newly created document and exit Word-
Perfect. Saving your data often while you work is a good habit—losing a few

minutes' worth of data is not so bad as losing an hour's worth. Let's look at the remaining options for saving an edited document.

Saving and Returning to the Document

■ *1*   Select File, Save or press F10. (If you have a dual floppy-disk system, you will see B:\MYLETTER as the document to be saved.)

```
Document to be saved: C:\WP51\MYLETTER
```

■*2*   ↵ to replace the file with the newly edited version. (Floppy-disk systems show B:\MYLETTER as the document to be replaced.)

```
Replace C:\WP51\MYLETTER? No (Yes)
```

To accept the file name displayed on the screen, place the pointer on the file name and double-click the left button.

2

■*3* Select **Y**es or press Y . The file is saved and
the letter remains on the screen, ready for
more editing.

```
Brian Knowles
Customer Service Manager
===============================================================================
OFFICE MEMO

TO: Sales Department

FROM: Charles Beck, CEO

SUBJECT: Monthly Sales

In the short time I've been with the company, our monthly sales
have increased over 50%. This is directly due to the outstanding
performance of the Sales Department.

Thank you for your fine efforts. Keep up the good work and everyone
will see a nice year-end bonus._



C:\WP51\MYLETTER                                 Doc 1 Pg 2 Ln 3.17" Pos 4.1"
```

Saving and Exiting from WordPerfect

■*1* Select **F**ile, **E**xit or press F7 .

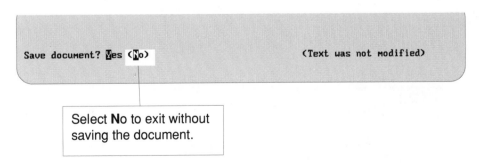

```
Save document? Yes (No)                          (Text was not modified)
```

Select **No** to exit without
saving the document.

Note If the "Text was not modified" message appears in the lower-right corner of your screen, it means no changes have been made in the document. Therefore, you do not have to save it since the original copy is already on the disk.

■*2* Select **Y**es or press [**Y**] or [↵]. (If using a floppy-disk, you will see B:\MYLETTER as the document to be saved.)

```
Document to be saved: C:\WP51\MYLETTER
```

Note If you want to give the file a new name, move the cursor to the file name and type the new name. Your file will be saved under that name. You will still have the old version of the file stored on the disk, without any of the new editing changes.

■*3* [↵] (Floppy-disk systems show B:\MYLETTER as the document to be replaced.)

```
Replace C:\WP51\MYLETTER? No (Yes)
```

To accept the file name, place the pointer on the file name displayed at the prompt, and double-click the left button.

2

■*4* Select **Y**es or press **Y**. The original version of
this file is overwritten by the newly edited one.

```
Exit WP? No (Yes)                          (Cancel to return to document)
```

> Press F1 (Cancel key) if you don't want
> to exit and want to keep the document
> on the screen.

■*5* Select **Y**es or press **Y** to exit WordPerfect
and return to the C prompt in DOS. (Floppy-
disk users will see the B prompt.)

```
C>_
```

Now that you're familiar with some of the many powerful features of Word-
Perfect 5.1, you may be wondering how you'll ever remember all this information. Remember that you have the WordPerfect menu bar, the Help template and
index, and this book for guidance. In fact, you already know enough to begin
creating many of the documents you'll need.

Move on to Chapter 3, where you'll learn some more editing techniques.

3

- More
- Editing
- Tools

■ **W**ordPerfect makes editing a document so easy, you'll wonder how you ever functioned without a word processor. In this chapter you'll learn how to use several powerful editing tools that will speed up the completion of your editing tasks and give you even more control over the look of your documents:

Revealing codes

Splitting and combining paragraphs

Repeating keystrokes

Working with multiple documents

Changing the date format

■ *Working with Hidden Codes*

Every character, space, tab, line return, and margin code in WordPerfect is actually an electronic signal that is created as you press a key on the keyboard, or is created automatically by WordPerfect's internal programming. Many of these electronic signals take the form of WordPerfect codes. WordPerfect uses the different codes to tell the printer how to print the document.

WordPerfect is designed so that these codes are invisible on the Edit screen. If you could see these codes, they would add a great deal of clutter to the screen, making it difficult to work with your document. However, when the need arises, you must be able to edit these hidden codes. Since it is difficult to edit the codes if you can't see them, WordPerfect provides a way to reveal them.

■ *1* Start WordPerfect, if you haven't already done so, and retrieve MYLETTER.

```
November 14, 1990


Mr. Randall Jones
2153 Ocean View Drive
Burbank, CA 91505
```

■ *2* Move the cursor to the *Y* in *Your*.

```
Dear Randall:

Your order of June 12 was traced to Nome, Alaska. It should arrive
at your office on August 25. Apparently, an incorrect address label
was placed on the container. This will not happen again. My
apologies. Mr. Beck, our new CEO, and I look forward to seeing you
at the convention in October.
```

3

■ 3   Select **E**dit, **R**eveal Codes or press ⟨ALT⟩⟨F3⟩.
Reveal Codes displays a portion of the document and the hidden codes on the bottom half
of the Edit screen.

```
Mr. Randall Jones
2153 Ocean View Drive
Burbank, CA 91505



Dear Randall:

Your order of June 12 was traced to Nome, Alaska. It should arrive
C:\WP51\MYLETTER                                    Doc 1 Pg 1 Ln 3" Pos 1"
                                                                }
[HRt]
Dear Randall:[HRt]
[HRt]
Your order of June 12 was traced to Nome, Alaska. It should arrive[SRt]
at your office on August 25. Apparently, an incorrect address label[SRt]
was placed on the container. This will not happen again. My[SRt]
apologies. Mr. Beck, our new CEO, and I look forward to seeing you[SRt]
at the convention in October.[HRt]
[HRt]
Best Regards,[HRt]

Press Reveal Codes to restore screen
```

The scale line shows the position of tabs (represented by triangles), margins that align with tabs ({ }), and margins that do not align with tabs ([ ]).

The hard return code is inserted when you press ⏎ at the end of a sentence or a paragraph, or when you insert a blank line in the text.

The soft return code is automatically inserted when word wrap moves the text to the next line.

> **Note** WordPerfect 5.1 allows you to change the size of the Reveal Codes window. You can make the window smaller if you want to see more of your document and fewer codes, or vice versa. (See "Using Screen in Reveal Codes" on page 763 of your WordPerfect documentation.)

Editing in Reveal Codes

WordPerfect 5.1 allows you to edit your document while in Reveal Codes. Let's do a little editing and watch how the document changes above and below the scale line. The cursor should still be on the *Y* in *Your*. Notice that the cursor in the Reveal Codes portion of the screen is actually a highlight. When you move the regular cursor in the Edit screen, the Reveal Codes highlight moves simultaneously, highlighting characters and codes.

■ *1* TAB to move the first sentence one tab stop.

Notice that the Tab code appears in the Reveal Codes part of the screen.

```
        Your order of June 12 was traced to Nome, Alaska. It should arrive
C:\WP51\MYLETTER                                Doc 1 Pg 1 Ln 3" Pos 1.5"
[
   }
[HRt]
Dear Randall:[HRt]
[HRt]
[Tab]Your order of June 12 was traced to Nome, Alaska. It should arrive[SRt]
at your office on August 25. Apparently, an incorrect address label[SRt]
```

■ *2* Move the cursor to the *J* in *June*.

```
        Your order of June 12 was traced to Nome, Alaska. It should
C:\WP51\MYLETTER                                Doc 1 Pg 1 Ln 3" Pos 2.9"
[
   }
[HRt]
Dear Randall:[HRt]
[HRt]
[Tab]Your order of June 12 was traced to Nome, Alaska. It should[SRt]
arrive at your office on August 25. Apparently, an incorrect address label[SRt]
```

3

■*3* `DEL` seven times.

```
       Your order of ▊was traced to Nome, Alaska. It should
C:\WP51\MYLETTER                           Doc 1 Pg 1 Ln 3" Pos 2.9"
{                                                               }
[HRt]
Dear Randall:[HRt]
[HRt]
[Tab]Your order of ▊was traced to Nome, Alaska. It should[SRt]
arrive at your office on August 25. Apparently, an incorrect address label[SRt]
was placed on the container. This will not happen again. My[SRt]
```

■*4* Type **July 25**.

```
       Your order of July 25▊was traced to Nome, Alaska. It should
C:\WP51\MYLETTER                           Doc 1 Pg 1 Ln 3" Pos 3.6"
{                                                               }
[HRt]
Dear Randall:[HRt]
[HRt]
[Tab]Your order of July 25▊was traced to Nome, Alaska. It should[SRt]
arrive at your office on August 25. Apparently, an incorrect address label[SRt]
```

■*5* `HOME` `←` to move the cursor back to the
Tab code.

```
       Your order of July 25 was traced to Nome, Alaska. It should
C:\WP51\MYLETTER                           Doc 1 Pg 1 Ln 3" Pos 1"
{                                                               }
[HRt]
Dear Randall:[HRt]
[HRt]
[Tab]Your order of July 25 was traced to Nome, Alaska. It should[SRt]
arrive at your office on August 25. Apparently, an incorrect address label[SRt]
```

■ 6 DEL The Tab code disappears and the first
sentence is no longer tabbed.

```
Your order of July 25 was traced to Nome, Alaska. It should arrive
C:\WP51\MYLETTER                              Doc 1 Pg 1 Ln 3" Pos 1"
[                                                              ]
[HRt]
Dear Randall:[HRt]
[HRt]
Your order of July 25 was traced to Nome, Alaska. It should arrive[SRt]
at your office on August 25. Apparently, an incorrect address label[SRt]
```

■ 7 Select **E**dit, **R**eveal Codes or press ALT F3 to
restore the Edit screen. You can no longer see
the codes, but they are still active.

```
Dear Randall:

Your order of July 25 was traced to Nome, Alaska. It should arrive
at your office on August 25. Apparently, an incorrect address label
was placed on the container. This will not happen again. My
apologies. Mr. Beck, our new CEO, and I look forward to seeing you
at the convention in October.
```

Note *Deleting Codes* Use Reveal Codes to delete and replace codes that may
format your document incorrectly. If you wish, you can delete hidden codes without
being in Reveal Codes. When you are in the process of deleting text and come upon
a hidden code, a prompt will appear in the lower-left corner of the screen asking if
you want to delete the code. (For a listing of WordPerfect codes, see the inside
covers of this book.)

Splitting and
■*Combining Paragraphs*

3

Have you ever written a paragraph and then realized that it should be split into two paragraphs? With WordPerfect you can do this very easily with ⏎. And, when you want to combine two paragraphs into one, you can do it just as easily with ⌫ or DEL.

■*1* Retrieve MYLETTER if it's not on the screen, and move the cursor to the *M* in *Mr. Beck.*

> Dear Randall:
>
> Your order of July 25 was traced to Nome, Alaska. It should arrive at your office on August 25. Apparently, an incorrect address label was placed on the container. This will not happen again. My apologies. Mr. Beck, our new CEO, and I look forward to seeing you at the convention in October.

■*2* ⏎ twice.

> Your order of July 25 was traced to Nome, Alaska. It should arrive at your office on August 25. Apparently, an incorrect address label was placed on the container. This will not happen again. My apologies.
>
> Mr. Beck, our new CEO, and I look forward to seeing you at the convention in October.

■ *3*   Select **E**dit, **R**eveal Codes or press ALT F3.

```
Mr. Beck, our new CEO, and I look forward to seeing you at the
C:\WP51\MYLETTER                              Doc 1 Pg 1 Ln 3.83" Pos 1"
{                                                                      }
was placed on the container. This will not happen again. My[SRt]
apologies. [HRt]
[HRt]
Mr. Beck, our new CEO, and I look forward to seeing you at the[SRt]
convention in October.[HRt]
```

Two hard return codes have been inserted, creating a second paragraph with a blank line inserted between it and the first paragraph.

If you want to combine two paragraphs, you simply delete the hard returns that separate them. Let's do this with our newly created paragraphs. Remain in Reveal Codes so you can see the [HRt] codes disappear. The cursor should still be on the *M* in *Mr. Beck.*

■ *1*   BKSP ← twice. You also can use DEL if you place
the Reveal Codes cursor on the [Hrt] codes;
you may have to delete a space or two in addition to the hard return codes.

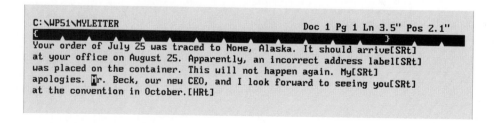

```
C:\WP51\MYLETTER                              Doc 1 Pg 1 Ln 3.5" Pos 2.1"
{                                                                      }
Your order of July 25 was traced to Nome, Alaska. It should arrive[SRt]
at your office on August 25. Apparently, an incorrect address label[SRt]
was placed on the container. This will not happen again. My[SRt]
apologies. Mr. Beck, our new CEO, and I look forward to seeing you[SRt]
at the convention in October.[HRt]
```

3

■*2*   Select **E**dit, **R**eveal Codes or press ⎡ALT⎤⎡F3⎤ to
     return to the Edit screen.

■*Repeating Keystrokes*

There may be times when you want to repeat a keystroke a specific number of
times; for example, you might need to place a line of dashes across the page,
move the cursor a precise number of times, or delete a specific number of words.
Instead of entering the needed keystrokes yourself, you can let the Repeat key
(⎡ESC⎤) do the work for you.

Retrieve MYLETTER if it's not already on the screen. Let's edit it using ⎡ESC⎤.

■*1*   Move the cursor to the space after the word
     bonus in the last line of the memo on page 2.

```
In the short time I've been with the company, our monthly sales
have increased over 50%. This is directly due to the outstanding
performance of the Sales Department.

Thank you for your fine efforts. Keep up the good work and everyone
will see a nice year-end bonus.█
C:\WP51\MYLETTER                              Doc 1 Pg 2 Ln 3.17" Pos 4.1"
```

■*2*   ⎡↵⎤ twice.

```
Thank you for your fine efforts. Keep up the good work and everyone
will see a nice year-end bonus.

█
C:\WP51\MYLETTER                              Doc 1 Pg 2 Ln 3.5" Pos 1"
```

■*3* 〔ESC〕

> Thank you for your fine efforts. Keep up the good work and everyone
> will see a nice year-end bonus.
>
> Repeat Value = **8**

The default value for the Repeat function is 8.

■*4* Type **64**. There are 64 spaces on each line
between the default left and right margins.

> Thank you for your fine efforts. Keep up the good work and everyone
> will see a nice year-end bonus.
>
> Repeat Value = **64**

■*5* Type **x**. The letter x is repeated 64 times.

> Thank you for your fine efforts. Keep up the good work and everyone
> will see a nice year-end bonus.
>
> xxxxxxxxxxxxxxxxxxxxxxxxxxxxxxxxxxxxxxxxxxxxxxxxxxxxxxxxxxxxxxxxxx■
> C:\WP51\MYLETTER   Doc 1 Pg 2 Ln 3.5" Pos 7.4"

3

■*6* **HOME** **←** to move the cursor to the beginning
of the line.

```
Thank you for your fine efforts. Keep up the good work and everyone
will see a nice year-end bonus.

▓xxxxxxxxxxxxxxxxxxxxxxxxxxxxxxxxxxxxxxxxxxxxxxxxxxxxxxxxxxxxxxxxxxx
C:\WP51\MYLETTER                                    Doc 1 Pg 2 Ln 3.5" Pos 1"
```

■*7* **ESC** and type **64**.

```
Thank you for your fine efforts. Keep up the good work and everyone
will see a nice year-end bonus.

xxxxxxxxxxxxxxxxxxxxxxxxxxxxxxxxxxxxxxxxxxxxxxxxxxxxxxxxxxxxxxxxxxx
Repeat Value = 64
```

■*8* **DEL** to erase the 64 x's.

```
Thank you for your fine efforts. Keep up the good work and everyone
will see a nice year-end bonus.

■
C:\WP51\MYLETTER                                    Doc 1 Pg 2 Ln 3.5" Pos 1"
```

You can see how this feature can speed up operations, such as creating forms or filling in form letters. Experiment a little, and then move on to the next lesson.

Working with
■*Multiple Documents*

WordPerfect provides a helpful feature that allows you to work with two documents at the same time, so you don't have to exit one document to retrieve a related document for editing or reference.

You activate this feature with the Switch command: **E**dit, **S**witch Document or [SHIFT][F3]. Let's practice using this command with MYLETTER. Make sure the letter is on the screen.

■*1*   Select **E**dit, **S**witch Document or press
[SHIFT][F3] to go into the Document 2 window.

                                                    Doc 2 Pg 1 Ln 1" Pos 1"

■*2*   Type this text in the second document
window inserting a hard return after each line:

PERSONAL REMINDERS
Notify Billing Dept. to update computerized customer address files.
Have Requisition Dept. order new Sybex WordPerfect computer books.

Note   You can move between Doc 1 and Doc 2 by using the Switch feature. This works great when you want to edit two documents at the same time.

■3 Select **F**ile, **E**xit, **Y**es or press `F7` `Y`, type
PERSONAL, and press `↵`.

```
Exit doc 2? No (Yes)                          (Cancel to return to document)
```

■4 Select **Y**es or press `Y` to exit Doc 2 and
return to Doc 1.

```
Thank you for your fine efforts. Keep up the good work and everyone
will see a nice year-end bonus.


C:\WP51\MYLETTER                              Doc 1 Pg 2 Ln 3.5" Pos 1"
```

The first document reappears.

Displaying Two Documents in Windows

At times you may find it easier to work on two documents when you can see
them both at the same time. This useful WordPerfect feature is called Windows.

With MYLETTER on the screen and the cursor at the end of the document, let's
activate the Windows feature and display the new file, PERSONAL.

■ *1*   Select **E**dit, **W**indow or press CTRL F3
      1 or **W**.

```
Best Regards,


Brian Knowles
Customer Service Manager
================================================================================
OFFICE MEMO

TO: Sales Department

FROM: Charles Beck, CEO

SUBJECT: Monthly Sales

In the short time I've been with the company, our monthly sales
have increased over 50%. This is directly due to the outstanding
performance of the Sales Department.

Thank you for your fine efforts. Keep up the good work and everyone
will see a nice year-end bonus.


Number of lines in this window: 24
```

The document uses 24 lines, the default line value.

■ *2*   Type **12**.

```
Thank you for your fine efforts. Keep up the good work and everyone
will see a nice year-end bonus.


Number of lines in this window: 12_
```

You also can press ↑ or ↓ instead of typing a new number. You will see the
scale line discussed in step 4. This line moves in the direction of the arrow.

■*3* ↵

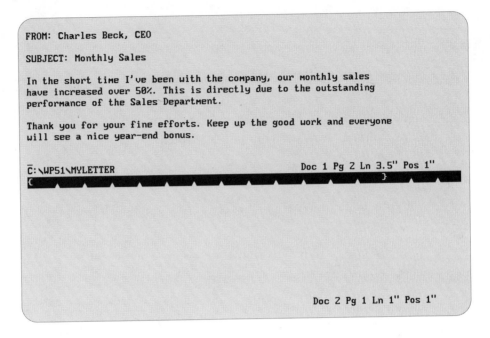

You now have portions of the Doc 1 and Doc 2 windows displayed on the screen.
The Doc 2 window is available for retrieving a file or typing in new text.

■*4* Select **E**dit, **S**witch Document or press
SHIFT F3. The cursor jumps to Doc 2 and
the triangles on the scale line now point
downward, indicating you are in Doc 2.

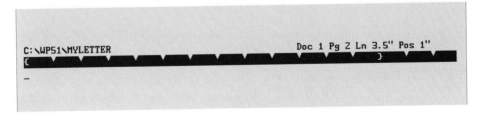

To use the mouse to switch between windows, place the pointer in the window you want and click the left button.

■**5**   Retrieve the file PERSONAL by selecting
File, Retrieve or pressing `SHIFT` `F10`. You can
now edit in either the Doc 1 or Doc 2 window
by using the Switch command: File, Switch
Document or `SHIFT` `F3`.

```
FROM: Charles Beck, CEO

SUBJECT: Monthly Sales

In the short time I've been with the company, our monthly sales
have increased over 50%. This is directly due to the outstanding
performance of the Sales Department.

Thank you for your fine efforts. Keep up the good work and everyone
will see a nice year-end bonus.


C:\WP51\MYLETTER                                   Doc 1 Pg 2 Ln 3.5" Pos 1"
{                                                              }
PERSONAL REMINDERS

Notify Billing Dept. to update computerized customer address files.

Have Requisition Dept. order new SYBEX WordPerfect computer books.



C:\WP51\PERSONAL                                   Doc 2 Pg 1 Ln 1" Pos 1"
```

Clearing the Windows

You clear a window by placing the cursor in it and exiting the document. Position the cursor in Doc 2; let's clear the lower window. (You didn't modify the text so you won't have to save the file.)

3

■ *1* Select **F**ile, **E**xit, **N**o, **Y**es or press
⌨F7, ⌨N ⌨Y.

> Thank you for your fine efforts. Keep up the good work and everyone
> will see a nice year-end bonus.
>
> C:\WP51\MYLETTER   Doc 1 Pg 2 Ln 3.5" Pos 1"

■ *2* Select **E**dit, **W**indow or press ⌨CTRL ⌨F3
1 or ⌨W, type **24**, then press ⌨↵ to clear
the Doc 2 window.

> FROM: Charles Beck, CEO
>
> SUBJECT: Monthly Sales
>
> In the short time I've been with the company, our monthly sales
> have increased over 50%. This is directly due to the outstanding
> performance of the Sales Department.
>
> Thank you for your fine efforts. Keep up the good work and everyone
> will see a nice year-end bonus.
>
> C:\WP51\MYLETTER   Doc 1 Pg 2 Ln 3.5" Pos 1"

> **Note** *Displaying the Scale Line* WordPerfect provides you with a full 24-line screen so you can see as much of your document as possible. However, if you don't mind seeing one less line of text on the display, you can have the scale line displayed at all times for referring to tab stops and margins.
>
> To display the scale line, select **Edit, W**indow or press `CTRL` `F3` `W`. Then type **23** and press `←`. You now will see all your tabs and margins. To delete the scale line, repeat the steps above, typing 24 instead of 23, and you return to 24 lines of text without the scale line displayed.

Changing
■ *the Date Format*

In Chapter 1 you selected the Date feature—Tools, Date **T**ext or `SHIFT` `F5`—to automatically insert the current date in MYLETTER. You also can use the Date command to change the format of the date, as long as you have an internal clock in your computer; if you don't, you must enter the date and time manually when you start your machine.

Start WordPerfect, or clear the screen if WordPerfect is running already by pressing `F7` `N` `N`. We'll create a new date format.

When you enter a date format, type in spaces as they appear in the examples. If you don't, the date on your screen will not have the correct spacing.

3

■ *1*   Select **T**ools, Date **F**ormat or press SHIFT F5
3 or **F**. Study the menu for a moment to
see all the available options.

```
Date Format

     Character  Meaning
         1      Day of the Month
         2      Month (number)
         3      Month (word)
         4      Year (all four digits)
         5      Year (last two digits)
         6      Day of the Week (word)
         7      Hour (24-hour clock)
         8      Hour (12-hour clock)
         9      Minute
         0      am / pm
        %,$     Used before a number, will:
                   Pad numbers less than 10 with a leading zero or space
                   Abbreviate the month or day of the week

     Examples:  3 1, 4      = December 25, 1984
                %6 %3 1, 4  = Tue Dec 25, 1984
                %2/%1/5 (6) = 01/01/85 (Tuesday)
                $2/$1/5 ($6) =  1/ 1/85 (Tue)
                8:90        = 10:55am

Date format: 3 1, 4
```

■ *2*   Type **2-1-5 (%6)**.

```
                %2/%1/5 (6)  = 01/01/85 (Tuesday)
                $2/$1/5 ($6) =  1/ 1/85 (Tue)
                8:90         = 10:55am


Date format: 2-1-5 (%6)_
```

■*3* ⏎

1 Date Text; 2 Date Code; 3 Date Format; 4 Outline; 5 Para Num; 6 Define: 0

■*4* 1 or T to place the current date on the
screen. This date format will be in effect until
you exit WordPerfect.

11-14-90 (Wed)_

When you restart WordPerfect at another time, the date format will revert to the
default format of 3 1,4 (month day,year).

Chapter 4 will show you some line formatting techniques that will enable you to
add some creative enhancements to your documents.

4

- *Line*
- *Formatting*
- *Techniques*

■ Like many people, you probably bought your computer in order to quickly and easily change the appearance and content of your documents. The arrangement of text on the page is called *formatting*. With a word processing program, you can change the tabs, line spacing, and margins and the entire document adjusts, or reformats, automatically.

Creating good-looking documents is important, whether they're for business or personal use. This chapter shows you these basic techniques to alter the format and improve the appearance of your documents:

Setting margins

Changing line spacing

Working with tabs

Hyphenating words

Justifying text

Eliminating widows and orphans

■*Setting Line Margins*

Try your hand at setting new margins. Start up WordPerfect and retrieve MYLETTER. The default left and right margins are 1 inch. You'll change them using the Format feature.

You can insert most WordPerfect formatting commands anywhere in a document—the change will take place onward from that location. Always position the cursor where you want the change to take place, and select the new settings.

You should begin the following exercise with the cursor located on the first letter of the date at the top of the document.

■ *1*   Select Layout, Line or press SHIFT F8 1 or L .

```
Format: Line

    1 - Hyphenation                         No

    2 - Hyphenation Zone - Left             10%
                          Right             4%

    3 - Justification                       Full

    4 - Line Height                         Auto

    5 - Line Numbering                      No

    6 - Line Spacing                        1

    7 - Margins - Left                      1"
                  Right                     1"

    8 - Tab Set                             Rel: -1", every 0.5"

    9 - Widow/Orphan Protection             No


Selection: 0
```

4

What you see are the current, default settings for line formatting. WordPerfect provides you with preset line, page, and document format settings. The following settings are in effect when you start WordPerfect:

Page Size:	8½" × 11", 66 lines per page, 54 typed lines
Top Margin:	1"
Bottom Margin:	1"
Left Margin::	1" (10 character positions from the left edge of the paper)
Right Margin:	1" (10 character positions from the right edge of the paper)
Page Numbering:	None
Line Spacing:	Single spacing, 6 lines per inch
Font:	Standard single-strike, 10 pitch (10 characters per inch)
Tabs:	Every 5 spaces (½")
Justification:	Full (printed on paper but not displayed on Edit screen)

You can change any or all of these settings to meet your needs.

■*2*   Select **Margins** or press **7** or **M**.

■*3*   Type **2** and press ⏎; type **2** again and press ⏎.

```
      6 - Line Spacing            1

      7 - Margins - Left          2"
                  Right           2"
```

■*4* Click the right mouse button or press F7,
then move the cursor down the page to adjust
the text. The letter is reformatted with 2-in.
margins. (See inside the book cover for a list
of cursor movements.)

```
                  November 14, 1990


                  Mr. Randall Jones
                  2153 Ocean View Drive
                  Burbank, CA 91505


                  Dear Randall:

                  Your order of June 12 was traced to Nome,
                  Alaska. It should arrive at your office on
                  August 25. Apparently, an incorrect address
                  label was placed on the container. This will
                  not happen again. My apologies. Mr. Beck, our
                  new CEO, and I look forward to seeing you at
                  the convention in October.

                  Best Regards,



C:\WP51\MYLETTER                          Doc 1 Pg 1 Ln 3.17" Pos 2"
```

Your screen shows
the cursor at Pos 2".

Editing Margin Codes

As you design a document layout, you may want to change your hidden formatting commands. Locate your newly created margin code using the Reveal Codes command, then change the margin using the Format feature.

■ *1*   HOME HOME ↑ to move the cursor to the beginning of the document.

■ *2*   Select **Edit**, **R**eveal Codes or press ALT F3.

```
            Dear Randall:
C:\WP51\MYLETTER                                    Doc 1 Pg 1 Ln 1" Pos 2"
                {                          }
[L/R Mar:2",2"]November 14, 1990[HRt]
[HRt]
[HRt]
```

■ *3*   Select **L**ayout, **L**ine **M**argins or press
SHIFT F8, 1 or L, 7 or M.

```
    6 - Line Spacing            1

    7 - Margins - Left          2"
                  Right         2"
```

■*4*   Type **1.4** and press ⏎ twice to set new left
and right margins. Click the right mouse button or press F7 to return to the document.

```
              Dear Randall:
C:\WP51\MYLETTER                                    Doc 1 Pg 1 Ln 1" Pos 1.4"
              {                                                              ]
[L/R Mar:2",2"][L/R Mar:1.4",1.4"]November 14, 1990[HRt]
[HRt]
[HRt]
```

You now see a second margin code. Your text will conform to the 1.4"
margins because this code is located *after* the 2" margin code.

■*5*   Select **E**dit, **R**eveal Codes or press ALT F3.
The Pos indicator shows that your cursor is
now at the new 1.4" left margin.

```
November 14, 1990


Mr. Randall Jones
2153 Ocean View Drive
Burbank, CA 91505


Dear Randall:

Your order of June 12 was traced to Nome, Alaska. It should
arrive at your office on August 25. Apparently, an
incorrect address label was placed on the container. This
will not happen again. My apologies. Mr. Beck, our new CEO,
and I look forward to seeing you at the convention in
October.

Best Regards,



Brian Knowles
C:\WP51\MYLETTER                              Doc 1 Pg 1 Ln 1" Pos 1.4"
```

Deleting Margin Codes

Pressing either ⌫ or DEL will delete a code from inside or outside the Reveal Codes screen. Make sure the cursor is still positioned on the first letter in the date.

■*1*   ⌫

Brian Knowles
Delete [L/R Mar:1.4",1.4"]? No (Yes)

> You are alerted that the 1.4" margin code is about to be deleted. Select **No** if you don't want to delete it.

■*2*   Select **Y**es or press Y . The margins shift back to 2".

November 14, 1990

Mr. Randall Jones
2153 Ocean View Drive
Burbank, CA 91505

Dear Randall:

Your order of June 12 was traced to Nome, Alaska. It should arrive at your office on August 25. Apparently, an incorrect address label was placed on the container. This will not happen again. My apologies. Mr. Beck, our new CEO, and I look forward to seeing you at the convention in October.

Best Regards,

C:\WP51\MYLETTER   Doc 1 Pg 1 Ln 1" Pos 2"

■*3*   Select **E**dit, **R**eveal Codes or press `ALT` `F3`.
The 1.4" margin code is gone.

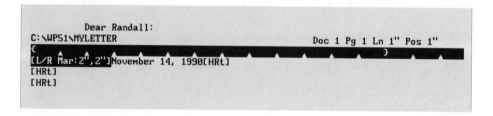

```
         Dear Randall:
C:\WP51\MYLETTER                                        Doc 1 Pg 1 Ln 1" Pos 2"
              {                                    }
[L/R Mar:2",2"]November 14, 1990[HRt]
[HRt]
[HRt]
```

■*4*   `←` to highlight the margin code.

```
         Dear Randall:
C:\WP51\MYLETTER                                        Doc 1 Pg 1 Ln 1" Pos 1"
{                                                  }
[L/R Mar:2",2"]November 14, 1990[HRt]
[HRt]
[HRt]
```

■*5*   `DEL` to delete the margin code.

```
Dear Randall:
C:\WP51\MYLETTER                                        Doc 1 Pg 1 Ln 1" Pos 1"
{                                                  }
November 14, 1990[HRt]
[HRt]
[HRt]
```

■*6* Select **Edit, R**eveal Codes or press `ALT` `F3` to
return to the document.

■*Setting Line Spacing*

4

You can set line spacing for any increment, in either decimal or fraction form.
The spacing increments will not always appear on your screen exactly the way
they should look, but they will print accurately.

Let's double-space the memo on page 2 of MYLETTER.

■*1* Press `PG DN` to move the cursor to the *O* in
OFFICE.

```
OFFICE MEMO

TO: Sales Department

FROM: Charles Beck, CEO
```

■*2* Select **Layout, L**ine or press `SHIFT` `F8`
`1` or `L`.

■*3* Select **L**ine **S**pacing or press `6` or `S`, type
2, then press `↵`.

```
   5 - Line Numbering              No

   6 - Line Spacing                2

   7 - Margins - Left              1"
               Right              1"
```

■ *4*   Click the right mouse button or press F7.

```
OFFICE MEMO


TO: Sales Department


FROM: Charles Beck, CEO


SUBJECT: Monthly Sales


In the short time I've been with the company, our monthly sales

have increased over 50%. This is directly due to the outstanding

performance of the Sales Department.


C:\WP51\MYLETTER                                    Doc 1 Pg 2 Ln 1" Pos 1"
```

If you like, experiment with some other line-spacing increments. When you've finished, delete all line-spacing codes to return the document to its original single-spaced version. Once you've done that, clear the screen for the next exercise by pressing F7 N N.

■ *Working with Tabs*

The tab is a fundamental tool in line formatting. On a typewriter, you mechanically set your tabs; with WordPerfect, you electronically set them.

WordPerfect provides four types of tab settings: Left, Centered, Right, and Decimal. You can also designate dot leaders (a series of periods) in the blank space between the tab and text. WordPerfect allows you to set tabs relative to the left margin. This means the distance between tab stops and the left margin remains the same whenever you change the margins. You can also set absolute tabs, which are measured from the left edge of the paper and maintain this distance regardless of where you set the margins.

The following exercises will show you how to use all four types of tabs.

Deleting Tab Stops and Setting New Ones

WordPerfect provides a default tab stop every half-inch. These default tabs are relative tabs. You first need to delete them to set your own tabs.

■ *1*   Select **L**ayout, **L**ine or press SHIFT F8
         1 or **L**.

■ *2*   Select **T**ab Set or press **8** or **T** to call up
         the Tab Set menu.

```
L...L...L...L....L....L....L....L....L....L....L....L....L....L....L....L....L...
|   ^   |   ^   |   ^   |   ^   |   ^   |   ^   |   ^   |   ^   |   ^
0"      +1"     +2"     +3"     +4"     +5"     +6"     +7"
Delete EOL (clear tabs); Enter Number (set tab); Del (clear tab);
Type; Left; Center; Right; Decimal; = Dot Leader; Press Exit when done.
```

Select **Type** to change between "Relative to margin" and "Absolute" tabs.

The *L* indicates where a left tab is set (an *R* indicates a right tab, a *C* indicates a center tab, and a *D* indicates a decimal tab).

■*3*   CTRL END to delete all tabs from the cursor to
the end of the line.

```
. . . . . . . . . . . . . . . . . . . . . . . . . . . . . . . . . . . . . . . . . . . . . . . . . . . . . . . . . . . . . . . . . . . . . . . . . . . . . . . . . . . . . . . . . . . . .
|     ^     |     ^     |     ^     |     ^     |     ^     |     ^     |     ^     |     ^
0"        +1"        +2"        +3"        +4"        +5"        +6"        +7"
Delete EOL (clear tabs); Enter Number (set tab); Del (clear tab);
Type; Left; Center; Right; Decimal; = Dot Leader; Press Exit when done.
```

To delete an individual tab, place the cursor on an L in the Tab Set menu and
press DEL.

The default tabs are cleared. Now set your own.

■*1*   → five spaces to the 0.5" mark and press
L to set a left tab.

```
. . . . L . . . . . . . . . . . . . . . . . . . . . . . . . . . . . . . . . . . . . . . . . . . . . . . . . . . . . . . . . . . . . . . . . . . . . . . . . . . . . . . . . . . . . . . . . . . . .
     ^           |     ^     |     ^     |     ^     |     ^     |     ^     |     ^
0"        +1"        +2"        +3"        +4"        +5"        +6"        +7"
Delete EOL (clear tabs); Enter Number (set tab); Del (clear tab);
Type; Left; Center; Right; Decimal; = Dot Leader; Press Exit when done.
```

You also can type **0.5** and press ⏎ and WordPerfect will place the tab for you.
If you want to type a fraction less than an inch, you have to place a zero in front
of the decimal point—typing a period alone places a dot leader in front of the
tabbed text. Typing tab entries only works with left tabs.

4

■2 → to the 3" mark and press C to set
a center tab.

```
.....L..............................C.....................................
 |    ^    |    ^    |    ^    |    ^    |    ^    |    ^    |    ^    |    ^
 0"       +1"       +2"       +3"       +4"       +5"       +6"       +7"
Delete EOL (clear tabs); Enter Number (set tab); Del (clear tab);
Type; Left; Center; Right; Decimal; = Dot Leader; Press Exit when done.
```

■3 → to the 5" mark and press R to set a
right tab.

```
.....L........................C................R.............................
 |    ^    |    ^    |    ^    |    ^    |    ^    |    ^    |    ^    |    ^
 0"       +1"       +2"       +3"       +4"       +5"       +6"       +7"
Delete EOL (clear tabs); Enter Number (set tab); Del (clear tab);
Type; Left; Center; Right; Decimal; = Dot Leader; Press Exit when done.
```

■4 → to the 6" mark. Press D and . (period)
to set a decimal tab and dot leader.

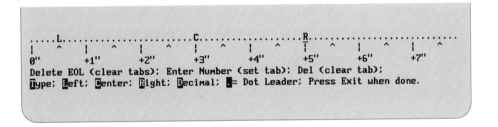

```
.....L........................C................R.........D..................
 |    ^    |    ^    |    ^    |    ^    |    ^    |    ^    |    ^    |    ^
 0"       +1"       +2"       +3"       +4"       +5"       +6"       +7"
Delete EOL (clear tabs); Enter Number (set tab); Del (clear tab);
Type; Left; Center; Right; Decimal; = Dot Leader; Press Exit when done.
```

Pressing ⌐ after you set a tab creates a dot leader before the tab stop. The *D* is highlighted because it has a dot leader before it.

■ *5*   Click the right mouse button twice or press
      F7 twice to return to the document.

Using Tab Stops

Let's create a short document and use the tab stops we just created.

■ *1*   TAB and type **Desks (2)**.

```
        Desks (2)_
```

■ *2*   TAB and type **(Boxed)**.

```
        Desks (2)        (Boxed)_
```

■ *3*   TAB and type **Dallas, Tx**.

```
        Desks (2)        (Boxed)    Dallas, Tx._
```

■ *4*   TAB and type **$1289.00**.

```
        Desks (2)        (Boxed)    Dallas, Tx. . . $1289.00_
```

■5 Using TAB, type two more lines, as shown in
the following screen:

```
Desks (2)          (Boxed)     Dallas, Tx. . . $1289.00
Table              (Oak)       Azusa, Ca. . . .$979.00
Chairs (14)        (Folding)   Elk, Ca. . . .$350.00_
```

■6 Select **E**dit, **R**eveal Codes or press ALT F3 to
show the codes for each tab command.

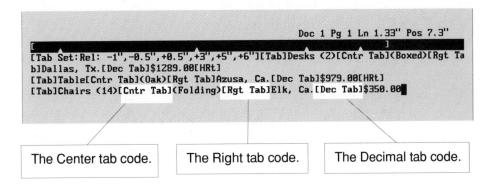

```
                                    Doc 1 Pg 1 Ln 1.33" Pos 7.3"
[                                                               ]
[Tab Set:Rel: -1",-0.5",+0.5",+3",+5",+6"][Tab]Desks (2)[Cntr Tab](Boxed)[Rgt Ta
b]Dallas, Tx.[Dec Tab]$1289.00[HRt]
[Tab]Table[Cntr Tab](Oak)[Rgt Tab]Azusa, Ca.[Dec Tab]$979.00[HRt]
[Tab]Chairs (14)[Cntr Tab](Folding)[Rgt Tab]Elk, Ca.[Dec Tab]$350.00
```

| The Center tab code. | The Right tab code. | The Decimal tab code. |

Remain in Reveal Codes, and go on to the next exercise.

Changing Tab Stops

Remember, all text that follows a particular formatting command is affected by
that command. If you want to change a tab setting, you must place the new setting in front of the text you want affected and then delete the old setting.

While in Reveal Codes, let's move the second column of text, created in the previous exercise, five spaces to the left.

■*1* Move the cursor to the [Tab] code on the first
line, in front of *Desks.*

```
                                              Doc 1 Pg 1 Ln 1" Pos 1"
[                                                          ]
[Tab Set:Rel: -1",-0.5",+0.5",+3",+5",+6"][Tab]Desks (2)[Cntr Tab](Boxed)[Rgt Ta
b]Dallas, Tx.[Dec Tab]$1289.00[HRt]
[Tab]Table[Cntr Tab](Oak)[Rgt Tab]Azusa, Ca.[Dec Tab]$979.00[HRt]
[Tab]Chairs (14)[Cntr Tab](Folding)[Rgt Tab]Elk, Ca.[Dec Tab]$350.00
```

■*2* Select **L**ayout, **L**ine, **T**ab Set or press
[SHIFT] [F8], [1] or [L], [8] or [T].

```
.....L..........................C...................R.........D................
├     ^       ├       ^       ├     ^     ├     ^     ├     ^    ├     ^     ├    ^
0"       +1"       +2"       +3"       +4"       +5"       +6"       +7"
Delete EOL (clear tabs); Enter Number (set tab); Del (clear tab);
Type; Left; Center; Right; Decimal; ▪= Dot Leader; Press Exit when done.
```

■*3* [→] to the 2.5" mark and press [C].

```
.....L.....................C...C...................R.........D................
├     ^       ├       ^     ├ ^ ├     ^     ├     ^     ├     ^    ├     ^     ├    ^
0"       +1"       +2"       +3"       +4"       +5"       +6"       +7"
Delete EOL (clear tabs); Enter Number (set tab); Del (clear tab);
Type; Left; Center; Right; Decimal; ▪= Dot Leader; Press Exit when done.
```

■ *4* To delete the old center tab, ⟶ to the 3"
center tab marker and press ⟨DEL⟩.

4

```
.....L.................C...................R........D.............
:   ^    :   ^    :   ^   :   ^    :   ^    :   ^    :   ^    :   ^
0"      +1"      +2"      +3"      +4"      +5"      +6"      +7"
Delete EOL (clear tabs); Enter Number (set tab); Del (clear tab);
Type; Left; Center; Right; Decimal; = Dot Leader; Press Exit when done.
```

■ *5* Click the right mouse button twice or press
⟨F7⟩ twice to return to Reveal Codes.

```
_     Desks (2)       (Boxed)        Dallas, Tx. . . $1289.00
      Table          (Oak)          Azusa, Ca. . . .$979.00
      Chairs (14)    (Folding)       Elk, Ca. . . .$350.00




                                          Doc 1 Pg 1 Ln 1" Pos 1"
[
[Tab Set:Rel: -1",-0.5",+0.5",+3",+5",+6"][Tab Set:Rel: -1",-0.5",+0.5",+2.5",+5
",+6"][Tab]Desks (2)[Cntr Tab](Boxed)[Rgt Tab]Dallas, Tx.[Dec Tab]$1289.00[HRt]
[Tab]Table[Cntr Tab](Oak)[Rgt Tab]Azusa, Ca.[Dec Tab]$979.00[HRt]
[Tab]Chairs (14)[Cntr Tab](Folding)[Rgt Tab]Elk, Ca.[Dec Tab]$350.00




Press Reveal Codes to restore screen
```

You now have two tab sets; the second one is the active one.

■ *6* ⬅ to the first tab set.

```
                                              Doc 1 Pg 1 Ln 1" Pos 1"
{                                             }
[Tab Set:Rel: -1",-0.5",+0.5",+3",+5",+6"][Tab Set:Rel: -1",-0.5",+0.5",+2.5",+5
",+6"][Tab]Desks (2)[Cntr Tab]<Boxed>[Rgt Tab]Dallas, Tx.[Dec Tab]$1289.00[HRt]
[Tab]Table[Cntr Tab]<Oak>[Rgt Tab]Azusa, Ca.[Dec Tab]$979.00[HRt]
[Tab]Chairs (14)[Cntr Tab]<Folding>[Rgt Tab]Elk, Ca.[Dec Tab]$350.00
```

■ *7* DEL to remove the unnecessary coding and to
reduce clutter on the Reveal Codes screen.

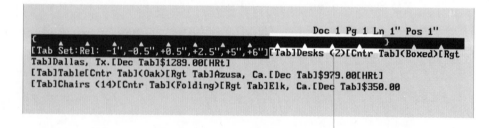

```
                                              Doc 1 Pg 1 Ln 1" Pos 1"
{                                             }
[Tab Set:Rel: -1",-0.5",+0.5",+2.5",+5",+6"][Tab]Desks (2)[Cntr Tab]<Boxed>[Rgt
Tab]Dallas, Tx.[Dec Tab]$1289.00[HRt]
[Tab]Table[Cntr Tab]<Oak>[Rgt Tab]Azusa, Ca.[Dec Tab]$979.00[HRt]
[Tab]Chairs (14)[Cntr Tab]<Folding>[Rgt Tab]Elk, Ca.[Dec Tab]$350.00
```

The scale line shows tabs
every five spaces because
any document created
before the tab-set coding
would have the default
tabs active. Move the cursor to the right, and watch
the tab indicators change.

Exit without saving this document if you want to take a break. If you want to
continue without exiting, clear the screen for the next exercise.

> **Note** *Replacing C, R, and D Tabs* If you want to replace a C, R, or D tab with another tab, you first must delete the hidden code in front of each text entry in the column. Next, you must go to the Tab menu and select new tab positions, then insert the new codes by placing the cursor at the correct locations and pressing [TAB].

4

■*Hyphenating Words*

Hyphenating words can improve the overall appearance of your text, especially when the text is fully justified. Without hyphenation there are often gaps between words, which can be very distracting to the reader.

WordPerfect provides three ways for you to hyphenate words: You can type hyphens at the keyboard, put them in manually with assistance from Word-Perfect, or have WordPerfect put them in automatically.

Start WordPerfect and try these exercises on the blank editing screen.

Keyboard Hyphenation

Words like *mother-in-law, half-truth,* and *baby-faced* require hyphens. Using the hyphen key by itself places a [-] code in the text. If such a word ever needs to be reformatted, word wrap will split the word at the hyphen.

■ *1*   Type this text:

```
Wilfred was 28 years old and experienced, but he still looked like
a baby-faced boy of 18._
```

■*2* Delete the words *he still*, and move the cursor down to adjust the text. Notice how *baby-faced* is hyphenated at the end of the line.

```
Wilfred was 28 years old and experienced, but looked like a baby-
faced boy of 18. _
```

You also can create a *hard hyphen*, which will ensure that a hyphenated word will not be broken if the text is reformatted. This hyphen is especially useful when you are typing formulas and equations that have minus signs and should not be split. To keep a word or equation together, press HOME and then press ⎯ .

■*1* Clear the screen by pressing F7 N N and type this text:

```
The most basic formula for the interdicting rate of change is _
```

■*2* Type the formula below, pressing HOME ⎯ to place the minus sign between *x* and *y*. The entire equation moves to the second line and does not break at the hyphen.

```
The most basic formula for the interdicting rate of change is
F(x-y)=M. _
```

A *soft hyphen* placed in a word does not appear on the screen unless word wrap needs to move part of that word onto the next line. (However, you can see the soft hyphen code in Reveal Codes.) Placing a soft hyphen in a long word at the end of a line tells WordPerfect to hyphenate the word at that hyphen, and you therefore have a smaller blank space at the end of the line. You use CTRL - .

Clear the screen and give it a try.

■ 1   Type this text:

```
WordPerfect offers a multitude of powerful features with _
```

■ 2   Type **hyphenation**, pressing CTRL - to place a soft hyphen after the first *n*. After you type the complete word, press the spacebar.

```
WordPerfect offers a multitude of powerful features with hyphen-
ation _
```

No hyphen appears when you initially press CTRL - . It does appear as soon as the rest of the word is wrapped to the next line.

Excellent work! If placing the hyphens yourself seems like too much trouble, you can have WordPerfect place the hyphens automatically. Clear your screen, and move on to the next exercise.

Automatic Hyphenation

When you use automatic hyphenation, you are asking WordPerfect to automatically place hyphens in words that require them. Let's see how it works.

■ *1* Select **L**ayout, **L**ine or press `SHIFT` `F8`
`1` or `L`

```
Format: Line

    1 - Hyphenation                No

    2 - Hyphenation Zone - Left    10%
                          Right    4%
```

■ *2* Select Hyphenation or press `1` or `Y`

```
Format: Line

    1 - Hyphenation                No (Yes)

    2 - Hyphenation Zone - Left    10%
                          Right    4%
```

■ *3* Select **Y**es or press `Y`

```
Format: Line

    1 - Hyphenation                Yes

    2 - Hyphenation Zone - Left    10%
                          Right    4%
```

■4　Click the right mouse button or press `F7` to return to the blank editing screen. If you select Reveal Codes, you'll see the [Hyph On] code.

■5　Type the text shown below. Be sure to press the spacebar after typing *irreversible*.

```
Robert sat in his favorite leather chair and made another irrevers-
ible _
```

Note　If you type a word that is not in the dictionary, WordPerfect prompts you to manually place the required hyphen. Use `←` and `→` to move the cursor where you want the hyphen to appear and press `ESC`. If you do not want to hyphenate the word, press `F1`.

Exit the screen, and go on to the next section.

■*Justifying Text*

WordPerfect allows you to choose from four kinds of justification: full, left, center, and right. *Fully justified text* is what you see in newspapers and most books (such as the text in this paragraph); the text aligns at both the left and right margins. When you type text on the screen, it will not appear justified, but when you print the text, or look at it with the View Document feature, it will be justified.

To fully justify text, WordPerfect adjusts the amount of space between words in sentences that extend along the entire width of the margins. Unless your printer has the ability to do proportional spacing, the spacing between words in justified text may distract the reader. The WordPerfect default is full justification.

This paragraph contains *left-justified text*. You'll notice the text is only aligned at the left margin. The right margin is uneven because the lines vary in length. You may remember that the documents you typed on your old typewriter (which has probably been gathering dust since you got your computer) were all left-justified and had a ragged right margin.

Right-justified text is aligned only at the right margin. The text along the left margin is unevenly aligned.

Center-justified text has an uneven left and right margin because each line is centered between the left and right margins. This is a handy feature if you need to center several lines of text, because you do not have to use the Center command for each line.

Let's take a look at how to select center and right justification.

■ *1* Select Layout, Line or press **SHIFT** **F8** **1** or **L** to call up the Format: Line menu.

■ *2* Select Justification or press **3** or **J**.

```
Justification: 1 Left; 2 Center; 3 Right; 4 Full: 0
```

■ *3* Select Center or press **2** or **C**.

```
    3 - Justification          Center

    4 - Line Height            Auto

    5 - Line Numbering         No
```

■4 Click the right mouse button or press [F7].
The cursor is centered halfway between the
left and right margins.

–

4

■5 Type the following text (as you type each
line, the text is automatically centered):

```
                    Moonlight Over the Rhine
                            by
                     Franz Reinhold_
```

Now let's select right justification, and type in some text.

■6 [↵] twice.

■7 Select **L**ayout, **L**ine, **J**ustification or press
[SHIFT][F8], [1] or [L], [3] or [J].

```
Justification: 1 Left; 2 Center; 3 Right; 4 Full: 0
```

■8 Select **R**ight or press [3] or [R].

```
    3 - Justification          Right

    4 - Line Height            Auto

    5 - Line Numbering         No
```

■ *9*   Click the right mouse button or press F7.
The cursor is now located at the right margin.

```
            Moonlight Over the Rhine
                     by
               Franz Reinhold


                                              _
```

■ *10*   Type this text:

```
            Moonlight Over the Rhine
                     by
               Franz Reinhold


                              Alpine Productions
                              Munich, Germany
                              21 November 1943_
```

If you go to the Reveal Codes screen, you'll see the [Just:Center] and [Just:Right] codes.

All the text entered after the insertion of the [Just:Center] code is centered on the screen. The text entered after the [Just:Right] code is aligned at the right margin. To cancel the existing justification, delete the appropriate codes.

Eliminating
■ *Widows and Orphans*

A *widow* is the first line of a paragraph that appears alone at the bottom of a page when a page break pushes the rest of the paragraph onto the next page. An *orphan* is the last line of a paragraph that appears alone at the top of a page when a page break splits a paragraph between two pages. With Widow/Orphan Protection turned on, WordPerfect ensures that any widow or orphan in your document is joined automatically by another line from that paragraph.

Do the following exercise on a clear screen.

4

■*1*   ⏎ 53 times to place the cursor at the beginning of the last line on the page (Ln 9.83").
Then, type the following text:

```
If you look closely, you will see this sentence is split apart by
----------------------------------------------------------------------
a WordPerfect soft page break.
      Turn the Widow/Orphan protection feature on to prevent the
single line at the top of page two from leading the life of an
orphan._
                                        Doc 1 Pg 2 Ln 1.5" Pos 1.7"
```

■*2*   HOME HOME ↑ to move to the top of the document. Select **L**ayout, **L**ine or press SHIFT F8 **1** or **L**

■*3*   Select **W**idow/Orphan Protection, **Y**es or press **9** or **W** **Y**

```
    9 - Widow/Orphan Protection        Yes


Selection: 0
```

■*4* Click the right mouse button or press F7 to
return to the document.

■*5* HOME HOME ↓ to move the cursor to the end
of the document. The last line from page 1 is
now located on page 2.

```
-------------------------------------------------------------------------------
If you look closely, you will see this sentence is split apart by
a WordPerfect soft page break.
    Turn the Widow/Orphan protection feature on to prevent the
single line at the top of page two from leading the life of an
orphan._


                                            Doc 1 Pg 2 Ln 1.67" Pos 1.7"
```

Widows will now be pushed onto the following page, and orphans will be joined
by a line from the previous page. This will change the line count on the page that
loses the line, so don't be concerned if you get one less line on a particular page
than you think you should get.

Let's go on to the next chapter and look at some ways to format individual
paragraphs.

5

- Paragraph
- Alignment
- Techniques

■ **T**his chapter shows you additional formatting features that will enhance the appearance of the paragraphs you create with WordPerfect 5.1. For example, you'll learn how to indent an entire paragraph to set it off from surrounding paragraphs and how to format your text so it is flush with the right margin or centered on the page. This chapter shows you the following techniques:

Indenting paragraphs

Centering text

Right-aligning text

▪*Indenting Paragraphs*

Let's look at three ways WordPerfect allows you to indent your text from the margins. You already know how to use TAB to indent the first line in a paragraph. Here you'll learn how to indent an entire paragraph from the left margin and from both the left and right margins, and how to create a *hanging indentation,* where all the lines after the first line in a paragraph are indented.

Indenting from the Left Margin

When you indent an entire paragraph from the left margin, you create a temporary left margin that remains in effect until you press ⏎.

Let's create a sample document using WordPerfect's default tabs. Start at the blank editing screen.

▪*1* Type this text:

```
Choosing the best viewpoint possible is very important when
attempting to take the perfect photograph. The best viewpoint must
be analyzed carefully and not rushed into._
```

▪*2* ⏎ twice, then select **L**ayout, **A**lign, **I**ndent-> or press F4. The Indent command moves the cursor to the next tab stop and creates a temporary margin at that position.

```
Choosing the best viewpoint possible is very important when
attempting to take the perfect photograph. The best viewpoint must
be analyzed carefully and not rushed into.

   _
```

■*3*   Type the text that appears on the screen
below. The text is indented ½ in. from the left
margin.

> Lighting is crucial to obtaining a clear, textured, detailed
> photograph. Choose a viewpoint that will avoid shooting
> directly into the sun._

■*4*   ⏎ twice, then select **L**ayout, **A**lign, **I**ndent->
twice or press **F4** twice.

> Lighting is crucial to obtaining a clear, textured, detailed
> photograph. Choose a viewpoint that will avoid shooting
> directly into the sun.
>
>   _

■*5*   Type the paragraph that appears on the screen
below. The text is indented 1 in.

> Perspective is the relationship of the subject to its
> surrounding objects. Change the viewpoint to provide
> added or diminished emphasis to your subject._

■*6*   ⏎ twice, then select **L**ayout, **A**lign, **I**ndent->
three times or press **F4** three times.

> Perspective is the relationship of the subject to its
> surrounding objects. Change the viewpoint to provide
> added or diminished emphasis to your subject.
>
>   _

■ 7   Type the paragraph that appears on the screen
below. The text is indented 1½ in.

> Framing the subject area through small changes of
> viewpoint allows you to include interesting details
> or colors that can enhance the photograph._

■ 8   [HOME] [HOME] [↑] to move the cursor to the top
of the page. Your document should look like
this:

> Choosing the best viewpoint possible is very important when
> attempting to take the perfect photograph. The best viewpoint must
> be analyzed carefully and not rushed into.
>
>   Lighting is crucial to obtaining a clear, textured, detailed
>   photograph. Choose a viewpoint that will avoid shooting
>   directly into the sun.
>
>   Perspective is the relationship of the subject to its
>   surrounding objects. Change the viewpoint to provide
>   added or diminished emphasis to your subject.
>
>   Framing the subject area through small changes of
>   viewpoint allows you to include interesting details
>   or colors that can enhance the photograph.
>
>   Doc 1 Pg 1 Ln 1" Pos 1"

■*9* Select **E**dit, **R**eveal Codes or press ⌨ALT⌨ ⌨F3⌨.
You can see the [->Indent] codes at the beginning of each paragraph.

```
                                        Doc 1 Pg 1 Ln 1" Pos 1"
[                                             ]
Choosing the best viewpoint possible is very important when[SRt]
attempting to take the perfect photograph. The best viewpoint must[SRt]
be analyzed carefully and not rushed into.[HRt]
[HRt]
[→Indent]Lighting is crucial to obtaining a clear, textured, detailed[SRt]
photograph. Choose a viewpoint that will avoid shooting[SRt]
directly into the sun.[HRt]
[HRt]
[→Indent][→Indent]Perspective is the relationship of the subject to its[SRt]
surrounding objects. Change the viewpoint to provide[SRt]


Press Reveal Codes to restore screen
```

■*10* Select **E**dit, **R**eveal Codes or press ⌨ALT⌨ ⌨F3⌨ to
exit Reveal Codes.

Indenting from the Left and Right Margins Simultaneously

Another kind of indentation uses temporary left *and* right margins. These new
margins are located at the default tab stops or at those tab stops that you have
selected. Let's try setting this kind of indentation, using the default tabs, which
are located every 0.5 in. Start at a clear editing screen by pressing ⌨F7⌨ ⌨N⌨ ⌨N⌨.

■*1* Type this text:

```
The lowest level of professional baseball in America is called
Single A. This is where many young players get their first
professional baseball experience._
```

■*2* ⏎ twice, then select **Layout**, **Align**, **Indent-> <-**
or press SHIFT F4. The Left/Right Indent com-
mand creates temporary left and right margins.

> The lowest level of professional baseball in America is called
> Single A. This is where many young players get their first
> professional baseball experience.
>
> –

■*3* Type the next paragraph. It is indented ½ in.
from the left and right margins.

> The second level is Double A. This league is more
> competitive than Single A, with greater demands placed on
> the players._

■*4* ⏎ twice, then select **Layout**, **Align**, **Indent-> <-**
twice or press SHIFT F4 twice.

> The second level is Double A. This league is more
> competitive than Single A, with greater demands placed on
> the players.
>
> –

■*5* Type the next paragraph. The paragraph is
indented 1 in. from both margins.

> The third level is Triple A. This is the final
> prep for the Major Leagues. The players are of
> a very high quality._

■6 ⏎twice, then select **L**ayout, **A**lign, **In**dent-> <-
three times or press SHIFT F4 three times.

```
       The third level is Triple A. This is the final
       prep for the Major Leagues. The players are of
       a very high quality.


           _
```

■7 Type the next paragraph. The text is indented
1½ in. from both margins.

```
           The Major Leagues is the highest
           level to which a player can progress
           and is the dream of all professional
           baseball players._
```

■8 HOME HOME ↑ to move the cursor to the top
of the page:

```
The lowest level of professional baseball in America is called
Single A. This is where many young players get their first
professional baseball experience.

    The second level is Double A. This league is more
    competitive than Single A, with greater demands placed on
    the players.

        The third level is Triple A. This is the final
        prep for the Major Leagues. The players are of
        a very high quality.

            The Major Leagues is the highest
            level to which a player can progress
            and is the dream of all professional
            baseball players.




                                    Doc 1 Pg 1 Ln 1" Pos 1"
```

5

■*9*   Select **Edit, R**eveal Codes or press ALT F3 .
Notice the [->Indent<-] codes.

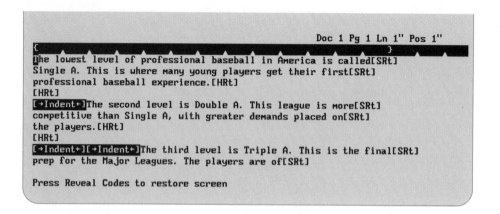

■*10*   Select **Edit, R**eveal Codes or press ALT F3 to
exit Reveal Codes.

Creating Hanging Indents

In a *hanging indent,* all the lines after the first line in a paragraph are indented.
This indent is often used in the numbered paragraphs of outlines or in other types
of numbered lists.

Clear your editing screen for the following exercise by pressing F7 N N .
You'll use the default tabs.

■*1*   Select **L**ayout, **A**lign, **I**ndent-> or press F4 ,
then select **L**ayout, **A**lign, **M**argin Rel<- or
press SHIFT TAB . Type **1.** and press TAB .

As you know, selecting Layout, Align, Indent-> or pressing [F4] creates a new, temporary left margin. When you select Layout, Align, Margin Rel<- or press [SHIFT][TAB] after indenting, you move the cursor one indent backward. You can select **Margin Rel** as many times as you select Indent->, depending on your formatting needs. When you enter text, the first line will "hang out" from all following lines that are word-wrapped.

■2   Type this text:

> 1.   Young men and women wishing to become professional airline
>   pilots must obtain the required aeronautical ratings. There
>   are three basic ratings that require flight training._

■3   [↵] twice, then select Layout, Align, Indent->
twice or press [F4] twice. Select Layout, Align,
Margin Rel<- or press [SHIFT][TAB]. Type **a.**

> 1.   Young men and women wishing to become professional airline
>   pilots must obtain the required aeronautical ratings. There
>   are three basic ratings that require flight training.
>
>   a._

■4   [TAB] and type this text:

> a.   Private Pilot - This is the primary aviator qualification
>   needed to pilot an aircraft._

5

■5   ⏎ twice, then select Layout, **Align, Indent->**
     twice or press **F4** twice. Select Layout, **Align,**
     Margin Rel<- or press **SHIFT TAB**. Type **b.**

```
    a.    Private Pilot - This is the primary aviator qualification
          needed to pilot an aircraft.

    b. _
```

■6   **TAB** and type the next paragraph:

```
    b.    Instrument Pilot - This is an advanced qualification
          necessary to pilot an aircraft only by the instruments._
```

■7   ⏎ twice, then select Layout, **Align, Indent->**
     twice or press **F4** twice. Select Layout, **Align,**
     Margin Rel<- or press **SHIFT TAB**. Type **c.**

```
    b.    Instrument Pilot - This is an advanced qualification
          necessary to pilot an aircraft only by the instruments.

    c. _
```

■8   **TAB** and type the next paragraph:

```
    c.    Commercial Pilot - This rating qualifies a pilot to fly
          an aircraft for compensation._
```

■*9*   Your screen should look like this:

1.   Young men and women wishing to become professional airline
     pilots must obtain the required aeronautical ratings. There
     are three basic ratings that require flight training.

     a.   Private Pilot - This is the primary aviator qualification
          needed to pilot an aircraft.

     b.   Instrument Pilot - This is an advanced qualification
          necessary to pilot an aircraft only by the instruments.

     c.   Commercial Pilot - This rating qualifies a pilot to fly
          an aircraft for compensation._

5

Note   If you wish to change the format you've created, simply delete the unwanted codes on the Reveal Codes screen and replace them with new ones. Your text will conform to the new codes.

■*Centering Text*

Centering text is a common word processing task. For example, titles and headings are frequently centered. In WordPerfect you can center text as you type it or center existing text without having to retype it.

> When you center text, WordPerfect inserts a hidden [Center] code in your document.

Centering New Text

Begin at a clear editing screen by pressing `F7` `N` `N`

■ *1*   Select Layout, Align, Center or press
`SHIFT` `F6`, and type the text on the screen
below. When you use the Center command,
the cursor moves to the center of the screen.

Personal Computing and You_

■ *2*   `↵` twice, select Layout, Align, Center or
press `SHIFT` `F6`, then type the next line:

Personal Computing and You

by_

■ *3*   `↵` twice, select Layout, Align, Center or
press `SHIFT` `F6`, then type the author's name:

Personal Computing and You

by

George Adams, II_

■*4* Select **E**dit, **R**eveal Codes or press `ALT` `F3`.

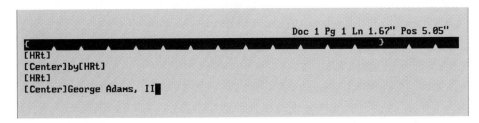

```
                                               Doc 1 Pg 1 Ln 1.67" Pos 5.05"
{                                              }
[HRt]
[Center]by[HRt]
[HRt]
[Center]George Adams, II█
```

Notice the [Center] code in front of each centered line. Any text that wraps to the next line will *not* be centered.

■*5* Select **E**dit, **R**eveal Codes or press `ALT` `F3` to exit Reveal Codes.

Centering Existing Text

There will be times when you want to center text you've already created. You don't have to delete the text and retype it to do this; you simply select the text as a *block,* then use the Center command. Clear the screen by pressing `F7` `N` `N` and try it (you'll learn more about blocking text in Chapter 8):

■*1* Type the following text, pressing `↵` after each of the first two lines:

```
WOK Wonders
by
Tsu Choi_
```

■*2*   Move the cursor to the *W* in *WOK*.

■*3*   Select **Edit**, **B**lock or press ALT F4, then press
         PG DN to highlight the text. Notice the "Block
         on" message in the lower-left corner of your
         screen.

The highlighting indicates that the text has been blocked. You'll block text often
while working with WordPerfect.

■*4*   Select **L**ayout, **A**lign, **C**enter or press
SHIFT F6.

    The [Just:Center] code indicates that all the blocked lines will be
centered as if they were initially typed with center justification turned on.

5

■*5*   Select **Y**es or press **Y**. (If the text disappears from your screen, press ⬆ until it
reappears.)

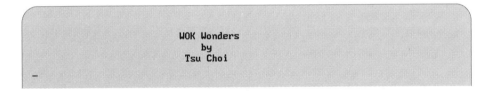

■*6*   Select **E**dit, **R**eveal Codes or press **ALT F3**.

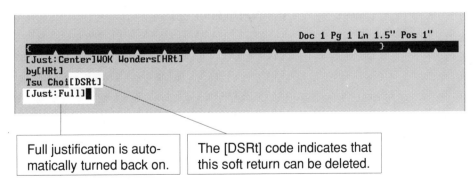

    Full justification is auto-
matically turned back on.

    The [DSRt] code indicates that
this soft return can be deleted.

■ *7*   Select **E**dit, **R**eveal Codes or press ALT F3 to
exit Reveal Codes.

Centering Text at the Cursor Location

WordPerfect allows you to position the cursor anywhere on the screen and center
your text at that location. Clear the screen and follow these steps:

■ *1*   Type **Apple Juice**, then press TAB four times.

```
Apple Juice                  _
```

■ *2*   Select **L**ayout, **A**lign, **C**enter or press
SHIFT F6, then type **12 Gallons**.

```
Apple Juice          12 Gallons_
```

■ *3*   Press ↵, type **Pear Juice**, then press TAB
four times.

```
Apple Juice          12 Gallons
Pear Juice                  _
```

■*4*   Select **Layout**, **Align**, **Center** or press
     `SHIFT` `F6`, then type **28 Pints**.

```
Apple Juice          12 Gallons
Pear Juice           28 Pints_
```

5

Aligning Text
■*Flush with the Right Margin*

Right alignment of text is often used for short columns of numbers, lists of data, items in an index, and so on. WordPerfect allows you to flush-right text that you have already entered, so you don't have to retype it.

Flush-Right Alignment of New Text

Work on this next exercise at a clear editing screen.

■*1*   Type **Automotive Labor Costs.**

```
Automotive Labor Costs_
```

■*2*   Select **L**ayout, **A**lign, **F**lush Right or press
        ALT F6.

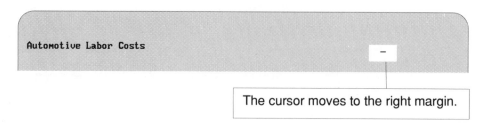

The cursor moves to the right margin.

■*3*   Type **Rebuild Carburetor**.

■*4*   ⏎, select **L**ayout, **A**lign, **F**lush Right or
        press ALT F6, then type **$95.00**.

■*5*   ⏎ twice, select **L**ayout, **A**lign, **F**lush Right
        or press ALT F6, then type **Brake Job.**

Automotive Labor Costs   Rebuild Carburetor
                                                   $95.00

                                          Brake Job_

■6 ⏎, select **L**ayout, **A**lign, **F**lush Right or
press ALT F6, then type **$55.00**.

```
Automotive Labor Costs                         Rebuild Carburetor
                                                         $95.00

                                                      Brake Job
                                                         $55.00_
```

■7 Select **E**dit, **R**eveal Codes or press ALT F3.
Notice the [Flsh Rgt] codes.

```
                                        Doc 1 Pg 1 Ln 1.67" Pos 7.5"
{                                                              }
[Flsh Rgt]$95.00[HRt]
[HRt]
[Flsh Rgt]Brake Job[HRt]
[Flsh Rgt]$55.00█
```

■8 Select **E**dit, **R**eveal Codes or press ALT F3 to
exit Reveal Codes.

Flush-Right Alignment of Existing Text

You can use the Block command with the Flush Right command to align existing
text on the right margin. Clear your editing screen to do the following exercise.

■1 Type this text:

```
8d Finishing nails - 28 boxes._
```

■*2* Move the cursor to *any* letter on the line.
(I chose the *n* in *finishing* for this example.)

■*3* Select **E**dit, **B**lock or press **ALT** **F4**, then press
END to block the text. (You can block *any*
amount of text.)

```
8d Finishing nails - 28 boxes.
```

■*4* Select **L**ayout, **A**lign, **F**lush Right or press
ALT **F6**.

```
[Just:Right]? No (Yes)
```

■*5* Select **Y**es or press **Y**. (If the text disappears from the screen, press **↑** to make it
reappear.)

```
                              8d Finishing nails - 28 boxes.
 -
```

You can also take a portion of an existing line of text and move it flush-right on that line. This is a very useful editing tool. For the following exercise, use the text on your screen from the previous exercise.

■ *1*   Select **E**dit, **R**eveal Codes or press ALT F3 .
Move the cursor to the *8* in *8d*.

■ *2*   BKSP ⬅ to delete the [Just:Right] code.

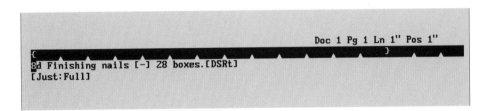

■ *3*   Move the cursor to the *2* in *28*.

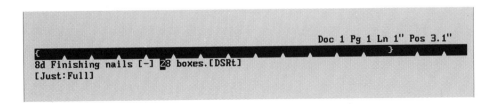

■4 Select **L**ayout, **A**lign, **F**lush Right or press
    ⌨ALT ⌨F6.

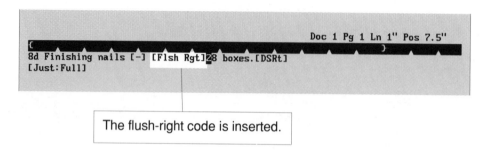

    Doc 1 Pg 1 Ln 1" Pos 7.5"

    8d Finishing nails [-] [Flsh Rgt]28 boxes.[DSRt]
    [Just:Full]

The flush-right code is inserted.

■5 Select **E**dit, **R**eveal Codes or press ⌨ALT ⌨F3 to
    exit Reveal Codes.

In the next chapter you'll learn about formatting features you can use when
working with an entire page.

6

- Page
- Formatting
- Techniques

■ In this chapter you'll learn to use some of the page formatting features WordPerfect 5.1 offers. For example, if you deal with many different forms and documents, you'll see how to print on paper other than the standard 8½" × 11" size. You'll learn how easy it is to number pages, placing the numbers where you want them. You'll also determine the amount of a page that will be used for text by learning how to select new top and bottom margins.

Setting top and bottom margins

Selecting paper size and type

Numbering pages

Setting Top ■*and Bottom Margins*

If you use the default top and bottom margins and the default font, you can fit 54 lines of text on an 8½" × 11" sheet of paper. From time to time you'll probably need to change the number of lines that fit on a page. This exercise will show you how to do this very quickly. Start with a clear screen by pressing [F7] [N] [N].

■ *1*   Select **Layout**, **P**age or press [SHIFT][F8]
        [2] or [P].

```
Format: Page

    1 - Center Page (top to bottom)      No

    2 - Force Odd/Even Page

    3 - Headers

    4 - Footers

    5 - Margins - Top                    1"
                  Bottom                 1"

    6 - Page Numbering

    7 - Paper Size                       8.5" x 11"
             Type                        Standard

    8 - Suppress (this page only)




Selection: 0
```

■*2*   Select **M**argins or press ⑤ or Ⓜ.

■*3*   Type **2** and press ⏎ to set the top margin
       at 2".

```
5 - Margins - Top              2"
            Bottom             1"

6 - Page Numbering
```

■*4*   Type **2** and press ⏎ to set the bottom margin
       at 2".

```
5 - Margins - Top              2"
            Bottom             2"

6 - Page Numbering
```

With these new margins, you will get 42 lines per page. The soft page break will occur automatically every 42 lines.

> Narrow your
> top and bottom
> margins to
> fit more text
> on a page.

■5 Click the right mouse button or press F7 to
return to the Edit screen.

Doc 1 Pg 1 Ln 2" Pos 1"

Notice the Ln 2" position indicator.

■6 Select **E**dit, **R**eveal Codes or press ALT F3.

All pages after the point in the document where you made the change will have
the new margins.

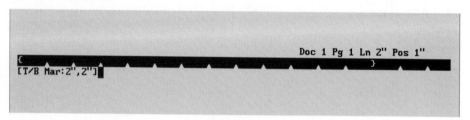

Doc 1 Pg 1 Ln 2" Pos 1"

[T/B Mar:2",2"]

■7 Select **E**dit, **R**eveal Codes or press ALT F3 to
exit Reveal Codes.

Defining Forms and
■Selecting Paper Size and Type

With WordPerfect you can select different paper sizes and types. The paper sizes
available to you are standard size (8.5" × 11"), legal size (8.5" × 14"), and en-
velope size (9.5" × 4"). You can also create your own special sizes. Paper types
range from standard and bond to envelopes and cardstock. You can also create
customized types.

Defining Forms

Before you select a new paper size and type, you must make sure the new form has been defined for your printer. WordPerfect uses the term *form* to describe the size and type of paper you print on. Clear your document screen by pressing [F7] [N] [N], and define a customized letterhead form.

■ *1* Select **Layout, Page** or press [SHIFT] [F8] [2] or [P].

6

■ *2* Select **Paper Size** or press [7] or [S].

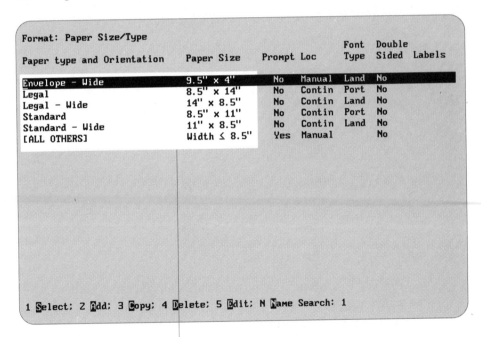

```
Format: Paper Size/Type
                                                          Font  Double
 Paper type and Orientation      Paper Size   Prompt Loc   Type  Sided  Labels

 Envelope - Wide                 9.5" x 4"     No   Manual  Land  No
 Legal                           8.5" x 14"    No   Contin  Port  No
 Legal - Wide                    14" x 8.5"    No   Contin  Land  No
 Standard                        8.5" x 11"    No   Contin  Port  No
 Standard - Wide                 11" x 8.5"    No   Contin  Land  No
 [ALL OTHERS]                    Width ≤ 8.5"  Yes  Manual        No




 1 Select; 2 Add; 3 Copy; 4 Delete; 5 Edit; N Name Search: 1
```

Don't be concerned if your list of forms is different; you have installed a different printer or printer file. You should still be able to do this exercise.

This is the list of forms currently defined for the HP LaserJet Series II printer. The [ALL OTHERS] option is for those paper sizes and types that do not fall into the list of defined forms.

■3   Select **Add** or press 2 or A to add a new form to the list of forms.

```
Format: Paper Type

     1 - Standard

     2 - Bond

     3 - Letterhead
```

■4   Select **Letterhead** or press 3 or H.

```
Format: Edit Paper Definition

          Filename           HPLASEII.PRS

     1 - Paper Size          8.5" x 11"

     2 - Paper Type          Letterhead
```

The Edit Paper Definition menu is the menu you use to define the form. You will define both the size and type of paper you'll be using in your printer.

■*5* Select Paper Size or press [1] or [S]. This
command defines the size of the customized
letterhead paper. You now see menu option O
for Other.

```
8 - A4                        (210mm x 297mm)

9 - A4 Landscape              (297mm x 210mm)

o - Other



Selection: 0
```

■*6* Select Other or press [O].

```
o - Other



Width: 0"        Height:
```

■*7* Type **8**, press [↵], type **10**, and press [↵].

```
1 - Paper Size           8" x 10"

2 - Paper Type           Letterhead

3 - Font Type            Portrait

4 - Prompt to Load       No

5 - Location             Continuous
```

■8   Select Location or press ⑤ or Ⓛ to specify
how the paper will be fed into the printer.

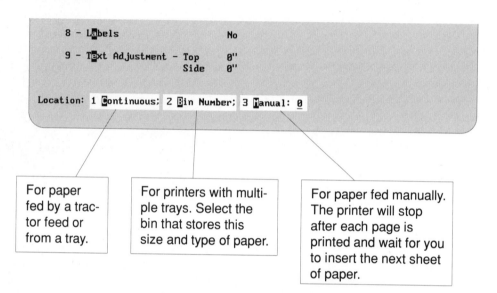

```
      8 - Labels                    No

      9 - Text Adjustment - Top     0"
                            Side    0"


   Location: 1 Continuous; 2 Bin Number; 3 Manual: 0
```

For paper
fed by a trac-
tor feed or
from a tray.

For printers with multi-
ple trays. Select the
bin that stores this
size and type of paper.

For paper fed manually.
The printer will stop
after each page is
printed and wait for you
to insert the next sheet
of paper.

If your printer is able
to print on both sides
of the page, you can
turn on Double Sided
Printing.

■ *9*   Select **C**ontinuous or press ⌐1⌐ or ⌐C⌐.

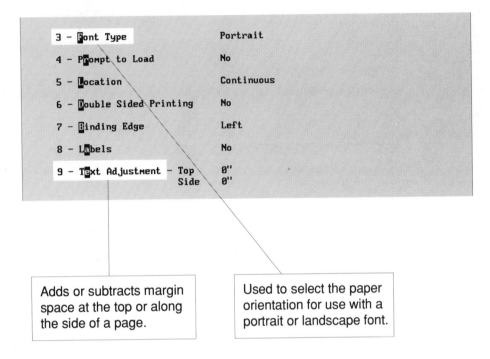

3 - **F**ont Type	Portrait
4 - P**r**ompt to Load	No
5 - **L**ocation	Continuous
6 - **D**ouble Sided Printing	No
7 - **B**inding Edge	Left
8 - L**a**bels	No
9 - T**e**xt Adjustment - Top	0"
Side	0"

Adds or subtracts margin space at the top or along the side of a page.

Used to select the paper orientation for use with a portrait or landscape font.

With the Font Type option, you can select *Portrait* to print pages that have the long edge oriented vertically. Select *Landscape* to print pages that have the long edge oriented horizontally.

Text adjustment is used for paper that has holes punched on the side or to compensate for printers that move the paper up before printing. You can move the text a specified distance upward, downward, to the left, or to the right on the page.

■ *10* Click the right mouse button or press [F7].
The new form is listed here:

```
Format: Paper Size/Type
                                                       Font  Double
Paper type and Orientation    Paper Size   Prompt Loc   Type  Sided  Labels

Envelope - Wide               9.5" x 4"    No  Manual  Land  No
Legal                         8.5" x 14"   No  Contin  Port  No
Legal - Wide                  14" x 8.5"   No  Contin  Land  No
Letterhead                    8" x 10"     No  Contin  Port  No
Standard                      8.5" x 11"   No  Contin  Port  No
Standard - Wide               11" x 8.5"   No  Contin  Land  No
[ALL OTHERS]                  Width ≤ 8.5" Yes Manual        No
```

■ *11* Click the right mouse button twice or press
[F7] twice to return to the Edit screen.

Selecting Paper Size and Type

Once you have defined the forms you will be using most often, you need to select them for use when you want them. Let's select the letterhead form you just defined.

■ *1* Select **L**ayout, **P**age or press [SHIFT][F8] [2]
or [P].

```
6 - Page Numbering

7 - Paper Size       8.5" x 11"
        Type         Standard

8 - Suppress (this page only)
```

■*2* Select Paper Size or press [7] or [S].

```
Format: Paper Size/Type

                                                     Font  Double
Paper type and Orientation   Paper Size   Prompt Loc  Type  Sided  Labels

Envelope - Wide              9.5" x 4"     No  Manual  Land  No
Legal                        8.5" x 14"    No  Contin  Port  No
Legal - Wide                 14" x 8.5"    No  Contin  Land  No
Letterhead                   8" x 10"      No  Contin  Port  No
Standard                     8.5" x 11"    No  Contin  Port  No
Standard - Wide              11" x 8.5"    No  Contin  Land  No
[ALL OTHERS]                 Width ≤ 8.5"  Yes Manual        No
```

6

■*3* Move the highlight to the letterhead paper size.

```
Format: Paper Size/Type

                                                     Font  Double
Paper type and Orientation   Paper Size   Prompt Loc  Type  Sided  Labels

Envelope - Wide              9.5" x 4"     No  Manual  Land  No
Legal                        8.5" x 14"    No  Contin  Port  No
Legal - Wide                 14" x 8.5"    No  Contin  Land  No
Letterhead                   8" x 10"      No  Contin  Port  No
Standard                     8.5" x 11"    No  Contin  Port  No
Standard - Wide              11" x 8.5"    No  Contin  Land  No
[ALL OTHERS]                 Width ≤ 8.5"  Yes Manual        No
```

■*4* Select Select or press [1], [S], or [↵].

```
      6 - Page Numbering

      7 - Paper Size              8" x 10"
              Type                Letterhead

      8 - Suppress (this page only)
```

■ *5*   Click the right mouse button or press F7.
Then, select **E**dit, **R**eveal Codes or press
ALT F3.

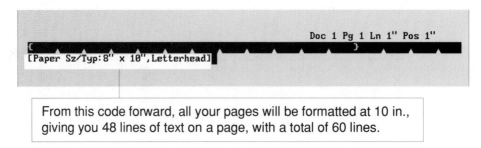

```
                                               Doc 1 Pg 1 Ln 1" Pos 1"
{                                                      }
[Paper Sz/Typ:8" x 10",Letterhead]
```

From this code forward, all your pages will be formatted at 10 in.,
giving you 48 lines of text on a page, with a total of 60 lines.

■ *6*   Select **E**dit, **R**eveal Codes or press ALT F3 to
exit Reveal Codes.

Selecting Sizes without Defining Forms

You can select a form smaller than the standard size without going through the
fairly long defining process. You may want to define a form that does not fit the
categories listed in the Paper Size/Type menu. Such a form can be no larger than
the Standard or All Others forms for your printer.

Let's assume you use 5" × 7" paper for thank-you notes. At a clear screen (press
F7 N N) do the following exercise:

You can define
a smaller than
standard
size form.

■ *1* Select **L**ayout, **P**age, Paper **S**ize or press
`SHIFT` `F8`, `2` or `P`, `7` or `S`.

```
Format: Paper Size/Type
                                                       Font  Double
Paper type and Orientation    Paper Size    Prompt Loc Type  Sided  Labels

Envelope - Wide               9.5" x 4"     No  Manual Land  No
Legal                         8.5" x 14"    No  Contin Port  No
Legal - Wide                  14" x 8.5"    No  Contin Land  No
Letterhead                    8" x 10"      No  Contin Port  No
Standard                      8.5" x 11"    No  Contin Port  No
Standard - Wide               11" x 8.5"    No  Contin Land  No
[ALL OTHERS]                  Width ≤ 8.5"  Yes Manual       No
```

■ *2* Select [ALL OTHERS].

■ *3* Select **O**ther or press `O`.

```
   o - Other



Width: 8"
```

■ *4* Type **5** and press `↵` to select a 5-in. form
width.

```
Height: 10"
```

■5   Type **7** and press ⏎ to select a 7-in. form
height.

```
Format: Paper Type

    1 - Standard

    2 - Bond

    3 - Letterhead
```

■6   Select **S**tandard or press 1 or S.

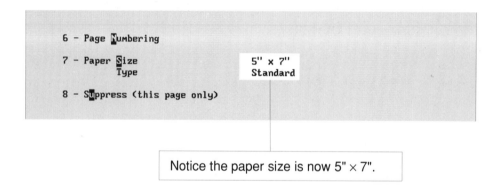

```
    6 - Page Numbering

    7 - Paper Size              5" x 7"
            Type                Standard

    8 - Suppress (this page only)
```

Notice the paper size is now 5" × 7".

■7   Click the right mouse button or press F7.
Then select **E**dit, **R**eveal Codes or press
ALT F3.

```
                                    Doc 1 Pg 1 Ln 1" Pos 1"
[                              ]
[Paper Sz/Typ:5" x 7",Standard]
```

Remember, you have 1-in. left, right, top, and bottom margins, leaving you only a 3" × 5" typing space. If you want to place more information on the thank-you note, simply change the margins.

■ *8*   Select **Edit**, **R**eveal Codes or press ALT F3 to exit Reveal Codes.

6

■*Numbering Pages*

WordPerfect will number pages only when you tell it to, and it's very easy to do so. You can choose from among eight locations on the page to put the number, and you can also tell WordPerfect to put odd numbers on the right side of the page and even numbers on the left side, as you see in this book. You can even suppress the page number on specified pages and then resume numbering on subsequent pages. There are three different styles of numbers to choose from: *Arabic* (1, 2, 3, etc.), *Lowercase Roman* (i, ii, iii, etc.), and *Uppercase Roman* (I, II, III, etc.).

Clear the screen by pressing F7 N N, and try the following exercise:

■ *1*   Select **Layout**, **P**age or press SHIFT F8 2 or P.

```
    6 - Page Numbering

    7 - Paper Size              8.5" x 11"
            Type                Standard

    8 - Suppress (this page only)
```

■*2*   Select Page **N**umbering or press **6** or **N**.

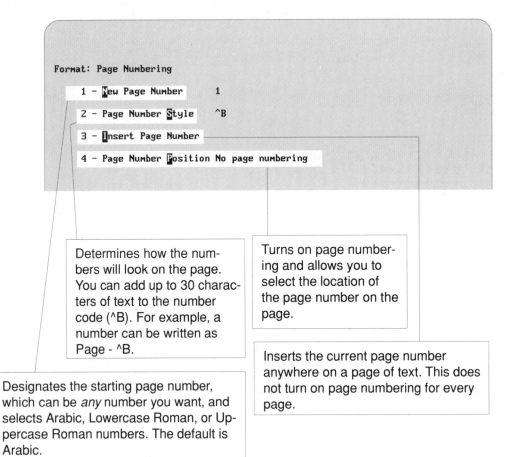

Format: Page Numbering

    1 - New Page Number   1

    2 - Page Number Style   ^B

    3 - Insert Page Number

    4 - Page Number Position No page numbering

Determines how the numbers will look on the page. You can add up to 30 characters of text to the number code (^B). For example, a number can be written as Page - ^B.

Turns on page numbering and allows you to select the location of the page number on the page.

Inserts the current page number anywhere on a page of text. This does not turn on page numbering for every page.

Designates the starting page number, which can be *any* number you want, and selects Arabic, Lowercase Roman, or Uppercase Roman numbers. The default is Arabic.

Note   You can tell WordPerfect to automatically place a page number at a specified location by placing the cursor where you want the number to appear and pressing CTRL B.

■*3* Select **N**ew Page Number or press `1` or `N`,
type **i** (Lowercase Roman), and press `⏎`.
(You can type **I** for Uppercase Roman, if
you wish.)

```
Format: Page Numbering

    1 - New Page Number        i

    2 - Page Number Style      ^B

    3 - Insert Page Number

    4 - Page Number Position No page numbering
```

6

■*4* Select Page Number **S**tyle or press `2` or
`S`, type **Intro -**, and press the spacebar after
you type the hyphen. Then press `CTRL` `B`
`⏎` to create the page number code, ^B.

```
Format: Page Numbering

    1 - New Page Number        i

    2 - Page Number Style      Intro - ^B

    3 - Insert Page Number

    4 - Page Number Position No page numbering
```

The ^B code must be placed in the Page Number Style entry in order for the page
number to appear on the page. If you forget to create the ^B code, WordPerfect
automatically places it at the end of the text.

■*5* Select Page Number **P**osition or press 4
or P.

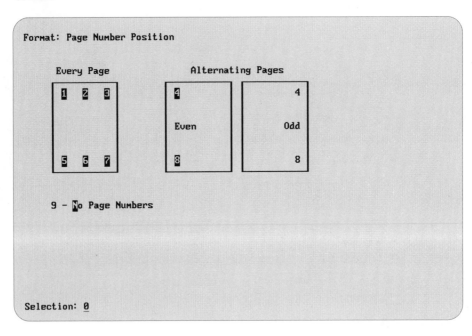

The numbers in the boxes represent the location of the page numbers on the printed pages.

■*6* Press 1 to select the top-left corner.

```
Format: Page Numbering

    1 - New Page Number        i

    2 - Page Number Style      Intro - ^B

    3 - Insert Page Number

    4 - Page Number Position Top Left
```

■ 7   Click the right mouse button or press F7.
Notice that you cannot see a number on
the screen.

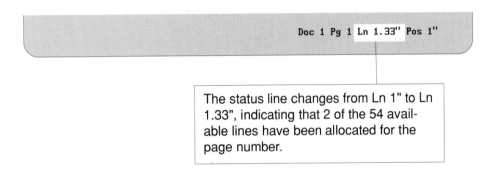

Doc 1 Pg 1 Ln 1.33" Pos 1"

The status line changes from Ln 1" to Ln
1.33", indicating that 2 of the 54 available lines have been allocated for the
page number.

6

■ 8   Select **F**ile, **P**rint, **V**iew Document or press
SHIFT F7, then press 6 or V to see the
page number on the View Document screen.

Keep this document on the screen for use in the next exercise.

Suppressing Page Numbers

Once you have selected page numbering, WordPerfect will number every page.
However, there may be times when you do not want a page numbered. Let's
create a title page and suppress the page number.

■ *1*   Press ⏎ 10 times and type the following
text, centering each line by selecting **Layout**,
Align, **Center** or pressing SHIFT F6:

```
                         THE TITLE PAGE
                              BY
                         MR. STU DENT_




                                          Doc 1 Pg 1 Ln 3.67" POS 4.85"
```

■ *2*   Select **Layout**, **Align**, Hard **P**age or press
CTRL ⏎ to enter a hard page break.

```
                      THE TITLE PAGE

                           BY

                      MR. STU DENT
==================================================================================
_
```

■*3* Move the cursor to the beginning of the document. Select **Layout**, **P**age or press [SHIFT][F8] [2] or [P].

■*4* Select S**u**ppress or press [8] or [U]. Take some time to look over the menu options.

Option 1 suppresses all the page numbers, headers, and footers that appear after the suppress code. Option 2 suppresses all the headers and footers in the document. Options 3–8 suppress the selected item on one page only.

6

```
Format: Suppress (this page only)

    1 - Suppress All Page Numbering, Headers and Footers

    2 - Suppress Headers and Footers

    3 - Print Page Number at Bottom Center    No

    4 - Suppress Page Numbering               No

    5 - Suppress Header A                      No

    6 - Suppress Header B                      No

    7 - Suppress Footer A                      No

    8 - Suppress Footer B                      No




Selection: 0
```

■*5* Select Suppress **P**age Numbering, **Y**es or press [4] or [P] [Y]. Selecting this option suppresses page numbering for this page only. The remaining pages will not be affected.

■ *6*   Click the right mouse button or press F7 to
return to the document.

■ *7*   Select **F**ile, **P**rint, **V**iew Document or press
SHIFT F7 6 or V to see your document on
the View Document screen—you'll notice
that the page number is no longer displayed
on page 1.

■ *8*   Press PG DN to see the number on page 2.

Keep this document on the screen for the next exercise.

Setting New Page Numbers

In the previous exercise, you suppressed the page number on page 1, but in
WordPerfect's internal numbering system, this page still counts as the first page
of your document. You want the page after the title page to be numbered *Intro - ii*,
not *Intro - i*. Here's how to set a new page number for the second page:

■ *1*   Press PG DN to ensure that the cursor is at the
top of page 2 (Intro - ii). You want this to be
page 1 (Intro - i).

```
                        MR. STU DENT
==================================================================================
_
```

■*2* Select **L**ayout, **P**age or press `SHIFT` `F8`
`2` or `P`.

■*3* Select Page **N**umbering or press `6` or `N`.

```
Format: Page Numbering

    1 - New Page Number        ii

    2 - Page Number Style      Intro - ^B

    3 - Insert Page Number

    4 - Page Number Position Top Left
```

6

■*4* Select **N**ew Page Number or press `1` or `N`,
type **i**, and press `↵`.

```
Format: Page Numbering

    1 - New Page Number        i

    2 - Page Number Style      Intro - ^B

    3 - Insert Page Number

    4 - Page Number Position Top Left
```

■5   Click the right mouse button or press `F7`.
Press `↑` and `↓` to move the cursor back
and forth between the title page and the
second page, and notice that Pg 1 on the
status line does not change.

```
                        MR. STU DENT
============================================================================
_




                                              Doc 1 Pg 1 Ln 1.33" Pos 1"
```

The page is now page 1 (which is printed
as Intro - i), even though you are on the
second page of the document.

At this point you know all the formatting features you need to know to get up
and running with any document you want to create. In the following chapters,
you'll learn about some special features. Though you may not use all of them,
these features are worth investigating—you never know when you might
need them.

7

- Formatting
- Characters and
- Using Fonts

■ In this chapter you will learn how to change the appearance of the characters you type on the screen. This process, often called *character formatting,* enables you to add variety and emphasis to your text. Some kinds of character formatting can be done with only a few keystrokes, while other kinds are a bit more involved and require working with fonts. In this chapter, you'll use these formatting techniques:

Typing boldfaced and underlined characters

Changing between uppercase and lowercase letters

Using fonts to change the size and appearance of your text

Typing Boldfaced and
■*Underlined Characters*

Boldfacing and underlining characters can add emphasis to text you want your reader to focus on.

■*1*   Select **F**ont, **A**ppearance, **B**old or press F6, and type this text: **Boldface is very effective.** Depending on the type of monitor you have, the Pos number on the status line will appear boldfaced or will change color.

■*2*   Select **F**ont, **N**ormal or press F6 or → to turn boldface off. Then select **E**dit, **R**eveal Codes or press ALT F3.

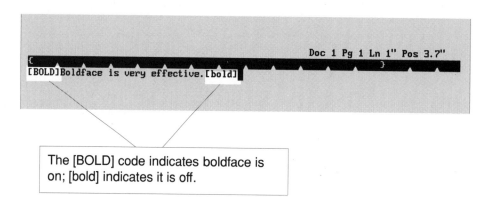

Doc 1 Pg 1 Ln 1" Pos 3.7"

[BOLD]Boldface is very effective.[bold]

The [BOLD] code indicates boldface is on; [bold] indicates it is off.

You'll underline the next sentence while remaining in Reveal Codes.

■*3*   [↵] twice to move down the screen. Select
Font, Appearance, Underline or press [F8],
and type this text: **See how easy it is to
underline.** The Pos number on the status line
may be underlined or in a different color on
some color monitors.

■*4*   Select Font, Normal or press [F8] or [→] to
turn underline off.

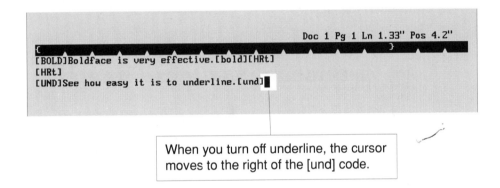

```
                                          Doc 1 Pg 1 Ln 1.33" Pos 4.2"
{                                                            }
[BOLD]Boldface is very effective.[bold][HRt]
[HRt]
[UND]See how easy it is to underline.[und]█
```

When you turn off underline, the cursor
moves to the right of the [und] code.

Remain in Reveal Codes and keep these sentences on the screen for use in the
next exercise.

Deleting Boldface and Underline

Suppose you've underlined parts of your document and now you want to return
the text to normal. You can delete the underlining by deleting the hidden codes.

Let's delete both the boldface and underline codes that you inserted in the last
exercise. You should still be in Reveal Codes; if not, display the codes by pressing [ALT][F3].

7

■ *1*   Move the cursor to either the [UND] or
[und] code.

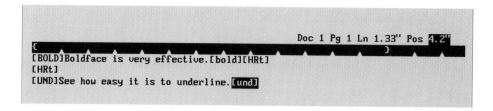

```
                                          Doc 1 Pg 1 Ln 1.33" Pos 4.2"
[                                                        }
[BOLD]Boldface is very effective.[bold][HRt]
[HRt]
[UND]See how easy it is to underline.[und]
```

■*2*   DEL   Check the status line; the Pos number is
no longer highlighted.

```
                                          Doc 1 Pg 1 Ln 1.33" Pos 4.2"
[                                                        }
[BOLD]Boldface is very effective.[bold][HRt]
[HRt]
See how easy it is to underline.█
```

Now let's delete the boldface code while not in Reveal Codes.

■*3*   Move the cursor to the *B* in *Boldface* and
select **E**dit, **R**eveal Codes or press ALT F3 to
exit Reveal Codes.

```
Boldface is very effective.

See how easy it is to underline.
```

The cursor is now located one character to the right of the hidden [BOLD] code.

■*4*　⌨ You are asked whether or not you want
to delete the [BOLD] code.

> Delete [BOLD]? No (Yes)

■*5*　Select **Y**es or press ⎡**Y**⎤.

Keep these two sentences on your screen; you'll use them again in the next exercise.

7

Boldfacing and Underlining Existing Characters

Once you have typed text on the screen, it cannot be boldfaced or underlined by simply selecting **B**oldface or **U**nderline—you must block the text first. Do the next exercise and you'll learn how to add these effects once you've typed the text.

■*1*　Move the cursor onto the letter *B* in *Boldface* and select **E**dit, **B**lock or press ⎡ALT⎤⎡F4⎤.

> **Block on**　　　　　　　　　　　　　　Doc 1 Pg 1 Ln 1" Pos ▌

You already used the Block feature in Chapter 5. (See Chapter 8 for more details on using Block.)

■*2* [END] to move the cursor to the end of the sentence and block the entire line.

> Boldface is very effective.
>
> See how easy it is to underline.

■*3* Select **F**ont, **A**ppearance, **B**oldface or press [F6] to boldface the sentence. (If you have difficulty seeing boldfaced text on your screen, try adjusting your monitor's contrast.)

■*4* Move the cursor to the *S* in the word *See*.

> Boldface is very effective.
>
> See how easy it is to underline.

■*5* Select **E**dit, **B**lock or press [ALT][F4], then press [.] (period) to highlight the sentence.

> Boldface is very effective.
>
> See how easy it is to underline.

■ *6*   Select **F**ont, **A**ppearance, **U**nderline or press
[F8] to underline the sentence.

Clear the screen by pressing [F7] [N] [N], and go to the next exercise.

Choosing an Underline Style

WordPerfect offers three kinds of underlining: *continuous* (underlines the word
and the spaces between words), *noncontinuous* (underlines only the words, not
the spaces), and *tab spaces* (underlines the blank space created by pressing [TAB]).
The default is single, continuous underlining. Let's change it to noncontinuous.
Begin at a clear screen.

■ *1*   Select **L**ayout, **O**ther or press [SHIFT][F8] [4]
or [O].

```
Format: Other

     1 - Advance

     2 - Conditional End of Page

     3 - Decimal/Align Character      .
         Thousands' Separator         ,

     4 - Language                     US

     5 - Overstrike

     6 - Printer Functions

     7 - Underline - Spaces           Yes
                     Tabs             No




Selection: 0
```

■*2* Select **U**nderline or press [**7**] or [**U**].

```
6 - Printer Functions

7 - Underline - Spaces        Yes (No)
                Tabs          No
```

■*3* Select **N**o or press [**N**] for noncontinuous,
and then select **N**o to not underline the tab
spaces.

```
6 - Printer Functions

7 - Underline - Spaces        No
                Tabs          No
```

■*4* Click the right mouse button or press [**F7**],
then select **F**ont, **A**ppearance, **U**nderline or
press [**F8**]. Type this sentence:

```
This is the non-continuous style.
```

■*5* Select **F**ile, **P**rint, **P**age or press [**SHIFT**][**F7**]
[**P**]. Your printout should look like this:

> This is the non-continuous style.

Changing between
■Uppercase and Lowercase

Sometimes it's necessary to change text you've typed with lowercase letters to uppercase, or vice versa. WordPerfect makes this a simple procedure when you use the Block feature and the Switch feature together. Begin at a clear screen.

7

■ 1   Type this text:

> Changing between uppercase and lowercase is easy._

■ 2   Select **E**dit, **B**lock or press `ALT` `F4`. `←` to the *h* in *Changing*.

> Changing between uppercase and lowercase is easy.

■ 3   Select **E**dit, **C**onvert Case, To **U**pper or press `SHIFT` `F3` `1` or `U`. (The Switch feature changes the editing screen between Doc 1 and Doc 2 when Block is off.)

> CHANGING BETWEEN UPPERCASE AND LOWERCASE IS EASY.

To change uppercase text back to lowercase, you simply block the text and select Edit, Convert Case, To Lower or press [SHIFT][F3] [2] or [L].

Changing the Size and Appearance of Your Text with Fonts

One of the advantages of producing documents with a word processor is the ease with which you can emphasize parts of your text, making it more interesting. In the following exercises, you'll learn how to use fonts to make characters smaller or larger than the normal, default size.

Changing Base Fonts

A *font* is a complete set of letters, numbers, and punctuation marks in a particular typeface and size. For example, the typeface used in the main text of this book is 11-point Times Roman. When you install your printer, WordPerfect automatically selects the default *base font*. The base font determines the size and style of the basic text. It also determines the fonts WordPerfect uses when you select options from the Size and Appearances menus.

You can select another base font to change the typeface and size, as long as your printer can handle different fonts. (Refer to your printer manual if you're not sure about your printer's capabilities.)

At a blank screen, try the following exercise. If you have an HP LaserJet Series II printer, you'll change the default base font to Line Printer. If you're using another printer, you may select any font your printer supports.

■ *1*   Select Font, Base Font or press `CTRL` `F8` `4`
        or `F`.

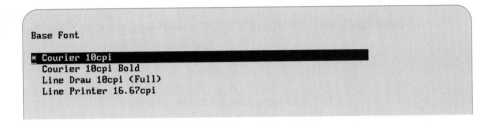

WordPerfect automatically supports these fonts on the HP LaserJet Series II
printer.

■ *2*   Place the highlight on *Line Printer 16.67cpi*
        (characters per inch) or on any font your
        printer supports.

```
Base Font

 × Courier 10cpi
   Courier 10cpi Bold
   Line Draw 10cpi (Full)
   Line Printer 16.67cpi
```

7

Choosing a lower
cpi font doesn't
allow as many
characters on
a line.

■*3* Select **S**elect or press 1 or S. Then select
Edit, **R**eveal Codes or press ALT F3.

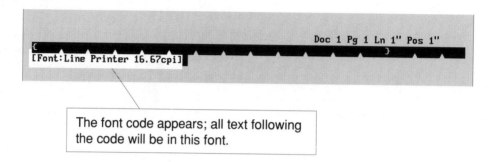

The font code appears; all text following
the code will be in this font.

The font code tells you how many characters per inch you may type; *16.67cpi*
indicates you can type 16.67 characters within every inch. The default font,
Courier 10cpi, prints 10 characters per inch. Whenever you choose a different
font size, the number of characters per inch changes and the Pos indicator
reflects the change.

■*4* Select **E**dit, **R**eveal Codes or press ALT F3 to
exit Reveal Codes.

Selecting Fonts

You may want to purchase additional fonts to expand your font selection.
Depending on which type of printer you have, you can use interchangeable font
cartridges, downloadable (soft) fonts, or a different print wheel. (*Downloadable*
fonts are stored on a disk and are not resident in the computer, which is why they
are also called *soft* fonts.) Refer to your printer manual for a list of available
cartridges and soft fonts. Before you can use soft fonts, you must install them.
Instructions for installation come with the font package.

Clear your screen; let's select a few fonts. For these exercises, I used the HP
LaserJet Series II printer and a series of Times Roman soft fonts. You may use
any printer or fonts that you have, using the same procedures.

■ *1*   Select **File**, **Print** or press SHIFT F7.

```
Options
        S - Select Printer                      HP LaserJet Series II
        B - Binding Offset                      0"
        N - Number of Copies                    1
        U - Multiple Copies Generated by        WordPerfect
        G - Graphics Quality                    Medium
        T - Text Quality                        High
```

7

■ *2*   Select **S**elect Printer or press S.

```
1 Select; 2 Additional Printers; 3 Edit; 4 Copy; 5 Delete; 6 Help; 7 Update: 1
```

Don't confuse the terms *typeface* and *font*. The face is the style of type (like Helvetica), and the font is the assortment of type in a particular face and size.

■3 Select **E**dit or press ③ or **E**.

Before proceeding to step 4, look at option 6 on this menu—the location of your fonts is displayed to the right. If your fonts are stored someplace other than your WP subdirectory, select **D**ownloadable or press ⑥ or **D** and type in the location.

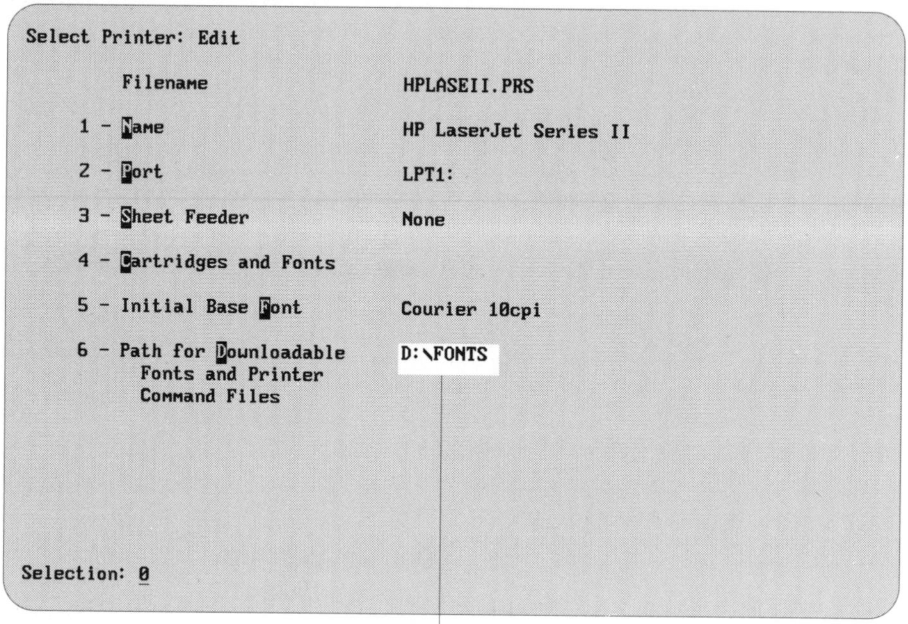

```
Select Printer: Edit

        Filename              HPLASEII.PRS

  1 - Name                    HP LaserJet Series II

  2 - Port                    LPT1:

  3 - Sheet Feeder            None

  4 - Cartridges and Fonts

  5 - Initial Base Font       Courier 10cpi

  6 - Path for Downloadable   D:\FONTS
      Fonts and Printer
      Command Files




Selection: 0
```

My fonts are on drive D in the FONTS subdirectory, so I typed in **D:\FONTS**.

■ *4*   Select Cartridges and Fonts or press ④
        or Ⓒ.

```
Select Printer: Cartridges and Fonts

Font Category                          Quantity        Available

Built-In
Cartridges                                  2               2
Soft Fonts                                350 K           350 K




NOTE: Most items listed under the Font Category (with the exception of Built-In)
are optional and must be purchased separately from your dealer or manufacturer.

In order to print soft fonts marked '*', you must run the Initialize Printer
option in WP each time you turn your printer on.

If soft fonts are not located in the same directory as your printer files, you
must specify a Path for Downloadable Fonts in the Select Printer: Edit menu.

1 Select: 2 Change Quantity: N Name search: 1
```

Select **Soft Fonts** to display a list of soft fonts supported by your printer.

Select **Built-In** to display a list of the fonts built into your printer. You do not have to select these fonts.

Select **Change Quantity** to change the amount of printer memory allocated for downloadable fonts.

Select **Cartridges** to display a list of font cartridges. (If you have a cartridge font, press ↵, move the cursor to the name of the cartridge you want to use, and mark it with an asterisk. Press F7 five times to save the selection and update the fonts.)

■5   Highlight the Soft Fonts menu option. Select
Select by pressing [1] or [S] or press [←]. (If
a Font Groups list is not displayed, move on
to step 7.)

```
Select Printer: Soft Fonts

Font Groups:

HP AC TmsRmn/Helv US
HP AD TmsRmn/Helv R8
HP AE TmsRmn/Helv US
HP AF TmsRmn/Helv R8
HP AG Helv Headlines PC-8
```

■6   Highlight the font group you want to use and
press [←].

```
Select Printer: Soft Fonts                     Quantity
                                     Total:     350 K
                                 Available:     350 K

HP AC TmsRmn/Helv US                                    Quantity Used

   (AC) Helv 06pt                                             8 K
   (AC) Helv 06pt (Land)                                      8 K
   (AC) Helv 06pt Bold                                        8 K
   (AC) Helv 06pt Bold (Land)                                 8 K
   (AC) Helv 06pt Italic                                      8 K
   (AC) Helv 06pt Italic (Land)                               8 K
   (AC) Helv 08pt                                             9 K
   (AC) Helv 08pt (Land)                                      9 K
   (AC) Helv 08pt Bold                                       11 K
   (AC) Helv 08pt Bold (Land)                                11 K
   (AC) Helv 08pt Italic                                     10 K
   (AC) Helv 08pt Italic (Land)                              10 K
   (AC) Helv 10pt                                            13 K
   (AC) Helv 10pt (Land)                                     13 K
   (AC) Helv 10pt Bold                                       13 K
   (AC) Helv 10pt Bold (Land)                                13 K

Mark:  * Present when print job begins          Press Exit to save
       + Can be loaded/unloaded during job     Press Cancel to cancel
```

> **Note** WordPerfect provides a font list based on the printer you selected, whether or not you've installed the actual fonts. So, if you select a font from the menu and it doesn't print, it may mean the font hasn't been installed. Check closely to see which fonts are actually included in your soft-font package.

■7 Place the cursor on each font you want to use and mark it with an asterisk or a plus sign. Select six fonts in ascending order. (I chose 6-, 8-, 10-, 12-, 14-, and 18-point fonts; advance to the next screen to see more sizes, if they are available.) You will use these fonts in the following exercises.

7

```
Select Printer: Soft Fonts                    Quantity
                                       Total:   350 K
                                   Available:   350 K

 HP AC TmsRmn/Helv US                             Quantity Used

 + (AC) Tms Rmn 06pt                                  8 K
   (AC) Tms Rmn 06pt (Land)                           8 K
   (AC) Tms Rmn 06pt Bold                             8 K
   (AC) Tms Rmn 06pt Bold (Land)                      8 K
   (AC) Tms Rmn 06pt Italic                           8 K
   (AC) Tms Rmn 06pt Italic (Land)                    8 K
 + (AC) Tms Rmn 08pt                                  9 K
   (AC) Tms Rmn 08pt (Land)                           9 K
   (AC) Tms Rmn 08pt Bold                            10 K
   (AC) Tms Rmn 08pt Bold (Land)                     10 K
   (AC) Tms Rmn 08pt Italic                          10 K
   (AC) Tms Rmn 08pt Italic (Land)                   10 K
 + (AC) Tms Rmn 10pt                                 12 K
   (AC) Tms Rmn 10pt (Land)                          12 K
   (AC) Tms Rmn 10pt Bold                            13 K
   (AC) Tms Rmn 10pt Bold (Land)                     13 K

 Mark: * Present when print job begins       Press Exit to save
       + Can be loaded/unloaded during job  Press Cancel to cancel
```

All fonts that you mark with an asterisk will be loaded into the printer when you give a print command. Be careful—doing this can sometimes cause you to exceed the amount of memory available in the printer or the quantity set in WordPerfect, and the fonts may not print. Fonts marked with a plus sign will be loaded *only* when they are needed for the print job. (I recommend using a plus sign if you use more than one or two fonts in a job.)

■*8* [F7] five times to update the fonts.

Changing the Size of a Base Font

If your printer supports different font sizes, you can change these sizes using one of two methods. One method is to select a new base font for each change, as you learned in the previous exercise. The other method is to use one of the Size options in the Font feature. This allows you to select a different size of type for a specific section of text and to return to the normal base font when you are finished. Your computer must support several font sizes for you to take advantage of this feature.

Turn your printer on, and clear your document screen by pressing [F7] [N] [N]. For this exercise, I selected 10-point Times Roman as my base font. You may use whatever fonts you have available. Select your base font with the Font command by pressing [CTRL] [F8] [F] before doing the exercise.

■*1* Select Font or press [CTRL] [F8] [1] or [S] to display one of the two Size menus provided by WordPerfect.

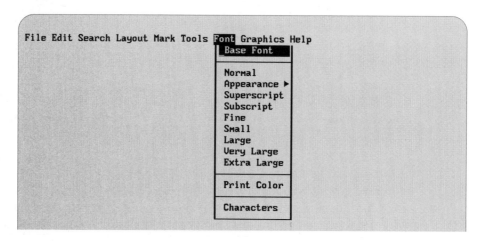

or

```
1 Su rscpt; 2 Su scpt; 3  ine; 4  mall; 5  arge; 6  ry Large; 7  xt Large: 0
```

Each of the Size menus offers *superscript* and *subscript* options. Subscript is used for numbers in formulas like H_2O. Superscript is used for exponential numbers in mathematical terms like x^2.

Also note the five other options ranging from Fine to Extra Large. The point sizes that you get when you choose these options depend on the sizes you designate in the "Select Printer: Cartridges and Fonts" menu and on the base font.

For example, in the previous exercise I chose font sizes of 6-, 8-, 10-, 12-, 14-, and 18-point. The *Extra Large* size changes the 10-point base font to 18-point size, *Very Large* changes it to 14-point size, and *Large* changes it to 12-point size. The *Small* size changes the font to 8-point size, and *Fine* changes it to 6-point size. These changes depend on the number of fonts available, your base font, and whether the fonts have been selected properly.

■2   Select **Large** or press 5 or L This produces 12-point type (based on the range of sizes you chose in the last exercise).

■3   Type the following text, centering it on the page by selecting **Layout**, **Align**, **Center** or by pressing SHIFT F6 :

```
                    Letter-Write Publications
```

■ *4*   Select Font, **N**ormal, or press `CTRL` `F8` `3` or
       `N`, or press `→`. The Large Size is turned
       off, and the font is returned to normal size—
       10-point Times Roman.

■ *5*   `⏎` twice and type the following text, center-
       ing each line:

```
                   Letter-Write Publications
                       234 Hill Street
                   Atlanta, Georgia 00000_
```

■ *6*   `⏎` twice. Select Font, **S**mall or press
       `CTRL` `F8`, `1` or `S`, `4` or `S`. This
       produces 8-point type.

■ *7*   Type the following text, centering it below
       the other text:

```
                   Letter-Write Publications
                       234 Hill Street
                   Atlanta, Georgia 00000
                       111-222-3333_
```

■*8*   Select **F**ont, **N**ormal, or press `CTRL` `F8` `3` or
`N`, or press `→`. The Small size is turned
off, and the font is returned to normal size.

■*9*   Select **F**ile, **P**rint, **P**age or press `SHIFT` `F7` `2`
or `P`. Your printout should look like this:

Letter-Write Publications

234 Hill Street

Atlanta, Georgia 00000

111-222-3333

7

Changing the Size of Existing Characters

Once a character is typed on the screen, its size and appearance cannot be altered until you mark the character using the Block command. Using the text from the last exercise, change the size of the company name.

■ *1* Move the cursor to the beginning of the company name.

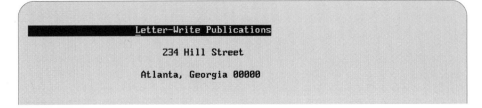

■ *2* Select **Edit**, **B**lock or press `ALT` `F4`, then press `END`.

■ *3* Select Font, **E**xtra Large or press `CTRL` `F8`, `1` or `S`, `7` or `E`. This produces 18-point bold type because that was my next largest font after 14-point.

■ *4* Select File, **P**rint, **P**age or press `SHIFT` `F7` `2` or `P`. Your printout should look like this:

Letter-Write Publications

234 Hill Street

Atlanta, Georgia 00000

111-222-3333

Changing the
■*Appearance of Characters*

To change the appearance of characters, you must have a printer that supports the appearance attributes WordPerfect offers. Check your printer manual for the attributes your printer supports or print a copy of the PRINTER.TST file, which is located on the WordPerfect 1 program disk or in the WP51 directory.

In this exercise, you'll learn how to select the double-underlining attribute. You can use the same steps to select other attributes. Turn on your printer, and clear your document screen by pressing F7 N N.

■.*1*   Select Font, Appearance or press CTRL F8 2 or A to display one of the two Appearance menus provided by WordPerfect.

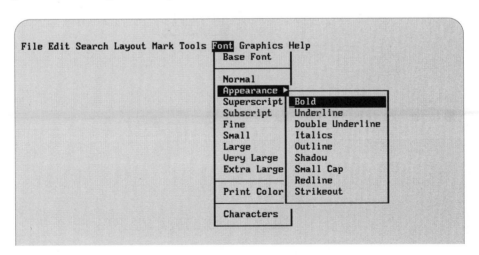

or

1 **B**old 2 **U**ndln 3 **D**bl Und 4 **I**talc 5 **O**utln 6 Sh**a**dw 7 S**m** **C**ap 8 **R**edln 9 **S**tkout: **0**

■2 Select **D**ouble Underline or press ⦗**3**⦘ or ⦗**D**⦘.

■3 Type the text on the screen below. (The text is highlighted on color monitors.)

This is double underline.

■4 Select **F**ont, **N**ormal, or press ⦗CTRL⦘⦗F8⦘ ⦗**3**⦘ or ⦗**N**⦘, or press ⦗→⦘ to turn double underlining off.

■5 Select **F**ile, **P**rint, **P**age or press ⦗SHIFT⦘⦗F7⦘ ⦗**2**⦘ or ⦗**P**⦘. Your printout should look like this:

This is double underline.

Combining Character Styles

If you wish, you can combine different character styles, such as double underlining and boldface. Clear the screen and try it.

■*1*   Select Font, Appearance, **B**old or press
[CTRL][F8], [2] or [A], [1] or [B].

■*2*   Select Font, Appearance, **D**ouble Underline
or press [CTRL][F8], [2] or [A], [3] or [D].

■*3*   Type this text:

> This is a combined style.

■*4*   Select Font, **N**ormal, or press [CTRL][F8] [3] or
[N], or press [→]. The Appearance feature is
now turned off.

■*5*   Select File, Print, **P**age or press [SHIFT][F7] [2]
or [P]. Your printout should look like this:

> ### This is a combined style.

7

You can boldface and underline characters at the same time by pressing both [F6] and [F8] before you type the text. If you want to do this to existing text, you have to block the text by pressing [ALT][F4], press [F6], block the text a second time, and press [F8]. Experiment with your own text, if you wish. When you're finished, clear the screen by pressing [F7] [N] [N].

This chapter is only an introduction to how you can use WordPerfect's many features to add interest and professional-looking effects to your printed pieces. Soon you'll begin using the graphics capabilities of WordPerfect 5.1 and learn the many ways you can illustrate documents.

8

Expanding Your
Editing Power with
the Block Feature

■ This chapter discusses one of the most powerful and impressive editing features of WordPerfect 5.1—block operations. You have already learned some simple block operations; now you'll learn these others:

Deleting, moving, and copying blocks

Moving blocks between documents

Printing blocks

Saving blocks

Appending blocks

Deleting, Moving, and Copying Blocks

A *block* is a designated group of one or more characters, including their formatting. Whenever you need to delete, move, or copy information in a document, you can use this feature. Let's do some exercises to see how blocking works.

Deleting Blocks

Let's delete a block using the Block command. First, you'll create a document that you will use in each exercise of this chapter.

■ *1* Start WordPerfect, and at a clear screen, type
these paragraphs:

```
In the United States there are four types of white flour which are
most commonly used in making breads:

White Flour - prepared by grinding and sifting cleaned wheat; not
amber durum or red durum.

Enriched Flour - white flour enriched by adding thiamine,
riboflavin, niacin and iron.

Bromated Flour - potassium bromate is added to white flour in order
to improve the baking qualities of the dough.

Whole Wheat - a more coarsely ground clean wheat: not red or amber
durum.

Self-rising Flour - a fine mixture of flour and baking powder (used
as a substitute for yeast).



                                              Doc 1 Pg 1 Ln 3.67" Pos 3.7"
```

■*2*   Move the cursor to the *W* in *Whole*.

■*3*   Select **E**dit, **B**lock or press ALT F4 to turn on
the Block feature.

■*4*   Move the cursor to the *S* in *Self-rising*.
(If you highlight beyond the S, simply move
back to the S and the highlighting disappears.)

```
Whole Wheat - a more coarsely ground clean wheat: not red or amber
durum.

Self-rising Flour - a fine mixture of flour and baking powder (used
as a substitute for yeast).
```

While in the Block mode, you can use the mouse or any of the cursor key movements to highlight desired text. Refer to the inside of the book cover for a list of
cursor key movements.

To block text with the mouse, place the cursor at the location where you want
to start blocking, press and hold down the left mouse button, and drag the cursor
across the text you want to block. After you block the text, do not move the mouse
before you click the right mouse button to activate the main menu. If you do, you'll
highlight the wrong text.

8

■ 5   Select **E**dit, **D**elete or press DEL . You are
        asked if you want to delete the highlighted
        text, including the [Hrt] codes after the
        sentence.

```
Delete Block? No (Yes)
```

■ 6   Select **Y**es or press Y to delete the block.

```
Bromated Flour - potassium bromate is added to white flour in order
to improve the baking qualities of the dough.

Self-rising Flour - a fine mixture of flour and baking powder (used
as a substitute for yeast).
```

Keep this document on the screen for the next exercise.

Moving Blocks

Let's move a block of text from the previous exercise using the Block and Move
commands.

■ 1   Move the cursor to the *B* in *Bromated Flour,*
        select **E**dit, **B**lock or press ALT F4 , and high-
        light the paragraph.

```
Bromated Flour - potassium bromate is added to white flour in order
to improve the baking qualities of the dough.

Self-rising Flour - a fine mixture of flour and baking powder (used
as a substitute for yeast).
```

Be sure to move the cursor to the *S* in *Self-rising* so that you can move the [Hrt] (hard return) codes that separate the paragraphs.

■*2*   Select **E**dit, **M**ove (Cut) or press `CTRL` `F4`,
`1` or `B`, `1` or `M`. The block disappears.

```
Move cursor; press Enter to retrieve.              Doc 1 Pg 1 Ln 2.5" Pos 1"
```

■*3*   Move the cursor to the *E* in *Enriched Flour.*

```
Enriched Flour - white flour enriched by adding thiamine,
riboflavin, niacin and iron.

Self-rising Flour - a fine mixture of flour and baking powder (used
as a substitute for yeast).
```

The blocked text will be moved to the cursor location.

■*4*   `↵` or double-click the left mouse button.
The *Bromated Flour* paragraph is inserted.

```
White Flour - prepared by grinding and sifting cleaned wheat; not
amber durum or red durum.

Bromated Flour - potassium bromate is added to white flour in order
to improve the baking qualities of the dough.

Enriched Flour - white flour enriched by adding thiamine,
riboflavin, niacin and iron.
```

8

Keep this document on the screen for the next exercise.

Copying Blocks

If you want to duplicate a section of text and insert it somewhere else in your document, you do so by making a copy of it. Try it using the text from the previous exercise.

■ *1* Move the cursor to the *w* in *white* in the first paragraph.

■ *2* Select **E**dit, **B**lock or press `ALT` `F4`. Move the cursor to the *f* in *flour* to include the space after *white*.

> In the United States there are four types of ▮white▮ flour which are most commonly used in making breads:
>
> White Flour - prepared by grinding and sifting cleaned wheat: not amber durum or red durum.

■ *3* Select **E**dit, **C**opy or press `CTRL` `F4`, `1` or `B`, `2` or `C`.

■ *4* Move the cursor to the *f* in *flour* in the next-to-last line.

> Enriched Flour - white flour enriched by adding thiamine, riboflavin, niacin and iron.
>
> Self-rising Flour - a fine mixture of ▮flour and baking powder (used as a substitute for yeast).

■*5* ⏎ or double-click the left mouse button.

```
Enriched Flour - white flour enriched by adding thiamine,
riboflavin, niacin and iron.

Self-rising Flour - a fine mixture of █hite flour and baking powder
(used as a substitute for yeast).
```

Keep this document on the screen for the next exercise.

Moving and Copying Blocks for Future Use

Whenever you copy or move a block of text, the text is stored in a retrieval area. Usually you will copy or move a block immediately, but there may be times when you want to store the text in the retrieval area and use it later in the editing process.

8

■*1* Select **Edit**, **B**lock or press ⒜⒧⒯ F4 and high-light *white flour* in the *Self-rising Flour* paragraph. Be sure to include the space after *flour*.

```
Enriched Flour - white flour enriched by adding thiamine,
riboflavin, niacin and iron.

Self-rising Flour - a fine mixture of ▐white flour▌and baking powder
(used as a substitute for yeast).
```

■*2* Select **Edit**, **M**ove or press CTRL F4, 1 or
B, 1 or M

■*3* **F1** (Cancel) or click the middle button on a three-button mouse or both buttons on a two-button mouse. The prompt "Move the cursor: press Enter to retrieve" disappears from the bottom of the screen. You are now free to edit. The block is stored in a retrieval area.

The retrieval area can hold only one block at a time. After you move or copy a block into the retrieval area, it remains there until you exit WordPerfect or move or copy another block.

■*4* Select **E**dit, **P**aste or press **CTRL** **F4**, **4** or **R**. (Make sure the cursor is still on the *a* in *and*.)

```
Retrieve: 1 Block; 2 Tabular Column; 3 Rectangle: 0
```

■*5* Select **B**lock or press **1** or **B** to insert the blocked text stored in the retrieval area.

```
Enriched Flour - white flour enriched by adding thiamine,
riboflavin, niacin and iron.

Self-rising Flour - a fine mixture of white flour and baking powder
(used as a substitute for yeast).
```

You can insert the stored block as many times as you want by following steps 4 and 5 above.

Keep this document on the screen for the next exercise.

Moving and Copying
■*Text between Documents*

The Block feature can also be used to move or copy text from one document to another. After you block text and activate the Move or Copy feature, you can immediately place the text at the cursor location by pressing ↵. Or, you can press F1 and, when ready, retrieve the text at a future time by selecting **E**dit, **P**aste.

In this exercise, let's immediately copy some text from the last exercise from Doc 1 to Doc 2.

8

■*1* Block the following paragraphs, which are in the Doc 1 window:

```
Bromated Flour - potassium bromate is added to white flour in order
to improve the baking qualities of the dough.

Enriched Flour - white flour enriched by adding thiamine,
riboflavin, niacin and iron.

Self-rising Flour - a fine mixture of white flour and baking powder
```

■*2* Select **E**dit, **C**opy or press CTRL F4,
1 or **B**, **2** or **C**.

■*3*   Select **E**dit, **S**witch Document or press
        SHIFT F3 to switch to the Doc 2 window.

Move cursor; press Enter to retrieve.   Doc 2 Pg 1 Ln 1" Pos 1"

In this case, the Doc 2 window is blank. At this point you could
also press F1 to store the blocked text temporarily, retrieve a
saved document into Doc 2, and then retrieve the blocked text
into the currently displayed document in Doc 2.

■*4*   Press ⏎ or double-click the left mouse but-
        ton. The selected text is copied into Doc 2.

Bromated Flour - potassium bromate is added to white flour in order
to improve the baking qualities of the dough.

Enriched Flour - white flour enriched by adding thiamine,
riboflavin, niacin and iron.

■*5*   Select **F**ile, E**x**it, **N**o, **Y**es or press F7 N
        Y to exit Doc 2 without saving the file.

Keep this document on the screen for the next exercise.

■*Printing Blocks*

For those times when you need to print a portion of a document, WordPerfect allows you to print a block of text. Let's print part of the text from the last exercise. Make sure your printer is turned on and ready to go.

■ *1*   Select **Edit, B**lock or press `ALT` `F4`, and highlight these paragraphs:

> Enriched Flour – white flour enriched by adding thiamine, riboflavin, niacin and iron.
>
> Self-rising Flour – a fine mixture of white flour and baking powder (used as a substitute for yeast).

■ *2*   Select **File, P**rint or press `SHIFT` `F7`.

> Print block? No (Yes)

■ *3*   Select **Y**es or press `Y`. The block is printed.

Headers, footers, and page numbers (if in use) will print along with the block. If you don't want them to print, save the block and print it as a separate file.

Keep this document on the screen for the next exercise.

8

■*Saving Blocks*

There may be portions of text you would like to use in another document or elsewhere in the same document. If so, you can block the text and save it as a separate file under a new name. (Do not use the name of an existing document or you will write over the original file with the blocked text.)

Using the text from the previous exercise, let's save a block of text.

■ *1*   Block the paragraphs on the screen below. Be sure to move the cursor to the *E* in *Enriched Flour.*

```
White Flour - prepared by grinding and sifting cleaned wheat; not
amber durum or red durum.

Bromated Flour - potassium bromate is added to white flour in order
to improve the baking qualities of the dough.

Enriched Flour - white flour enriched by adding thiamine,
```

■ *2*   Select **F**ile, **S**ave or press F10 and type this file name for the blocked text, **flour1**.

```
Block name: flour1
```

■ *3*   ⏎ The blocked text is now stored in the file named FLOUR1.

Keep this document on the screen for the next exercise.

Appending Blocks to
■*Existing Files*

Many writers like to have a notes file where they record spur-of-the moment ideas or text that they may want to use later. With the Block feature, you can append text to the end of an existing document file that is stored on the disk.

Use your document from the previous exercise.

8

■*1*   Block the last two paragraphs.

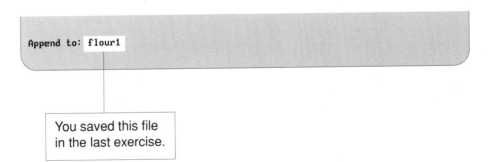

> Enriched Flour – white flour enriched by adding thiamine, riboflavin, niacin and iron.
>
> Self-rising Flour – a fine mixture of white flour and baking powder (used as a substitute for yeast).

■*2*   Select **E**dit, **A**ppend, **T**o **F**ile or press
    CTRL F4, 1 or B, 4 or A and type
    flour1.

Append to: flour1

You saved this file
in the last exercise.

■*3*   ⏎ to append the block to the FLOUR1 file.

Now retrieve the FLOUR1 file to the Doc 2 window.

■*4*   Select **E**dit, **S**witch Document or press
SHIFT F3 to change to Doc 2.

■*5*   Select **F**ile, **R**etrieve or press SHIFT F10.

Document to be retrieved: (List Files)

■*6*   Type **flour1** (the document file name) and
press ⏎. The paragraphs are appended to
the FLOUR1 file, which already contains the
first two paragraphs.

```
White Flour - prepared by grinding and sifting cleaned wheat; not
amber durum or red durum.

Bromated Flour - potassium bromate is added to white flour in order
to improve the baking qualities of the dough.

Enriched Flour - white flour enriched by adding thiamine,
riboflavin, niacin and iron.

Self-rising Flour - a fine mixture of white flour and baking powder
(used as a substitute for yeast).
```

The Block feature is a powerful editing tool that you will use for many word
processing tasks. The following chapter will show you how to create footnotes
and headers to make your documents easier to read.

9

- Using Headers and
- Footers,Footnotes
- and Endnotes

■ **M**any large documents are structurally complicated, requiring you to use special tools to organize and present information. For example, you
may want to create headers or footers that guide
the reader through a book, or generate endnotes or
footnotes that provide additional material of interest. In this chapter you'll learn how to use these
features to make your document well organized
and easy to read:

Adding headers and footers to a page

Creating footnotes and endnotes

■*Creating Headers and Footers*

A *header* is the information you find printed at the top of a page; a *footer* is the information you find printed at the bottom of a page. They are separate from the main body of text. A header can be the title or chapter title of a book, or a page number. A footer can contain the same information but usually is a page number. WordPerfect can automatically place headers and footers in your document.

Creating a Header

WordPerfect starts a header at the first usable text line (not in the top margin). The header can be as many lines long as you wish, but keep in mind that it will take up space that could be used by main document text lines.

A header *must* be created on the first line of the document page in order to appear on that page. A header does not appear on the screen, unless you view it with the View Document option by selecting **F**ile, **P**rint, **V**iew Document or pressing `SHIFT` `F7` `6` or `V`, but it does print out.

Let's create a header that will appear on the first line of every page. Start at a clear screen by pressing `F7` `N` `N`.

> Using headers and footers reduces the amount of text lines on your pages. If overused they give documents a cluttered look.

■ *1* Select **L**ayout, **P**age or press SHIFT F8
2 or **P**.

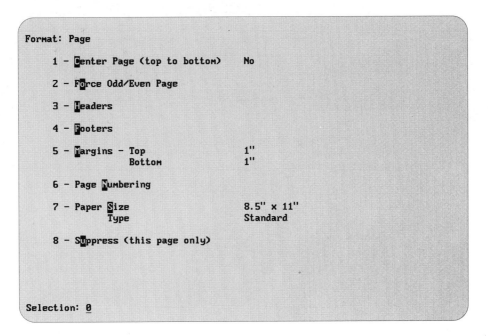

■ *2* Select **H**eaders or press **3** or **H**.

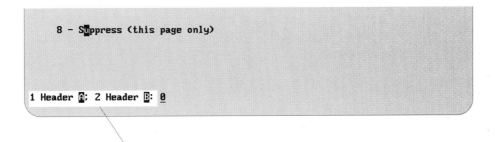

You can put two headers in place.

9

■*3* Select Header **A** or press 1 or A.

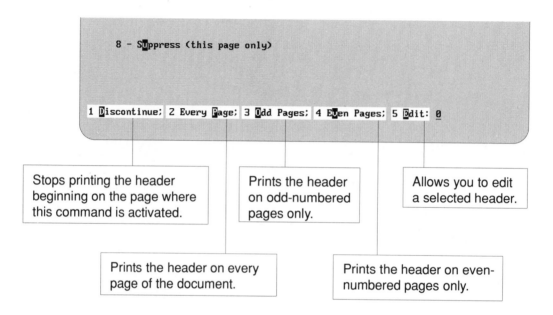

8 - S**u**ppress (this page only)

1 **D**iscontinue; 2 Every **P**age; 3 **O**dd Pages; 4 E**v**en Pages; 5 **E**dit: **0**

Stops printing the header beginning on the page where this command is activated.

Prints the header on odd-numbered pages only.

Allows you to edit a selected header.

Prints the header on every page of the document.

Prints the header on even-numbered pages only.

■*4* Select Every **P**age or press 2 or P, then type this header text on the screen: **Creating Headers and Footers**.

WordPerfect automatically inserts a blank line between the header and the main body of the text.

■5   **F7** to return to the Format: Page menu.

> 2 - F**o**rce Odd/Even Page
>
> 3 - **H**eaders   HA Every page
>
> 4 - **F**ooters

Keep this menu on the screen for the next exercise.

Creating a Footer

WordPerfect places the footer above the bottom margin on the last text line of the page. Here you'll create a footer that will contain a page number that appears on every page. You should have the Page menu displayed from the previous exercise.

9

■1   Select **F**ooters or press **4** or **F**.

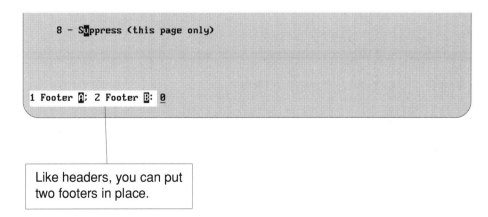

> 8 - S**u**ppress (this page only)
>
> 1 Footer **A**: 2 Footer **B**: **0**

Like headers, you can put two footers in place.

■*2* Select Footer **A** or press `1` or `A`. The
menu choices are the same as those in the
Headers menu.

```
    8 - Suppress (this page only)



1 Discontinue; 2 Every Page; 3 Odd Pages; 4 Even Pages; 5 Edit: 0
```

■*3* Select Every **P**age or press `2` or `P`.

```
Footer A:  Press Exit when done                          Ln 1" Pos 1"
```

■*4* Select **L**ayout, **A**lign, **F**lush Right or press
`ALT` `F6` to place the cursor flush-right.

■*5* Type **Page**, then press the spacebar.

```
                                                        Page _
```

■*6* `CTRL` `B`

```
                                                        Page ^B_
```

WordPerfect recognizes the ^B code as the current page number and will print the page number at that location. (^B is also used to place page numbers in headers.)

■ 7   [F7] to return to the Format: Page menu.

```
3 - Headers              HA Every page

4 - Footers              FA Every page

5 - Margins - Top        1"
            Bottom       1"
```

■ 8   Click the right mouse button or press [F7] to return to the editing screen.

9

```
                                           Doc 1 Pg 1 Ln 1.33" Pos 1"
```

> The status line indicates that the first line of text will be placed on Ln 1.33". The header and one blank line will appear at the top of each page of text.

■ 9   Select **E**dit, **R**eveal Codes or press [ALT][F3].

Both the header and footer codes are displayed.

```
                                           Doc 1 Pg 1 Ln 1.33" Pos 1"
{                                                         }
[Header A:Every page;Creating Headers and Footers][Footer A:Every page;[Flsh Rgt
]Page ^B]
```

■ *10* Select **F**ile, **P**rint, **V**iew Document or press
[SHIFT][F7] [6] or [V] to see what the header
and footer look like.

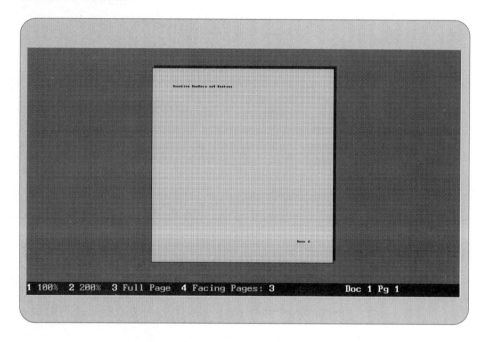

    1 100% 2 200% 3 Full Page 4 Facing Pages: 3   Doc 1 Pg 1

■ *11* Click the right mouse button or press [F7].
Then, select **E**dit, **R**eveal Codes or press
[ALT][F3] to return to the editing screen.

Keep this document on the screen for the next exercise.

Editing and Deleting
■*Headers and Footers*

WordPerfect makes it easy to change headers and footers. Let's edit the header
and delete the footer that we created in the last exercise.

■ *1* Select **L**ayout, **P**age or press `SHIFT` `F8`
       `2` or `P`.

■ *2* Select **H**eaders, Header **A** or press `3` or `H`
       `1` or `A`.

■ *3* Select **E**dit or press `5` or `E`, and press
       `END` to move the cursor to the end of the line.

> Creating Headers and Footers_

9

■ *4* Select **L**ayout, **A**lign, **F**lush Right or press
       `ALT` `F6`.

> Creating Headers and Footers   _

■ *5* Type **Page**, press the spacebar, then press
       `CTRL` `B`.

> Creating Headers and Footers   Page ^B_

■ *6* `F7` twice to return to the editing screen.

■ 7   Select **E**dit, **R**eveal Codes or press `ALT` `F3`.

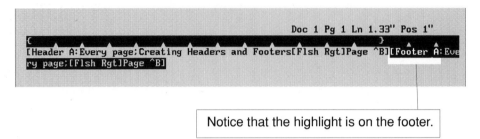

Doc 1 Pg 1 Ln 1.33" Pos 1"
{
[Header A:Every page;Creating Headers and Footers[Flsh Rgt]Page ^B][Footer A:Eve
ry page;[Flsh Rgt]Page ^B]

Notice that the highlight is on the footer.

■ 8   `DEL` to delete the footer code.

Doc 1 Pg 1 Ln 1.33" Pos 1"
{
[Header A:Every page;Creating Headers and Footers[Flsh Rgt]Page ^B]

■ 9   Select **E**dit, **R**eveal Codes or press `ALT` `F3` to
return to the editing screen.

If you wish, you can view your edited header on the View Document screen by
selecting **F**ile, **P**rint, **V**iew Document or pressing `SHIFT` `F7` `6` or `V`.

Working with Footnotes and
■*Endnotes*

Footnotes and endnotes are an integral part of many technical books and reports
and academic papers. WordPerfect simplifies the task of keeping track of numbers used to identify such notes by assigning the numbers automatically and

adjusting them when notes are added or deleted. This means that you don't have to type the numbers, just the text of the note.

If, in your editing, you move text that has been footnoted to another page, the footnote also moves to the new page. Footnotes are always tied to their source reference by a WordPerfect code.

Creating Footnotes

Let's create a paragraph with a footnote. Start at a clear screen.

■ *1* Type this paragraph:

```
The primary factor that determines the health of the economy is
"confidence." Without confidence in the nation's financial
institutions, any major downward fluctuation in the economy can
engender fear, with the next step being panic._
```

9

■ *2* Select **L**ayout, **F**ootnote, **C**reate or press <kbd>CTRL</kbd> <kbd>F7</kbd>, <kbd>1</kbd> or <kbd>F</kbd>, <kbd>1</kbd> or <kbd>C</kbd> The footnote number is displayed automatically, ready for you to enter the text of the footnote.

```
    1_
```

■*3* Press the spacebar twice, and type the following footnote text:

> 1 This is precisely why the stock market crash of 1929 was so
> ruinous. If people had had confidence in the economy's ability to
> recover, there would not have been the massive bank runs and wild
> stock selling._

■*4* [F7] to return to the editing screen.

> The primary factor that determines the health of the economy is
> "confidence." Without confidence in the nation's financial
> institutions, any major downward fluctuation in the economy can
> engender fear, with the next step being panic.**1**

> The footnote number appears in the text.

If you wish, you can view your footnote on the View Document screen—select
File, **P**rint, **V**iew Document or press [SHIFT][F7] [6] or [V].

Keep this document on the screen for the next exercise.

Editing Footnotes

Let's edit the footnote from the last exercise. You can be anywhere in your document when you do the editing, as long as you know the number of the footnote.
(However, if you use the footnote option that allows you to begin each page's footnotes with the number 1, you will have to return to that specific page to make your edits. The exercise in the section "Formatting Footnotes" explains this and other footnote options.)

■ *1*   `HOME` `HOME` `↑` to return the cursor to the beginning of the document.

■ *2*   Select **Layout, Footnote, Edit** or press `CTRL` `F7`, `1` or `F`, `2` or `E`. Type in the number of the footnote you want to edit, if it is different from the one WordPerfect displays here.

```
Footnote number? 1
```

■ *3*   `↵` to call up footnote number 1.

```
_     1  This is precisely why the stock market crash of 1929 was
so ruinous. If people had had confidence in the economy's ability
to recover, there would not have been the massive bank runs and
wild stock selling.
```

■ *4*   Move the cursor to the end of the paragraph and type **The Great Depression resulted.**

```
      1  This is precisely why the stock market crash of 1929 was so
ruinous. If people had had confidence in the economy's ability to
recover, there would not have been the massive bank runs and wild
stock selling. The Great Depression resulted._
```

■ *5*   `F7` to return to the editing screen.

Keep this document on the screen for the next exercise.

9

Formatting Footnotes

There are nine footnote options that affect the format of footnotes. The footnote format you saw in the previous exercises uses the default options. In this exercise you'll change one of the options. Experiment with the remaining options to find those that best suit your needs. Let's use the footnote from the previous exercise and change the footnote citation from a number to a letter.

■ *1*   HOME HOME ↑ to move the cursor to the beginning of the document.

■ *2*   Select **L**ayout, **F**ootnote, **O**ptions or press CTRL F7, **1** or **F**, **4** or **O**. Take a moment to read this menu.

```
Footnote Options

     1 - Spacing Within Footnotes          1
              Between Footnotes             0.167"

     2 - Amount of Note to Keep Together    0.5"

     3 - Style for Number in Text           [SUPRSCPT][Note Num][suprscpt]

     4 - Style for Number in Note                   [SUPRSCPT][Note Num][suprscpt

     5 - Footnote Numbering Method          Numbers

     6 - Start Footnote Numbers each Page   No

     7 - Line Separating Text and Footnotes 2-inch Line

     8 - Print Continued Message            No

     9 - Footnotes at Bottom of Page        Yes



Selection: 0
```

The Numbers method is currently in use.

■*3* Select Footnote Numbering **M**ethod or press
[5] or [M].

9 - Footnotes at **B**ottom of Page   Yes

1 **N**umbers; 2 **L**etters; 3 **C**haracters: **0**

Enables you to use other keyboard characters, such as asterisks
or dollar signs, instead of numbers or letters. For example, footnotes 1, 2, and 3 could be represented by *, **, and ***.

9

■*4* Select **L**etters or press [2] or [L].

5 - Footnote Numbering **M**ethod   Letters

6 - Start Footnote Numbers each **P**age   No

7 - **L**ine Separating Text and Footnotes 2-inch Line

■*5* [F7] to return to the editing screen.

■*6* [HOME] [HOME] [↓] to move the cursor to the end
of the document and activate the option
change. The citation now has the letter *a*.

The primary factor that determines the health of the economy is
"confidence." Without confidence in the nation's financial
institutions, any major downward fluctuation in the economy can
engender fear, with the next step being panic.**a**_

Keep this document on the screen for the next exercise.

Creating and Editing Endnotes

An *endnote* is a footnote placed at the end of a document. Endnotes are created and edited in the same way as footnotes, except that you select the Endnote menu options.

In this exercise, you'll add an endnote to the text created in the previous exercise.

■ *1* Move the cursor to the end of the first sentence, after *"confidence."*

■ *2* Select **L**ayout, **E**ndnote, **C**reate or press `CTRL` `F7`, `2` or `E`, `1` or `C`. Endnotes, by default, are not superscripted.

```
   1. _
```

■ *3* `TAB` to place space between the number and the endnote. Type the text of the endnote as shown below; underline *The Reality of 20th Century Economics* by pressing `F8`.

```
   1.   Diane Woodward, The Reality of 20th Century Economics (Los
Angeles, California, 1979) pp. 123-128._
```

■*4* [F7] to return to the editing screen.

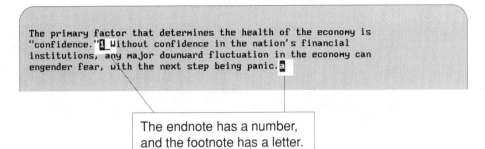

The primary factor that determines the health of the economy is "confidence."1 Without confidence in the nation's financial institutions, any major downward fluctuation in the economy can engender fear, with the next step being panic.a

The endnote has a number, and the footnote has a letter.

■*5* Select **F**ile, **P**rint, **V**iew Document or press [SHIFT] [F7] [6] or [V].

9

You only see the footnote. Endnotes are automatically placed on the page following the last page of the document.

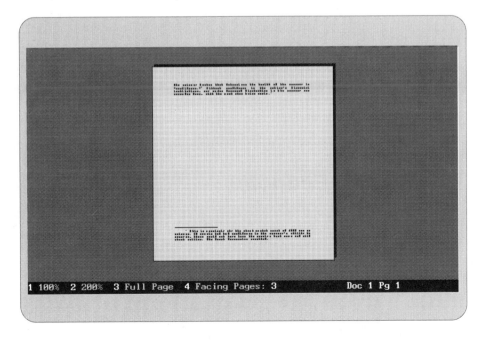

■ 6   [PG DN] to view the endnote page.

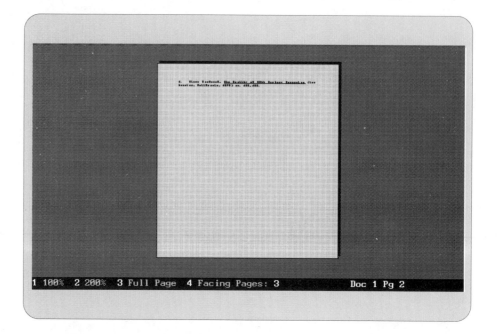

■ 7   Click the right mouse button or press [F7] to
return to the editing screen.

Keep this document on the screen for the next exercise.

Formatting Endnotes

You can change the format of endnotes, just like you can with footnotes. If you
wish to change the default format of an endnote, you have five options to choose
from. Let's look at one of these options. You should still have the text from the
previous exercise on the screen.

■ *1* Select **L**ayout, **E**ndnote, **O**ptions or press
`CTRL` `F7`, `2` or `E`, `4` or `O`.

```
Endnote Options

    1 - Spacing Within Endnotes          1
              Between Endnotes           0.167"

    2 - Amount of Endnote to Keep Together  0.5"

    3 - Style for Numbers in Text         [SUPRSCPT][Note Num][suprscpt]

    4 - Style for Numbers in Note         [Note Num].

    5 - Endnote Numbering Method          Numbers
```

The default style places a period after the note number.

9

■ *2* Select Style for Numbers in **N**ote or press
`4` or `N`.

```
Replace with: [Note Num].
```

■ *3* `END` `BKSP`, then type a closed parenthesis,).

```
Replace with: [Note Num])_
```

■*4* ⏎

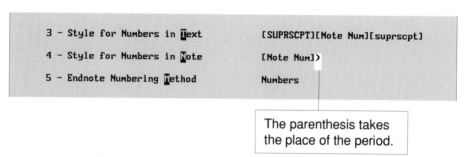

> 3 - Style for Numbers in ▊ext   [SUPRSCPT][Note Num][suprscpt]
>
> 4 - Style for Numbers in ▊ote   [Note Num]〉
>
> 5 - Endnote Numbering ▊ethod   Numbers

The parenthesis takes
the place of the period.

■*5* Click the right mouse button or press F7 to
return to the editing screen.

If you wish, you can select **File**, **Print**, **View** Document or press SHIFT F7 6 or
V, and press PG DN to see the edited endnote. (Select 100% or 200% to see the
endnote number more clearly.)

Note *Deleting Footnotes and Endnotes* To delete a footnote or endnote, place
the cursor on the number of the note in the document, then press DEL. Press Y at the
prompt that appears and the note is deleted. Notes that follow the deleted note are
automatically renumbered.

In the next chapter, you'll learn about some additional editing features, including
the Speller and the Thesaurus.

10

- *Using Document*
- *Summaries and*
- *Comments,*
- *Search and Replace,*
- *the Speller,*
- *and the Thesaurus*

■ As you have seen, WordPerfect 5.1 offers a great many features to meet your word processing needs. In this chapter you'll learn about some editing features that will make it easier for you to produce useful and accurate documents.

Creating document summaries and comments

Using Search and Replace

Using the Speller and Thesaurus

Creating Document
■*Summaries and Comments*

Document summaries and comments can be very helpful in managing and editing your documents. When you've accumulated a large number of documents, the file name alone often is not descriptive enough to identify them. To remedy this problem, you can add a brief, nonprinting summary at the beginning of your file. You can also insert nonprinting comments throughout your document that can serve as reminders or notes.

Creating a Document Summary

Sometimes it's difficult to remember what information a particular file contains, especially if it has a generic name like LETTER1. This is where a *document summary* can be very handy. A document summary is nonprinting text placed at the beginning of a file that briefly describes the contents of that document.

Let's create a document summary for MYLETTER, the document you created in Chapter 1.

> Document summaries can contain many different types of information.

■ *1*   Retrieve the file MYLETTER by selecting
File, **R**etrieve or pressing `SHIFT` `F10`.

November 14, 1990

Mr. Randall Jones
2153 Ocean View Drive
Burbank, CA 91505

Dear Randall:

Your order of June 12 was traced to Nome, Alaska. It should arrive
at your office on August 25. Apparently, an incorrect address label
was placed on the container. This will not happen again. My
apologies. Mr. Beck, our new CEO, and I look forward to seeing you
at the convention in October.

Best Regards,

Brian Knowles
Customer Service Manager
C:\WP51\MYLETTER   Doc 1 Pg 1 Ln 1" Pos 1"

10

■ *2*   Move the cursor to Ln 2.33" (midway
between the address and the greeting) and
type **RE: Misplaced Parts Shipment**.

Mr. Randall Jones
2153 Ocean View Drive
Burbank, CA 91505

RE: Misplaced Parts Shipment_

Dear Randall:

The text entered after the *RE:* entry will be used later when you create a subject
description in the document summary.

■*3* Select **L**ayout, **D**ocument or press SHIFT F8
3 or **D** (You can select **F**ill, Su**m**mary
also, which displays the Document Summary
screen shown in step 4.)

```
Format: Document

     1 - Display Pitch - Automatic Yes
                         Width     0.1"

     2 - Initial Codes

     3 - Initial Base Font          Courier 10cpi

     4 - Redline Method             Printer Dependent

     5 - Summary




Selection: 0
```

Remember, docu-
ment summaries
appear only on
the screen—they
don't print out.

■ *4*   Select **S**ummary or press ⑤ or Ⓢ. You see
a screen with seven options.

```
Document Summary

        Revision Date   09-28-90 02:54p

    1 - Creation Date   11-14-90 02:23p

    2 - Document Name
        Document Type

    3 - Author
        Typist

    4 - Subject

    5 - Account

    6 - Keywords

    7 - Abstract




Selection: 0                    (Retrieve to capture; Del to remove summary)
```

This prompt line offers the options of
automatically creating document summary fields or deleting existing fields.
(See the Subject and Abstract options
listed below for a description of summary fields.)

10

The following list explains the options on the Document Summary screen:

■ Revision Date —Last date the document was edited (based on the
  computer's current date and time).

■ Creation Date—Date the summary was created.

- Document Name—Up to 68 characters (see Chapter 14 for a description of Long Document Names).
- Document Type—Up to 20 characters.
- Author/Typist—Author's or typist's name.
- Subject—Manually type in a document subject (up to 160 characters), or automatically enter a document subject (up to 39 characters) when you capture the document summary field (the text entered after an RE: entry).
- Account—Account number or information that will help identify the document (up to 160 characters).
- Keywords—Help locate the file when using the Find option in List Files (up to 160 characters).
- Abstract—Manually type in a document description (up to 780 characters), or automatically enter a document description when you capture the document summary field (the first 400 characters of the document).

Let's enter a Subject and Abstract description in the document summary by capturing the document summary fields.

■*5* Select Retrieve or press SHIFT F10 as noted in the prompt line to automatically capture the document summary fields.

```
   7 - Abstract




Capture Document Summary Fields? No (Yes)
```

■ *6*   Select **Y**es or press **Y**.

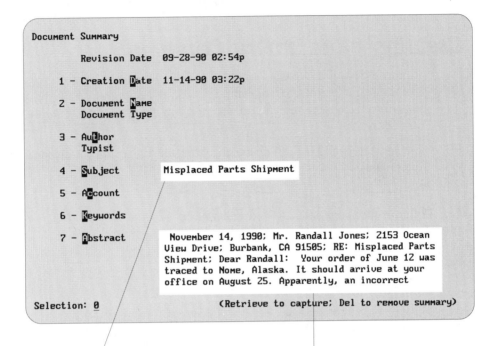

```
Document Summary

         Revision Date   09-28-90 02:54p

     1 - Creation Date   11-14-90 03:22p

     2 - Document Name
         Document Type

     3 - Author
         Typist

     4 - Subject         Misplaced Parts Shipment

     5 - Account

     6 - Keywords

     7 - Abstract            November 14, 1990; Mr. Randall Jones; 2153 Ocean
                          View Drive; Burbank, CA 91505; RE: Misplaced Parts
                          Shipment; Dear Randall:  Your order of June 12 was
                          traced to Nome, Alaska. It should arrive at your
                          office on August 25. Apparently, an incorrect

  Selection: 0                    (Retrieve to capture; Del to remove summary)
```

The Subject summary field displays the text following the RE: entry that appears within the first 400 characters of the document.

The Abstract summary field displays the first 400 characters of the document.

You can edit the captured Subject and Abstract descriptions by selecting the relevant menu option and using WordPerfect's editing tools.

■ *7*   Click the right mouse button or press **F7** to return to the editing screen.

Keep this document on the screen for use in the next exercise.

The document summary will not print and does not appear on the screen. To see it, you must save the document, select **F**ile, **L**ist Files or press F5, then select **L**ook or press 6 or L. (You can also perform steps 3 and 4 above.)

Creating a Document Comment

Comments can be created anywhere in a document and refer to specific portions of the text. Comments, like summaries, do not print out. However, you can see them on the editing screen. You can insert as many comments as you like within a document.

Using the MYLETTER document from the last exercise, create a document comment.

■ *1*   Move the cursor after the word *October.*

■ *2*   Select **E**dit, **C**omment, **C**reate or press
CTRL F5, 4 or C, 1 or C.

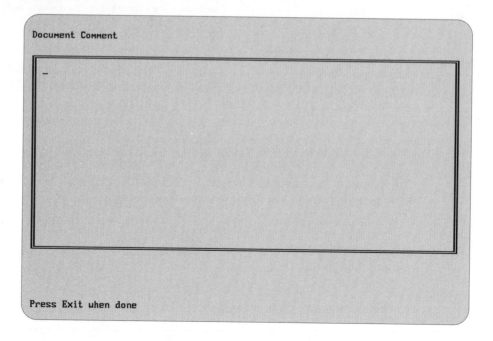

■*3*   Type the following comment:

Document Comment

Call airlines and reserve two first class seats for Mr. Beck._   _

■*4*   Click the right mouse button or press F7 to
return to the editing screen.

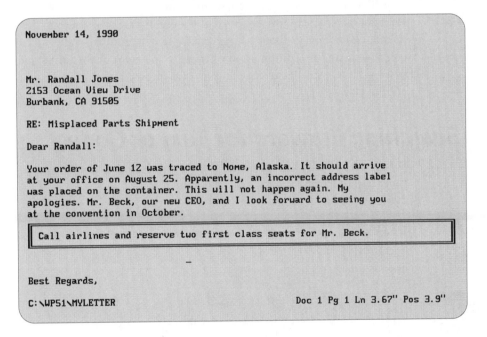

November 14, 1990

Mr. Randall Jones
2153 Ocean View Drive
Burbank, CA 91505

RE: Misplaced Parts Shipment

Dear Randall:

Your order of June 12 was traced to Nome, Alaska. It should arrive
at your office on August 25. Apparently, an incorrect address label
was placed on the container. This will not happen again. My
apologies. Mr. Beck, our new CEO, and I look forward to seeing you
at the convention in October.

Call airlines and reserve two first class seats for Mr. Beck.

                                  _

Best Regards,

C:\WP51\MYLETTER   Doc 1 Pg 1 Ln 3.67" Pos 3.9"

10

Keep this document on the screen for use in the next exercise.

> **Note** *Editing and Deleting Comments* To edit a comment, select **Edit**, Com-
> ment, **Edit** or press [CTRL][F5], [4] or [C], [2] or [E]. To delete a comment, select **Edit**,
> **Reveal Codes** or press [ALT][F3] and delete the [Comment] code.

Using Search to
■Find and Replace Data

WordPerfect allows you to search forward or backward through any document
for a character (including spaces and punctuation marks), number, word, phrase,
or hidden code and, if necessary, replace it with another code or character. You
can search for any combination of these items from one to fifty-nine characters
in length. As you'll see, this feature can be invaluable in speeding up certain
editing tasks.

Searching Forward for Text or Codes

Using the MYLETTER document from the previous exercise, let's search for
some text with the forward Search command.

■ *1*   [HOME] [HOME] [↑] to return the cursor to the
beginning of the document.

■ *2*   Select **S**earch, **F**orward or press [F2]. The ->
in the lower-left corner of the screen indicates
a forward search.

```
Best Regards,

-> Srch: _
```

■*3* Type **Randall**, but do not press ⏎. (See the
Note in this section for rules about typing
search text.)

```
Best Regards,

-> Srch: Randall_
```

■*4* Click the right mouse button or press F2 to
begin the search. WordPerfect finds the first
occurrence of *Randall*.

```
Mr. Randall Jones
2153 Ocean View Drive
Burbank, CA 91505

RE: Misplaced Parts Shipment
```

10

■*5* Select Search, Next or press F2 twice to continue the search. A second *Randall* is located.
(If the search word does not exist, the message "*Not found*" appears in the lower-left
corner of the screen.)

```
RE: Misplaced Parts Shipment

Dear Randall:

Your order of June 12 was traced to Nome, Alaska. It should arrive
at your office on August 25. Apparently, an incorrect address label
was placed on the container. This will not happen again. My
```

Now let's search for the [Comment] code for the document comment you created previously. After you activate the Search feature, you must type in the appropriate command-key sequence until the code appears on the search line. You cannot use the pull-down menus to select the code. This procedure works for all hidden codes.

■ *6*   Select **S**earch, **F**orward or press `F2` and press `CTRL` `END` to delete *Randall*.

```
Best Regards,

-> Srch: _
```

■ *7*   `CTRL` `F5` and select **C**omment or press `1` or `C` to select the [Comment] code.

```
Best Regards,

-> Srch: [Comment]_
```

■ *8*   Click the right mouse button or press `F2`.
WordPerfect finds the document comment.

```
apologies. Mr. Beck, our new CEO, and I look forward to seeing you
at the convention in October.

 ┌──────────────────────────────────────────────────────────────┐
 │ Call airlines and reserve two first class seats for Mr. Beck.  │
 └──────────────────────────────────────────────────────────────┘

                                 -

Best Regards,
```

Keep this document on the screen for use in the next exercise.

Note When you enter search text, characters you type in lowercase will match both lowercase and uppercase characters. Those typed in uppercase will match *only* those in uppercase. Search words that form *part* of another word will be located during a search. For example, the word *so* will be found in *sofa* and *also*. To prevent this, type a space before and after the word you are searching for.

Searching Backward through a Document

A backward search works the same as a forward search, except that it goes from the cursor location toward the beginning of the document. Try it using the text from the previous exercise.

■ *1* Move the cursor after the word *Manager* on the last line.

■ *2* Select **S**earch, **B**ackward or press SHIFT F2.

10

```
Best Regards,


Brian Knowles
Customer Service Manager
<- Srch: [Comment]
```

Indicates a backward search. The code is still listed from the previous search, so you don't need to type in new search text.

■*3* Click the right mouse button or press F2 to
begin the backward search.

```
Dear Randall:

Your order of June 12 was traced to Nome, Alaska. It should arrive
at your office on August 25. Apparently, an incorrect address label
was placed on the container. This will not happen again. My
apologies. Mr. Beck, our new CEO, and I look forward to seeing you
at the convention in October.
┌──────────────────────────────────────────────────────────────────────┐
│ Call airlines and reserve two first class seats for Mr. Beck.          │
└──────────────────────────────────────────────────────────────────────┘

                              ‑

Best Regards,



Brian Knowles
Customer Service Manager
================================================================================
OFFICE MEMO

TO: Sales Department

C:\WP51\MYLETTER                              Doc 1 Pg 1 Ln 3.67" Pos 3.9"
```

■ *4* Select **S**earch, **N**ext or press SHIFT F2 to repeat the backward search.

Keep this document on the screen for use in the next exercise.

Replacing Text

The *replace* option of the Search and Replace feature allows you to search for every occurrence of a character, word, phrase, or hidden code and replace it with another. You can also specify whether or not you want to confirm each replacement before it is made. When you use Search and Replace, it's easy to unknowingly replace portions of words that don't need changing. For example, replacing the word *it* with *house* makes the word *itself* become *houseself.* I recommend confirming each replacement unless you're positive there will be no incorrect replacements.

Let's replace some text in the MYLETTER document from the last exercise.

■ *1* HOME HOME ↑ to move the cursor to the beginning of the document.

■ *2* Select **S**earch, **R**eplace or press ALT F2. The prompt for confirmation appears. If you choose **Y**es, the cursor stops at every occurrence of the search item, and you are asked for permission to replace it at every location.

```
Best Regards,

w/Confirm? No (Yes)
```

10

■*3*   Select **Y**es or press Y.

> Best Regards,
>
> -> Srch: [Comment]

The previous search item is still active.

■*4*   DEL to delete the current choice. Type **not** as
the search text.

> Best Regards,
>
> -> Srch: not_

■*5*   Click the right mouse button or press F2.
Type **never** as the replacement text.

> Best Regards,
>
> Replace with: never_

■ *6*   Click the right mouse button or press F2.

```
Mr. Randall Jones
2153 Ocean View Drive
Burbank, CA 91505

RE: Misplaced Parts Shipment

Dear Randall:

Your order of June 12 was traced to Nome, Alaska. It should arrive
at your office on August 25. Apparently, an incorrect address label
was placed on the container. This will not happen again. My
apologies. Mr. Beck, our new CEO, and I look forward to seeing you
at the convention in October.

┌─────────────────────────────────────────────────────────────────┐
│  Call airlines and reserve two first class seats for Mr. Beck.    │
└─────────────────────────────────────────────────────────────────┘


Best Regards,


Brian Knowles
Confirm? No (Yes)
```

You are prompted to confirm the replacement.

10

■ *7*   Select **Y**es or press **Y** to replace *not* with
never.

```
Your order of June 12 was traced to Nome, Alaska. It should arrive
at your office on August 25. Apparently, an incorrect address label
was placed on the container. This will never happen again. My
apologies. Mr. Beck, our new CEO, and I look forward to seeing you
at the convention in October.
```

Using the Speller
■*and the Thesaurus*

Have you ever found yourself using the same words over and over again in the same document? And how about spelling? Are you always accurate? I know I'm not, and I also make lots of typos when I type. WordPerfect provides a spelling checker to help you avoid embarrassing misspellings and a thesaurus to guide you to just the right word.

Spell-Checking a Document

If you're using a dual floppy-disk system, the Speller dictionary is too large to fit on your WordPerfect program disk, so you'll have to do a little disk-swapping when you want to spell-check a document. The process is a little simpler if you have a hard disk.

Start WordPerfect and make sure you're at the blank editing screen. If you have a dual floppy-disk system, you should place a document disk in drive B.

■ *1*   Type this sentence (including the mistakes):

```
Using the spell cheker (Ctrl-F2) is the best thingthat ever
happened to me or my my fellow office workers._
```

If you have floppy disks, follow steps 2–4. If you have a hard disk, go to step 5.

■*2*   Remove the document disk from drive B and
replace it with the Speller disk. Select **T**ools,
Sp**e**ll or press `CTRL` `F2` to begin spell-checking.

```
WP{WP}US.LEX not found: 1 Enter Path; 2 Skip Language; 3 Exit Spell: 3
```

Allows you to skip spell-checking and
check the total number of words in the text.

■*3*   Select Enter **P**ath or press `1` or `P` and
type **B:**.

```
Temporary dictionary path: B:\_
```

10

■*4*   `↵` You see the Speller menu. Read the ex-
planation of this menu in step 5, then go to
step 6.

```
Check: 1 Word; 2 Page; 3 Document; 4 New Sup. Dictionary; 5 Look Up; 6 Count: 0
```

■*5*   Select **T**ools, Sp**e**ll or press `CTRL` `F2`. You
see the Speller menu.

```
Check: 1 Word; 2 Page; 3 Document; 4 New Sup. Dictionary; 5 Look Up; 6 Count: 0
```

The functions of the Speller menu options are described below.

- Word—Spell-checks the word that the cursor is currently located on.
- Page—Spell-checks one page of a document.
- Document—Spell-checks the entire document.
- New Sup. Dictionary—See the section "Creating a Supplemental Dictionary."
- Look Up—Looks up the spelling of a word prior to entering the word.
- Count—Tells you the total number of words in your document.

■*6*   Select **D**ocument or press ⟨**3**⟩ or ⟨**D**⟩. Word-Perfect highlights the unrecognized word and offers a list of alternatives. The menu provides six options for dealing with the word, described below.

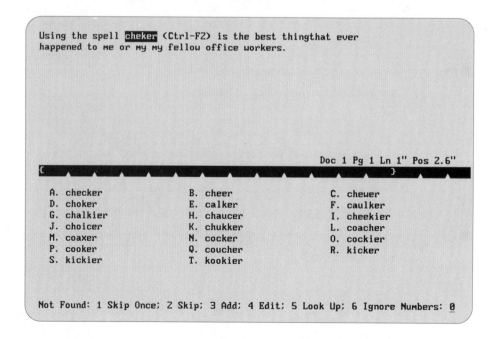

- Skip Once—Leaves the word spelled "as is" this time only. Repeats of this spelling will be highlighted.
- Skip—Accepts all occurrences of this spelling in the entire document.
- Add—Places the word in the supplemental dictionary. (For correctly spelled words not in WordPerfect's main dictionary, see "Creating a Supplemental Dictionary.")
- Edit—Allows you to edit the word. Use ⬅ and ➡ to position the cursor before editing. Press F7 to return to spell-checking.
- Look Up—Looks up the correct spelling prior to entering the word. See "Using Wildcards" later in this chapter.
- Ignore Numbers—Skips words containing numbers.

Note You can press the letter located in front of any correctly spelled word displayed on the lower portion of your spell-check screen and that word will automatically replace the incorrectly spelled word.

10

■ 7 Press **A** to replace *cheker* with *checker* from the list of alternatives. The cursor immediately advances to the next unrecognizable word.

```
Using the spell checker (Ctrl-F2) is the best thingthat ever
happened to me or my my fellow office workers.
```

■ 8 Select Skip Once or press **1**, or select Skip or press **2**, or select Ignore Numbers or press **6** to bypass this spelling since it is correct.

```
Using the spell checker (Ctrl-F2) is the best thingthat ever
happened to me or my my fellow office workers.
```

■ *9*   Select Edit or press ⎣**4**⎦ or press ⎣**→**⎦ to edit
the text.

> Using the spell checker (Ctrl-F2) is the best thing that ever
> happened to me or my my fellow office workers.

The highlighting disappears; now you can insert a space between *thing* and *that*.

■ *10*   Click the right mouse button or press ⎣**F7**⎦ to
exit Edit mode. The spell-checker highlights
the double occurrence of the word *my* and
provides a new menu.

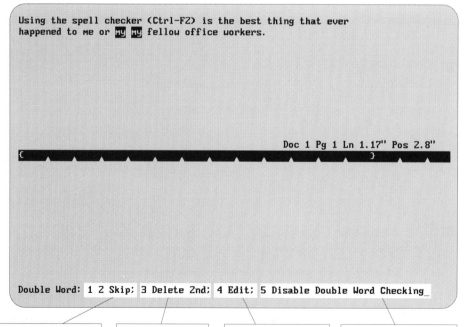

Accepts this occur-
rence because both
words are needed.

Deletes the
second word.

Allows you to
manually make
a correction.

Double words no
longer flagged.

■ *11* Select Delete 2nd or press ⟨**3**⟩ to delete the second *my.*

```
Using the spell checker (Ctrl-F2) is the best thing that ever
happened to me or my fellow office workers._
```

When spell-checking is completed, a word count appears in the lower-left corner of the screen.

■ *12* Press any key to return to the editing screen.

Creating a Supplemental Dictionary

10

The main WordPerfect dictionary contains over 100,000 words, which is substantial, but it doesn't have every word you may use in your document. This is especially true if you use technical terminology. So WordPerfect has made it possible for you to add words to a supplemental dictionary.

If the WordPerfect Speller finds a word it doesn't recognize, it will search the main dictionary first, then the supplemental dictionary.

WordPerfect provides a default supplemental dictionary to which it automatically adds new words. However, you can create another supplemental dictionary with its own name. To create a named supplemental dictionary, you type a list of words you want included and save this list under a file name of your choice. It is important to note that you can recall a named or default supplemental dictionary (WP{WP}EN.SUP) to the editing screen, add or delete words, and save the file.

Here's how to tell WordPerfect to check your named supplemental dictionary:

■ *1* Select **T**ools, **S**pell or press CTRL F2.

```
Check: 1 Word; 2 Page; 3 Document; 4 New Sup. Dictionary; 5 Look Up; 6 Count: 0
```

■ *2* Select **N**ew Sup. Dictionary or press 4
or N.

```
Supplemental dictionary name: _
```

■ *3* At the prompt "Supplemental dictionary
name:", type in a name for the supplemental
dictionary file. (Remember, a file name can
have no more than eight total characters, with
a three-character extension.)

■ *4* ⏎ The words in your document will be
compared with those in the supplemental dictionary as well as those in the main dictionary.

Note *Adding Words to the Supplemental Dictionary* When you activate your
supplemental dictionary, all unrecognizable words that you select for addition to the
dictionary are added to the supplemental dictionary.

Using Wildcards

There will be many times when you *almost* know how to spell a word, but not quite. Using *wildcards* when searching the dictionary is very helpful in this kind of situation. A wildcard is a character that stands for an undetermined character or group of characters. WordPerfect's Speller uses the asterisk (*) and question mark (?) as wildcards. The asterisk represents any number of characters; the question mark represents only one character.

This may be a little confusing, so let's try an exercise. At a clear screen, do the following:

■ *1* Type this text: **This truck was**

■ *2* Select **T**ools, Spell or press CTRL F2.

> Check: 1 **W**ord; 2 **P**age; 3 **D**ocument; 4 **N**ew Sup. Dictionary; 5 **L**ook Up; 6 **C**ount: 0

10

We want to know how to spell the word *camouflaged*. Let's look up some possible spellings using the asterisk wildcard.

■ *3* Select **L**ook Up or press **5** or **L** and type **cam*d**.

> Word or word pattern: cam*d_

There are a number of ways you can use the wildcard here. You can type *cam* or *camo*ed,* for example. The more you restrict the parameters of the search, the quicker WordPerfect is able to find the word.

■*4* ↵Choice C shows the correct spelling.

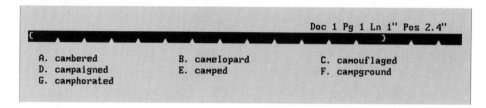

Doc 1 Pg 1 Ln 1" Pos 2.4"

```
A. cambered        B. camelopard      C. camouflaged
D. campaigned      E. camped          F. campground
G. camphorated
```

■*5* F7 twice to return to the editing screen, then
type **camouflaged**.

```
This truck was camouflaged_
```

You use the question mark as a wildcard in a similar way. For example, if you aren't sure whether *occurrence* is spelled with an *e* or an *a*, you can type **occurr?nce** and the Speller looks for all words that are similar. You can also use an asterisk and a question mark in the same search.

Using the Thesaurus

Using the WordPerfect Thesaurus is an easy way to find just the right words to express your thoughts or to avoid using the same word over and over again. The Thesaurus provides synonyms and antonyms for approximately 10,000 words.

Clear your screen; let's see how it works.

■*1* Type this text: **The fog lay like a veil**

If you have floppy disks, place the Thesaurus disk in drive B.

■*2*   Place the cursor on or at the end of the
word *veil,* and select **T**ools, **Th**esaurus or
press ALT F1. The Thesaurus screen displays
synonyms and antonyms for *veil.*

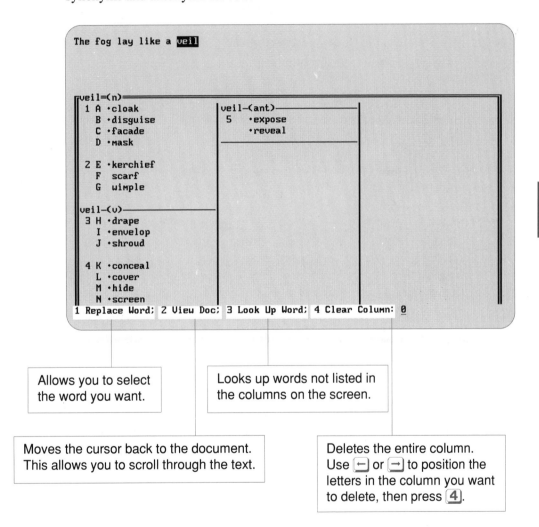

Allows you to select the word you want.

Looks up words not listed in the columns on the screen.

Moves the cursor back to the document. This allows you to scroll through the text.

Deletes the entire column. Use ← or → to position the letters in the column you want to delete, then press 4.

10

Note that the definitions are organized into parts of speech. Words with similar definitions are put into numbered subgroups.

To use the list of synonyms and antonyms for a word, press the letter key that precedes the word. Make sure the word you want is preceded by a letter. If not, use ← or → to move the letters between columns.

■ 3   [C] to select the word *facade*. A thesaurus listing for *facade* appears in column 2.

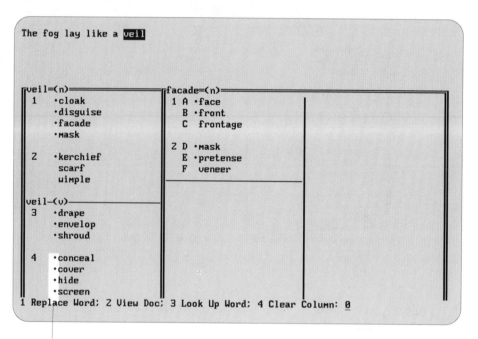

The dot in front of a word designates a *headword*. Only headwords provide listings of synonyms and antonyms.

Note that the list of antonyms that was in column 2 in the previous step has now moved to the bottom of column 1 (off-screen). Columns can be scrolled with ⬆ and ⬇, with ⊝ and ⊕ on the numeric keypad, and with ᴾᴳ ᵁᴾ and ᴾᴳ ᴰᴺ. You can go to a specific subgroup by pressing ᶜᵀᴿᴸ ᴴᴼᴹᴱ (Go To), followed by the subgroup number and ⏎.

Let's select *veneer* to replace *veil.*

■*4* Select Replace Word or press ①.

```
  4      •conceal
         •cover
         •hide
         •screen
Press letter for word _
```

■*5* Ｆ to replace *veil* with *veneer.*

```
The fog lay like a veneer_
```

10

The Speller and Thesaurus are fun to experiment with, so take a few moments to look up words from one of your own documents. Then move on to Chapter 11, where you'll see how to use some of the powerful printing options available in WordPerfect.

11

- *Getting the*
- *Most from*
- *Your Printer*

■ **W**ordPerfect provides several options that add versatility to your printing. In Chapter 1, you became familiar with some basic print options. In this chapter you'll learn these others:

Printing a document from disk

Printing portions of a document

Setting binding margins

Selecting number of copies

Controlling printing quality

Monitoring your printing jobs

■*Printing from Disk*

You can print a document without first recalling it into RAM. WordPerfect can read a stored document straight off the disk and print all or part of it, as long as the file has been saved with its WordPerfect formatting codes. A document saved with Fast Save (see "Taking a Look at the Setup Menu" in Chapter 14) will not print from the disk. You must recall the document to the screen and print it from there.

Turn your printer on, and print your MYLETTER file directly from the disk. From a clear screen, do the following:

■ *1*   Select **F**ile, **P**rint or press <kbd>SHIFT</kbd><kbd>F7</kbd>.

```
Print

        1 - Full Document
        2 - Page
        3 - Document on Disk
        4 - Control Printer
        5 - Multiple Pages
        6 - View Document
        7 - Initialize Printer


Options

        S - Select Printer                          HP LaserJet Series II
        B - Binding Offset                          0"
        N - Number of Copies                        1
        U - Multiple Copies Generated by            WordPerfect
        G - Graphics Quality                        Medium
        T - Text Quality                            High




Selection: 0
```

■*2*   Select **D**ocument on Disk or press ③ or Ⓓ
and type **MYLETTER** (lowercase letters
also can be used). If you have a dual floppy-
disk system, you may have to type
B:\MYLETTER.

> Document name: MYLETTER_

■*3*

> Page(s): (All)

You now must specify how much of the document you want to print. We'll print
all of MYLETTER; see the note "Printing Portions of a Document."

11

■*4*   to select the (All) option and begin
printing.

Note *Printing Portions of a Document* WordPerfect provides a number of options for printing all or just parts of a file. The instructions below are entered at the prompt "Page(s): All" seen in step 3.

Print Option	Instruction
All pages	
Single page	Type the page number and press ⏎.
Several pages	Type each of the page numbers, separated by commas (1,5,9).
Consecutive pages	Type the page numbers, separated by a hyphen (12-24).
From a specified page to the end of the document	Type the page number
From the first page to a specified page	Type a hyphen, followed by the ending page number (-5).

■ *Setting Margins for Binding*

If you need to print pages that will be bound, you probably need to allow extra space along the edge of the paper for the binding—the left and right page margins will be slightly different. Rather than selecting a separate margin for each page, you can set one margin and select a binding width to take into account the extra space you need. Clear your screen and try it.

■ *1* Select **F**ile, **P**rint or press SHIFT F7.

```
Options

     S - Select Printer                    HP LaserJet Series II
     B - Binding Offset                     0"
     N - Number of Copies                   1
     U - Multiple Copies Generated by       WordPerfect
     G - Graphics Quality                   Medium
     T - Text Quality                       High
```

■*2* Select **B**inding Offset or press **B**, then type
.5 and press ⏎.

```
Options

    S - Select Printer              HP LaserJet Series II
    B - Binding Offset              0.5"
    N - Number of Copies            1
    U - Multiple Copies Generated by  WordPerfect
    G - Graphics Quality            Medium
    T - Text Quality                High
```

WordPerfect will now automatically add a half-inch to the left margin of odd-numbered pages and a half-inch to the right margin of even-numbered pages.

■*3* Retrieve one of your own documents with
two or more pages, and Select **F**ile, **P**rint,
View Document or press SHIFT F7 6 or V
to see the effect.

Selecting Binding Offset does not insert a permanent formatting code in the document. It will, however, affect any document you print until you exit Word-Perfect or change the Binding Offset setting.

■*Selecting the Number of Copies*

This option allows you to print more than one copy of the document with a single print command. At a clear screen, do the following:

11

■*1*   Select **F**ile, **P**rint, **N**umber of Copies or press
SHIFT F7 N

```
Options

        S - Select Printer              HP LaserJet Series II
        B - Binding Offset              0.5"
        N - Number of Copies            1
        U - Multiple Copies Generated by  WordPerfect
        G - Graphics Quality            Medium
        T - Text Quality                High
```

■*2*   Type **4** and press ↵. All your print jobs will
now produce four copies.

```
Options

        S - Select Printer              HP LaserJet Series II
        B - Binding Offset              0.5"
        N - Number of Copies            4
        U - Multiple Copies Generated by  WordPerfect
        G - Graphics Quality            Medium
        T - Text Quality                High
```

The number of copies you select remains in effect for all documents printed until
you exit WordPerfect or change the Number of Copies option.

Selecting Graphics
■*and Text Print Quality*

The Graphics Quality and Text Quality options on the Print menu work exactly
the same; you use them to set the print quality of your graphics and text at any

one of four different levels of resolution. The option you select remains in effect for any print job until you select another option or exit WordPerfect.

I'll show you how to select Graphics Quality options. (The same steps apply to the Text Quality options.) Start at a blank editing screen.

■ *1*   Select **F**ile, **P**rint, **G**raphics Quality or press `SHIFT` `F7` `G`.

Graphics Quality: 1 Do Not Print; 2 Draft; 3 Medium; 4 High: 3

The graphic will not print. For printers without enough memory to print text and graphics together, first print the text. Next, turn graphics back on and turn text off, then reprint the same sheet the text was printed on. The graphics will print in the proper places.

Prints draft copies at a lower resolution at the fastest speed.

Prints copies at a higher resolution with more detail than draft copies; the printing speed is slightly slower.

Prints high-resolution copies of excellent quality, but the document takes longer to print.

11

■ *2*   Select Do **N**ot Print or press `1` or `N`.

```
Options

    S - Select Printer                      HP LaserJet Series II
    B - Binding Offset                      0.5"
    N - Number of Copies                    4
    U - Multiple Copies Generated by        WordPerfect
    G - Graphics Quality                    Do Not Print
    T - Text Quality                        High
```

▪*Controlling Your Printing Jobs*

The Control Printer menu options allow you to monitor the current status of any print job in operation. They allow you to select more than one document for printing and place them in a printing order called a *print queue* so that you can specify the next document for printing while another document is being printed. Let's take a look at the Control Printer menu. Start at a blank editing screen.

▪ *1*   Select File, Print or press SHIFT F7.

```
Print

     1 - Full Document
     2 - Page
     3 - Document on Disk
     4 - Control Printer
     5 - Multiple Pages
     6 - View Document
     7 - Initialize Printer
```

The print queue
comes in handy
when printing long
documents.

■*2*   Select Control Printer or press **4** or **C**.

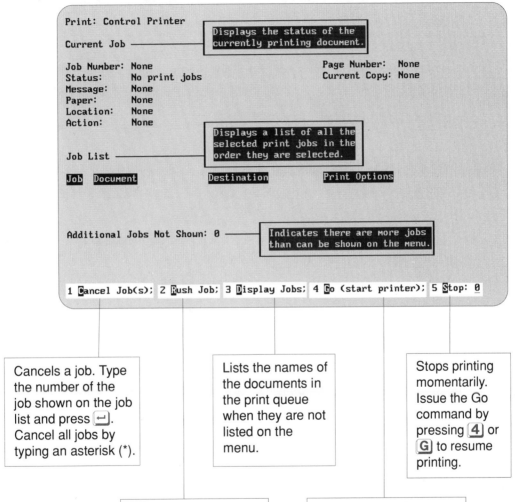

```
Print: Control Printer
                              Displays the status of the
Current Job ———————           currently printing document.

Job Number: None                        Page Number:  None
Status:     No print jobs               Current Copy: None
Message:    None
Paper:      None
Location:   None
Action:     None
                              Displays a list of all the
                              selected print jobs in the
Job List ————————             order they are selected.

Job Document              Destination          Print Options




Additional Jobs Not Shown: 0 ———   Indicates there are more jobs
                                   than can be shown on the menu.



1 Cancel Job(s); 2 Rush Job; 3 Display Jobs; 4 Go (start printer); 5 Stop: 0
```

11

Cancels a job. Type the number of the job shown on the job list and press ↵. Cancel all jobs by typing an asterisk (*).

Lists the names of the documents in the print queue when they are not listed on the menu.

Stops printing momentarily. Issue the Go command by pressing **4** or **G** to resume printing.

Moves a job to the front of the print queue. Type in the job number and it will print after the document currently being printed.

Issues the Print command after stopping the print procedure. Also used to manually feed paper into the printer one sheet at a time.

> **Note** *Using the Print Buffer* Printers that have a *print buffer,* a sort of holding area for text waiting to be printed, store data until the printer calls for it and continue to print after a stop or cancel command is given. When the buffer is empty, the printer stops. If you do not want the printer to continue printing, turn off the printer and turn it on again to delete data in the buffer. You will see the prompt "Cancel all print jobs? No (Yes)" if you try to exit prior to the printing of the documents. You can cancel the jobs and exit, or not cancel and remain in WordPerfect until the jobs are printed.

■*3*   Click the right mouse button or press F7 to return to the editing screen.

In Chapter 12 you will learn how to place text in columns, a very useful Word-Perfect skill.

12

- Working
- with
- Columns

■ **W**ordPerfect can handle text columns quickly and easily, which is especially useful when you're working with desktop publishing projects. You can create up to 24 columns on a single page, with much of the tedious work of figuring out the column widths done for you automatically. And, unlike some other word processors, WordPerfect shows you the columns side by side on the screen, as well as on the printed copy.

WordPerfect uses two types of columns: newspaper and parallel. *Newspaper columns* contain text that flows from column to column, exactly the way text in a newspaper flows. *Parallel columns* contain text that does not flow from one column to the next.

In this chapter you'll learn these skills:

Creating newspaper columns

Creating parallel columns

Changing column definitions

Editing and moving between columns

▪*Creating Newspaper Columns*

When working with newspaper columns, as you fill one column with text, the continuing text automatically spills into the next column. When a page is full, the text flows onto the next page, starting at the lefthand column.

Before you can type text into columns, you need to tell WordPerfect what you want these columns to look like, a process WordPerfect calls *defining* columns. Let's define three newspaper columns with three-quarters of an inch between each one. Start at a clear editing screen.

▪*1* Select **L**ayout, **C**olumns, **D**efine or press
[ALT][F7], [1] or [C], [3] or [D].

```
Text Column Definition

    1 - Type                          Newspaper

    2 - Number of Columns             2

    3 - Distance Between Columns

    4 - Margins

    Column   Left    Right   Column   Left    Right
      1:     1"      4"        13:
      2:     4.5"    7.5"      14:
      3:                       15:
      4:                       16:
      5:                       17:
      6:                       18:
      7:                       19:
      8:                       20:
      9:                       21:
     10:                       22:
     11:                       23:
     12:                       24:

Selection: 0
```

Newspaper is the default option. The default number of columns is 2.

■2   Select **N**umber of Columns or press `2`
or `N`.

```
Text Column Definition
    1 - Type                          Newspaper
    2 - Number of Columns             2
    3 - Distance Between Columns
```

■3   Type **3** and press `↵`. Three columns are automatically assigned margins within the designated left and right page margins.

```
    2 - Number of Columns             3
    3 - Distance Between Columns
    4 - Margins
```

12

■4   Select **D**istance Between Columns or press
`3` or `D`.

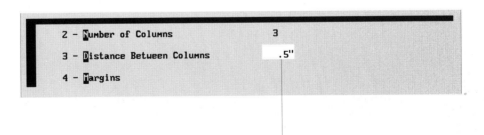

```
    2 - Number of Columns             3
    3 - Distance Between Columns      .5"
    4 - Margins
```

The default distance between columns is always ½ in.

■*5*   Type **.75** and press ⏎.

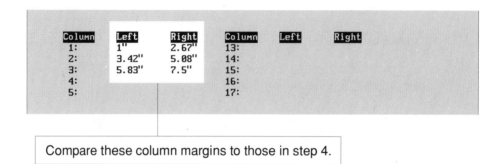

Column	Left	Right	Column	Left	Right
1:	1"	2.67"	13:		
2:	3.42"	5.08"	14:		
3:	5.83"	7.5"	15:		
4:			16:		
5:			17:		

Compare these column margins to those in step 4.

You've defined the columns. Now turn on the columns so you can type text into them.

■*6*   Click the right mouse button or press F7 to display the Columns menu line.

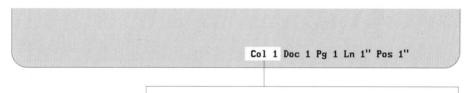

Columns: 1 On; 2 Off; 3 Define: 0

■*7*   Select **On** or press 1 or O.

Col 1 Doc 1 Pg 1 Ln 1" Pos 1"

The status line indicates that the Columns feature is turned on and that the cursor is in the first column.

■ *8*   Select **E**dit, **R**eveal Codes or press ALT F3.

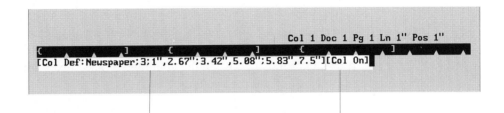

```
                                              Col 1 Doc 1 Pg 1 Ln 1" Pos 1"
{            ]          {             ]        {            ]
[Col Def:Newspaper;3;1",2.67";3.42",5.08";5.83",7.5"][Col On]
```

This code displays the type and number of columns you've defined and the margins of each column.	This code indicates that the column format is turned on and all that you type will appear in the columns.

Once you have completed entering the text in the columns, you must turn the column formatting off. This is very simple to do.

■ *9*   Select **L**ayout, **C**olumns, **O**ff or press
ALT F7 , **1** or **C** , **2** or **F** .

12

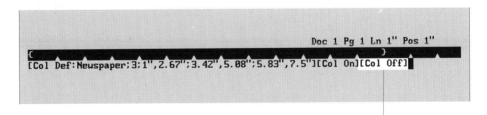

```
                                          Doc 1 Pg 1 Ln 1" Pos 1"
{                                        }
[Col Def:Newspaper;3;1",2.67";3.42",5.08";5.83",7.5"][Col On][Col Off]
```

The [Col Off] code is inserted in the document.

Note   While you are in column mode, you are not able to select new page margins. When Columns is turned off, the regular margins return.

∎*Creating Parallel Columns*

Parallel columns work well when you need short columns across a page, such as for use in conference or meeting agendas, schedules, or any document where the columns must align in a particular way. When you type into parallel columns, you have to manually move between the columns to place text within them—text does not flow from column to column.

Let's work through a short exercise in setting up parallel columns. As with newspaper columns, you must first define the columns. Start at a clear editing screen.

∎*1* Select **L**ayout, **C**olumns, **D**efine or press
[ALT][F7], [1] or [C], [3] or [D].

```
Text Column Definition

    1 - Type                        Newspaper

    2 - Number of Columns           2
```

∎*2* Select **T**ype or press [1] or [T].

```
    10:                    22:
    11:                    23:
    12:                    24:

Column Type: 1 Newspaper; 2 Parallel; 3 Parallel with Block Protect: 0
```

> **Note**   *Using Block Protection* You can choose the Parallel with Block Protect option in the Column Type menu to block-protect your column text. A protected block of columns is not split between two pages when a page break occurs in the middle of the columns; it is moved onto the following page. Unprotected parallel columns are split across the page break, placing part of the columns on the following page.

■ *3*   Select **P**arallel or press **2** or **P**.

```
Text Column Definition

    1 - Type                        Parallel

    2 - Number of Columns           2
```

■ *4*   Select **M**argins or press **4** or **M**, type **2**,
and press **↵** to define a 2" left margin for
the first column.

12

Column	Left	Right	Column	Left	Right
1:	2"	4"	13:		
2:	4.5"	7.5"	14:		
3:			15:		
4:			16:		
5:			17:		

■5 ⏎ twice to accept 4" for the right margin
and 4.5" for the left margin of the second
column.

Column	Left	Right	Column	Left	Right
1:	2"	4"	13:		
2:	4.5"	7.5"	14:		
3:			15:		
4:			16:		
5:			17:		

■6 Type **6.5** and press ⏎ to define a 6.5" right
margin for the second column.

Column	Left	Right	Column	Left	Right
1:	2"	4"	13:		
2:	4.5"	6.5"	14:		
3:			15:		
4:			16:		
5:			17:		

■7 Click the right mouse button or press **F7** to
leave the column-definition screen, then
select **O**n or press **1** or **O** to turn on the
columns.

■8 Type the following text in the first column,
pressing ⏎ after the first line:

```
        Carlsbad Caverns
        SE N. Mexico
```

■*9* CTRL ← to move to the second column, then type the text seen in column 2 below—do not press ← after each line. (In Columns, CTRL ← does not insert a hard page break but moves the cursor between columns.)

```
Carlsbad Caverns          60 million years
SE N. Mexico              old. Among the
                         largest caverns in
                         the world and a home
                         for millions of bats.
```

■*10* CTRL ← to return to the first column, then type the text seen below in that column. (Press ← after the first line.)

```
Carlsbad Caverns          60 million years
SE N. Mexico              old. Among the
                         largest caverns in
                         the world and a home
                         for millions of
                         bats.

Channel Islands
S. California
```

12

> Use parallel
> columns when
> you want com-
> plete control over
> the alignment
> of text.

■ *11*   CTRL ↵ and type the text seen below in the second column. (Do not press ↵ after each line.)

```
Carlsbad Caverns     60 million years
SE N. Mexico         old. Among the
                     largest caverns in
                     the world and a home
                     for millions of
                     bats.

Channel Islands      Islands off the
S. California        coast of Santa
                     Barbara, CA. Known
                     for large rookeries
                     of sea lions, sports
                     fishing, and many
                     varieties of sea
                     birds.
```

■ *12*   Select **L**ayout, **C**olumns, **O**ff or press ALT F7, **1** or **C**, **2** or **F** to turn off the columns and return to the regular document margins.

Keep this document on the screen for use in the next exercise.

Note   If you need to use these columns again, there is no need to redefine them—simply turn them on again by selecting **L**ayout, **C**olumns, **O**n or by pressing ALT F7, **1** or **C**, **1** or **O** at any location in the document beyond the point at which you defined them.

Changing Existing
■*Column Definitions*

What if you are unhappy with the look of your columns? Just redefine the columns and then delete the old [Col Def] code. You do not have to delete the text in the columns.

Let's redefine the parallel columns created in the last exercise and make the second column wider.

■*1* Select **E**dit, **R**eveal Codes or press ⌊ALT⌋⌊F3⌋.

■*2* ⌊HOME⌋ ⌊HOME⌋ ⌊↑⌋ to return the cursor to the beginning of the document.

```
                                    Barbara, CA. Known
                                    for large rookeries
                                           Col 1 Doc 1 Pg 1 Ln 1" Pos 2"
         {                 }  {                 }
[Col Def:Parallel;2;2",4";4.5",6.5"][Col On]Carlsbad Caverns [HRt]
SE N. Mexico[HPg]
60 million years[SRt]
```

■*3* ⌊←⌋ to place the highlight on the [Col On] code.

```
                                    Barbara, CA. Known
                                    for large rookeries
                                           Doc 1 Pg 1 Ln 1" Pos 1"
{
[Col Def:Parallel;2;2",4";4.5",6.5"][Col On]Carlsbad Caverns [HRt]
SE N. Mexico[HPg]
60 million years[SRt]
```

12

You will now insert the new column-definition coding between the old [Col Def] code and the [Col On] code.

■4　Select **L**ayout, **C**olumns, **D**efine or press
ALT F7, 1 or C, 3 or D.

```
4 - Margins

Column    Left      Right     Column    Left      Right
  1:      2"        4"          13:
  2:      4.5"      6.5"        14:
  3:                            15:
  4:                            16:
```

■5　Select **M**argins or press 4 or M, then press
⏎ three times to accept three of the current
margin settings.

```
3 - Distance Between Columns

4 - Margins

Column    Left      Right     Column    Left      Right
  1:      2"        4"          13:
  2:      4.5"      6.5"        14:
  3:                            15:
  4:                            16:
```

■6 Type **7** and press ⏎. The second column is
now a half-inch wider.

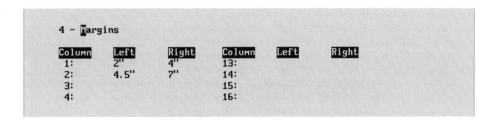

■7 Click the right mouse button twice or press
F7 twice to return to the editing screen.

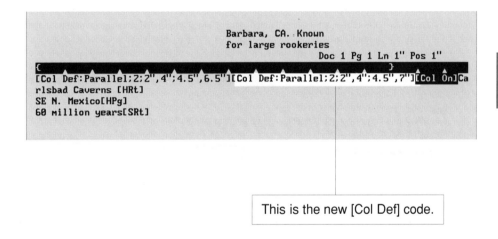

This is the new [Col Def] code.

■8 ⏎ twice and delete the old column code.
Only the new code will be active.

12

■9 Select **E**dit, **R**eveal Codes or press ALT F3.

The text adjusts to the new column margins.

```
Carlsbad Caverns          60 million years old.
SE N. Mexico              Among the largest caverns
                          in the world and a home
                          for millions of bats.

Channel Islands           Islands off the coast of
S. California             Santa Barbara, CA. Known
                          for large rookeries of
                          sea lions, sports
                          fishing, and many
                          varieties of sea birds.
```

Keep this document on the screen for use in the next exercise.

It is possible to have more than one style of column in a document. The [Col Def] code immediately preceding any [Col On] code is the one in effect.

Editing and Moving the ■*Cursor between Columns*

Once you have placed text in columns, nearly all the editing features of Word-Perfect can be used. The exceptions are CTRL END (deletes to the end of a line), which, when used with columns, deletes to the end of a column; ← and →, which move the cursor only within a column; and HOME ← and HOME →, which move the cursor to the end of the line in a column you're currently working in.

When editing text in columns, there is a handy shortcut for moving the cursor from one column to another: Select **S**earch, **G**oto or press CTRL HOME. Using the columns from the previous exercise, let's try it.

■ *1*   [HOME] [HOME] [↑] to position the cursor at the
beginning of the document.

```
Carlsbad Caverns        60 million years old.
SE N. Mexico            Among the largest caverns
                        in the world and a home
                        for millions of bats.

Channel Islands         Islands off the coast of
S. California           Santa Barbara, CA. Known
                        for large rookeries of
                        sea lions, sports
                        fishing, and many
                        varieties of sea birds.
```

■ *2*   Select **S**earch, **G**oto or press [CTRL] [HOME].

```
Go to
```

12

■ *3*   [→] to move the cursor to column 2.

```
Carlsbad Caverns        60 million years old.
SE N. Mexico            Among the largest caverns
                        in the world and a home
                        for millions of bats.

Channel Islands         Islands off the coast of
S. California           Santa Barbara, CA. Known
                        for large rookeries of
                        sea lions, sports
                        fishing, and many
                        varieties of sea birds.
```

Notice that the status line indicates you are now in Col 2.

■*4*   Select **S**earch, **G**oto or press `CTRL` `HOME` and
press `←` to return to column 1.

```
    Carlsbad Caverns        60 million years old.
    SE N. Mexico            Among the largest caverns
                            in the world and a home
                            for millions of bats.

    Channel Islands         Islands off the coast of
    S. California           Santa Barbara, CA. Known
                            for large rookeries of
                            sea lions, sports
                            fishing, and many
                            varieties of sea birds.
```

If you wish, select **F**ile, **P**rint, **V**iew Document or press `SHIFT` `F7`, `6` or `V` to
see how the columns will appear on the printed page.

Chapter 13 discusses the most exciting new feature in WordPerfect 5.1: Tables.
I think you'll find Tables very useful in many of your word processing projects.
Once you experience the versatility of Tables, you'll wonder how you ever got
along without this feature.

13

- **Creating and**
- **Working with**
- **Tables**

■ The ability to create tables is one of the most impressive and innovative features offered by WordPerfect 5.1. You can use tables to create many different kinds of documents, including calendars, inventories, and invoices. There is almost no limit to the documents you can create with the tables feature. There are so many aspects of this feature that I will not be able to cover them all in this entry-level book. In this chapter, you'll learn how to create and change a table structure, enter and format text in the table, format the lines that form the table and cell borders, and use the Math feature to calculate numerical data.

Creating the table structure

Changing the table structure

Placing text in a table

Using math with tables

Formatting data in a table

■*Creating the Table Structure*

A table is structured like a spreadsheet, with *columns* and *rows* that intersect to form *cells*. You can structure a table that has up to 32 columns and 32,765 rows of data. Let's create a table that has four columns and five rows. Begin at a clear editing screen.

■*1*   Select **L**ayout, **T**ables, **C**reate or press
ALT F7, 2 or T, 1 or C. You are
prompted to enter the number of columns
you want.

```
Number of Columns: 3
```

■*2*   Type **4** and press ↵. You are prompted to
enter the number of rows you want.

```
Number of Rows: 1
```

■*3*   Type **5** and press ⏎. The table structure appears on the Table Edit screen. Twenty cells are formed by the intersection of four columns (A through D) and five rows (1 through 5). At the bottom of the screen is the Table Edit menu.

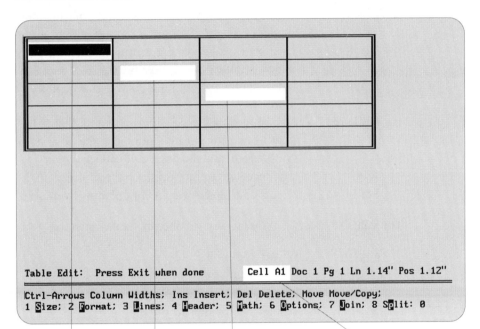

```
Table Edit:  Press Exit when done        Cell A1 Doc 1 Pg 1 Ln 1.14" Pos 1.12"

Ctrl-Arrows Column Widths; Ins Insert; Del Delete; Move Move/Copy;
1 Size; 2 Format; 3 Lines; 4 Header; 5 Math; 6 Options; 7 Join; 8 Split: 0
```

13

The cell highlight appears in cell A1. The highlight designates the cell or cells you are currently working with.

This is cell B2, the intersection of column B and row 2.

This is cell C3, the intersection of column C and row 3.

The status line displays the cell in which the highlight is located.

You can change the structure and format of a table with these options on the Table Edit menu:

Size	Increases and/or decreases the number of rows and/or columns.
Format	Changes the font attributes and alignment of text within cells and columns, the width of columns, and the height of cells.
Lines	Changes the types of lines used for the table and cell borders (you can choose to have no lines at all).
Header	Designates rows of text to be repeated on pages where a table continues over a page break.
Math	Adds, subtracts, multiplies, and divides columns or rows of numbers.
Options	Determines the space between text and table lines, the display format for negative numbers, the position of the table on the page, and the degree of cell shading.
Join	Connects two or more cells so they form a single cell.
Split	Creates multiple rows and/or columns from a single cell.

You can also use the following keys to alter the table structure:

CTRL →	Widen a column.
CTRL ←	Narrow a column.
INS	Insert a blank row or column.
DEL	Delete an existing row or column.

■*4* F7 and select **E**dit, **R**eveal Codes or press
ALT F3. The width of each column (1.63") is
displayed in the [Tbl Def:] (table definition)
code. This information is helpful for checking
the width measurement for each column.

■*5*

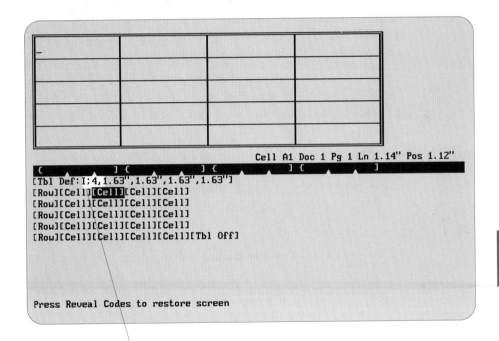

The *I;4* part of the code means this is table
roman numeral I, and it consists of 4 columns.

Select **E**dit, **R**eveal Codes or press ALT F3 to
return to the Table Edit screen.

Keep the Table Edit screen displayed for use in the next exercise.

13

■*Changing the Table Structure*

If your word processing projects are like mine, I'm sure you'll agree that they are evolutionary in nature, requiring many changes before the finished product is turned out. When you work with tables, you will often want to change the table structure to suit the amount and type of information that is placed in the table. WordPerfect provides an extensive array of formatting tools to assist you in this endeavor. Unfortunately, I cannot cover all the various ways to change a table's structure; however, I will show you some basics that will provide you with enough information to get started creating your own tables immediately.

In this exercise, you will learn how to join four cells into a single cell, how to increase the size of the table, and how to change the type of lines used to draw the borders of the table cells.

Joining Cells

You can block two or more cells and combine them to form a larger cell.

■ *1*   If you are not at the Table Edit screen, select **Layout, Tables, Edit** or press **ALT F7** to display it.

> If a table cell is too small for your needs, join it with another one.

■*2*   With the highlight in cell A1, press `ALT` `F4`
(Block) and `→` three times (or `END` once)
to highlight cells A1, B1, C1, and D1 (the
top row).

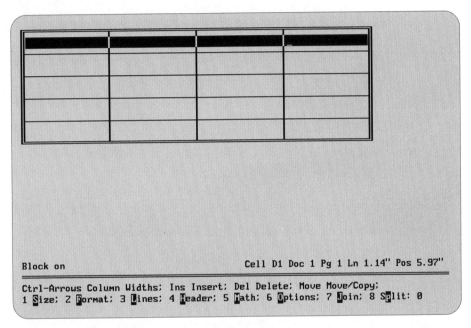

Block on   Cell D1 Doc 1 Pg 1 Ln 1.14" Pos 5.97"

Ctrl-Arrows Column Widths; Ins Insert; Del Delete; Move Move/Copy;
1 Size; 2 Format; 3 Lines; 4 Header; 5 Math; 6 Options; 7 Join; 8 Split: 0

13

■*3*   Select Join or press `7` or `J`. You are asked
if you want to join the cells.

k on   Cell D1 Doc 1 Pg 1 Ln 1.14" Pos 6"

-Arrows Column Widths; Ins Insert; Del Delete; Move Move/Copy;
cells? No (Yes)

■*4*   Select **Y**es or press ⟨**Y**⟩. Cells A1 through D1
become a single cell, A1.

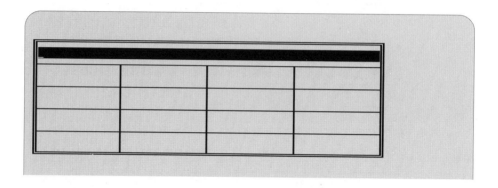

Changing Table Size

WordPerfect makes it easy for you to enlarge your table.

■*5*   Select **S**ize or press ⟨**1**⟩ or ⟨**S**⟩

```
Table Edit:   Press Exit when done      Cell A1 Doc 1 Pg 1 Ln 1.14" Pos 1.12"

Table Size: 1 Rows: 2 Columns: 0
```

■6   Select **R**ows or press ⎡1⎦ or ⎡R⎦ to change the
     number of rows. (You would select **C**olumns
     to change the number of columns.)

```
Table Edit:  Press Exit when done          Cell A1 Doc 1 Pg 1 Ln 1.14" Pos 1.12"
════════════════════════════════════════════════════════════════════════════════

Number of Rows: 5
```

■7   Type **7** and press ⎣←⎦. The table now has two
     more rows.

```
Table Edit:  Press Exit when done          Cell A7 Doc 1 Pg 1 Ln 2.82" Pos 1.12"
────────────────────────────────────────────────────────────────────────────────
Ctrl-Arrows Column Widths: Ins Insert: Del Delete: Move Move/Copy:
1 Size: 2 Format: 3 Lines: 4 Header: 5 Math: 6 Options: 7 Join: 8 Split: 0
```

13

Changing Line Size and Type

You can also change the size and type of lines that border the cells in the table.

■ *8*   ⬆ five times to move the highlight to
cell A2.

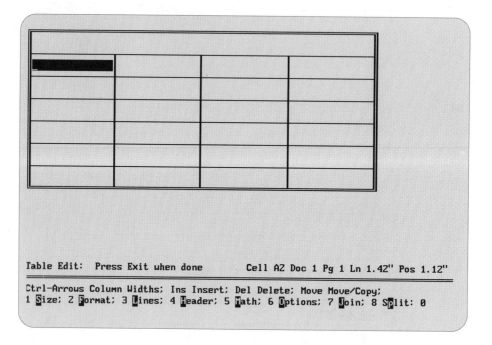

```
Table Edit:   Press Exit when done        Cell A2 Doc 1 Pg 1 Ln 1.42" Pos 1.12"

Ctrl-Arrows Column Widths; Ins Insert; Del Delete; Move Move/Copy;
1 Size; 2 Format; 3 Lines; 4 Header; 5 Math; 6 Options; 7 Join; 8 Split: 0
```

To move the highlight with the mouse, place the mouse pointer in the cell where you want the highlight and click the left button. You also can use the mouse to block cells—press ALT F4, place the pointer in the most distant cell of the group of cells you want to block, and click the left button.

■ *9*   ALT F4   Then press → three times to move
the highlight to cell D2, or place the mouse
pointer in cell D2 and click the left button.

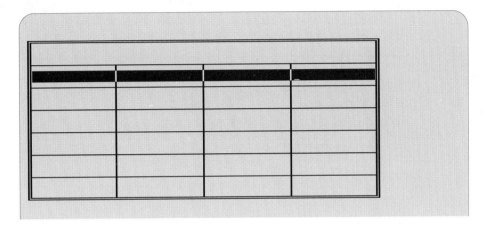

■ *10* Select **L**ines or press **3** or **L**.

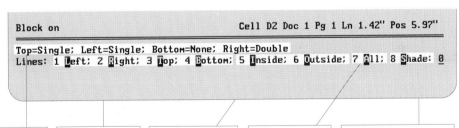

13

Displays the status of the lines that form the high-lighted cells.

Selects the specific line you want to change.

Changes the lines in-side the highlighted cells and the outside border lines of the high-lighted cells.

Changes all the in-side and outside bor-der lines of the high-lighted cells.

Turns cell shad-ing on or off (the amount of shad-ing is set with the **O**ptions menu option on the Table Edit menu).

> **Note** When you first create them, all the cells, except those in the far-right column and bottom row, have only the top and left borders defined with a single or a double line. There is no right or bottom border assigned. The right and bottom borders of each cell, except those in the far-right column and the bottom row, are actually the top and left border lines of the adjacent cell. If you assign right and bottom lines to these cells, you increase the thickness of the left or top line of the adjacent cells.

■ *11* Select **T**op or press 3 or T to display menu options for selecting the type of line you want for the highlighted cells.

```
Block on                              Cell D2 Doc 1 Pg 1 Ln 1.42" Pos 5.97"

Top=Single; Left=Single; Bottom=None; Right=Double
1 None; 2 Single; 3 Double; 4 Dashed; 5 Dotted; 6 Thick; 7 Extra Thick: 0
```

■ *12* Select **D**ouble or press 3 or D. The top line for cells A2 through D2 is changed to a double line.

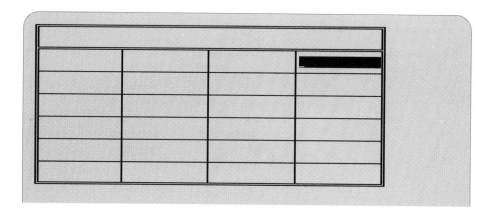

■*13* ⬇️ to move the highlight to cell D6. Then
press [ALT][F4] (the Block command), and
press ⬅️ three times to move the highlight to
cell A6. (Remember, you can use the mouse
to position the highlight.)

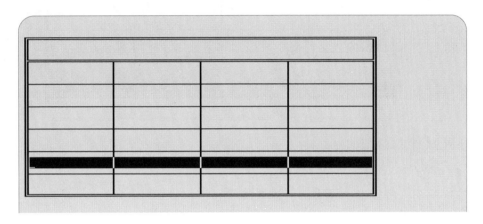

■*14* Select **L**ines, **T**op, **T**hick or press [3] or [L],
[3] or [T], [6] or [T] to place a thick line between rows 5 and 6.

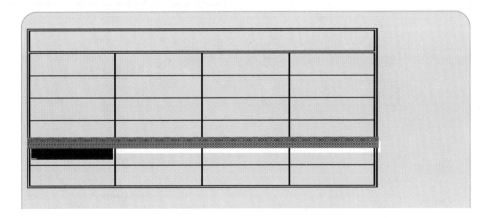

13

■ *15* [ALT][F4] Press [→] three times to move the
highlight back to cell D6.

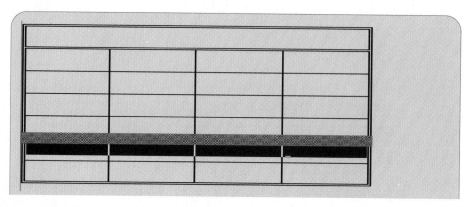

■ *16* Select **L**ines, **B**ottom, **T**hick or press
[3] or [L], [4] or [B], [6] or [T] to place a
thick line between rows 6 and 7.

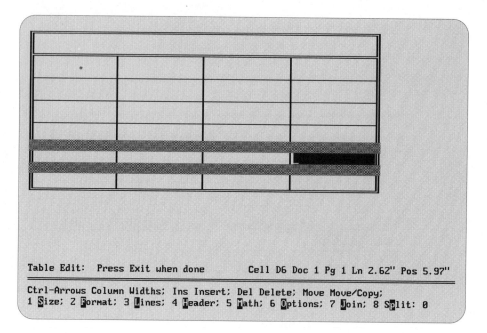

■ *17* F7 to exit the Table Edit screen and return to the editing screen.

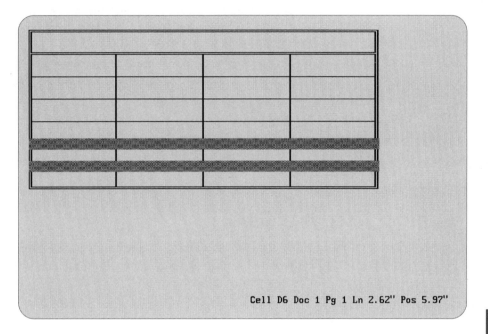

Cell D6 Doc 1 Pg 1 Ln 2.62" Pos 5.97"

13

The thick, shaded lines will print as regular lines, only thicker. The cursor location is displayed on the status line (in this case, cell D6).

If you wish, select File, Print, View Document or press SHIFT F7 6 or V to view the table on your screen. (Select 100% or press 1 to see how the borders will appear on the printed page.) When finished, press F7 and return the cursor to cell D6.

Keep this document on the screen for use in the next exercise.

■*Placing Text in the Table*

There is more than one way to place text in a table. You can type text directly in a cell, block and copy or move text from anywhere in the document and place it in a cell (not covered in this section), or retrieve a file stored on disk directly into a cell. In other words, you can enter and manipulate text in a table using the same WordPerfect tools you use when working with any other kind of document.

To move the cursor between columns containing data, press `TAB` to move from left to right and `SHIFT` `TAB` to move from right to left. Press `→` or `←` to move between columns without data. To move the cursor between rows, press `↓` to move downward and `↑` to move upward. Also, as on the Table Edit screen, you can move the cursor with the mouse.

Typing Text in a Cell

Let's type some text in the table you created in the previous exercises. I will give you keyboard instructions for moving the cursor within the table, but remember, you can also use the mouse.

■ *1*   `↑` five times to move the cursor to cell A1;

select **L**ayout, **A**lign, **C**enter or press `SHIFT` `F6`.

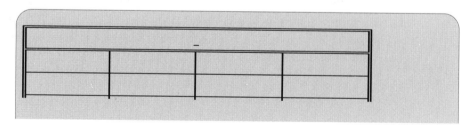

■*2* Type **Airline Ticket Sales**.

```
┌──────────────────────────────────────────────────┐
│              Airline Ticket Sales                  │
├──────────┬──────────┬──────────┬──────────────────┤
│          │          │          │                  │
├──────────┼──────────┼──────────┼──────────────────┤
│          │          │          │                  │
└──────────┴──────────┴──────────┴──────────────────┘
```

■*3* Use the cursor keys or the mouse to move the cursor to the correct cells, and type the text shown on this screen:

Airline Ticket Sales			
	January	February	March
Europe	$23,000.00	$7,000.00	$20,000.00
South America	$42,000.00	$102,000.00	$40,000.00
Asia	$31,000.00	$27,000.00	$7,000.00
Monthly Sales Totals			

13

> When you type more than one line of text in a cell, the cell height automatically adjusts downward so that the additional text fits.

Keep this document on the screen for use in the next exercise.

Retrieving a File into a Cell

To retrieve a file into a cell, you place the cursor in the desired cell, select **F**ile, **R**etrieve, type the file name, and press ⏎. The size of the cell adjusts to accept the data in the file. Before we retrieve a file into the table, let's join cells A7, B7, C7, and D7 to form a single cell.

■ *1* Select **L**ayout, **T**ables, **E**dit or press ALT F7.
Then press ↓ one time to move the high-
light to cell A7.

Airline Ticket Sales			
	January	February	March
Europe	$23,000.00	$7,000.00	$20,000.00
South America	$42,000.00	$102,000.00	$40,000.00
Asia	$31,000.00	$27,000.00	$7,000.00
Monthly Sales Totals			

Table Edit: Press Exit when done   Cell A7 Doc 1 Pg 1 Ln 3.13" Pos 1.12"

Ctrl-Arrows Column Widths; Ins Insert; Del Delete; Move Move/Copy;
1 Size; 2 Format; 3 Lines; 4 Header; 5 Math; 6 Options; 7 Join; 8 Split: 0

■*2* ⟦ALT⟧⟦F4⟧ Move the highlight to cell D7. Then
select **J**oin, **Y**es or press ⟦7⟧ or ⟦J⟧, ⟦Y⟧ to
create a single cell.

```
                    Airline Ticket Sales
              ┌──────────────┬──────────────┬──────────────┐
              │ January      │ February     │ March        │
┌─────────────┼──────────────┼──────────────┼──────────────┤
│Europe       │ $23,000.00   │ $7,000.00    │ $20,000.00   │
│South America│ $42,000.00   │ $102,000.00  │ $40,000.00   │
│Asia         │ $31,000.00   │ $27,000.00   │ $7,000.00    │
│Monthly Sales│              │              │              │
│Totals       │              │              │              │
└─────────────┴──────────────┴──────────────┴──────────────┘


Table Edit:  Press Exit when done        Cell A7 Doc 1 Pg 1 Ln 3.13" Pos 1.12"

Ctrl-Arrows Column Widths; Ins Insert; Del Delete; Move Move/Copy;
1 Size; 2 Format; 3 Lines; 4 Header; 5 Math; 6 Options; 7 Join; 8 Split: 0
```

13

■*3* ⟦F7⟧ to return to the editing screen. The cursor
should be in cell A7, which now extends
across the entire width of the table.

■*4*   Select **File**, **R**etrieve, type **PERSONAL**, and
press ⏎ to retrieve the file PERSONAL,
which you created in Chapter 3. Cell A7 is
enlarged so that the text will fit.

Airline Ticket Sales			
	January	February	March
Europe	$23,000.00	$7,000.00	$20,000.00
South America	$42,000.00	$102,000.00	$40,000.00
Asia	$31,000.00	$27,000.00	$7,000.00
Monthly Sales Totals			

PERSONAL REMINDERS

Notify Billing Dept. to update computerized customer address files.

Have Requisition Dept. order new Sybex WordPerfect Computer books.

Cell A7 Doc 1 Pg 1 Ln 3.13" Pos 1.12"

Keep this document on the screen for use in the next exercise.

■*Working with Numerical Entries*

Using WordPerfect's Table Math feature allows you to calculate tabulated numbers. Table Math can perform only addition, subtraction, multiplication, and division. The operators for these functions are as follows:

Addition   +
Subtraction   –
Multiplication   *
Division   /

The easiest method for performing calculations in a table is to create a *formula* that consists of the *cell addresses* that contain the numbers to be used in the calculation. If you have experience using a spreadsheet program, creating formulas is a familiar task. If not, these next exercises will show you how to work with some simple formulas.

Creating Formulas

Let's create a formula that will add up the ticket sales for each month. The table you created in the previous exercises should still be on the screen.

13

■ *1*   Select **L**ayout, **T**ables, **E**dit or press `ALT` `F7` to activate the Table Edit screen. Then move the highlight to cell B6.

	January	February	March
Europe	$23,000.00	$7,000.00	$20,000.00
South America	$42,000.00	$102,000.00	$40,000.00
Asia	$31,000.00	$27,000.00	$7,000.00
Monthly Sales Totals			
PERSONAL REMINDERS			

We will now place formulas in cells B6, C6, and D6.

■2 Select Math or press **5** or **M**.

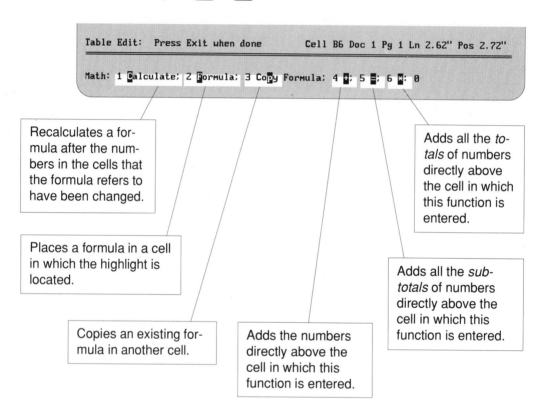

Recalculates a formula after the numbers in the cells that the formula refers to have been changed.

Adds all the *totals* of numbers directly above the cell in which this function is entered.

Places a formula in a cell in which the highlight is located.

Adds all the *subtotals* of numbers directly above the cell in which this function is entered.

Copies an existing formula in another cell.

Adds the numbers directly above the cell in which this function is entered.

■3 Select Formula or press **2** or **F**.

```
Table Edit:  Press Exit when done        Cell B6 Doc 1 Pg 1 Ln 2.62" Pos 2.72"


Enter formula:
```

■*4* Type **B3+B4+B5** (lowercase letters also
work). These cells contain the numbers that,
when added together, give the total dollar
amount of sales in January.

```
Table Edit:  Press Exit when done      Cell B6 Doc 1 Pg 1 Ln 2.62" Pos 2.72"
_____

Enter formula: B3+B4+B5
```

■*5* ⏎ The total appears in cell B6.

	January	February	March
Europe	$23,000.00	$7,000.00	$20,000.00
South America	$42,000.00	$102,000.00	$40,000.00
Asia	$31,000.00	$27,000.00	$7,000.00
Monthly Sales Totals	96,000.00		

PERSONAL REMINDERS

Notify Billing Dept. to update computerized customer address
files.

Have Requisition Dept. order new Sybex WordPerfect computer
books.

```
=B3+B4+B5                              Cell B6 Doc 1 Pg 1 Ln 2.62" Pos 3.62"
Ctrl-Arrows Column Widths; Ins Insert; Del Delete; Move Move/Copy;
1 Size; 2 Format; 3 Lines; 4 Header; 5 Math; 6 Options; 7 Join; 8 Split: 0
```

13

The formula is displayed on the Table Edit
screen to remind you of the Math function.

Keep the Table Edit screen displayed for use in the next exercise.

Copying Formulas to Other Cells

Now let's copy the formula in cell B6 to cells C6 and D6. When copying a formula, you must place the highlight in the cell that contains the formula to be copied. The highlight should still be in cell B6.

■ *1*   Select **M**ath, **C**opy Formula or press
       [5] or [M], [3] or [P].

```
Have Requisition Dept. order new Sybex WordPerfect computer
books.
=B3+B4+B5                        Cell B6 Doc 1 Pg 1 Ln 2.62" Pos 3.62"


Copy Formula To: 1 Cell; 2 Down; 3 Right: 0
```

| Copies the formula to a specific cell; you place the highlight in the cell where you want the formula copied and press [↵]. | Copies the formula to a cell or cells directly below the cell containing the formula. | Copies the formula to a cell or cells directly to the right of the cell containing the formula. |

■ *2*   Select **R**ight or press [3] or [R].

```
Have Requisition Dept. order new Sybex WordPerfect computer
books.
=B3+B4+B5                        Cell B6 Doc 1 Pg 1 Ln 2.62" Pos 3.62"


Number of times to copy formula: 1
```

■*3*   Type **2** and press ⏎. The formula is automatically placed in cells C6 and D6, and the totals are calculated.

```
┌─────────────────────────────────────────────────────────────┐
│                     Airline Ticket Sales                      │
├───────────────┬─────────────┬──────────────┬─────────────────┤
│               │ January     │ February     │ March           │
├───────────────┼─────────────┼──────────────┼─────────────────┤
│ Europe        │ $23,000.00  │ $7,000.00    │ $20,000.00      │
├───────────────┼─────────────┼──────────────┼─────────────────┤
│ South America │ $42,000.00  │ $102,000.00  │ $40,000.00      │
├───────────────┼─────────────┼──────────────┼─────────────────┤
│ Asia          │ $31,000.00  │ $27,000.00   │ $7,000.00       │
├───────────────┼─────────────┼──────────────┼─────────────────┤
│ Monthly Sales │ 96,000.00   │ 136,000.00   │ 67,000.00       │
│ Totals        │             │              │                 │
└───────────────┴─────────────┴──────────────┴─────────────────┘
```

PERSONAL REMINDERS

Notify Billing Dept. to update computerized customer address
files.

Have Requisiton Dept. order new SYBEX WordPerfect computer
books.

`=B3+B4+B5`   Cell B6 Doc 1 Pg 1 Ln 2.62" Pos 2.72"

Ctrl-Arrows Column Widths; Ins Insert; Del Delete; Move Move/Copy;
1 **S**ize; 2 **F**ormat; 3 **L**ines; 4 **H**eader; 5 **M**ath; 6 **O**ptions; 7 **J**oin; 8 S**p**lit: 0

13

■*4*   F7 to return to the editing screen.

Keep the editing screen displayed for use in the next exercise.

Recalculating Formulas

Let's change the number in cell B4 (the dollar amount for South America in
January) and recalculate the total.

■ *1* Move the cursor to cell B4 and edit the number so it reads *$36,000.00*. Then select **L**ayout, **T**ables, **E**dit or press `ALT` `F7` to call up the Table Edit menu.

Airline Ticket Sales			
	January	February	March
Europe	$23,000.00	$7,000.00	$20,000.00
South America	$36,000.00	$102,000.00	$40,000.00
Asia	$31,000.00	$27,000.00	$7,000.00
Monthly Sales Totals	96,000.00	136,000.00	67,000.00

PERSONAL REMINDERS

■ *2* Select **M**ath, **C**alculate or press `5` or `M` `1` or `C`. The total in cell B6 (total sales for January) changes from *$96,000.00* to *$90,000.00*.

Airline Ticket Sales			
	January	February	March
Europe	$23,000.00	$7,000.00	$20,000.00
South America	$36,000.00	$102,000.00	$40,000.00
Asia	$31,000.00	$27,000.00	$7,000.00
Monthly Sales Totals	90,000.00	136,000.00	67,000.00

PERSONAL REMINDERS

Keep the Table Edit screen displayed for use in the next exercise.

■*Formatting Cells*

Once data is entered in cells, it may need to be formatted differently. For example, you may want to realign text or decimal-align numbers to improve the appearance of a document. Let's make some basic formatting changes to the document created in the previous exercises.

■ *1*   Place the highlight in cell B2 (the January heading), press ᴬᴸᵀ F4, and move the highlight to cell D2 (the March heading).

```
                         Airline Ticket Sales
              │January      │February      │March
Europe        │$23,000.00   │$7,000.00     │$20,000.00
South America │$36,000.00   │$102,000.00   │$40,000.00
Asia          │$31,000.00   │$27,000.00    │$7,000.00
Monthly Sales │90,000.00    │136,000.00    │67,000.00
Totals
PERSONAL REMINDERS

Notify Billing Dept. to update computerized customer address
files.

Have Requisiton Dept. order new SYBEX WordPerfect computer
books.
Block on                          Cell D2 Doc 1 Pg 1 Ln 1.45" Pos 5.97"

Ctrl-Arrows Column Widths; Ins Insert; Del Delete; Move Move/Copy;
1 Size; 2 Format; 3 Lines; 4 Header; 5 Math; 6 Options; 7 Join; 8 Split: 0
```

13

■*2*   Select **F**ormat or press **2** or **F**.

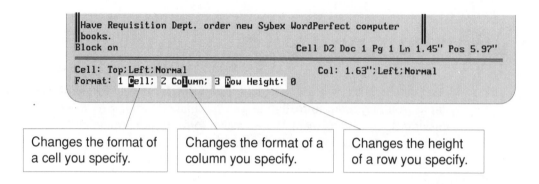

```
Have Requisition Dept. order new Sybex WordPerfect computer
books.
Block on                                    Cell D2 Doc 1 Pg 1 Ln 1.45" Pos 5.97"

Cell: Top;Left;Normal                           Col: 1.63";Left;Normal
Format: 1 Cell; 2 Column; 3 Row Height: 0
```

Changes the format of a cell you specify.	Changes the format of a column you specify.	Changes the height of a row you specify.

■*3*   Select **C**ell or press **1** or **C**.

```
Have Requisition Dept. order new Sybex WordPerfect computer
books.
Block on                                    Cell D2 Doc 1 Pg 1 Ln 1.45" Pos 5.97"

Cell: Top;Left;Normal                           Col: 1.63";Center;Normal
Cell: 1 Type; 2 Attributes; 3 Justify; 4 Vertical Alignment; 5 Lock: 0
```

■*4*   Select **J**ustify or press **3** or **J**.

```
Have Requisition Dept. order new Sybex WordPerfect computer
books.
Block on                                    Cell D2 Doc 1 Pg 1 Ln 1.45" Pos 5.97"

Cell: Top;Left;Normal                           Col: 1.63";Center;Normal
Justification: 1 Left; 2 Center; 3 Right; 4 Full; 5 Decimal Align; 6 Reset: 1
```

■*5* Select **C**enter or press **2** or **C** to center the
three month titles.

Airline Ticket Sales			
	January	February	March
Europe	$23,000.00	$7,000.00	$20,000.00

■*6* Move the highlight to cell B3 (sales total for
Europe in January), press **ALT F4**, and move
the highlight to cell D6 (monthly sales total
for monthly March) to block all the dollar
amounts on the screen.

Airline Ticket Sales			
	January	February	March
Europe	$23,000.00	$7,000.00	$20,000.00
South America	$36,000.00	$102,000.00	$40,000.00
Asia	$31,000.00	$27,000.00	$7,000.00
Monthly Sales Totals	90,000.00	136,000.00	67,000.00
PERSONAL REMINDERS			

13

■*7* Select **F**ormat, **C**ell or press **2** or **F**
1 or **C**.

■ *8*   Select **J**ustify, **D**ecimal Align or press
[3] or [J], [5] or [D] to align all the numbers on the decimal point.

```
┌──────────────────────────────────────────────────────────────┐
│  ┌─────────────────────────────────────────────────────────┐  │
│  │               Airline Ticket Sales                      │  │
│  ├───────────────┬─────────────┬─────────────┬─────────────┤  │
│  │               │   January   │  February   │    March    │  │
│  ├───────────────┼─────────────┼─────────────┼─────────────┤  │
│  │ Europe        │ $23,000.00  │  $7,000.00  │ $20,000.00  │  │
│  ├───────────────┼─────────────┼─────────────┼─────────────┤  │
│  │ South America │ $36,000.00  │ $102,000.00 │ $40,000.00  │  │
│  ├───────────────┼─────────────┼─────────────┼─────────────┤  │
│  │ Asia          │ $31,000.00  │ $27,000.00  │  $7,000.00  │  │
│  ├───────────────┼─────────────┼─────────────┼─────────────┤  │
│  │ Monthly Sales │ 90,000.00   │ 136,000.00  │  67,000.00  │  │
│  │ Totals        │             │             │             │  │
│  └───────────────┴─────────────┴─────────────┴─────────────┘  │
│  PERSONAL REMINDERS                                            │
│                                                                │
│  Notify Billing Dept. to update computerized customer address │
│  files.                                                        │
│                                                                │
│  Have Requisiton Dept. order new SYBEX WordPerfect computer    │
│  books.                                                        │
│  =D3+D4+D5 Align char = .           Cell D6 Doc 1 Pg 1 Ln 2.62" Pos 6.48" │
│                                                                │
│  Ctrl-Arrows Column Widths; Ins Insert; Del Delete; Move Move/Copy; │
│  1 Size; 2 Format; 3 Lines; 4 Header; 5 Math; 6 Options; 7 Join; 8 Split: 0 │
└──────────────────────────────────────────────────────────────┘
```

■ *9*   [F7] to return to the editing screen.

■ *10*   Insert a dollar sign ($) in front of each of the
monthly sales totals.

■ *11* Select **F**ile, **P**rint, **P**age or press `SHIFT` `F7` `2` or `P` to print the table. Your printout should look like this:

Airline Ticket Sales			
	January	February	March
Europe	$23,000.00	$7,000.00	$20,000.00
South America	$36,000.00	$102,000.00	$40,000.00
Asia	$31,000.00	$27,000.00	$7,000.00
Monthly Sales Totals	$90,000.00	$136,000.00	$67,000.00

PERSONAL REMINDERS

Notify Billing Dept. to update computerized customer address files.

Have Requisition Dept. order new Sybex WordPerfect computer books.

13

Chapter 14 introduces you to the WordPerfect features that allow you to change the default settings. You will also learn how to create macros, which are small programs that automatically execute a series of keystrokes. You will find these two subjects very helpful in streamlining many of your everyday tasks.

14

- *Customizing*
- *WordPerfect with*
- *the Setup Feature*
- *and Macros*

■ **In** this chapter you will learn how to change WordPerfect's start-up defaults and how to create macros. Both are extremely useful skills that will help you streamline your work and enable you to customize WordPerfect to meet your needs.

Default values are the formatting codes and commands that are automatically in operation when you start WordPerfect. *Macros* automate those operations where you use the same keystrokes over and over again. Macros are handy time-savers when you need to write the same word or phrase many times in one or more documents. You'll learn the following in this chapter:

Creating automatic backup files

Changing default formatting codes

Changing units of measurement

Creating long document names

Creating and using macros

■ *Taking a Look at the Setup Menu*

So far you have learned to change the WordPerfect default options only for the document you are creating. When you exit the program, the changes you have made revert to the original defaults. But what if you always need hyphenation turned on and justification turned off, instead of the default, which is the other way around? With the Setup menu you can change the defaults to new values that will be in place every time you start WordPerfect. Setup also has an automatic backup option that saves your work at intervals you specify—an almost effortless method you can use to avert the irritation caused by lost data due to a power failure.

WordPerfect's Setup menu offers six choices with which you can customize your operations. Although there are many more Setup options than I can discuss in a book of this nature, we'll take a quick look at each one of these options first, then explore more deeply a few that are important for your most immediate needs.

Start at a blank editing screen.

Tired of always having to go through a series of menus to change margins and spacing? Don't despair—you can use the Setup menu to change the defaults.

■ *1*   Select File, Setup or press SHIFT F1.

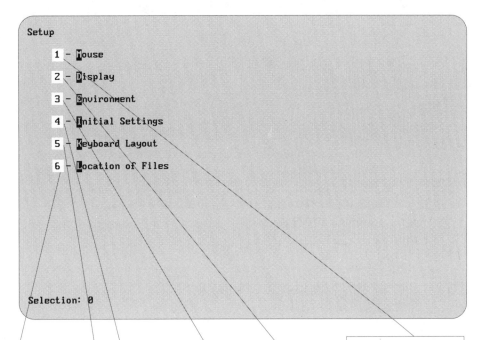

```
Setup
     1 - Mouse
     2 - Display
     3 - Environment
     4 - Initial Settings
     5 - Keyboard Layout
     6 - Location of Files




Selection: 0
```

Tells WordPerfect which directories contain your document files, graphics, printer files, macros, and the dictionary and thesaurus lexicons.

Sets operating parameters that govern the operation of the mouse.

Changes the appearance, color, size, and presentation style of the screen display.

Changes the use of the function keys; for example, the F1 key can be changed to function as the Esc key and the Esc key as the Cancel key.

Permanently sets initial formatting values, the date format, print options, and other advanced options.

Structures the WordPerfect operating environment: cursor speed, automatic backup, units of measurement, hyphenation rules, etc.

14

▪ *Creating Automatic Backup Files*

WordPerfect uses two methods to create backups of your documents—timed backups and original backups.

Timed Backups

The timed backup feature has saved me from losing my work on countless occasions. Remember, if you haven't saved the text you see on your monitor and your computer loses power for any reason, you will lose all the work you've done up to that point. With Timed Backup operating, your document is automatically saved at a time interval of your choice. These timed backups are automatically erased when you exit WordPerfect by selecting **F**ile, **E**xit or pressing [F7].

▪ *1*   Select **F**ile, **S**etup, **E**nvironment or press
[SHIFT][F1][3] or [E].

```
Setup: Environment

    1 - Backup Options

    2 - Beep Options
```

■*2*   Select **B**ackup Options or press 1 or B.

```
Setup: Backup
        Timed backup files are deleted when you exit WP normally.  If you
        have a power or machine failure, you will find the backup file in the
        backup directory indicated in Setup: Location of Files.

        Backup Directory

     1 - Timed Document Backup              Yes
         Minutes Between Backups           30


     Original backup will save the original document with a .BK! extension
     whenever you replace it during a Save or Exit.

     2 - Original Document Backup           No




Selection: 0
```

■*3*   Select **T**imed Document Backup or press 1
or T.

14

```
        Backup Directory

     1 - Timed Document Backup             Yes (No)
         Minutes Between Backups          30
```

■*4*   Select **Y**es or press ⦋**Y**⦌, type the number of
minutes you want for the time interval (*10* in
this example), then press ⦋↵⦌.

```
        Backup Directory

   1 - ▊imed Document Backup              Yes
       Minutes Between Backups            10
```

■*5*   ⦋**F7**⦌ or click the right mouse button to return
to the document screen.

The Timed Backup setting is saved and is activated every time you start Word-
Perfect. WordPerfect makes a copy of your document in the time interval you
specify (10 minutes in this example). By default, the copy will be located in the
WordPerfect subdirectory unless you designate a separate directory (as described in the next exercise).

Since you can't do anything at the keyboard while WordPerfect creates the automatic backup, I suggest not having too small a time interval, especially when
you work on large documents. Creating a backup every minute or two will interfere with your work and slow you down.

Creating an Auxiliary Backup-File Directory

WordPerfect saves your timed backup files in the WordPerfect subdirectory unless you designate another location. It is helpful, but not necessary, to create such
an auxiliary backup-file directory to make it easier to locate and retrieve your
timed backups.

Let's store backup files in another location. First, however, you must create a
subdirectory under the WordPerfect subdirectory, which in this case is named
WP51. Exit WordPerfect to the C:\WP51 prompt. Type **MD BACKUPS** and
press ⦋↵⦌. Restart WordPerfect. (The following exercise is for a computer with a
hard disk, but the same principles apply if you are using a dual floppy-disk
machine.)

■ *1* Select **F**ile, **S**etup, **L**ocation of Files or press
SHIFT F1 6 or L .

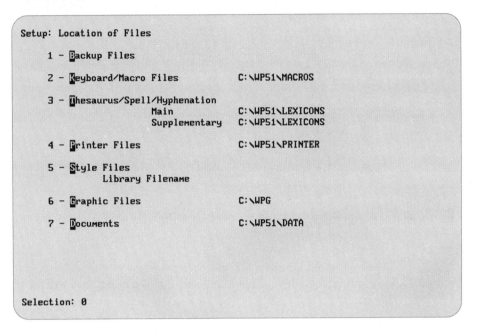

```
Setup: Location of Files

    1 - Backup Files

    2 - Keyboard/Macro Files          C:\WP51\MACROS

    3 - Thesaurus/Spell/Hyphenation
                        Main          C:\WP51\LEXICONS
                        Supplementary C:\WP51\LEXICONS

    4 - Printer Files                 C:\WP51\PRINTER

    5 - Style Files
            Library Filename

    6 - Graphic Files                 C:\WPG

    7 - Documents                     C:\WP51\DATA




Selection: 0
```

This screen displays a complete list of the types of auxiliary files to which you can assign directory locations. Notice that I have created directories for several of these file groupings as an example to guide you in creating your own.

14

■*2* Select **B**ackup Files or press 1 or B , then
type **C:\WP51\BACKUPS**. Press ⏎ . (If
you have dual floppy disks, you can enter *B:*
instead of *C:* so that the backup files will be
saved to your document disk in drive B.)

```
Setup: Location of Files

    1 - Backup Files                   C:\WP51\BACKUPS

    2 - Keyboard/Macro Files           C:\WP51\MACROS
```

■*3* F7 or click the right mouse button to return
to the document screen.

All your backup files will automatically be stored in the BACKUPS directory.
This makes it much easier to locate your backup files when you need them, because you don't have to comb through all your WordPerfect program files.

Original Backups

When you retrieve a document to edit it, there is still a copy of it stored on the
disk. When you save the edited document, it overwrites the old, stored version.
The old version, however, may contain needed information that is not in the new
version. With the Original Backup feature, you can copy the old version to a
backup file before the new version overwrites it. You can then retrieve the old
version if you need to refer to it.

It is very easy to activate this feature. Let's do it.

■ *1* Select **F**ile, **Se**tup, **E**nvironment, **B**ackup Options or press [SHIFT][F1], [3] or [E], [1] or [B].

> Original backup will save the original document with a .BK! extension
> whenever you replace it during a Save or Exit.
>
> 2 - Original Document Backup   No

■ *2* Select **O**riginal Document Backup, **Y**es or
press [2] or [O], [Y].

> Original backup will save the original document with a .BK! extension
> whenever you replace it during a Save or Exit.
>
> 2 - Original Document Backup   Yes

■ *3* [F7] or click the right mouse button to save
the new setting and return to the document
screen.

Original backups are always stored in the same directory that the original file is
stored in.

14

Note Original backups take up disk space equal to the size of the original file. If
your disk is nearly full, you may not be able to save a large file. If this happens, turn
off the Original Backup feature or use another disk.

Retrieving Backup Files

The original backup file has the same name as the newly saved file, except the extension changes to .BK!. For example, the backup file for a file called SETUP is named SETUP.BK!. You locate a backup file by selecting **F**ile, **L**ist **F**iles or by pressing **F5**. You give the backup file a new name by selecting **M**ove/Rename or by pressing **3** or **M** and retrieving it like any other file.

Timed backups are saved under the name WP{WP}.BK1 if you are using the Doc 1 screen, or WP{WP}.BK2 if you are using Doc 2. You can recall these files using these file names, but it is more convenient if you rename them before you recall them. When you have to restart WordPerfect because of a power failure, a malfunction, or because you did not exit properly, you see this prompt:

Are other copies of WordPerfect currently running? (Y/N)

Press **N** for No; the computer thinks WordPerfect is already in operation because there are timed backup files. Before starting to work, follow these steps to delete or rename the files:

- Select **F**ile, **L**ist **F**iles or press **F5**.
- Locate the backup files.
- Select **D**elete or press **2** or **D** to delete the files; select **M**ove/Rename or press **3** or **M** to rename them.

Remember to look in your auxiliary backup subdirectory if you've created one.

If you do not rename or delete these timed backup files before you begin working, you see the following prompt when WordPerfect tries to make the first timed backup of your next document:

Old backup file exists. 1 **R**ename; 2 **D**elete:

Select **R**ename or press **1** or **R** to rename the old timed-backup files. Select **D**elete or press **2** or **D** to delete them.

Changing
■*Default Formatting Codes*

WordPerfect automatically assigns default margins, tabs, line spacing, justification, and more—the list is quite long. If you consistently use formatting settings that are different from these defaults, you can change the defaults.

For example, suppose you want to change justification so it is turned off instead of on every time you start WordPerfect. You can follow the procedure below; the method is basically the same for changing other defaults.

■ *1*   Select **F**ile, **S**etup or press (SHIFT)(F1).

```
Setup

      1 - Mouse

      2 - Display

      3 - Environment

      4 - Initial Settings

      5 - Keyboard Layout

      6 - Location of Files













Selection: 0
```

14

■*2*   Select Initial Settings or press **4** or **I**.

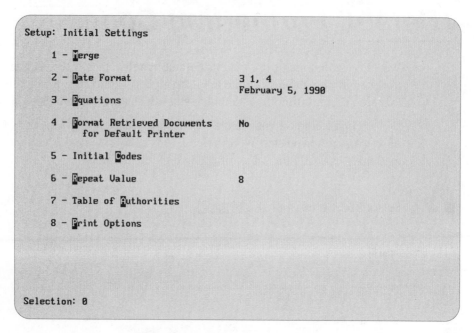

```
Setup: Initial Settings

     1 - Merge

     2 - Date Format                    3 1, 4
                                        February 5, 1990
     3 - Equations

     4 - Format Retrieved Documents     No
           for Default Printer

     5 - Initial Codes

     6 - Repeat Value                   8

     7 - Table of Authorities

     8 - Print Options




Selection: 0
```

■*3*   Select Initial Codes or press **5** or **C**. You
see a Reveal Codes screen.

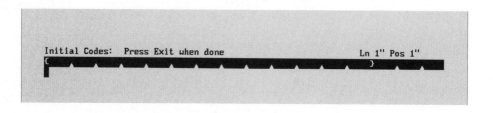

```
Initial Codes:  Press Exit when done                          Ln 1" Pos 1"
```

■*4* Select **L**ayout, **L**ine or press SHIFT F8
1 or **L**.

```
     3 - Justification              Full

     4 - Line Height                Auto

     5 - Line Numbering             No
```

■*5* Select **J**ustification, **L**eft or press **3** or **J**
1 or **L**.

```
     3 - Justification              Left

     4 - Line Height                Auto

     5 - Line Numbering             No
```

■*6* **F7** or click the right mouse button. The
newly selected default-formatting codes will
appear below the tab ruler.

14

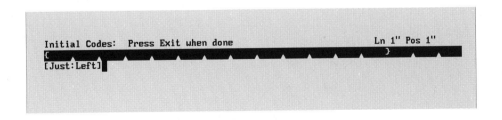

```
Initial Codes:   Press Exit when done                        Ln 1" Pos 1"
[                                                             }
[Just:Left]
```

■*7* [F7] twice to return to your document.

From now on, your document will automatically be left-justified when you start WordPerfect. To select another justification alignment, follow the above steps to select the new alignment, then delete the [Just:Left] code when you get to the Reveal Codes screen. Keep in mind that any formatting codes changed with the Setup command are not displayed on a normal document Reveal Codes screen.

■*Changing Units of Measurement*

The line (Ln) and position (Pos) indicator default measurements are in inches, which is great for doing most page layouts, especially those with graphics. There are other units of measurement available, such as centimeters, points, and WordPerfect 4.2 units. (WordPerfect 4.2 is the software version that came out prior to WordPerfect 5 and 5.1. It uses *units* (one unit = 1 character space) as its system of measurement.)

Follow these steps to change the system of measurement to WordPerfect 4.2 units:

■*1* Select **F**ile, **S**etup, **E**nvironment or press
[SHIFT] [F1] [3] or [E].

```
6 - Hyphenation              External Dictionary/Rules

7 - Prompt for Hyphenation   When Required

8 - Units of Measure
```

■*2*   Select **U**nits of Measure or press `8` or `U`.

```
Setup: Units of Measure

     1 - Display and Entry of Numbers              "
            for Margins, Tabs, etc.

     2 - Status Line Display                       "


Legend:

     " = inches
     i = inches
     c = centimeters
     p = points
     w = 1200ths of an inch
     u = WordPerfect 4.2 Units (Lines/Columns)
```

```
Selection: 0
```

■*3*   Select **D**isplay and Entry of Numbers or press
       `1` or `D`. Press `U`.

```
Setup: Units of Measure

     1 - Display and Entry of Numbers              u
            for Margins, Tabs, etc.

     2 - Status Line Display                       "
```

14

■*4*   Select Status Line Display or press
       [2] or [S]. Press [U].

```
Setup: Units of Measure

    1 - Display and Entry of Numbers            u
          for Margins, Tabs, etc.

    2 - Status Line Display                     u
```

■*5*   [F7] or click the right mouse button to return
       to the editing screen.

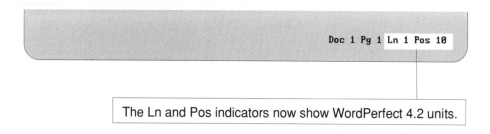

```
                                    Doc 1 Pg 1 Ln 1 Pos 10
```

The Ln and Pos indicators now show WordPerfect 4.2 units.

■*6*   Before moving on to the next exercise, reset
       Units of Measure to inches by repeating steps
       1–5 above and selecting the " = *inches* option.

Setting Up WordPerfect ■*to Use Long File Names*

As you know, DOS limits the length of a file name to eight characters with a three-character extension. A file name of this length often does not provide enough information to effectively describe the contents of the file. Word-Perfect 5.1 provides a method of naming files with longer, more descriptive file names. To activate this feature, you must use WordPerfect's Setup feature.

At a clear editing screen, do the following:

■*1* Select **F**ile, Set**u**p, **E**nvironment or press SHIFT F1 3 or E.

```
    4 - Document Management/Summary

    5 - Fast Save (unformatted)          No

    6 - Hyphenation                      External Dictionary/Rules
```

■*2* Select **D**ocument Management/Summary or press 4 or D.

14

```
Setup: Document Management/Summary

    1 - Create Summary on Save/Exit    No

    2 - Subject Search Text            RE:

    3 - Long Document Names            No

    4 - Default Document Type
```

■*3* Select **L**ong Document Names, **Y**es or press
③ or Ⓛ, Ⓨ.

```
Setup: Document Management/Summary
    1 - Create Summary on Save/Exit    No
    2 - Subject Search Text            RE:
    3 - Long Document Names            Yes
    4 - Default Document Type
```

■*4* F7 or click the right mouse button to return
to the editing screen.

The Long Document Name option is now active. Let's retrieve a file to the
screen and save it under a long file name.

■*5* Select **F**ile, **R**etrieve or press SHIFT F10, and
retrieve the file MYLETTER.

```
November 14, 1990


Mr. Randall Jones
2153 Ocean View Drive
Burbank, CA 91505
```

■*6*   Select **F**ile, **S**ave or press `F10`, and type
Randall Jones - lost shipment.

```
Brian Knowles
Customer Service Manager
Long Document Name: Randall Jones - lost shipment
```

■*7*   `↵` and type **LETTER**. The Document Type
description is used simply to categorize certain types of documents to help you find files.

```
Brian Knowles
Customer Service Manager
Long Document Type: LETTER
```

■*8*   `↵` You are prompted for the regular eight-
character file name. You can edit the displayed file name, if you wish.

14

```
Brian Knowles
Customer Service Manager
Document to be saved: C:\WP51\DATA\MYLETTER
```

■ *9* ⏎ and select **Y**es or press **Y** to accept
the current MYLETTER name and return
to the editing screen.

■ *10* Select **F**ile, List **F**iles or press **F5**, and
press ⏎. Then move the highlight down
the list of files until you reach the
MYLETTER file.

```
02-05-90  01:45p              Directory C:\WP51\DATA\*.*
Document size:    1,355   Free:  3,510,272 Used:     819,286    Files:    66
Descriptive Name                  Type     Filename      Size    Revision Date

                                           LETTER   .FIL    1,665   06-27-89 09:41a
                                           LETTER1  .WP5      873   04-05-89 05:00p
                                           LETTERWP.          761   11-14-89 04:51p
                                           MACRO    .LST    1,170   05-23-88 09:46a
                                           NEWFMT   .       11,141   07-23-89 06:00p
                                           PERSONAL.          474   01-19-90 02:33p
                                           PROSE    .DOC    5,086   06-11-88 10:11a
                                           PROSE    .PRC    1,462   06-11-88 10:12a
Randall Jones - lost shipment  LETTER      MYLETTER.        1,297   02-05-90 01:37p
                                           READING  .LST    2,523   02-14-89 09:50a
                                           RECORDS  .WP5      567   12-13-88 02:22p
                                           RESUME   .DOC    4,243   07-27-88 02:17p
                                           RESUME1  .DOC      617   07-26-88 01:04p
                                           RICHARD  .LTR    1,774   01-19-90 04:09p
                                           SALES    .       18,139   12-08-89 12:54p
                                           SCRIPT   .FMT      571   07-19-88 04:35p
                                           SPLCHK   .W51      412   01-31-90 08:48p
                                           STUEVAL  .DIA    5,996   07-27-88 04:05p

 1 Retrieve; 2 Delete; 3 Move/Rename; 4 Print; 5 Short/Long Display;
 6 Look; 7 Other Directory; 8 Copy; 9 Find; N Name Search: 6
```

Long file names and
types appear on the
left side of the screen.

Regular file names
appear on the right
side of the screen.

Switches between
long document
names and the
shorter, eight-
character names.

■ *11* [F7] or click the right mouse button to return
to the editing screen.

■*Creating Macros*

A *macro* contains stored information that is recalled to the screen by pressing
keys that you assign, linking those keys to the macro. When you use a macro,
you place the cursor where you want the information to appear and press the
macro keys that you assigned. For example, instead of having to type in an
often-used chemical formula each time, you can create a macro that contains the
text

CH_3CH_2OH

and assign it to the key Alt-C. Whenever you need to use that formula, you can
press Alt-C instead of typing in the text.

There are three types of macros in WordPerfect: *named* macros, *Alt* macros (like
the one in the example), and *temporary* macros. Let's create one of each. First,
clear your screen.

Named Macros

14

A named macro can have one to eight characters in its name.

■ *1* Select **T**ools, **Ma**cro, **D**efine or
press [CTRL][F10] and type **AUTO** for
the macro's name.

```
Define macro: AUTO
```

■*2* ⏎ and type **automotive parts**.

Description: automotive parts

You see this description only when you edit the macro. You can use up to 39 characters to describe what the macro contains, but it is not necessary to type a description.

■*3* ⏎ You can now create the contents of the macro.

Macro Def   Doc 1 Pg 1 Ln 1" Pos 1"

Note *Mistakes and Corrections in Macros* All keystrokes, including corrections of mistakes you make when creating a macro, become a part of that macro. These correction keystrokes are executed when the macro is activated. (However, since macros execute very rapidly, you don't actually see the corrections being made on the screen.) You may not notice a mistake until you activate the macro for the first time, in which case you'll want to fix it. WordPerfect offers a sophisticated macro-editing feature for advanced WordPerfect users. Your first macros will probably be fairly simple, so I recommend you re-create them from scratch if you make any mistakes.

■ *4* Select **L**ayout, **A**lign, **C**enter or press
⌗SHIFT⌗⌗F6⌗, and type **Automotive Parts Requi-sition**.

> Automotive Parts Requisition

■ *5* Select **T**ools, **Ma**cros, **D**efine or press
⌗CTRL⌗⌗F10⌗ to close the macro. You see the con-tents of the macro displayed on the screen.

Whenever you need to use this text, just select **T**ools, **Ma**cro, **E**xecute or press ⌗ALT⌗⌗F10⌗, and type in the macro name. You'll use this macro in just a moment.

A copy of the macro's contents always appears on your document screen as you create the macro, unless the macro consists entirely of hidden codes. Such a macro cannot be seen on the editing screen, but these codes do become a part of your document and can adversely affect the look of the document from their location onward. To avoid this problem, I suggest that you create macros on a blank document screen, not in your document.

Alt Macros

Alt macros are two-key macros consisting of the Alt key and a letter key, like ⌗ALT⌗⌗A⌗. Clear the screen and do the following:

■ *1* Select **T**ools, **Ma**cro, **D**efine or press
⌗CTRL⌗⌗F10⌗.

> Define macro:

14

■*2*   ALT A

Description:

You press ALT A to activate this macro. Alt macros are easier to remember if you name them mnemonically. (I chose ALT A because this is a macro for auto parts, and *automobile* begins with the letter *A*.)

■*3*   ↵ Do not enter a description for this macro.

Macro Def   Doc 1 Pg 1 Ln 1" Pos 1"

■*4*   Type **Engine** to begin creating the macro's contents.

Engine

■*5*   TAB three times and type **Electrical**.

Engine   Electrical

■*6* ⌨TAB three times, type **Suspension**, and
press ⏎

```
Engine              Electrical              Suspension
-
```

■*7* Select **T**ools, **M**acro, **D**efine or press CTRL F10
The macro is closed and completed.

We will use this macro in a moment.

Temporary Macros

A temporary macro is saved under the name WP{WP}.WPM. You can only
create a single temporary macro. To activate a temporary macro, you simply
press ⏎—you do not have to type in the macro name.

At a clear screen, do the following:

■*1* Select **T**ools, **M**acro, **D**efine or press CTRL F10

```
Define macro:
```

■*2* ⏎

```
Macro Def                          Doc 1 Pg 1 Ln 1" Pos 1"
```

You cannot assign a special name to a temporary macro.

■*3*   Press ⬚ (asterisk) as many times as is necessary to create a line of asterisks running to the right margin, which should be set at Pos 7.5".

****

■*4*   Select **T**ools, **M**acro, **D**efine or press `CTRL``F10` to close the macro.

■*Using Macros*

You have created each type of macro: named, Alt, and temporary. Let's activate them in a document. At a clear screen, do the following:

■*1*

Select **T**ools, **M**acro, **E**xecute or press `ALT``F10`. Type **AUTO**. This is the name of the named macro.

```
Macro: AUTO
```

■*2*   ⏎ The AUTO macro appears on your
screen.

```
              Automotive Parts Requisition
```

■*3*   ⏎ four times. Press ALT A to activate the
Alt macro.

```
              Automotive Parts Requisition


   Engine            Electrical           Suspension
```

You do not have to select menu options or press ALT F10 to activate Alt macros.

■*4*   Select **T**ools, **M**acro, **E**xecute or
press ALT F10.

■*5*   ⏎ to activate the temporary macro.

```
              Automotive Parts Requisition


   Engine            Electrical           Suspension
   xxxxxxxxxxxxxxxxxxxxxxxxxxxxxxxxxxxxxxxxxxxxxxxxxxxxxx
```

14

If you often have to deal with large mailing lists, read Chapter 15, which introduces you to WordPerfect's powerful mail-merge feature.

15

- *Creating Form*
- *Letters with the*
- *Merge Feature*

■ **A**nyone who has a mailbox knows what form letters are. Businesses and other organizations constantly mail these letters to promote their products, solicit support, or inform their audience. And because your name is on the letter, it all seems so personal. My favorites are sweepstakes mailings. They always spell my name correctly, speak of my town as though they lived next door, and try hard to make me feel like I'm special— but I never win. I always wondered how they did this until I discovered merge operations on the computer. It was a bit disillusioning; now I don't feel so special, and I still don't win. WordPerfect's Merge feature is powerful and versatile and can be used for personal projects as well as for business. You'll explore these merge operations:

Creating the primary file

Creating the secondary file

Merging the primary and secondary files

Merging to the printer

■*Creating the Primary File*

The *primary file* is the main document you use in a merge operation. You create this document once and then merge names, addresses, phone numbers, or other specific data with it from a *secondary file* (which will be covered in the exercise following this one). This merge is done by placing *field codes* in the primary document that are tied to data located in the secondary file. A *field* represents a specific category of information that is common to each letter. For example, name, address, and phone number may be a common item in each letter, but the information will be different for each letter. So, the letter in the primary file has a *field code* for each of these categories and pulls the different names, addresses, and phone numbers from the secondary file.

> **Note**   WordPerfect 5.1 allows you to insert merge codes in headers, footers, endnotes, footnotes, and graphic boxes that contain text.

Let's create a primary file containing the necessary field codes. Form letters are the most common type of merge document, so we'll create one of these. Start WordPerfect with a clear screen.

■*1*   Select **T**ools, **D**ate Code or press `SHIFT` `F5` `2` or `C` to place the current date code in the document.

```
August 10, 1990
```

> **Note** The date code, while placing the current date in your document when you first create it, in the future, tells WordPerfect to automatically place the new current date, according to your computer system clock, in the document when you retrieve it. The new date replaces the previous date.

■2   ⏎ twice and select **T**ools, **Me**rge Codes, **F**ield or press `SHIFT` `F9` `1` or `F` to place the first field code in your form letter.

```
Enter Field:
```

■3   Type **1** and press ⏎.

```
August 10, 1990

{FIELD}1~
```

> You must assign each field a number. {Field}1~ will contain names that come from the secondary file you will create.

■4   ⏎ and select **T**ools, **Me**rge Codes, **F**ield or press `SHIFT` `F9` `1` or `F`.

```
Enter Field:
```

15

■5   Type **2** and press ⏎ to create the second
field, which will contain addresses.

```
August 10, 1990

{FIELD}1~
{FIELD}2~
```

■6   ⏎ and select **T**ools, **Mer**ge Codes, **F**ield or
press SHIFT F9 1 or F.

```
Enter Field:
```

■7   Type **3** and press ⏎ to create the third field,
which will contain phone numbers.

```
August 10, 1990

{FIELD}1~
{FIELD}2~
{FIELD}3~
```

■8   ⏎ twice, type **Dear**, and press the spacebar.
All your letters will contain this greeting.

```
August 10, 1990

{FIELD}1~
{FIELD}2~
{FIELD}3~

Dear _
```

■9   Select **T**ools, **M**erge Codes, **F**ield or press
        [SHIFT] [F9] [1] or [F], type **1**, and press [↵] (to
        create another name field).

```
August 10, 1990

{FIELD}1~
{FIELD}2~
{FIELD}3~

Dear {FIELD}1~
```

■10   Press [,] (comma), then press [↵] twice and
        type the body of the letter as shown below.

```
August 10, 1990

{FIELD}1~
{FIELD}2~
{FIELD}3~

Dear {FIELD}1~,

This is just a reminder that our annual holiday fund raising
projects are getting into full swing. Your generous support in the
past has been greatly appreciated and we hope you are able to
contribute again this year. Have a Happy Holiday!

Sincerely,


Hap E. Day
Projects Manager



                                              Doc 1 Pg 1 Ln 4" Pos 2.6"
```

15

This is your primary file.

■ *11*   Select File, Exit, **Y**es or press F7 Y to
save your document. (Make sure the Long
Document Name option is turned off.) Name it
LETTER1, press ↵, and select **N**o or press N
to remain in WordPerfect with a clear screen.

Note   You can create as many fields as your disk is capable of storing. You can
use these fields as often as you want throughout the document.

■*Creating the Secondary File*

The *secondary file* consists of *records*. A record contains data that form the fields
that are merged into the primary document. A grouping of one name, address,
and phone number makes up one record, a second grouping of a name, address, and phone number makes up a second record, and so on. If you plan to
send letters to 100 people, you must create 100 records containing the field data
for each individual.

Let's create a secondary file that consists of three records. You'll use three
records to create three copies of the form letter that you typed in the last exercise. Start at a clear screen.

■ *1*   Type **Mr. Dan Fox**, the name in the first record. This is Field 1. Do not press ↵. This
data will appear wherever you placed a
{Field}1~ code in the primary file.

```
Mr. Dan Fox
```

■2   [F9] to place an {END FIELD} code at the
end of the first field.

```
Mr. Dan Fox{END FIELD}
```

■3   Type the address below. Press [↵] after the
street name and [F9] after the zip code. This is
Field 2.

```
Mr. Dan Fox{END FIELD}
213 Ash St.
Waco, TX 71121{END FIELD}
```

■4   Type the phone number below and press [F9].
This is Field 3.

```
Mr. Dan Fox{END FIELD}
213 Ash St.
Waco, TX 71121{END FIELD}
(902)555-0711{END FIELD}
```

15

■5 Select **T**ools, **M**erge Codes, **E**nd Record or
press `SHIFT` `F9` `2` or `E`.

```
Mr. Dan Fox{END FIELD}
213 Ash St.
Waco, TX 71121{END FIELD}
(902)555-0711{END FIELD}
{END RECORD}
================================================================================
```

The {END RECORD} code, followed by
the hard page-break lines, indicates the
end of the record.

■6 Add the remaining two records to the file, as
shown below; just follow the steps above.

```
Mr. Dan Fox{END FIELD}
213 Ash St.
Waco, TX 71121{END FIELD}
(902)555-0711{END FIELD}
{END RECORD}
================================================================================
Ms. Amy Walsh{END FIELD}
22 Kling St.
Banks, CO 81444{END FIELD}
(728)555-6720{END FIELD}
{END RECORD}
================================================================================
Mr. Bob Thomas{END FIELD}
4316 Elm St.
Dayton, OH 21141{END FIELD}
(310)555-6789{END FIELD}
{END RECORD}
================================================================================




Field: 1                                           Doc 1 Pg 4 Ln 1" Pos 1"
```

You've completed the secondary merge file. Now save the file.

■ 7   Select **F**ile, **E**xit, **Y**es or press F7 Y, name
      the file **RECORDS**, then press ⏎ and select
      No or press N to remain in WordPerfect.

Note   *Merging Records with Missing Fields*   WordPerfect reads each record in the secondary file and inserts fields in the primary document in the listed numerical order. Therefore, each secondary-file record *must* have at least the same number of fields as called for in the primary document. If you have a record that is missing a field, such as a phone number, you must still place the {END RECORD} code in that field slot by pressing F9. This causes a blank space to appear in the printed document unless you recall the primary document and place a question mark between the field number and the tilde, as in *{END FIELD}3?~*. Use the normal editing keys to insert the question mark in an existing field, or, when creating a new field, select **T**ools, **M**erge Codes, **F**ield or press SHIFT F9 1 or F, then type the field number and the question mark before pressing ⏎.

Merging the Primary
■*and Secondary Files*

15

Now that you have completed the primary and secondary files, it's time to merge them. If you merge the two files you created in the last two exercises (LETTER1 and RECORDS), you will have three letters, each with a different name and address. Be careful if you have large files—you may run out of computer memory or room on your disk. If that happens, break your files into smaller files and then merge them.

Once you've merged the information, you are free to edit the documents, print them from the screen, or save them on the disk in their merged form. (You'll learn more about this in a moment.)

Let's merge LETTER1 and RECORDS on the document screen. Start at a clear screen.

■ *1* Select **T**ools, **M**erge or press CTRL F9
1 or **M**.

Primary file: (List Files)

■ *2* Type **LETTER1** and press ↵.

Secondary file: (List Files)

> When entering the names of primary and secondary files for margins, be sure to include the path name if the file is on a different disk.

■*3*   Type **RECORDS** and press ⏎.

You may now edit, save, or print the letters.

```
Hap E. Day
Projects Manager
=================================================================================
August 10, 1990

Mr. Bob Thomas
4316 Elm St.
Dayton, OH 21141
(310)555-6789

Dear Mr. Bob Thomas,

This is just a reminder that our annual holiday fund raising
projects are getting into full swing. Your generous support in the
past has been greatly appreciated and we hope you are able to
contribute again this year. Have a Happy Holiday!

Sincerely,



Hap E. Day
Projects Manager
                                                  Doc 1 Pg 3 Ln 4.17" Pos 2.6"
```

The final letter appears on your screen after the files have been merged.

■*Merging Directly to the Printer*   15

You may wish to print your documents as they are merged. If so, there are a few important codes that must be added to the primary file.

Let's try it. Clear your screen, and retrieve the LETTER1 file.

■ *1*   [HOME] [HOME] [↓]

```
Sincerely,


Hap E. Day
Projects Manager
```

■ *2*   Select **T**ools, Me**r**ge Codes, **P**age Off or
press [SHIFT] [F9] [4] or [P].

```
Sincerely,


Hap E. Day
Projects Manager{PAGE OFF}
```

The {PAGE OFF} code prevents a blank sheet of paper from
being sent through the printer after each page is printed.

■ *3*   Select **T**ools, Me**r**ge Codes, **M**ore or press
[SHIFT] [F9] [6] or [M].

```
August 10, 1990              ┌──────────────────────────────────────┐
                             │{ASSIGN}var~expr~                       │
{FIELD}1~                    │{BELL}                                  │
{FIELD}2~                    │{BREAK}                                 │
{FIELD}3~                    │{CALL}label~                            │
                             │{CANCEL OFF}                            │
Dear {FIELD}1~,              │{CANCEL ON}                             │
                             │{CASE}expr~cs1~lb1~...csN~lbN~~          │
This is just a reminder that our annual │{CASE CALL}expr~cs1~lb1~...csN~lbN~~    │
projects are getting into full swing. Yo│{CHAIN MACRO}macroname~          (^G)  │
past has been greatly appreciated and we│{CHAIN PRIMARY}filename~                │
contribute again this year. Have a Happy└──────────────────────────────────────┘
```

The majority of these codes are used for creating advanced
macros and will not be covered in this entry-level book.

■*4* ⬇ to move the highlight through the list and
place it on the {PRINT} code.

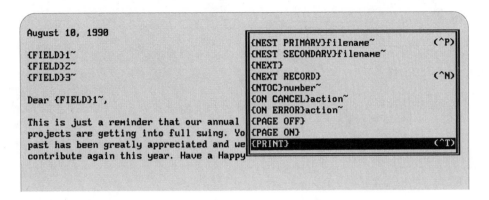

```
August 10, 1990

{FIELD}1~
{FIELD}2~
{FIELD}3~

Dear {FIELD}1~,

This is just a reminder that our annual
projects are getting into full swing. Yo
past has been greatly appreciated and we
contribute again this year. Have a Happy
```

```
{NEST PRIMARY}filename~          (^P)
{NEST SECONDARY}filename~
{NEXT}
{NEXT RECORD}                    (^N)
{NTOC}number~
{ON CANCEL}action~
{ON ERROR}action~
{PAGE OFF}
{PAGE ON}
{PRINT}                          (^T)
```

■*5* ⏎ to place the {PRINT} code in the
document.

```
Sincerely,


Hap E. Day
Projects Manager{PAGE OFF}{PRINT}
```

15

■*6* Select **File, Exit, Y**es or press F7 Y to save
the file with the new merge codes.

Now you're ready to print the letters. Turn on your printer, and follow these steps:

■ *1*   Select **T**ools, **M**erge or press `CTRL` `F9`
`1` or `M`.

Primary file: (List Files)

■ *2*   Type **LETTER1** and press `⏎`.

Secondary file: (List Files)

When you merge direct-
ly to the printer, you
can stop printing by
selecting File, Print,
Control Printer, Stop or
by pressing `SHIFT` `F7`, `4`
or `C`, `5` or `S`.

■*3*   Type **RECORDS** and press ⏎. All three letters print out.

August 10, 1990

Ms. Amy Walsh
22 Kling St.
Banks, CO 81444
(728)555-6720

Dear Ms. Amy Walsh

This is just a re
projects are getti
past has been gre
contribute again t

Sincerely,

August 10, 1990

Mr. Dan Fox
213 Ash St.
Waco, TX 71121
(902)555-0711

Dear Mr. Dan Fox,

This is just a reminder that our annual holiday fun

August 10, 1990

Mr. Bob Thomas
4316 Elm St.
Dayton, OH 21141
(310)555-6789

Dear Mr. Bob Thomas,

This is just a reminder that our annual holiday fund raising projects are getting into full swing. Your generous support in the past has been greatly appreciated and we hope you are able to contribute again this year. Have a Happy Holiday!

Sincerely,

Hap E. Day
Projects Manager

15

Chapter 16 introduces you to one of the most powerful features of WordPerfect: graphics. If you are interested in doing desktop publishing projects, such as creating informative and attractive newsletters and fliers, this chapter is indispensable. Don't miss it!

16

- *Working*
- *with*
- *Graphics*

■ **T**his chapter covers the most dynamic of all the features of WordPerfect 5.1: graphics. Using this feature, you can create vertical and horizontal lines of various widths and shades. You can also create boxes into which you can import and edit text, charts, diagrams, and graphics files. WordPerfect will accept graphics from a multitude of sources, including Lotus 1-2-3, PC Paintbrush, AutoCAD, GEM Paint, and TIFF files from scanners.

The full power of WordPerfect's graphics capability goes beyond the scope of this book, but I'll show you the basics to get you started right away.

In this chapter you'll learn about these features:

Creating horizontal and vertical lines

Creating boxes for text and graphics

Placing text in boxes

Importing WordPerfect graphics into a document

Importing graphics from other programs

Creating Horizontal ■and Vertical Lines

In the following exercises, you'll create a horizontal and a vertical line to add graphic interest to a book title page. As you work through the exercises in this section, keep in mind that you cannot see graphics on your document screen; you must print them or use the View Document screen by selecting File, Print, View Document or by pressing SHIFT F7 6 or V.

Drawing Horizontal Lines

Start WordPerfect with a clear editing screen.

■ *1*   Select **L**ayout, **A**lign, **C**enter or press
SHIFT F6. Type the following title:

```
                    GRAPHICS FOR WORDPERFECT 5.1_
```

```
                                              Doc 1 Pg 1 Ln 1" POS 5.65"
```

■*2* ⏎ to position the cursor on the line where you'll create a horizontal line. Then, select **Graphics, Line, Create H**orizontal or press ⎇ F9, 5 or L, 1 or H.

```
Graphics: Horizontal Line

    1 - Horizontal Position      Full

    2 - Vertical Position        Baseline

    3 - Length of Line

    4 - Width of Line            0.013"

    5 - Gray Shading (% of black) 100%
```

You'll use the options on this menu to specify how your line will look and where it will begin and end.

Let's create a shaded line centered below the title, 3 in. long and 0.1 in. wide. The line will be shaded at 50%. Shading percentage determines how dark a line appears when it is printed. Fifty-percent shading is half as dark as 100%, which is black.

■*3* Select **H**orizontal Position or press 1 or H.

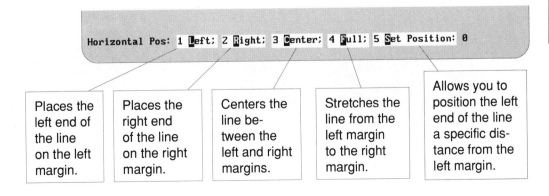

```
Horizontal Pos: 1 Left; 2 Right; 3 Center; 4 Full; 5 Set Position: 0
```

16

Places the left end of the line on the left margin.

Places the right end of the line on the right margin.

Centers the line between the left and right margins.

Stretches the line from the left margin to the right margin.

Allows you to position the left end of the line a specific distance from the left margin.

■ *4*   Select **C**enter or press **3** or **C** to center the
line under the title.

```
Graphics: Horizontal Line
     1 - Horizontal Position      Center
     2 - Vertical Position        Baseline
```

■ *5*   Select **L**ength of Line or press **3** or **L**.
Type **3** and press ⏎ to make the line 3 in.
long. (You don't have to type the inch mark.)

```
     3 - Length of Line           3"
     4 - Width of Line            0.013"
     5 - Gray Shading (% of black) 100%
```

■ *6*   Select **W**idth of Line or press **4** or **W**.
Type **.1** and press ⏎ to make the line
0.1 in. wide.

```
     3 - Length of Line           3"
     4 - Width of Line            0.1"
     5 - Gray Shading (% of black) 100%
```

■ 7   Select **G**ray Shading or press ⑤ or Ⓖ.
Type **50** and press ⏎ to shade the line at
50 percent of black.

```
3 - Length of Line          3"

4 - Width of Line           0.1"

5 - Gray Shading (% of black) 50%
```

■ 8   F7 or click the right mouse button to return
to the document screen. You can use the View
Document command to see the results prior to
printing.

Keep this document on the screen for use in the next exercise.

Note   *Determining Your Printer's Graphics Capabilities*   The type of printer you
have governs how many graphics features you can take advantage of. If you're not
sure what your printer will do, print the PRINTER.TST file that comes with Word-
Perfect. It includes special printing effects, such as underscore, subscripts, and
graphics. To print the PRINTER.TST file, select **F**ile, **R**etrieve or press SHIFT F10, type
PRINTER.TST, and retrieve the file to the editing screen. Then select **F**ile, **P**rint,
Full Document or press SHIFT F7 ① or Ⓕ. When the document prints, you will see
right away which features your printer cannot accommodate.

16

Drawing Vertical Lines

Using the document produced in the last exercise, let's create a vertical line.

■ *1*

Select **Graphics**, **Line**, Create **Vertical** or press `ALT``F9`, `5` or `L`, `2` or `V`.

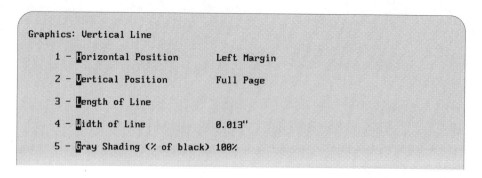

```
Graphics: Vertical Line

     1 - Horizontal Position      Left Margin

     2 - Vertical Position        Full Page

     3 - Length of Line

     4 - Width of Line            0.013"

     5 - Gray Shading (% of black) 100%
```

Let's create a line that will be 5 in. long and .05 in. wide. It will start 2.5 in. from the left margin and will be centered between the top and bottom margins.

■ *2*

Select **H**orizontal Position or press `1` or `H`.

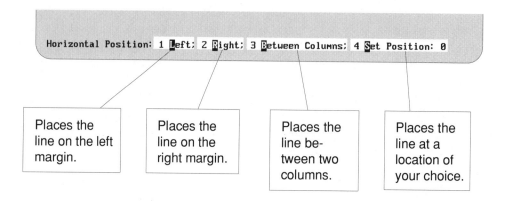

```
Horizontal Position: 1 Left; 2 Right; 3 Between Columns; 4 Set Position: 0
```

Places the line on the left margin.

Places the line on the right margin.

Places the line between two columns.

Places the line at a location of your choice.

■*3*   Select **S**et Position or press **4** or **S**.
Type **2.5** and press ⏎ to start the line 2.5 in.
from the left margin.

```
Graphics: Vertical Line

    1 - Horizontal Position      2.5"

    2 - Vertical Position        Full Page
```

■*4*   Select **V**ertical Position or press **2** or **V**.

```
Vertical Position: 1 Full Page; 2 Top; 3 Center; 4 Bottom; 5 Set Position: 0
```

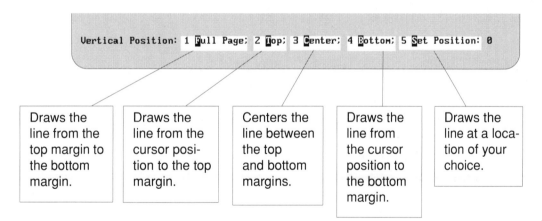

| Draws the line from the top margin to the bottom margin. | Draws the line from the cursor position to the top margin. | Centers the line between the top and bottom margins. | Draws the line from the cursor position to the bottom margin. | Draws the line at a location of your choice. |

■*5*   Select **C**enter or press **3** or **C**.

```
Graphics: Vertical Line

    1 - Horizontal Position      2.5"

    2 - Vertical Position        Center
```

16

■6 Select **L**ength of Line or press ③ or Ⓛ.
Type **5** and press ⏎.

```
3 - Length of Line          5"
4 - Width of Line           0.013"
5 - Gray Shading (% of black) 100%
```

■7 Select **W**idth of Line or press ④ or Ⓦ.
Type **.05** and press ⏎.

```
3 - Length of Line          5"
4 - Width of Line           0.05"
5 - Gray Shading (% of black) 100%
```

■8 [F7] or click the right mouse button to return
to the document screen.

You can use a
vertical line to
emphasize the
left or right
margin.

■*9*   ⏎ 20 times and type the credit lines below,
pressing Tab four times before typing each
line and placing five blank lines between
each line.

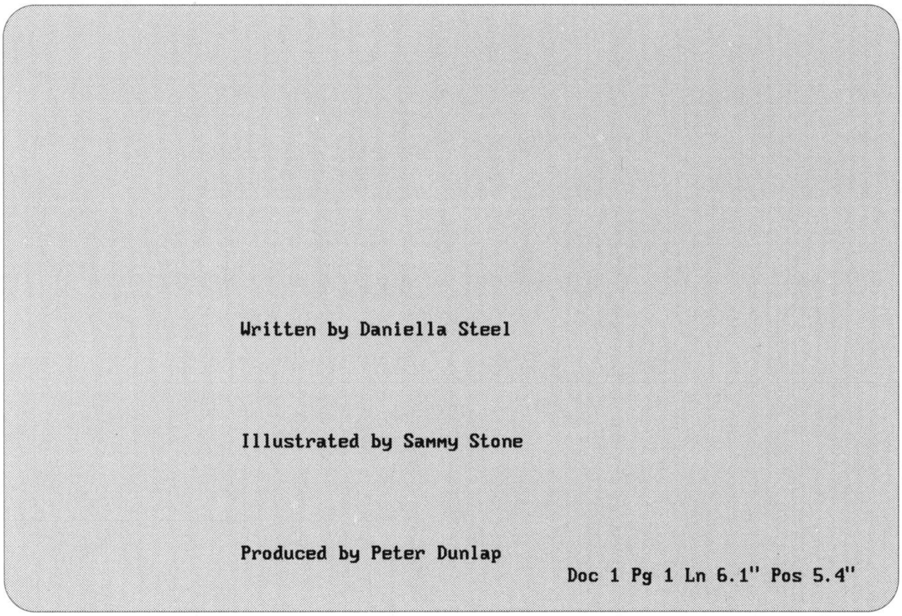

```
                         Written by Daniella Steel



                         Illustrated by Sammy Stone



                         Produced by Peter Dunlap
                                          Doc 1 Pg 1 Ln 6.1" Pos 5.4"
```

■*10* Select **F**ile, **P**rint, **P**age or press SHIFT F7
2 or **P**.

16

The printout of your title page should look like this:

```
        Written by Daniella Steel



        Illustrated by Sammy Stone



        Produced by Peter Dunlap
```

Creating Boxes
■*for Text and Graphics*

WordPerfect allows you to create boxes in which you can place text or graphics to make your documents look more interesting. There are five types of boxes: figure, table, text, user, and equation. These box titles help you organize your files. For example, a figure box is used for illustrations that support your text, a table box contains graphs or lists of data, a text box contains textual material, and a user box can contain anything not covered by the other three. (An equation box contains mathematical equations; WordPerfect's Equation Editor feature is not covered in this book.) However, you can place text in a figure box, an illustration in a table box, a graph in a text box, or any combination you wish—the mechanics of placing data in one type of box work exactly the same for any other type of box. The boxes are named differently simply to help you organize your files more efficiently.

Creating a Box Style

There are several box options that determine effects like the style and shading of the borders, where the caption (if there is one) is located, and the numbering style when a figure number is used.

Let's create a Text Box style with a double border. In a later exercise, you'll determine the location and size of the box, then place some text inside it. Start at a clear screen.

If you want to use a combination of numbered tables and figures, put your tables in table boxes and your figures in figure boxes—WordPerfect will keep track of your numbers automatically.

16

■*1* Select **G**raphics, Text **B**ox, **O**ptions or
press ALT F9, 3 or B, 4 or O.

```
Options: Text Box

        1 - Border Style
                Left                    None
                Right                   None
                Top                     Thick
                Bottom                  Thick
        2 - Outside Border Space
                Left                    0.167"
                Right                   0.167"
                Top                     0.167"
                Bottom                  0.167"
        3 - Inside Border Space
                Left                    0.167"
                Right                   0.167"
                Top                     0.167"
                Bottom                  0.167"
        4 - First Level Numbering Method   Numbers
        5 - Second Level Numbering Method  Off
        6 - Caption Number Style        [BOLD]1[bold]
        7 - Position of Caption         Below box, Outside borders
        8 - Minimum Offset from Paragraph  0"
        9 - Gray Shading (% of black)   10%

Selection: 0
```

The styles you see here are offered for all four types. The border-space dimensions are a measure of the distance between the text and the box border.

■*2* Select **B**order Style or press 1 or B.

```
        6 - Caption Number Style        [BOLD]1[bold]
        7 - Position of Caption         Below box, Outside borders
        8 - Minimum Offset from Paragraph  0"
        9 - Gray Shading (% of black)   10%

1 None; 2 Single; 3 Double; 4 Dashed; 5 Dotted; 6 Thick; 7 Extra Thick: 0
```

■*3*   Select **D**ouble four times or press ③ or Ⓓ
four times to place the same border style on
all sides of the box.

```
Options: Text Box

    1 - Border Style
            Left                    Double
            Right                   Double
            Top                     Double
            Bottom                  Double
```

■*4*   F7 or click the right mouse button to return
to the editing screen. Then, select **E**dit,
Reveal Codes or press ALT F3.

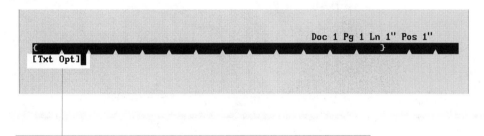

```
                                              Doc 1 Pg 1 Ln 1" Pos 1"
{                                                         }
[Txt Opt]
```

Every text box created after this code will have double
borders.

16

■*5*   Select **E**dit, **R**eveal Codes or press ALT F3 to
return to the editing screen, then go to the
next exercise.

■*Placing Text in Boxes*

Now that you've selected your box options and have determined how your box will look, let's fill it with text. You can type text directly in a box or retrieve a file from disk and place it in the box. You'll use the box option (double border) that you selected in the last exercise. First, you'll tell WordPerfect where you want the box positioned on the page, then you'll type text in it.

■ *1*   Select **G**raphics, Text **B**ox, **C**reate or
press `ALT``F9`, `3` or `B`, `1` or `C`.

```
Definition: Text Box

    1 - Filename

    2 - Contents         Text

    3 - Caption

    4 - Anchor Type       Paragraph

    5 - Vertical Position  0"

    6 - Horizontal Position  Right

    7 - Size              3.25" wide x 0.625" (high)

    8 - Wrap Text Around Box  Yes

    9 - Edit


Selection: 0
```

Prevents text outside a box from printing over the contents of the box.

Allows you to set a vertical page position in relation to the text or the top margin.

You select this option and type in a file name if you want to import a file or graphic into a box. The box size adjusts automatically to the size of the file.

■*2*   Select Anchor **T**ype or press [4] or [T] to
select what type of box this will be.

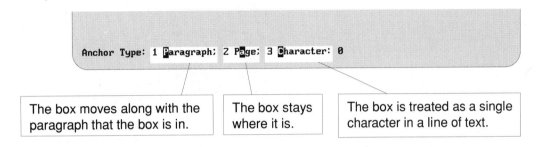

Anchor Type: 1 **P**aragraph; 2 P**a**ge; 3 **C**haracter: 0

| The box moves along with the paragraph that the box is in. | The box stays where it is. | The box is treated as a single character in a line of text. |

■*3*   Select P**a**ge or press [2] or [A] to select a
page type.

Number of pages to skip: 0

If you want to place the text box on a page after the page where the cursor is
located, enter after "Number of pages to skip" the number of pages you would
need to skip over to reach the page where you want the text box. For example, if
you want the box to appear five pages from the current page, you type *4* and
press [↵].

■*4*   [↵] to accept the current page.

16

4 - Anchor **T**ype   Page

5 - **V**ertical Position   Top

6 - **H**orizontal Position Margin, Right

■5　Select **H**orizontal Position or press **6** or **H**.

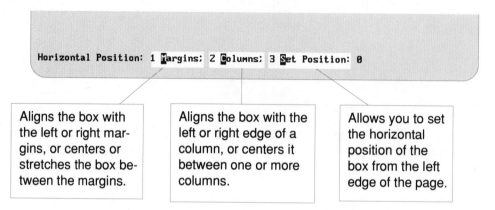

Horizontal Position: 1 **M**argins; 2 **C**olumns; 3 **S**et Position: 0

Aligns the box with the left or right margins, or centers or stretches the box between the margins.

Aligns the box with the left or right edge of a column, or centers it between one or more columns.

Allows you to set the horizontal position of the box from the left edge of the page.

■6　Select **M**argins or press **1** or **M**.

Horizontal Position: 1 **L**eft; 2 **R**ight; 3 **C**enter; 4 **F**ull: 0

■7　Select Center or press **3** or **C** to center the box between the margins.

```
6 - Horizontal Position  Margin, Center

7 - Size                 3.25" wide x 0.625" (high)

8 - Wrap Text Around Box Yes

9 - Edit
```

■*8*   Select **S**ize or press **7** or **S**.

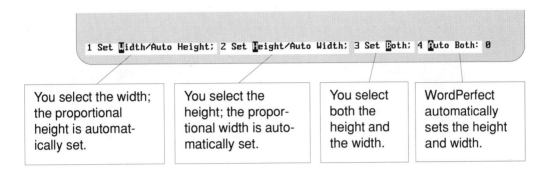

1 Set **W**idth/Auto Height; 2 Set **H**eight/Auto Width; 3 Set **B**oth; 4 **A**uto Both: 0

You select the width; the proportional height is automatically set.

You select the height; the proportional width is automatically set.

You select both the height and the width.

WordPerfect automatically sets the height and width.

You'll use the options mentioned above to create a box 3 in. wide and 1 in. high.

■*9*   Select Set **B**oth or press **3** or **B**.

Width = 3.25"

■*10*   Type **3** and press ⏎.

Height = 0.167"

16

■ *11* Type **1** and press ⏎

```
    7 - Size              3" wide x 1" high

    8 - Wrap Text Around Box Yes

    9 - Edit
```

Now that you've set the size and location of your box, you can add text to it.

■ *12* Select **Edit** or press **9** or **E** Then, select **Layout**, **Align**, **Center** or press **SHIFT** **F6** and type the following text:

```
    CONGRATULATIONS

This was the best sales
month in company history!
```

■ *13* **F7** twice, then ⏎ nine times. The TXT 1 notation shows the location of your text box in the document.

```
  ┌TXT 1─────────────────────
  │                          │
  │                          │
  │                          │
  │                          │
  └──────────────────────────┘


 _
```

■ *14* Select **F**ile, **P**rint, **V**iew Document or
press `SHIFT` `F7` `6` or `V` to see how the
box will print.

■ *15* `F7` or click the right mouse button to return
to the editing screen.

■ *16* Type the following text, or any other text you
wish, in your document. When you're done,
you will import a graphic into the document
and see how the text adjusts.

At long last the Rubber Baby Bumper Division has surpassed the
sales record set in 1962 by that legendary salesman Lance "tell 'em
you got it from me" Lassiter. Our mysterious accountant associates
from Wiggins, Watkins and Whong emerged into the sunlight yesterday for the first time since WWII and delivered the good news to the
Chief. The Chief was so pleased he authorized unlimited use of
the photocopier for one hour today between the hours of 11 P.M. and
midnight. I know it's Friday night, but break those dates, don't leave
town for the weekend, and let's meet at the photocopier at 10:50 P.M.
so we don't lose a minute of this valuable time. Bring your own soft
drinks; the soda dispenser is still broken. The Chief said if next
month's sales are better than this month's, he'll get the thing fixed by
summer.

So, keep up the good work, and remember, a baby-buggy without a
rubber baby bumper is a baby-buggy without a rubber baby bumper.

Keep smiling, and see you next month!

Keep this document on the screen for use in the next exercise.

16

Importing WordPerfect ■*Graphics into a Document*

WordPerfect provides 30 graphics files with your program. These images are included on the PTR Program/Graphics disks. Illustrations of these images are provided in your WordPerfect documentation in the "Graphic Images" section in Appendix B. The graphics files have a .WPG extension and can be easily located using the List Files feature.

Let's create a figure box below the text box you just created. Then we'll import a WordPerfect graphic into the box.

■ *1* Position the cursor on the first letter in the first sentence of the main body of text. If you are using the text from the last exercise, it will be the letter *A* in *At.*

WordPerfect offers
a panoply of
graphic images
that you can use
to enliven your
documents.

■2   Select **G**raphics, **F**igure, **O**ptions or press
ALT F9 , 1 or F , 4 or O .

```
Options: Figure

        1 - Border Style
                Left                        Single
                Right                       Single
                Top                         Single
                Bottom                      Single
        2 - Outside Border Space
                Left                        0.167"
                Right                       0.167"
                Top                         0.167"
                Bottom                      0.167"
        3 - Inside Border Space
                Left                        0"
                Right                       0"
                Top                         0"
                Bottom                      0"
        4 - First Level Numbering Method    Numbers
        5 - Second Level Numbering Method   Off
        6 - Caption Number Style            [BOLD]Figure 1[bold]
        7 - Position of Caption             Below box, Outside borders
        8 - Minimum Offset from Paragraph   0"
        9 - Gray Shading (% of black)       0%

Selection: 0
```

■3   Select **I**nside Border Space or press 3 or
I . Then type **.1** four times, pressing ⏎
after each entry.

```
        3 - Inside Border Space
                Left                        0.1"
                Right                       0.1"
                Top                         0.1"
                Bottom                      0.1"
        4 - First Level Numbering Method    Numbers
        5 - Second Level Numbering Method   Off
```

16

■*4* [F7] or click the right mouse button to return
to the editing screen.

Your figure box will have a single-line border with 0.1 in. between the border
and the graphic.

■*5* Select **G**raphics, **F**igure, **C**reate or
press [ALT][F9], [1] or [F], [1] or [C].

```
Definition: Figure
    1 - Filename
    2 - Contents         Empty
    3 - Caption
```

■*6* Select **F**ilename or press [1] or [F].

```
Enter filename:                                              (List Files)
```

■ *7*   Type **TROPHY.WPG** for the name of the
graphic you want to use, and press ⏎.
TROPHY.WPG is a graphic supplied by
WordPerfect. (If you are prompted that the
file cannot be found, include the path when
you enter the file name, for example,
C:\GRAPHICS\TROPHY.WPG.)

```
Definition: Figure

     1 - Filename          TROPHY.WPG

     2 - Contents          Graphic

     3 - Caption
```

■ *8*   Select Caption or press 3 or C. The figure
box number appears at the top of the screen.
It will print under the box if you need a caption number.

```
Figure 1
```

■ *9*   BKSP⏎ one time to delete *Figure 1*. Type the following caption for the graphic:

```
AWARD WINNING
```

16

■ *10* [F7] to return to the menu.

```
3 - Caption              AWARD WINNING

4 - Anchor Type          Paragraph

5 - Vertical Position    0"

6 - Horizontal Position  Right

7 - Size                 3.25" wide x 2.49" (high)
```

■ *11* Select Size or press [7] or [S].

```
1 Set Width/Auto Height; 2 Set Height/Auto Width; 3 Set Both; 4 Auto Both: 0
```

■ *12* Select Set **Width**/Auto Height or press
[1] or [W]. Type **1.5** for the width and
press [↵].

```
7 - Size                 1.5" wide x 1.18" (high)

8 - Wrap Text Around Box Yes

9 - Edit
```

The box height is automatically adjusted.

■ *13* [F7] or click the right mouse button to return to the document screen. Then press [↓] once to adjust the text.

```
┌TXT 1─────────────────────────────┐
│                                   │
│                                   │
│                                   │
│                                   │
│                                   │
└───────────────────────────────────┘

At long last the Rubber Baby Bumper Division has      ┌FIG 1───────────┐
surpassed the sales record set in 1962 by that        │                │
legendary salesman Lance "tell 'em you got it from     │                │
me" Lassiter. Our mysterious accountant associates     │                │
from Wiggins, Watkins and Whong emerged into the       │                │
sunlight yesterday for the first time since WWII       │                │
and delivered the good news to the Chief. The          │                │
Chief was so pleased he authorized unlimited use       │                │
of the photocopier for one hour today between the      │                │
hours of 11 P.M. and midnight. I know it's Friday      └────────────────┘
night, but break those dates, don't leave town for the weekend, and
let's meet at the photocopier at 10:50 P.M. so we don't lose a
minute of this valuable time. Bring your own soft drinks; the soda
dispenser is still broken. The Chief said if next month's sales are
better than this month's, he'll get the thing fixed by summer.
                                            Doc 1 Pg 1 Ln 2.67" Pos 1"
```

■ *14* Select **F**ile, **P**rint, **P**age or press [SHIFT][F7][2] or [P]. (Make sure you have full justification turned on.)

16

Your printout should look like this:

```
                    CONGRATULATIONS

               This was the best sales
               month in company history!
```

At long last the Rubber Baby Bumper Division has
surpassed the sales record set in 1962 by that
legendary salesman Lance "tell 'em you got it from
me" Lassiter. Our mysterious accountant associates
from Wiggins, Watkins and Whong emerged into the
sunlight yesterday for the first time since WWII
and delivered the good news to the Chief. The
Chief was so pleased he authorized unlimited use
of the photocopier for one hour today between the
hours of 11 P.M. and midnight. I know it's Friday   AWARD WINNING
night, but break those dates, don't leave town for the weekend, and
let's meet at the photocopier at 10:50 P.M. so we don't lose a
minute of this valuable time. Bring your own soft drinks; the soda
dispenser is still broken. The Chief said if next month's sales are
better than this month's, he'll get the thing fixed by summer.

So, keep up the good work, and remember, a baby-buggy without a
rubber baby bumper is a baby-buggy without a rubber baby bumper.

Keep smiling, and see you next month!

Importing Graphics
∎*from Other Programs*

If you work with programs that contain or generate their own graphics, you can retrieve those graphics into WordPerfect documents. The procedure is exactly the same as described in the previous section, "Importing WordPerfect Graphics into a Document." The only difference is in the file name you type in for the graphic you want to import. Graphics programs assign very specific extensions to their graphic file names. When you type in the file name, you must be sure to include the extension or the file will not load.

Here are some common graphics formats that can be used with WordPerfect, along with their extensions:

CGM	Computer Graphics Metafile
DHP	Dr. Halo PIC
DXF	AutoCAD
EPS	Encapsulated PostScript
GEM	GEM Draw Format
HPGL	Hewlett-Packard Graphics Language Plotter File
IMG	GEM Paint Format
MSP	Microsoft Windows Paint
PCX	PC Paintbrush Format
PIC	Lotus 1-2-3 PIC Format
PNTG	Macintosh Paint Format
PPIC	PC Paint Plus Format
TIFF	Tagged Image File Format
WPG	WordPerfect Graphics Format

16

If you have a graphics program that is not included in this list, consult its documentation to see whether the format is compatible with WordPerfect.

A

- **Formatting**
- **Floppy**
- **Disks**

■ **If** you try to use a new floppy disk straight out of the box, you'll find your computer can't read it or write information onto it. First, you have to format it using the Disk Formatting command (FORMAT.COM).

Formatting a floppy disk places magnetic recording tracks on it, similar to the grooves on a phonograph record. These grooves accept information generated within your computer's operating system. When this formatting is done, you can write data to the disk and read data off it. If you want to use the disk in a computer using a different operating system (such as an Apple), however, it will not work, and you will have to reformat it for use with that system. When you reformat a disk for use in another system, you erase any information currently recorded on the disk.

Before you begin the exercises in this appendix, please read the Introduction—it contains some important information on how to use this book.

■*Formatting Your Floppy Disks*

There are two sizes of floppy disk used in the majority of personal computers: 5¼" and 3½". Both of these disk sizes are available with two amounts of storage capacity: double-density (often referred to as low-density) and high-density. A 5¼" double-density disk can store 360K of data, and the high-density disk can store 1.2Mb of data. A double-density 3½" disk can hold 720K of data, and the high-density disk can hold 1.44 Mb. To give you an idea of how large a document can fit on a disk, a 360K double-density disk can hold approximately 250 double-spaced pages of text.

If you have a double-density disk drive, you cannot format or use high-density disks in it. However, you can format and use double-density disks in a high-density disk drive.

Formatting disks on a dual floppy-disk computer is different from formatting them on a hard disk computer. Let's look at both methods.

■*Dual Floppy-Disk Computers*

■ *1*　Insert the DOS disk in drive A, and start your computer. Ignore the date and time requests by pressing ⏎.

```
Current date is Wed 11-21-90
Enter new date (mm-dd-yy):
Current time is 12:20:45.08
Enter new time:


A>_
```

■*2* Type **format b:** and press ⏎.

```
A>format b:
Insert new diskette for drive B:
and strike ENTER when ready_
```

■*3* Insert the new floppy disk in drive B and
press ⏎.

```
Format complete

   1213952 bytes total disk space
   1213952 bytes available on disk

Format another (Y/N)?_
```

For a high-density disk drive (1.2 Mb). A double-density drive shows 360K.

■*4* **Y** to format another disk, then press ⏎.

■*5* Repeat steps 3 and 4 until all the blank disks
are formatted.

■*6* **N** ⏎ to return to the DOS prompt.

```
   1213952 bytes total disk space
   1213952 bytes available on disk

Format another (Y/N)?n
A>_
```

A

■*Hard Disk Computers*

Before you can format a new disk on a hard disk computer, you must locate the FORMAT.COM file. You'll find FORMAT.COM in your DOS program files. Many times DOS is located in the root directory, but it is often installed in its own directory. If it is, you have to change to that directory before attempting to format a disk.

■ *1*   Start your computer. Ignore the date and time
requests (if they appear on your computer
screen) by pressing ⏎.

```
Current date is Wed 11-21-90
Enter new date (mm-dd-yy):
Current time is 12:20:45.08
Enter new time:

C>_
```

If you are not
familiar with
DOS, please
read the Intro-
duction to this
book.

■2 Type **dir/w** and press ⏎ to check the
contents of the root directory. Look for
FORMAT.COM. If you do not find this file,
go to step 3. Otherwise, move on to step 4.

```
C>dir/w

 Volume in drive C is TURBO
 Directory of  C:\

.                 ..               4201     CPI   5202     CPI   ANSI     SYS
APPEND   EXE   ASSIGN   COM   ATTRIB   EXE   AUTOEXEC BAT   CAPTURE  COM
CHKDSK   COM   COMMAND  COM   COMP     COM   CONFIG   SYS   COPY
COUNTRY  SYS   DEBUG    COM   DISKCOMP COM   DISKCOPY COM   DISPLAY  SYS
DRIVER   SYS   EDLIN    COM   EGA      CPI   EXE2BIN  EXE   FASTOPEN EXE
FC       EXE   FDISK    COM   FIND     EXE   FORMAT   COM   GRAFTABL COM
GRAPHICS COM   GWBASIC  EXE   JOIN     EXE   KEYB     COM   KEYBOARD SYS
LABEL    COM   LCD      CPI   LINK     EXE   MODE     COM   MORE     COM
NLSFUNC  EXE   PRINT    COM   PRINTER  SYS   RAMDRIVE SYS   RECOVER  COM
REPLACE  EXE   RESTORE  COM   SELECT   COM   SHARE    EXE   SORT     EXE
SUBST    EXE   SYS      COM   TREE     COM   XCOPY    EXE
       54 File(s)    4278272 bytes free

C>_            _
```

■3 Type **cd**, followed by the directory name, to
change to the directory containing DOS. My
DOS files are located in my DOS directory,
so I have to type *cd\dos* and press ⏎ to
change directories.

```
C>cd\dos

C>_
```

Do not put a space between the \ and the directory name.

A

■4   Type **dir/w** to see a file listing similar to that
in step 2. Locate the FORMAT.COM file.

■5   Once you're in the proper directory, type
format a: and press ⏎ . If you have a high-
density (1.2Mb) disk drive and want to format a
low-density (360K) disk, type **format a:/4** and
press ⏎ .

```
C>format a:
Insert new diskette for drive A:
and strike ENTER when ready_
```

■6   Insert a new disk in drive A and press ⏎ .

```
Format complete

    1213952 bytes total disk space
    1213952 bytes available on disk

Format another (Y/N)?_
```

For a high-density
disk drive (1.2Mb). A
double-density drive
shows 360K.

■ 7   [Y] and repeat steps 5 and 6 to format as
many disks as you need. When you're
finished, press [N] [↵].

```
   1213952 bytes total disk space
   1213952 bytes available on disk

Format another (Y/N)?n
C>_
```

A

B

- *Installing*
- *WordPerfect 5.1*

■ **W**hen you install WordPerfect 5.1, you prepare the program to run on your particular computer system. The WordPerfect disks are low-density (360K) disks that contain *compressed* (archived) program and data files needed to operate WordPerfect. You must expand these files to run WordPerfect on your computer. WordPerfect provides an installation program that *must* be run to expand the program files. It's a simple process requiring only a few minutes of your time.

Before you can execute WordPerfect, you also need to make a small change in your CONFIG.SYS file, if you have one, and you're ready to go. CONFIG.SYS stands for "configure system." It is a file on your DOS disk that your computer automatically reads when you turn on the computer. Instructions are placed in this file that tell the computer to configure the system to operate in a designated way. WordPerfect needs many files to run the program, and an entry that allows for 20 files is needed in the CONFIG.SYS file to facilitate the use of the WordPerfect files. The CONFIG.SYS file is automatically created or updated when you run the installation program.

Please read the Introduction before proceeding. It contains valuable information on how to use this book.

■*Installation on Dual Floppy Disks*

Your WordPerfect program disks contain compressed files that will not run on your computer unless you run the installation program first. Your disk drives must be 720K or more to run WordPerfect 5.1. WordPerfect will not run on low-density (360K) drives. If you're not sure of the size of your disk drives, consult your computer manual or the dealer that sold you the computer.

Before you start, you will need at least 10 blank, formatted disks, onto which the installation program will copy the WordPerfect program files. (See Appendix A for instructions on formatting new floppy disks.)

■*1* Insert your DOS disk in drive A, and start your computer. Ignore the date and time requests by pressing ⏎.

```
Current date is Wed 11-14-90
Enter new date (mm-dd-yy):
Current time is 12:16:45.08
Enter new time:

A>_
```

■*2* Remove the DOS disk, and place the Install/Learn/Utilities 1 disk in drive A.

■*3*   At the A prompt, type **install** and press ⏎.

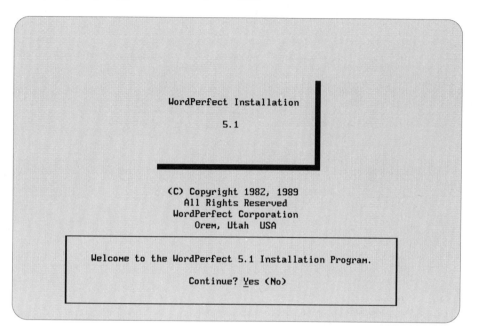

```
              WordPerfect Installation

                      5.1



              (C) Copyright 1982, 1989
                 All Rights Reserved
              WordPerfect Corporation
                  Orem, Utah  USA

      Welcome to the WordPerfect 5.1 Installation Program.

                   Continue? Yes (No)
```

■*4*   Y to continue.

```
          Installing to a Hard Disk? Yes (No)
```

B

∎5   [N] Label your 10 floppy disks with the label
names displayed on the screen.

```
To install WordPerfect, you need at least 10 blank formatted diskettes.
Please label the diskettes as follows:

        WordPerfect 1              Speller
        WordPerfect 2              Thesaurus
        Install/Utilities          PTR Program
        Learning/Images            Fonts/Graphics
        Macros/Keyboards           Printer (.ALL) Files














WordPerfect will not run on low density (360K) 5¼" diskettes.
Do you have at least 10 blank formatted diskettes available? Yes (No)
```

If you do not have 10 formatted disks, press [N] again and you will return to the
DOS prompt. Follow the instructions in Appendix A for formatting new floppy
disks. When you are finished, begin with step 3 above, and run the installation
program again.

■ *6*   If you have 10 blank, formatted disks, press **Y**.
Six installation options are displayed. As an
entry-level user, you need only be concerned
with option 1 (Basic).

```
Installation

    1 - Basic        Perform a standard installation to B:.

    2 - Custom       Perform a customized installation.  (User selected
                     directories.)

    3 - Network      Perform a customized installation onto a network.
                     (To be performed by the network supervisor.)

    4 - Printer      Install updated Printer (.ALL) File.

    5 - Update       Install updated program file(s).

    6 - Copy Disks   Install every file from an installation diskette to a
                     specified location.  (Useful for installing all the
                     Printer (.ALL) Files.)




Selection: 1
```

■ *7*   **1** or **B**

```
              Do you want to install the Utility Files? Yes (No)
              The files will be installed to B:\



              The Utility Files contain a variety of programs.
              This includes the convert, spell, and installation programs.
```

B

■*8* [Y]

```
Insert the diskette you labeled Install/Utilities into B:\
                                   <Enter = Continue  F1 = Cancel>_
```

■*9* Place the correctly labeled, blank, formatted
disk in drive B and press [↵] Follow the instructions on your screen until you've made
copies of all your WordPerfect disks.

If you do not want to use the WordPerfect workbook or tutorial, you can press
[N] to skip the installation of the Learn files.

When the installation process is complete, you see the following screen concerning the creation or updating of your CONFIG.SYS file:

```
Check A:\CONFIG.SYS file

This file contains information about your system configuration.


Insert your DOS boot disk into drive A:\,
which is the disk you use when you turn your computer on.

Press any key to continue_
```

■*10* Insert your DOS disk in drive A, and press
any key.

■ 11 You see one of the following three screen
displays:

```
A:\CONFIG.SYS not found.   Create it? Yes (No)
```

If you see this screen, you do not have a
CONFIG.SYS file and need to create one.
Press Y You are then asked whether you
want to add FILES=20 to your CONFIG.SYS
file. Go to step 12 to continue the installation.

```
Your CONFIG.SYS file allows for only 10 files.
It needs to allow for at least 20 files for WordPerfect to run.
Would you like to have it changed? Yes (No)
```

If you see this screen, you have an existing
CONFIG.SYS file, but it does not allow
enough files to run WordPerfect. Press Y
Go to step 12 to continue the installation.

```
It allows for 20 files.
No changes are necessary.
```

If you see this screen, you have enough files
to run WordPerfect. Go to step 12 to continue
the installation.

B

■ *12* ⏎ Press any key, if prompted to do so on
the bottom of your screen.

```
Check A:\AUTOEXEC.BAT file

This file contains a "batch" of commands that are automatically executed each
time you start your computer.

Insert your DOS boot disk into drive A:\,
which is the disk you use when you turn your computer on.

Press any key to continue_
```

Note   The AUTOEXEC.BAT file instructs your computer to "*auto*matically *ex-
ec*ute" certain actions whenever you start your computer. WordPerfect's installation
program creates an AUTOEXEC.BAT file or updates an existing AUTOEXEC.BAT
file. You are then able to start WordPerfect from any directory.

■ *13* Press any key.

■ *14* You see one of the following three screen
displays:

```
A:\AUTOEXEC.BAT not found.   Create it? Yes (No)
```

If you see this screen, you do not have an
AUTOEXEC.BAT file. I recommend that you
create one.

Press Y. You are then asked whether you
want to create a path command and add B:\ to
the AUTOEXEC.BAT file. Go to step 15 to
continue the installlation.

```
Adding WordPerfect to the path allows WordPerfect to be run from any directory.
Create a path command and add "B:\" to the path? Yes (No)
```

If you see this screen, you can add B:\ to the
AUTOEXEC.BAT file. Go to step 15.

```
Check A:\AUTOEXEC.BAT file

This file contains a "batch" of commands that are automatically executed each
time you start your computer.

Insert your DOS boot disk into drive A:\,
which is the disk you use when you turn your computer on.

Press any key to continue

Your AUTOEXEC.BAT contains B:\ in the path.
```

If you see this screen, you already have B:\ in
your AUTOEXEC.BAT file. Press any key.
Go to step 16.

B

■ *15* Ⓨ

```
          The Printer master diskette is the Printer 1
          diskette or an additional Printer diskette
          numbered 5 or greater.




  Insert the Printer master diskette into A:\
                                             (Enter = Continue   F1 = Cancel)_
```

WordPerfect supports dozens of different printer models. After you insert the Printer master disk, you see a long list of them on your screen.

■ *16* Insert your WordPerfect Printer 1 disk in drive A and press ⏎. You see a partial list of names for different printer models. (Your list may not be exactly the same as the one below.)

```
 1  Acer LP-76                        Printers marked with '*' are
 2  AEG Olympia Compact RO            not included with shipping
 3  AEG Olympia ESW 2000              disks.  Select printer for
 4  AEG Olympia Laserstar 6           more information.
 5  AEG Olympia Startype
 6  Alphacom Alphapro 101
 7  Alps Allegro 24
 8 *Alps Allegro 24 (Additional)
 9  Alps ALQ200 (18 pin)
10  Alps ALQ200 (24 pin)
11  Alps ALQ224e
12 *Alps ALQ224e (Additional)
13  Alps ALQ300 (18 pin)
14  Alps ALQ300 (24 pin)
15  Alps ALQ324e
16 *Alps ALQ324e (Additional)
17  Alps P2000
18 *Alps P2000 (Additional)
19  Alps P2100
20 *Alps P2100 (Additional)
21  Alps P2400C
22  Amstrad DMP 4000

N Name Search; PgDn More Printers; PgUp Previous Screen; F3 Help; F7 Exit;
Selection: 0
```

Locates a specific printer by name. Press **N** and type the name.

Displays the next screen of printer names.

Displays the previous screen of printer names.

If you do not find your printer on the list, press **F3** for help about how to obtain a printer disk from WordPerfect that lists your particular printer model.

B

■ *17* [PG DN] to scroll through the list of printer
names until you find your printer. If you do
not find your printer name, check your printer
manual to see whether there is a printer on
this list that emulates your printer, and then
select that name.

■ *18* Type the number for your printer and
press [↵]. I chose the HP LaserJet Series II.

```
 1 *HP LaserJet 500+ (Additional)        Printers marked with '*' are
 2  HP LaserJet IID                      not included with shipping
 3 *HP LaserJet IID (Additional)         disks.  Select printer for
 4  HP LaserJet IIP                      more information.
 5 *HP LaserJet IIP (Additional)
 6  HP LaserJet Series II
 7 *HP LaserJet Series II (Additional)
 8  HP LaserJet+
 9 *HP LaserJet+ (Additional)
10  HP PaintJet
11  HP PaintJet XL
12 *HP PaintJet XL (Additional)
13  HP QuietJet
14  HP QuietJet Plus
15  HP RuggedWriter
16 *HP RuggedWriter (Additional)
17  HP ThinkJet
18  IBM 3812 Pageprinter
19  IBM 4216-30 Personal Page Printer II
20  IBM 4216-31 Personal Page Printer II
21  IBM 4216 Personal Pageprinter
22  IBM 5216 Wheelprinter


Select printer HP LaserJet Series II? Yes (No)
```

■ *19* [Y] to select the printer. (WPH1.ALL is for
the HP LaserJet Series II.)

```
Do you want to install the Printer (.ALL) File? Yes (No)
The files will be installed to B:\


The Printer File WPHP1.ALL will be copied.
This file is necessary to create or update the
HP LaserJet Series II printer file (.PRS).
```

■ *20* [Y] to install the printer file for your printer.

```
Insert the diskette you labeled Printer (.ALL) Files into B:\
                                       (Enter = Continue  F1 = Cancel)_
```

■ *21* Insert the hand-labeled Printer (.ALL) Files
disk in drive B and press ⏎.

```
Insert the Printer 2 master diskette into A:\
                               (Enter = Continue   F1 = Cancel)_
```

You may see another printer disk displayed here.

B

■*22* Remove the Printer 1 disk from drive A and
replace it with the appropriate printer disk.
Press ⏎.

```
Do you want to install another printer? No (Yes)
```

■*23* If you want to install more than one printer,
press **Y** and go through the printer selection
procedure again. When you are finished,
press **N**; your printer file is installed. (If
relevant for your printer, some helps and hints
may be displayed.)

```
Insert the diskette you labeled Install/Utilities into B:\
                                        (Enter = Continue  F1 = Cancel)_
```

■*24* Insert your hand-labeled Install/Utilities disk
in drive B and press ⏎.

```
Insert the diskette you labeled WordPerfect 2 into B:\
                                        (Enter = Continue  F1 = Cancel)_
```

■*25* Insert your hand-labeled WordPerfect 2 disk
in drive B and press⏎ .

```
Insert the diskette you labeled WordPerfect 1 into B:\
                                        (Enter = Continue  F1 = Cancel)_
```

■*26* Insert your hand-labeled WordPerfect 1 disk
in drive B and press⏎ .

```
Insert the diskette you labeled Printer (.ALL) Files into A:\
                                        (Enter = Continue  F1 = Cancel)_
```

After you gain some
experience, you may
want to change some
of WordPerfect's
default settings.
Chapter 14 shows
you how.

B

■27 Insert your hand-labeled Printer (.ALL) Files
disk in drive A and press ⏎. Your printer file
is installed, and the WordPerfect program
starts automatically.

Doc 1 Pg 1 Ln 1" Pos 1"

■*28* F7 N Y to exit WordPerfect and end the
installation procedures.

```
Installation Complete.
If you changed the AUTOEXEC.BAT file or the CONFIG.SYS file,
you will need to re-boot your computer before running WordPerfect.

README files which contain information about recent changes to the program
are contained on the following disks:

Program 1                README.WP
Spell/Thesaurus 1        README.SPL
PTR Program/Graphics 1   README.PTR
Install/Learn/Utilities 1  README.UTL

To run WordPerfect, insert the WordPerfect 1 diskette,
type WP, and press Enter

A:\>_
```

Congratulations! WordPerfect 5.1 is now installed and ready to go. You may
now proceed with Chapter 1.

___

Note Put your original WordPerfect disks in a safe place; they are your backup
disks. If you damage any of the working disks you created with the installation procedure, use the originals and rerun the installation program to reinstall the programs
on new, undamaged floppy disks.

B

∎*Installation on a Hard Disk*

If your computer has a hard disk, you must run the installation program to copy the WordPerfect program and printer files onto your hard disk. The compressed WordPerfect files are expanded during this installation process.

This program creates the WP51 subdirectory, the LEARN subdirectory, and the proper CONFIG.SYS and AUTOEXEC.BAT files. The WordPerfect files are automatically copied into the WP51 subdirectory, and the tutorial files are copied into the LEARN subdirectory.

∎*1*

Start your computer. Ignore the date and time requests (if they appear on your screen) by pressing ⏎.

```
Current date is Wed 11-14-90
Enter new date (mm-dd-yy):
Current time is 12:16:45.08
Enter new time:

C>_
```

WordPerfect's installation procedure is quick and easy to use.

■*2*   Insert the Install/Learn/Utilities 1 disk in
drive A. Type **a:install** and press ⏎.

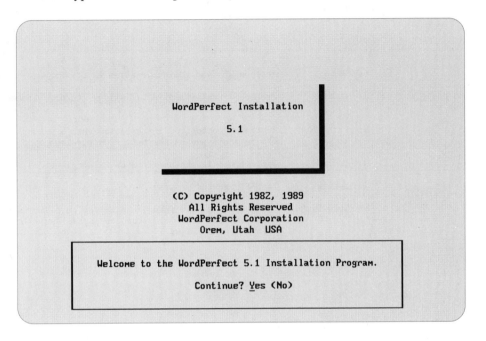

```
                    WordPerfect Installation

                            5.1




                    (C) Copyright 1982, 1989
                      All Rights Reserved
                    WordPerfect Corporation
                       Orem, Utah  USA

  ┌─────────────────────────────────────────────────────────┐
  │ Welcome to the WordPerfect 5.1 Installation Program.      │
  │                                                           │
  │                  Continue? Yes (No)                       │
  └─────────────────────────────────────────────────────────┘
```

■*3*   Ⓨ to continue.

```
  ┌─────────────────────────────────────────────────────────┐
  │                                                           │
  │              Installing to a Hard Disk? Yes (No)          │
  │                                                           │
  │                                                           │
  └─────────────────────────────────────────────────────────┘
```

B

■*4*   **Y** Six installation options are displayed on your screen. As an entry-level user, you need only be concerned with option 1 (Basic).

```
Installation

    1 - Basic        Perform a standard installation to C:\WP51.

    2 - Custom       Perform a customized installation.  (User selected
                     directories.)

    3 - Network      Perform a customized installation onto a network.
                     (To be performed by the network supervisor.)

    4 - Printer      Install updated Printer (.ALL) File.

    5 - Update       Install updated program file(s).

    6 - Copy Disks   Install every file from an installation diskette to a
                     specified location.  (Useful for installing all the
                     Printer (.ALL) Files.)




Selection: 1
```

■*5*   **1** or **B**

```
          Do you want to install the Utility Files? Yes (No)
          The files will be installed to C:\WP51\




          The Utility Files contain a variety of programs.
          This includes the convert, spell, and installation programs.
```

You are offered the choice of installing or not installing the Utilities files. I recommend that you install them. You may need these files in the future.

■*6* [Y]

```
Insert the Install/Learn/Utilities 1 master diskette into A:\
                                        (Enter = Continue  F1 = Cancel)_
```

■*7* The Install/Learn/Utilities 1 disk is already in drive A. Press [↵] to continue with the installation.

Follow the instructions displayed on your screen for installing the WordPerfect program disks. If you do not want to use the WordPerfect workbook or tutorial, you can skip installing the Learning files by pressing [N] when prompted.

■*8* When the installation process is complete, you see one of the following three screens concerning the creation of your CONFIG.SYS file:

```
Check C:\CONFIG.SYS file

This file contains information about your system configuration.

C:\CONFIG.SYS not found.  Create it? Yes (No)
```

B

If you see this screen, you do not have a
CONFIG.SYS file and need to create one.
Press ⓨ. You are asked whether or not you
want to add FILES=20 to your CONFIG.SYS
file. Press ⓨ. Then press any key. Go to step
9 to continue the installation.

```
Check C:\CONFIG.SYS file

This file contains information about your system configuration.

Your CONFIG.SYS file allows for only 10 files.
It needs to allow for at least 20 files for WordPerfect to run.
Would you like to have it changed? Yes (No)
```

If you see a screen similar to this one, you
have an existing CONFIG.SYS file, but it
does not allow enough files to run Word-
Perfect. Press ⓨ. Then press any key. Go to
step 9 to continue the installation.

```
Check C:\CONFIG.SYS file

This file contains information about your system configuration.

It allows for 20 files.
No changes are necessary.
```

If you see a screen similar to this one, your
CONFIG.SYS file allows enough files to run
WordPerfect. Press any key. Go to step 9.

> **Note** The AUTOEXEC.BAT file instructs your computer to "*auto*matically *ec*ute" certain actions whenever you start your computer. WordPerfect's installation program creates an AUTOEXEC.BAT file or updates an existing AUTOEXEC.BAT file. You are then able to start WordPerfect from any directory.

■*9* You will see one of the following three screens:

```
This file contains a "batch" of commands that are automatically executed each
time you start your computer.

C:\AUTOEXEC.BAT not found.   Create it? Yes (No)
```

If you see this screen, you do not have an AUTOEXEC.BAT file. I recommend that you create one. Press [Y]. You are asked whether or not you want to create a path command and add C:\WP51\ to the path. Press [Y]. If you are not asked to reboot your computer, go to step 10.

If you are asked to reboot (restart) your computer, follow the instructions on the screen carefully. Be *sure* to open the latch on drive A before you press [CTRL][ALT][DEL]. After you see

B

the C prompt, close the latch on drive A. Type
a:install and press ⏎. Press **Y**. You'll see
the screen shown in step 10.

Go to step 11.

```
Check C:\AUTOEXEC.BAT file

This file contains a "batch" of commands that are automatically executed each
time you start your computer.

Adding WordPerfect to the path allows WordPerfect to be run from any directory.
Add "C:\WP51\" to the path? Yes (No)
```

If you see this screen, you have an existing
AUTOEXEC.BAT file. I recommend that you
add C:\WP51\ to the path. Press **Y**. If you
are asked to reboot (restart) your computer,
follow the rebooting instructions above. If
you are not asked to reboot, go to step 10.

```
Check C:\AUTOEXEC.BAT file

This file contains a "batch" of commands that are automatically executed each
time you start your computer.

Your AUTOEXEC.BAT contains C:\WP51\ in the path.
```

If you see this screen, you already have
C:\WP51\ in your AUTOEXEC.BAT file.
Press any key. Go to step 10.

■ *10* After you complete the installation of your
CONFIG.SYS and AUTOEXEC.BAT files,
you see this screen:

```
                    The Printer master diskette is the Printer 1
                    diskette or an additional Printer diskette
                    numbered 5 or greater.




Insert the Printer master diskette into A:\
                                          (Enter = Continue   F1 = Cancel)_
```

B

■ *11* Insert the Printer 1 disk in drive A and press
⏎. You see a partial list of names for different printer models. (Your list may not be
exactly the same as the one below.)

```
 1   Acer LP-76                          Printers marked with '*' are
 2   AEG Olympia Compact RO              not included with shipping
 3   AEG Olympia ESW 2000                disks.  Select printer for
 4   AEG Olympia Laserstar 6             more information.
 5   AEG Olympia Startype
 6   Alphacom Alphapro 101
 7   Alps Allegro 24
 8  *Alps Allegro 24 (Additional)
 9   Alps ALQ200 (18 pin)
10   Alps ALQ200 (24 pin)
11   Alps ALQ224e
12  *Alps ALQ224e (Additional)
13   Alps ALQ300 (18 pin)
14   Alps ALQ300 (24 pin)
15   Alps ALQ324e
16  *Alps ALQ324e (Additional)
17   Alps P2000
18  *Alps P2000 (Additional)
19   Alps P2100
20  *Alps P2100 (Additional)
21   Alps P2400C
22   Amstrad DMP 4000

N Name Search; PgDn More Printers; PgUp Previous Screen; F3 Help; F7 Exit;
Selection: 0
```

Locates a specific printer by name. Press **N** and type the name.

Displays the next screen of printer names.

Displays the previous screen of printer names.

If you do not find your printer on the list, press F3 for help about how to obtain a printer disk from WordPerfect that lists your particular printer model.

■ *12* PG DN to scroll through the list of printer
names until you find your printer. If you do
not find your printer name, check your printer
manual to see whether there is a printer on
this list that emulates your printer, and then
select that name.

■ *13* Type the number for your printer and press
↵. I chose the HP LaserJet Series II.

```
 1 *HP LaserJet 500+ (Additional)       Printers marked with 'x' are
 2  HP LaserJet IID                      not included with shipping
 3 *HP LaserJet IID (Additional)         disks.  Select printer for
 4  HP LaserJet IIP                      more information.
 5 *HP LaserJet IIP (Additional)
 6  HP LaserJet Series II
 7 *HP LaserJet Series II (Additional)
 8  HP LaserJet+
 9 *HP LaserJet+ (Additional)
10  HP PaintJet
11  HP PaintJet XL
12 *HP PaintJet XL (Additional)
13  HP QuietJet
14  HP QuietJet Plus
15  HP RuggedWriter
16 *HP RuggedWriter (Additional)
17  HP ThinkJet
18  IBM 3812 Pageprinter
19  IBM 4216-30 Personal Page Printer II
20  IBM 4216-31 Personal Page Printer II
21  IBM 4216 Personal Pageprinter
22  IBM 5216 Wheelprinter


Select printer HP LaserJet Series II? Yes (No)
```

B

■ *14* (Y) to select the printer. (WPH1.ALL is for
the HP LaserJet Series II.)

Do you want to install the Printer (.ALL) File? <u>Y</u>es (No)
The files will be installed to C:\WP51\

The Printer File WPHP1.ALL will be copied.
This file is necessary to create or update the
HP LaserJet Series II printer file (.PRS).

■ *15* (Y) to install the printer file for your printer.

Insert the Printer 2 master diskette into A:\
(Enter = Continue F1 = Cancel)_

You may see another printer disk displayed here.

■ *16* Insert the appropriate printer disk in drive A
and press (↵).

Do you want to install another printer? <u>N</u>o (Yes)

■ *17* If you want to install more than one printer, press ⟨**Y**⟩ and go through the printer selection procedure again.

■ *18* When you are finished, press ⟨**N**⟩; your printer file is installed. If relevant for your printer, some helps and hints may be displayed.

```
Printer Helps and Hints:  HP LaserJet Series II

11/6/89
Initializing the printer will delete all soft fonts in printer memory and
those fonts marked with an asterisk (*) will be downloaded.

Additional font support for this printer is available on a supplementary
diskette.  To order this diskette, call WP Corp. at (801)225-5000.  There
will be a $10 shipping fee.
```

■ *19* Press any key to exit to the DOS prompt.

Congratulations! WordPerfect 5.1 is installed on your hard disk. You are now ready to run WordPerfect and go to work creating professional-quality documents. Turn to Chapter 1 and get started. I think you'll enjoy learning how to use all the powerful word-processing features WordPerfect has to offer.

B

C

- *Using a Mouse*
- *with*
- *WordPerfect 5.1*

■ One of the most exciting enhancements made to WordPerfect in the new version is the addition of mouse support. If you have read the Introduction, you know you can use the mouse to activate an extensive pull-down menu system and select menu options. You can also use the mouse to scroll through your documents, reposition your cursor, and block text. Using the mouse and the keyboard together allows you to select Word-Perfect features quickly and efficiently, decreasing the time you spend formatting and editing your documents.

This appendix provides you with some general hints for installing the mouse and shows you the different modifications you can make to the mouse settings using WordPerfect's Setup feature. If you haven't read the Introduction, I recommend you do so before you read on.

■*Hints for Installing Your Mouse*

When installing your mouse, follow the manufacturer's documentation very carefully. There are several steps you may have to follow to correctly connect the mouse so it will work in your system. You most likely will have to install a mouse card in one of the expansion slots inside your computer. The mouse is connected to this mouse card to send the signals that move the cursor on your screen.

You will also receive software that contains the mouse driver, which enables the mouse to communicate with your computer. The mouse driver must be installed on the hard or floppy disk that you use to start your computer or in the directory where WP.EXE (the WordPerfect program file) is stored. MOUSE.COM, MOUSE.SYS, MSMOUSE.COM, and MSMOUSE.SYS are four common mouse drivers.

To determine the location of your mouse driver, type *DIR/W* at the DOS prompt for the root directory (C:\ for a hard disk and A:\ for a floppy disk) or the Word-Perfect directory (C:\WP51), and look for the driver file name.

If your mouse driver is MOUSE.COM or MSMOUSE.COM, you can activate the mouse by moving to the DOS prompt for the directory that contains the mouse driver, typing *MOUSE* or *MSMOUSE* (depending on the name of your driver) and pressing ⏎. The mouse program is loaded into memory. When you run WordPerfect, the mouse will be available for use.

If your mouse driver is MOUSE.SYS or MSMOUSE.SYS, you need to add the entry *device=mouse.sys* or *device=msmouse.sys* to your CONFIG.SYS file. Follow these steps to add the mouse device to your CONFIG.SYS file:

■ *1* If you have a hard disk, turn on your computer. If you have floppy disks, insert your DOS disk in drive A and turn on your computer. Ignore any date and time requests by pressing ⏎. Hard disk systems display the C prompt; floppy disk systems display the A prompt.

■*2* You should be at the DOS prompt where your
CONFIG.SYS file is located. Normally it's
located in the root directory for a hard disk
(type **cd** and press ⏎ to ensure that you are
at the root directory) and on the DOS system
disk for a floppy disk.

■*3* Type **type config.sys** and press ⏎.

```
C>type config.sys
files=20
buffers=20
C>_
```

The contents of your CONFIG.SYS file are displayed on the screen. If you do
not see the mouse-device entry described above, you will have to add it.

■*4* At the DOS prompt, type

copy config.sys+con config.sys

■*5* ⏎

```
C>type config.sys
files=20
buffers=20
C>copy config.sys+con config.sys
CONFIG.SYS
CON
_
```

C

■6 Type **device=mouse.sys** or
device=msmouse.sys and press ⏎.

```
C>type config.sys
files=20
buffers=20
C>copy config.sys+con config.sys
CONFIG.SYS
CON
device=mouse.sys
_
```

■7 CTRL Z ⏎

```
C>type config.sys
files=20
buffers=20
C>copy config.sys+con config.sys
CONFIG.SYS
CON
device=mouse.sys
^Z
        1 File(s) copied

C>_
```

■8 CTRL ALT DEL to reboot your computer.
This allows the computer to read the new
CONFIG.SYS file and activate the mouse.

You are now able to use the mouse when you start WordPerfect.

■*Setting Up Your Mouse*

WordPerfect supports many different kinds of mouse drivers, so you have to tell WordPerfect the type of mouse you are using. WordPerfect automatically selects the mouse driver called MOUSE.COM when you install the program. If your mouse uses a driver other than MOUSE.COM, you must tell WordPerfect the name of that driver before the mouse will work with the program. To do this, you use WordPerfect's Setup feature.

To select a different mouse driver, follow these steps:

■*1*   Start your computer and run WordPerfect.
(Refer to Chapter 1 for assistance.)

■*2*   At a blank WordPerfect screen, press
SHIFT F1 (the Setup key). The Setup menu
appears.

```
Setup

    1 - Mouse

    2 - Display

    3 - Environment

    4 - Initial Settings

    5 - Keyboard Layout

    6 - Location of Files
```

C

■*3* 〔1〕 or 〔M〕 to see the mouse options.

```
Setup: Mouse

   1 - Type                              Mouse Driver (MOUSE.COM)

   2 - Port

   3 - Double Click Interval (1 = .01 sec) 70

   4 - Submenu Delay Time (1 = .01 sec)    15

   5 - Acceleration Factor                 24

   6 - Left-Handed Mouse                    No

   7 - Assisted Mouse Pointer Movement      Yes
```

The following mouse options provide you with the means to customize your mouse according to your needs:

Type	Provides a list of different mouse types. (See the Note at the end of this section if your mouse is not listed.)
Port	Allows you to select the port to which your mouse is connected.
Double Click Interval	Allows you to change the amount of time in which you can click the mouse button twice to carry out a particular mouse command.
Submenu Delay Time	Allows you to change the amount of time that passes between the time you select a menu option that has a submenu and the time the submenu appears.
Acceleration Factor	Allows you to adjust how fast the pointer moves when you move the mouse. You can choose a number from 1 to 1200. The lower the number, the slower the pointer moves; the higher the number, the faster the pointer moves.

Left-Handed Mouse	Allows you to switch the functions of the mouse buttons if you want to use your left hand. The right mouse button performs the functions the left button did and vice versa. The default is set to No, which is for right-handed users.
Assisted Mouse Pointer Movement	Automatically moves the mouse pointer to the menu bar when the menu bar is activated.

■ *4* ⬚1 or ⬚T to display a list of mouse drivers supported by WordPerfect.

```
Setup: Mouse Type

   CH Products Roller Mouse (PS/2)
   CH Products Roller Mouse (Serial)
   IBM PS/2 Mouse
   Imsi Mouse, 2 button (Serial)
   Imsi Mouse, 3 button (Serial)
   Kensington Expert Mouse (PS/2)
   Kensington Expert Mouse (Serial)
   Keytronic Mouse (Bus)
   Keytronic Mouse (Serial)
   Logitech Mouse (Bus)
   Logitech Mouse (Serial)
   Logitech Mouse, 3 button (PS/2)
   Microsoft Mouse (Bus)
   Microsoft Mouse (Serial)
 * Mouse Driver (MOUSE.COM)
   Mouse Systems Mouse, 3 button (Serial)
   MSC Technology PC Mouse 2 (Serial)
   Numonics Mouse (Serial)
   PC-Trac Trackball (Serial)




1 Select; 2 Auto-select; 3 Other Disk; N Name Search: 1
```

■ *5* ⬚↓ to move the highlight to the name of your mouse.

C

■ *6*   **1** or **S** to select the name. You are ready
to go.

Note   If your mouse is not listed on the screen, contact WordPerfect Corporation.
They frequently update their software and may have included your mouse driver in
a later release of the WordPerfect program. (See page 481 of the WordPerfect
documentation for the address to write to.)

Index

Selections from The SYBEX Library

WORD PROCESSING

The ABC's of Microsoft Word (Third Edition)
Alan R. Neibauer
461pp. Ref. 604-9

This is for the novice WORD user who wants to begin producing documents in the shortest time possible. Each chapter has short, easy-to-follow lessons for both keyboard and mouse, including all the basic editing, formatting and printing functions. Version 5.0.

The ABC's of WordPerfect
Alan R. Neibauer
239pp. Ref. 425-9

This basic introduction to WordPefect consists of short, step-by-step lessons—for new users who want to get going fast. Topics range from simple editing and formatting, to merging, sorting, macros, and more. Includes version 4.2

The ABC's of WordPerfect 5
Alan R. Neibauer
283pp. Ref. 504-2

This introduction explains the basics of desktop publishing with WordPerfect 5: editing, layout, formatting, printing, sorting, merging, and more. Readers are shown how to use WordPerfect 5's new features to produce great-looking reports.

The ABC's of WordPerfect 5.1
Alan R. Neibauer
352pp. Ref. 672-3

Neibauer's delightful writing style makes this clear tutorial an especially effective learning tool. Learn all about 5.1's new drop-down menus and mouse capabilities that reduce the tedious memorization of function keys.

Advanced Techniques in Microsoft Word (Second Edition)
Alan R. Neibauer
462pp. Ref. 615-4

This highly acclaimed guide to WORD is an excellent tutorial for intermediate to advanced users. Topics include word processing fundamentals, desktop publishing with graphics, data management, and working in a multiuser environment. For Versions 4 and 5.

Advanced Techniques in MultiMate
Chris Gilbert
275pp. Ref. 412-7

A textbook on efficient use of MultiMate for business applications, in a series of self-contained lessons on such topics as multiple columns, high-speed merging, mailing-list printing and Key Procedures.

Advanced Techniques in WordPerfect 5
Kay Yarborough Nelson
586pp. Ref. 511-5

Now updated for Version 5, this invaluable guide to the advanced features of Word-Perfect provides step-by-step instructions and practical examples covering those specialized techniques which have most perplexed users—indexing, outlining, foreign-language typing, mathematical functions, and more.

The Complete Guide to MultiMate
Carol Holcomb Dreger
208pp. Ref. 229-9

This step-by-step tutorial is also an excellent reference guide to MultiMate features and uses. Topics include search/replace, library and merge functions, repagination, document defaults and more.

Encyclopedia WordPerfect 5.1

Greg Harvey
Kay Yarborough Nelson

1100pp. Ref. 676-6

This comprehensive, up-to-date Word-Perfect reference is a must for beginning and experienced users alike. With complete, easy-to-find information on every WordPerfect feature and command -- and it's organized by practical functions, with business users in mind.

Introduction to WordStar

Arthur Naiman

208pp. Ref. 134-9

This all time bestseller is an engaging first-time introduction to word processing as well as a complete guide to using WordStar—from basic editing to blocks, global searches, formatting, dot commands, SpellStar and MailMerge. Through Version 3.3.

Mastering DisplayWrite 4

Michael E. McCarthy

447pp. Ref. 510-7

Total training, reference and support for users at all levels—in plain, non-technical language. Novices will be up and running in an hour's time; everyone will gain complete word-processing and document-management skills.

Mastering Microsoft Word on the IBM PC (Fourth Edition)

Matthew Holtz

680pp. Ref. 597-2

This comprehensive, step-by-step guide details all the new desktop publishing developments in this versatile word processor, including details on editing, formatting, printing, and laser printing. Holtz uses sample business documents to demonstrate the use of different fonts, graphics, and complex documents. Includes Fast Track speed notes. For Versions 4 and 5.

Mastering MultiMate Advantage II

Charles Ackerman

407pp. Ref. 482-8

This comprehensive tutorial covers all the capabilities of MultiMate, and highlights the differences between MultiMate Advantage II and previous versions—in pathway support, sorting, math, DOS access, using dBASE III, and more. With many practical examples, and a chapter on the On-File database.

Mastering WordPerfect

Susan Baake Kelly

435pp. Ref. 332-5

Step-by-step training from startup to mastery, featuring practical uses (form letters, newsletters and more), plus advanced topics such as document security and macro creation, sorting and columnar math. Through Version 4.2.

Mastering WordPerfect 5

Susan Baake Kelly

709pp. Ref. 500-X

The revised and expanded version of this definitive guide is now on WordPerfect 5 and covers wordprocessing and basic desktop publishing. As more than 200,000 readers of the original edition can attest, no tutorial approaches it for clarity and depth of treatment. Sorting, line drawing, and laser printing included.

Mastering WordPerfect 5.1

Alan Simpson

1050pp. Ref. 670-7

The ultimate guide for the WordPerfect user. Alan Simpson, the "master communicator," puts you in charge of the latest features of 5.1: new dropdown menus and mouse capabilities, along with the desktop publishing, macro programming, and file conversion functions that have made WordPerfect the most popular word processing program on the market.

Mastering WordStar Release 5.5

Greg Harvey
David J. Clark

450pp. Ref. 491-7

This book is the ultimate reference book for the newest version of WordStar. Readers may use Mastering to look up any word processing function, including the new Version 5 and 5.5 features and enhancements, and find detailed instructions for fundamental to advanced operations.

Microsoft Word Instant Reference for the IBM PC
Matthew Holtz
266pp. Ref. 692-8

Turn here for fast, easy access to concise information on every command and feature of Microsoft Word version 5.0 -- for editing, formatting, merging, style sheets, macros, and more. With exact keystroke sequences, discussion of command options, and commonly-performed tasks.

Practical WordStar Uses
Julie Anne Arca
303pp. Ref. 107-1

A hands-on guide to WordStar and MailMerge applications, with solutions to comon problems and "recipes" for day-to-day tasks. Formatting, merge-printing and much more; plus a quick-reference command chart and notes on CP/M and PC-DOS. For Version 3.3.

Understanding Professional Write
Gerry Litton
400pp. Ref. 656-1

A complete guide to Professional Write that takes you from creating your first simple document, into a detailed description of all major aspects of the software. Special features place an emphasis on the use of different typestyles to create attractive documents as well as potential problems and suggestions on how to get around them.

Understanding WordStar 2000
David Kolodney
Thomas Blackadar
275pp. Ref. 554-9

This engaging, fast-paced series of tutorials covers everything from moving the cursor to print enhancements, format files, key glossaries, windows and MailMerge. With practical examples, and notes for former WordStar users.

Visual Guide to WordPerfect
Jeff Woodward
457pp. Ref. 591-3

This is a visual hands-on guide which is ideal for brand new users as the book shows each activity keystroke-by-keystroke. Clear illustrations of computer screen menus are included at every stage. Covers basic editing, formatting lines, paragraphs, and pages, using the block feature, footnotes, search and replace, and more. Through Version 5.

WordPerfect 5 Desktop Companion
SYBEX Ready Reference Series
Greg Harvey
Kay Yarborough Nelson
1006pp. Ref. 522-0

Desktop publishing features have been added to this compact encyclopedia. This title offers more detailed, cross-referenced entries on every software features including page formatting and layout, laser printing and word processing macros. New users of WordPerfect, and those new to Version 5 and desktop publishing will find this easy to use for on-the-job help.

WordPerfect Instant Reference
SYBEX Prompter Series
Greg Harvey
Kay Yarborough Nelson
254pp. Ref. 476-3, 4 ¾" × 8"

When you don't have time to go digging through the manuals, this fingertip guide offers clear, concise answers: command summaries, correct usage, and exact keystroke sequences for on-the-job tasks. Convenient organization reflects the structure of WordPerfect. Through Version 4.2.

WordPerfect 5 Instant Reference
SYBEX Prompter Series
Greg Harvey
Kay Yarborough Nelson
316pp. Ref. 535-2, 4 ¾" × 8"

This pocket-sized reference has all the program commands for the powerful WordPerfect 5 organized alphabetically for quick access. Each command entry has the exact key sequence, any reveal codes, a list of available options, and option-by-option discussions.

WordPerfect 5.1 Instant Reference
Greg Harvey
Kay Yarborough Nelson
252pp. Ref. 674-X

Instant access to all features and commands of WordPerfect 5.0 and 5.1, highlighting the newest software features. Complete, alphabetical entries provide exact key sequences, codes and options, and step-by-step instructions for many important tasks.

WordPerfect 5 Macro Handbook
Kay Yarborough Nelson
488pp. Ref. 483-6

Readers can create macros customtailored to their own needs with this excellent tutorial and reference. Nelson's expertise guides the WordPerfect 5 user through nested and chained macros, macro libraries, specialized macros, and much more.

WordPerfect 5.1 Tips and Tricks (Fourth Edition)
Alan R. Neibauer
675pp. Ref. 681-2

This new edition is a real timesaver. For on-the-job guidance and creative new uses, this title covers all versions of WordPerfect up to and including 5.1—streamlining documents, automating with macros, new print enhancements, and more.

WordStar Instant Reference
SYBEX Prompter Series
David J. Clark
314pp. Ref. 543-3, 4 ¾" × 8"

This quick reference provides reminders on the use of the editing, formatting, mailmerge, and document processing commands available through WordStar 4 and 5. Operations are organized alphabetically for easy access. The text includes a survey of the menu system and instructions for installing and customizing WordStar.

DESKTOP PUBLISHING

The ABC's of the New Print Shop
Vivian Dubrovin
340pp. Ref. 640-4

This beginner's guide stresses fun, practicality and original ideas. Hands-on tutorials show how to create greeting cards, invitations, signs, flyers, letterheads, banners, and calendars.

The ABC's of Ventura
Robert Cowart
Steve Cummings
390pp. Ref. 537-9

Created especially for new desktop publishers, this is an easy introduction to a complex program. Cowart provides details on using the mouse, the Ventura side bar, and page layout, with careful explanations of publishing terminology. The new Ventura menus are all carefully explained. For Version 2.

Mastering COREL DRAW!
Steve Rimmer
403pp. Ref. 685-5

This four-color tutorial and user's guide covers drawing and tracing, text and special effects, file interchange, and adding new fonts. With in-depth treatment of design principles. For version 1.1.

Mastering PageMaker on the IBM PC (Second Edition)
Antonia Stacy Jolles
384pp. Ref. 521-2

A guide to every aspect of desktop publishing with PageMaker: the vocabulary and basics of page design, layout, graphics and typography, plus instructions for creating finished typeset publications of all kinds.

Mastering Ventura
(Second Edition)
Matthew Holtz
613pp. Ref. 581-6
A complete, step-by-step guide to IBM PC desktop publishing with Xerox Ventura Publisher. Practical examples show how to use style sheets, format pages, cut and paste, enhance layouts, import material from other programs, and more. For Version 2.

Understanding PFS:
First Publisher
Gerry Litton
310pp. Ref. 616-2
This complete guide takes users from the basics all the way through the most complex features available. Discusses working with text and graphics, columns, clip art, and add-on software enhancements. Many page layout suggestions are introduced. Includes Fast Track speed notes.

DESKTOP PRESENTATION

Mastering Harvard Graphics
Glenn H. Larsen
318pp. Ref. 585-9
Here is a solid course in computer graphing and chart building with the popular software package. Readers can create the perfect presentation using text, pie, line, bar, map, and pert charts. Customizing and automating graphics is easy with these step-by-step instructions. For Version 2.1.

OPERATING SYSTEMS

The ABC's of DOS 4
Alan R. Miller
275pp. Ref. 583-2
This step-by-step introduction to using DOS 4 is written especially for beginners. Filled with simple examples, *The ABC's of DOS 4* covers the basics of hardware,

software, disks, the system editor EDLIN, DOS commands, and more.

ABC's of MS-DOS
(Second Edition)
Alan R. Miller
233pp. Ref. 493-3
This handy guide to MS-DOS is all many PC users need to manage their computer files, organize floppy and hard disks, use EDLIN, and keep their computers organized. Additional information is given about utilities like Sidekick, and there is a DOS command and program summary. The second edition is fully updated for Version 3.3.

DOS Assembly Language
Programming
Alan R. Miller
365pp. 487-9
This book covers PC-DOS through 3.3, and gives clear explanations of how to assemble, link, and debug 8086, 8088, 80286, and 80386 programs. The example assembly language routines are valuable for students and programmers alike.

DOS Instant Reference
SYBEX Prompter Series
Greg Harvey
Kay Yarborough Nelson
220pp. Ref. 477-1, 4 ¾" × 8"
A complete fingertip reference for fast, easy on-line help:command summaries, syntax, usage and error messages. Organized by function—system commands, file commands, disk management, directories, batch files, I/O, networking, programming, and more. Through Version 3.3.

DOS User's Desktop Companion
SYBEX Ready Reference Series
Judd Robbins
969pp. Ref. 505-0
This comprehensive reference covers DOS commands, batch files, memory enhancements, printing, communications and more information on optimizing each user's DOS environment. Written with step-by-step instructions and plenty of examples, this volume covers all versions through 3.3.

Encyclopedia DOS
Judd Robbins

1030pp. Ref. 699-5

A comprehensive reference and user's guide to all versions of DOS through 4.0. Offers complete information on every DOS command, with all possible switches and parameters -- plus examples of effective usage. An invaluable tool.

Essential OS/2
(Second Edition)
Judd Robbins

445pp. Ref. 609-X

Written by an OS/2 expert, this is the guide to the powerful new resources of the OS/2 operating system standard edition 1.1 with presentation manager. Robbins introduces the standard edition, and details multitasking under OS/2, and the range of commands for installing, starting up, configuring, and running applications. For Version 1.1 Standard Edition.

Essential PC-DOS
(Second Edition)
Myril Clement Shaw
Susan Soltis Shaw

332pp. Ref. 413-5

An authoritative guide to PC-DOS, including version 3.2. Designed to make experts out of beginners, it explores everything from disk management to batch file programming. Includes an 85-page command summary. Through Version 3.2.

Graphics Programming
Under Windows
Brian Myers
Chris Doner

646pp. Ref. 448-8

Straightforward discussion, abundant examples, and a concise reference guide to graphics commands make this book a must for Windows programmers. Topics range from how Windows works to programming for business, animation, CAD, and desktop publishing. For Version 2.

Hard Disk Instant Reference
SYBEX Prompter Series
Judd Robbins

256pp. Ref. 587-5, 4 ¾" × 8"

Compact yet comprehensive, this pocket-sized reference presents the essential information on DOS commands used in managing directories and files, and in optimizing disk configuration. Includes a survey of third-party utility capabilities. Through DOS 4.0.

The IBM PC-DOS Handbook
(Third Edition)
Richard Allen King

359pp. Ref. 512-3

A guide to the inner workings of PC-DOS 3.2, for intermediate to advanced users and programmers of the IBM PC series. Topics include disk, screen and port control, batch files, networks, compatibility, and more. Through Version 3.3.

Inside DOS: A Programmer's
Guide
Michael J. Young

490pp. Ref. 710-X

A collection of practical techniques (with source code listings) designed to help you take advantage of the rich resources intrinsic to MS-DOS machines. Designed for the experienced programmer with a basic understanding of C and 8086 assembly language, and DOS fundamentals.

Mastering DOS
(Second Edition)
Judd Robbins

722pp. Ref. 555-7

"The most useful DOS book." This seven-part, in-depth tutorial addresses the needs of users at all levels. Topics range from running applications, to managing files and directories, configuring the system, batch file programming, and techniques for system developers. Through Version 4.

MS-DOS Advanced
Programming
Michael J. Young

490pp. Ref. 578-6

Practical techniques for maximizing performance in MS-DOS software by making best use of system resources. Topics include functions, interrupts, devices, multitasking, memory residency and more, with examples in C and assembler. Through Version 3.3.

MS-DOS Handbook
(Third Edition)
Richard Allen King
362pp. Ref. 492-5

This classic has been fully expanded and revised to include the latest features of MS-DOS Version 3.3. Two reference books in one, this title has separate sections for programmer and user. Multi-DOS partitons, 3 ½-inch disk format, batch file call and return feature, and comprehensive coverage of MS-DOS commands are included. Through Version 3.3.

MS-DOS Power User's Guide,
Volume I
(Second Edition)
Jonathan Kamin
482pp. Ref. 473-9

A fully revised, expanded edition of our best-selling guide to high-performance DOS techniques and utilities—with details on Version 3.3. Configuration, I/O, directory structures, hard disks, RAM disks, batch file programming, the ANSI.SYS device driver, more. Through Version 3.3.

Programmers Guide to
the OS/2 Presentation Manager
Michael J. Young
683pp. Ref. 569-7

This is the definitive tutorial guide to writing programs for the OS/2 Presentation Manager. Young starts with basic architecture, and explores every important feature including scroll bars, keyboard and mouse interface, menus and accelerators, dialogue boxes, clipboards, multitasking, and much more.

Programmer's Guide to
Windows
(Second Edition)
David Durant
Geta Carlson
Paul Yao
704pp. Ref. 496-8

The first edition of this programmer's guide was hailed as a classic. This new edition covers Windows 2 and Windows/386 in depth. Special emphasis is given to over fifty new routines to the Windows interface, and to preparation for OS/2 Presentation Manager compatibility.

Understanding DOS 3.3
Judd Robbins
678pp. Ref. 648-0

This best selling, in-depth tutorial addresses the needs of users at all levels with many examples and hands-on exercises. Robbins discusses the fundamentals of DOS, then covers manipulating files and directories, using the DOS editor, printing, communicating, and finishes with a full section on batch files.

Understanding Hard Disk
Management on the PC
Jonathan Kamin
500pp. Ref. 561-1

This title is a key productivity tool for all hard disk users who want efficient, error-free file management and organization. Includes details on the best ways to conserve hard disk space when using several memory-guzzling programs. Through DOS 4.

Up & Running
with Your Hard Disk
Klaus M Rubsam
140pp. Ref. 666-9

A far-sighted, compact introduction to hard disk installation and basic DOS use. Perfect for PC users who want the practical essentials in the shortest possible time. In 20 basic steps, learn to choose your hard disk, work with accessories, back up data, use DOS utilities to save time, and more.

Up & Running with Windows
286/386
Gabriele Wentges
132pp. Ref. 691-X

This handy 20-step overview gives PC users all the essentials of using Windows -- whether for evaluating the software, or getting a fast start. Each self-contained lesson takes just 15 minutes to one hour to complete.

Understanding PostScript Programming (Second Edition)
David A. Holzgang
472pp. Ref. 566-2

In-depth treatment of PostScript for programmers and advanced users working on custom desktop publishing tasks. Hands-on development of programs for font creation, integrating graphics, printer implementations and more.

Ventura Instant Reference SYBEX Prompter Series
Matthew Holtz
320pp. Ref. 544-1, 4 ¾" × 8"

This compact volume offers easy access to the complex details of Ventura modes and options, commands, side-bars, file management, output device configuration, and control. Written for versions through Ventura 2, it also includes standard procedures for project and job control.

Ventura Power Tools
Rick Altman
318pp. Ref. 592-1

Renowned Ventura expert, Rick Altman, presents strategies and techniques for the most efficient use of Ventura Publisher 2. This includes a power disk with DOS utilities which is specially designed for optimizing Ventura use. Learn how to soup up Ventura, edit CHP files, avoid design tragedies, handle very large documents, and improve form.

Your HP LaserJet Handbook
Alan R. Neibauer
564pp. Ref. 618-9

Get the most from your printer with this step-by-step instruction book for using LaserJet text and graphics features such as cartridge and soft fonts, type selection, memory and processor enhancements,

PCL programming, and PostScript solutions. This hands-on guide provides specific instructions for working with a variety of software.

COMMUNICATIONS

Mastering Crosstalk XVI (Second Edition)
Peter W. Gofton
225pp. Ref. 642-1

Introducing the communications program Crosstalk XVI for the IBM PC. As well as providing extensive examples of command and script files for programming Crosstalk, this book includes a detailed description of how to use the program's more advanced features, such as windows, talking to mini or mainframe, customizing the keyboard and answering calls and background mode.

Mastering PROCOMM PLUS
Bob Campbell
400pp. Ref. 657-X

Learn all about communications and information retrieval as you master and use PROCOMM PLUS. Topics include choosing and using a modem; automatic dialing; using on-line services (featuring CompuServe) and more. Through Version 1.1b; also covers PROCOMM, the "shareware" version.

Mastering Serial Communications
Peter W. Gofton
289pp. Ref. 180-2

The software side of communications, with details on the IBM PC's serial programming, the XMODEM and Kermit protocols, non-ASCII data transfer, interrupt-level programming and more. Sample programs in C, assembly language and BASIC.

TO JOIN THE SYBEX MAILING LIST OR ORDER BOOKS
PLEASE COMPLETE THIS FORM

NAME _____ COMPANY _____

STREET _____ CITY _____

STATE _____ ZIP _____

☐ PLEASE MAIL ME MORE INFORMATION ABOUT **SYBEX** TITLES

ORDER FORM (There is no obligation to order)

PLEASE SEND ME THE FOLLOWING:

TITLE	QTY	PRICE
_____	____	____
_____	____	____
_____	____	____
_____	____	____

TOTAL BOOK ORDER _____ $_____

SHIPPING AND HANDLING PLEASE ADD $2.00 PER BOOK VIA UPS _____

FOR OVERSEAS SURFACE ADD $5.25 PER BOOK PLUS $4.40 REGISTRATION FEE _____

FOR OVERSEAS AIRMAIL ADD $18.25 PER BOOK PLUS $4.40 REGISTRATION FEE _____

CALIFORNIA RESIDENTS PLEASE ADD APPLICABLE SALES TAX _____

TOTAL AMOUNT PAYABLE _____

☐ CHECK ENCLOSED ☐ VISA
☐ MASTERCARD ☐ AMERICAN EXPRESS

ACCOUNT NUMBER _____

EXPIR. DATE _____ DAYTIME PHONE _____

CUSTOMER SIGNATURE _____

CHECK AREA OF COMPUTER INTEREST:

☐ BUSINESS SOFTWARE

☐ TECHNICAL PROGRAMMING

☐ OTHER: _____

THE FACTOR THAT WAS MOST IMPORTANT IN YOUR SELECTION:

☐ THE SYBEX NAME

☐ QUALITY

☐ PRICE

☐ EXTRA FEATURES

☐ COMPREHENSIVENESS

☐ CLEAR WRITING

☐ OTHER _____

OTHER COMPUTER TITLES YOU WOULD LIKE TO SEE IN PRINT:

_____

_____

OCCUPATION

☐ PROGRAMMER   ☐ TEACHER

☐ SENIOR EXECUTIVE   ☐ HOMEMAKER

☐ COMPUTER CONSULTANT   ☐ RETIRED

☐ SUPERVISOR   ☐ STUDENT

☐ MIDDLE MANAGEMENT   ☐ OTHER:

☐ ENGINEER/TECHNICAL   _____

☐ CLERICAL/SERVICE

☐ BUSINESS OWNER/SELF EMPLOYED

CHECK YOUR LEVEL OF COMPUTER USE

☐ NEW TO COMPUTERS

☐ INFREQUENT COMPUTER USER

☐ FREQUENT USER OF ONE SOFTWARE

   PACKAGE:

   NAME _____

☐ FREQUENT USER OF MANY SOFTWARE

   PACKAGES

☐ PROFESSIONAL PROGRAMMER

OTHER COMMENTS:

_____

_____

_____

_____

_____

_____

_____

PLEASE FOLD, SEAL, AND MAIL TO SYBEX

SYBEX, INC.
2021 CHALLENGER DR. #100
ALAMEDA, CALIFORNIA USA
        94501

SEAL

■ WordPerfect 5.1 Command Summary

Command	Keystrokes	Menu Selections
Append Block	Ctrl-F4 B A	Edit ➤ Append
Base Font	Ctrl-F8 F	Font ➤ Base Font
Block Text	Alt-F4 *or* F12	Edit ➤ Block
Case Conversion (Block)	Shift-F3	Edit ➤ Convert Case Center Page
(Top/Bottom)	Shift-F8 P C	Layout ➤ Page C
Center Text	Shift-F6	Layout ➤ Align ➤ Center
Columns On/Off	Alt-F7 C O	Layout ➤ Columns ➤ On
Control Printer	Shift-F7 C	File ➤ Print C
Copy Block	Ctrl-F4 B C	Edit ➤ Copy
Copy Sentence/Paragraph/Page	Ctrl-F4	Edit ➤ Select
Date Code	Shift-F5 C	Tools ➤ Date Code
Date Format	Shift-F5 F	Tools ➤ Date Format
Date Text	Shift-F5 T	Tools ➤ Date Text
Define Columns	Alt-F7 C D	Layout ➤ Columns ➤ Define
Document Comment	Ctrl-F5 C	Edit ➤ Comment
Document Summary	Shift-F8 D S	File ➤ Summary
DOS (Temporary Exit)	Ctrl-F1	File ➤ Goto DOS
Endnote	Ctrl-F7 E	Layout ➤ Endnote
Exit WordPerfect	F7	File ➤ Exit
Flush Right	Alt-F6	Layout ➤ Align ➤ Flush Right
Font (Size)	Ctrl-F8 S	Font
Footers	Shift-F8 P F	Layout ➤ Page F
Footnote	Ctrl-F7 F	Layout ➤ Footnote
Fully Justify Text	Shift-F8 L J F	Layout ➤ Justify ➤ Full
Go to Character/Page	Ctrl-Home	Search ➤ Goto
Graphics Print Quality	Shift-F7 G	File ➤ Print G
Headers	Shift-F8 P H	Layout ➤ Page H
Help	F3	Help
Hyphenation On/Off	Shift-F8 L Y	Layout ➤ Line Y
Indent →	F4	Layout ➤ Align ➤ Indent →
→Indent←	Shift-F4	Layout ➤ Align ➤ Indent →←
Left-Justify Text	Shift-F8 L J L	Layout ➤ Justify ➤ Left
Line (Graphic)	Alt-F9 L	Graphics ➤ Line
Line Spacing	Shift-F8 L S	Layout ➤ Line S
List Files	F5	File ➤ List Files
Macro (Define)	Ctrl-F10	Tools ➤ Macro ➤ Define
Macro (Execute)	Alt-F10	Tools ➤ Macro ➤ Execute